Radiopharmaceutical Therapy and Dosimetry

Medical Physics Monograph No. 40

Radiopharmaceutical Therapy and Dosimetry

Robert F. Hobbs, Joe O'Donoghue, and Jessica Clements

Editors

American Association of Physicists in Medicine
2023 Summer School Proceedings

University of Minnesota
Minneapolis, Minnesota
June 2–7, 2023

Published for the
American Association of Physicists in Medicine
by Medical Physics Publishing, Inc.

To order American Association of Physicists in Medicine (AAPM) publications, contact:

Medical Physics Publishing, Inc.
4555 Helgesen Dr.
Madison, WI 53718
Phone: (800) 442-5778 or (608) 224-4508
Fax: (608) 224-5016
Email: mpp@medicalphysics.org
Web: medicalphysics.org

First printing June 2023

Published by:
Medical Physics Publishing, Inc.
Madison, Wisconsin

Published for:
American Association of Physicists in Medicine (AAPM)
1631 Prince St.
Alexandria, VA 22314
Phone: (571) 298-1300
Fax: (571) 298-1301

Library of Congress Control Number: 2023934451

ISBN hardcover: 978-1-951134-22-8
ISBN eBook: 978-1-951134-23-5

Printed in the United States of America

Contents

List of Contributors

Diane Alvarez, MSc, DABR
Department of Radiation Oncology
Miami Cancer Institute at Baptist Hospital
Miami, FL

David Barbee, Ph.D.
Clinical Associate Professor
NYU Langone Health
New York, NY

Rachel Barbee, Ph.D., DABR
Associate Professor
Department of Imaging Physics
MD Anderson Cancer Center
Houston, TX

Bryan Bednarz, Ph.D.
Department of Medical Physics, Human Oncology, Radiology, and Engineering Physics
University of Wisconsin–Madison
Madison, WI

Denis E. Bergeron, Ph.D.
Radiation Physics Division
National Institute of Standards and Technology
Gaithersburg, MD

Wesley E. Bolch, Ph.D., PE, FAAPM
Director, Advanced Laboratory for Radiation Dosimetry Studies
J. Crayton Pruitt Family Department of Biomedical Engineering
University of Florida
Gainesville, FL

Jessica Clements, MS, DABR, FAAPM
Imaging Physicist and Radiation Safety Officer
Department of Radiology
The University of Vermont Medical Center
Burlington, VT

Chaitanya Divgi, MD
CEO
Divgi Consulting, LLC
Meadowbrook, PA

Alireza Ehtenham, Ph.D.
Postdoctoral Research Associate
Department of Radiation Oncology
Washington University School of Medicine
St. Louis, MO

Eric C. Frey, Ph.D., FSNMMI
Professor
Division of Radiological Physics
Russell H. Morgan Department of Radiology and Radiological Science
Baltimore, MD
and Founder and CEO
Rapid, LLC

Jonathan I. Gear, Ph.D.
Joint Department of Physics
Institute of Cancer Research and
Royal Marsden NHS Foundation Trust
United Kingdom

Stephen A. Graves, Ph.D., DABR
Assistant Professor of Radiology
Division of Nuclear Medicine
University of Iowa
Iowa City, IA

Robert F. Hobbs, Ph.D., DABR
Associate Professor and Medical Physicist
Department of Radiation Oncology
Johns Hopkins Medical Institute
Baltimore, MD

Roger W. Howell, Ph.D., FSNMMI
ICRU Commissioner and
Chief, Division of Radiation Research
Department of Radiology
Rutgers New Jersey Medical School
Newark, NJ

Adam Kesner, Ph.D., DABR
Associate Attending, Nuclear Imaging Physics
Department of Medical Physics
Memorial Sloan Kettering Cancer Center
New York, NY

Nichole M. Maughan, Ph.D, DABR
Clinical Physicist, Division of Medical Physics
Department of Radiation Oncology
Washington University School of Medicine, St. Louis, MO
and Clinical Physicist, Department of Radiation Oncology
Intermountain Health, Provo, UT

Joe O'Donoghue, Ph.D.
Attending Physicist
Department of Medical Physics
Memorial Sloan Kettering Cancer Center
New York, NY

Katie Tapp, Ph.D.
Medical Radiation Safety Team
U.S. Nuclear Regulatory Commission
Washington, D.C.

Russell Tarver, M.S., DABR, FAAPM
Chief Physicist
Department of Radiation Oncology
Center for Cancer and Blood Disorders
Fort Worth, TX

Gerald White, M.S., FAAPM
Colorado Associates in Medical Physics
Colorado Springs, CO

Brian E. Zimmerman, Ph.D.
Radiation Physics Division
National Institute of Standards and Technology
Gaithersburg, MD

Chapter 1

Current FDA-Approved RPTs and Their Uses

Nichole M. Maughan[1], Ph.D., DABR
and Alireza Ehtenham[2], Ph.D.

[1]Clinical Physicist
Department of Radiation Oncology
Washington University School of Medicine
St. Louis, MO
and Clinical Physicist
Department of Radiation Oncology
Intermountain Health
Provo, UT

[2]Postdoctoral Research Associate
Department of Radiation Oncology
Washington University School of Medicine
St. Louis, MO

1.1 Introduction

This chapter will provide a brief introduction of the currently FDA-approved *radiopharmaceutical therapies* (RPT). The goal is to provide a quick reference guide for currently available FDA-approved RPTs, their uses, and basic treatment information, such as typical prescriptions, administration route, and radiation safety precautions that are specific to the RPT. Greater safety recommendations as pertaining to regulatory aspects, release criteria and waste management are provided in chapters 2, 5, and 6, respectively. As more RPTs are approved, we recommend the reader studies the appropriate literature and follows manufacturer and local regulatory guidelines for each drug/isotope.

1.2 Iodine-131 Sodium Iodide

Sodium iodide-131 ([^{131}I]NaI) or radioiodine-131 is a white non-metallic solid that is soluble in water. Sodium iodide-131, in liquid or capsule form, can be used for the management and treatment of hyperthyroidism and *differentiated thyroid carcinomas* (DTC) as the thyroid and most thyroid cancers naturally take up iodine in the sodium iodide symporters present in the cell membrane plasma of thyroid tissue. Some key recommendations are provided below. For more detailed recommendations, please refer to SNMMI Practice Guideline for ^{131}I Therapy for Thyroid Disease (Silberstein et al. 2012).

1.2.1 Treatment of Hyperthyroidism

Indication

In 1941, Saul Hertz pioneered the use and administration of radioactive material for medicinal purposes when *radioactive iodine ablation* (RAI) was first used to treat Grave's disease, a common cause of hyperthyroidism. RAI remains a treatment option for hyperthyroidism today. The thyroid preferentially takes up iodine when used as ^{131}I (Abraham 2010; RSNA 2022).

Imaging and Diagnosis

Diagnosis and categorization of hyperthyroidism consists of analysis of *thyroid-stimulating hormone level* (TSH) as well as diagnostic scans of radioiodine uptake (1–4 mCi or 3.7–15 MBq), as measured either by probes or planar imaging. For these, ^{123}I is typically used, as the half-life is shorter (13.3 h); ^{123}I decays via electron capture, so no energetic electrons are emitted, and there is an 83% branching ratio for a 159-keV photon.

Typical Prescription

The initial required dosage is 4–10 mCi (148–370 MBq) by mouth, but for toxic nodular goiters and other serious thyroid conditions, larger dosages may be required (Medscape). In particular, ablative dosages typically run in the 10–30 mCi range (Radswiki).

Route of Administration

Sodium iodine-131 is available as a stabilized aqueous solution or a solid capsule that can be administered orally. Due to radiation safety concerns as described below, the capsule form is preferred over the liquid form. Once absorbed by the gastrointestinal tract, the [^{131}I]NaI is redistributed to concentrate in the thyroid (Mandel 2003).

1.2.2 Treatment of Carcinoma of the Thyroid

Indication

Sodium iodine-131 is effective for treating well-differentiated thyroid carcinomas. About 90 percent of thyroid cancers are well-differentiated, which selectively take up radioactive iodine (Palot et al. 2022). Patients are classified in three main categories: low risk, intermediate-risk, and high-risk categories (Palot et al. 2022). More details on treating these three categories with [^{131}I]NaI are provided by Silberstein et al. (2012). A low-iodine diet should be followed for 10–14 days prior to administration, and a high level of TSH present in the system improves the uptake to the thyroid or thyroid disease. Often recombinant human TSH is administered to the patient prior to the [^{131}I]NaI.

Typical Prescription

Prescription values for the different disease stages are a subject of debate. Typical prescriptions for thyroid carcinoma are 100–150 mCi (3700–5550 MBq) and 50 mCi (1850 MBq) for post-operative ablation of normal thyroid tissue, while 200 mCi (7400 MBq) is a standard value for disseminated disease. As with hyperthyroidism, diagnosis of thyroid cancer may include blood level of TSH as well as imaging methods that include radioactive iodine uptake. Activity uptake of ^{123}I may inform the prescribed therapeutic activity of ^{131}I. In some cases, ^{124}I may be used for diagnostic or pretherapeutic imaging, as it has a better sensitivity and quantitative interpretation (Bockisch et al. 2006; Jentzen et al. 2010; Hobbs et al. 2009). Iodine-124 is a b^+ emitter (24% branching ratio) with a half-life of 4.18 d, which allows for PET imaging and multi-timepoint analysis for better pharmacokinetic extrapolation to the ^{131}I half-life (FDA-3; Radswiki).

1.2.3 Production and Decay Scheme

Iodine-131 decays by β^- emission (~90% of local irradiation) and associated gamma emission (~10% of local irradiation), with a physical half-life of 8.04 days (UW Hos-

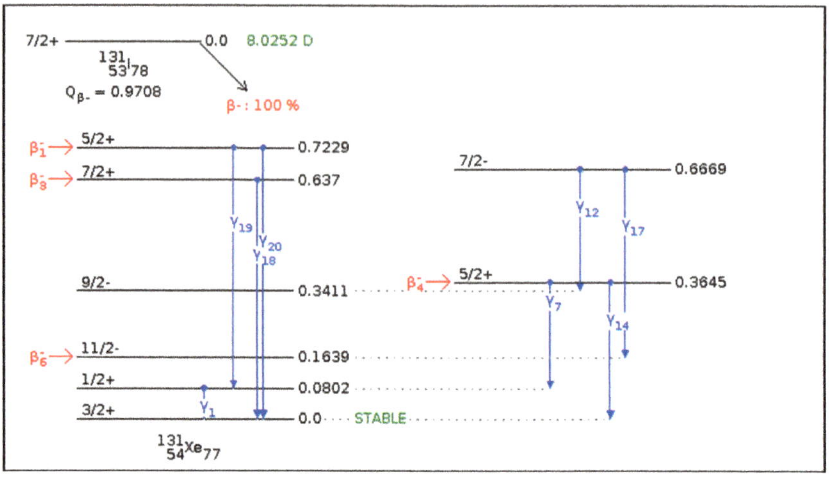

Figure 1.1 Decay scheme for ^{131}I. Transitional energies are shown in MeV. (Figure from Medical Internal Radiation Dose.)

pitals and Clinics). The maximum energy of beta particles is 0.61 MeV, with an average energy of 0.192 MeV. There is an 82% branching ratio for 364-keV gamma rays that allows for imaging via *single photon computed tomography* (SPECT) but can also present radiation safety concerns. The full decay scheme is presented in Figure 1.1.

Most of the medicinal ^{131}I is produced by bombarding ^{130}Te in powder form with neutrons in a nuclear facility. Several methods for extracting the ^{131}I exist. Iodine-131 may also be produced in nuclear reactors as a by-product of ^{235}U fission:

$$^{130}_{52}Te(n,\gamma)\,^{131}_{52}Te \rightarrow ^{131}_{53}I. \tag{1.1}$$

1.2.4 Precautions

Radiation Safety

As with all radioactive products, appropriate *personal protective equipment* (PPE) should be utilized for both staff and the patient. Specifically, maximum precautions should be taken to avoid spills, drips, and splatters when dealing with the liquid form of [^{131}I]NaI, Liquid [^{131}I]NaI is highly volatile in air and at high risk of localizing within staff members' thyroids when inhaled (Radswiki et al.). The use of a charcoal filter is recommended to reduce the amount of volatile ^{131}I, which can escape the container. It is also recommended that thyroid bioassays of staff be taken to assess the amount of ^{131}I that could be absorbed from this volatile iodine. As with all RPTs, be sure to survey the entire area thoroughly when releasing a room/area.

Sodium iodine-131 is typically administered as an outpatient procedure; however, high activities may require an inpatient stay until the patient is considered "releasable" according to regulatory requirements. Even performed as an outpatient procedure, patients and caregivers should exercise appropriate precautions to reduce exposure to themselves and others and provide take-home instructions to patients to protect caregivers and visitors.

Shielding

- Half-Value Layer (HVL / Lead): 0.09 inch = 0.23 cm
- Tenth-Value Layer (TVL / Lead): 0.28 inch = 0.70 cm
- Half-Value Layer (HVL / Water or Tissue) 2.50 inch = 6.30 cm (University of Michigan)

Biological Concerns

The effective half-life (biological + physical) for [^{131}I]NaI for the typical adult is about 5.5 days (about 132 h). There is a faster rate of biological clearance of [^{131}I]NaI for patients who have had surgical resection of their thyroids (Ravichandran et al. 2010). Potential toxicity from ^{131}I administration is bone marrow toxicity or hematotoxicity. Additionally, salivary uptake of radioactive iodine may lead to sialadenitis, dry mouth, and other oral sequelae, and it is a serious quality-of-life concern (Mandel et al. 2003). Metastases in the lungs may lead to concern for pulmonary toxicity (Hobbs et al. 2009).

1.3 Iodine-131 Iobenguane

1.3.1 Treatment

Indication

Iobenguane-iodine-131, commercially known as AZEDRA® (Progenics Pharmaceuticals), is indicated for the treatment of pheochromocytoma or paraganglioma. Typical candidates for iobenguane ^{131}I therapy (CardinalHealth) are adults and children (12+ years old) with iobenguane scan positive, unresectable, locally advanced or metastatic pheochromocytoma or paraganglioma who require systemic anticancer therapy. Pediatrics under 12 years of age have been treated off-label.

Typical Prescription

The recommended planar or SPECT diagnostic/planning dosages for patients:

- Greater than 50 kg: 185 to 222 MBq (5 to 6 mCi)
- 50 kg or less: 3.7 MBq/kg (0.1 mCi/kg)

The recommended therapeutic dosages for patients (up to 2 cycles):

- Greater than 62.5 kg: 18,500 MBq (500 mCi)
- 62.5 kg or less: 296 MBq/kg (8 mCi/kg)

If needed, Azedra therapeutic dosages can be adjusted based on absorbed dose estimated from the diagnostic/planning SPECT scan (FDA-2).

Route of Administration

Iobenguane I-131 is administered intravenously. It is provided as a clear solution in a vial with concentrations of 555 MBq/mL (15 mCi/ml) at *time of calibration* (TOC).

1.3.2 Precautions

Radiation Safety

Due to the high activities of ^{131}I that these patients receive, patients are typically not considered "releasable" and are kept as radioactive inpatients (3–5 days). Staff who are caring for the patient during the inpatient stay should be trained in handling radioactive patients, including appropriate use of PPE, shielding, and wearing radiation monitoring badges. Pediatric patients typically also require careful monitoring with video surveillance, bladder catheterization, and anesthesia for some younger patients. Also, pediatric patients have a higher risk of long-term radiation effects/toxicity than adults (FDA-2).

Biological Concerns

The mean biological clearance for Azedra (iobenguane-^{131}I) is reported as 62 ± 24 mL/hr/kg by the manufacturer, in a unit that is difficult to interpret, and the mean terminal blood half-life is 35 ± 14 hours (FDA-2). The primary toxicity concerns are for hematological toxicity, both acute and out of concern for myeloplastic syndrome. Activity should be adjusted downwards or the treatment discontinued as a function of platelet and neutrophil counts. Renal toxicity is also a concern from the clearance of the activity through the kidneys, as is salivary gland toxicity or hypothyroidism from any free ^{131}I. Patients take iodine tablets prior to therapy to saturate the thyroid and salivary glands in iodine and lower the risk of such toxicities.

Azedra is the only FDA-approved RPT that recommends dosimetry with the goal of limiting the amount of activity prescribed as a function of the projected absorbed doses for the dose-limiting normal organs. The recommendation is to perform organ-level dosimetry from 3 planar images and adjust as a function of the given absorbed dose limits provided (12 Gy, 16.5 Gy, 18 Gy, 31 Gy, and 40 Gy for the bone marrow, lungs, kidneys, liver, and small intestine, respectively). Caveats to the recommended methodology are that SPECT/CT would provide more accurate information and the absorbed dose limits seem to be taken from external beam experience.

1.4 Radium-223 Dichloride

1.4.1 Treatment

Indication

Radium-223 dichloride, [^{223}Ra]RaCl$_2$, commercially known as XOFIGO® (Bayer), is a radiopharmaceutical used for the treatment of patients with *castration-resistant prostate cancer* (mCRPC), specifically those with symptomatic bone metastases and no known visceral metastatic disease. In 2013, Xofigo was the first FDA-approved radionuclide for the treatment of patients with mCRPC (CardinalHealth; Morgan et al. 2021). Other indications may be forthcoming. Radium is in the same chemical family as calcium, thus it localizes to areas of calcium uptake, notably bone formations.

Typical Prescription

The recommended prescription for Xofigo is 55 kBq/kg (1.49 μCi/kg), which is given for 6 cycles at 4-week intervals (FDA-4). Because of the concern for myelotoxicity, contra-indications exist for low blood counts, both for the initial cycle as well as for succeeding cycles.

Route of Administration

Xofigo is provided as a clear/colorless solution in a patient-specific syringe. It is administered via slow intravenous injection through a syringe. The line should be primed and flushed with saline before and after administration, respectively (Bayer).

1.4.2 Decay Scheme

Radium-223 (half-life 11.4 days), is primarily an alpha-emitter with a Q-value of 5.98 MeV (most alpha-particle energies in the 5.5–5.7 MeV range) and small fractions of gamma emissions (in particular, a 14% branching ratio for 269-keV photons). Alpha particles from RPT have ranges on the order of 50–100 μm, which makes the energy deposition very conformal. However, ^{223}Ra also has radioactive daughters, decaying to ^{219}Rn, which decays via α-decay with a half-life of 4 s and a Q-value of 6.95 MeV to ^{215}Po. Polonium-215 decays by α-particle decay to ^{211}Pb with a half-life of 1.78 ms and a Q-value of 7.53 MeV. Lead-211 is a β-emitter with a half-life of 36.1 min and its daughter ^{211}Bi then provides another α-particle to the decay chain (with a half-life of 2.14 min), either directly to ^{207}Tl or with a 0.3% branching ratio of β-decay to ^{211}Po, which then decays with an alpha-particle. In all, four alpha-particles are emitted in the decay chain. The full decay scheme is provided in Figure 1.2 over the next several pages (FDA-4). In principle, photons from ^{223}Ra and ^{211}Pb can be imaged for activity quantification; however, due to the low amount of administered therapeutic activity, this has not been feasible until recently (Benabdallah et al. 2021). From a dosimetric standpoint, relocalization of the daughters—in particular the ^{211}Pb with a 36.1-min half-life—should be considered. Imaging and preclinical studies have shown that for Xofigo, the daughters tend to stay where the parent isotope decays (Henriksen et al. 2003).

Radium-223 is most commonly generated from naturally occurring ^{226}Ra. The ^{226}Ra is bombarded with neutrons to form ^{227}Ra, which decays to ^{227}Ac, then to ^{227}Th, which then decays in turn to ^{227}Ac. Since ^{227}Ac has a half-life of 21.8 years, once sufficient quantities of ^{227}Ac have been produced, the ^{223}Ra may be eluted from the ^{227}Ac generator:

$$^{226}_{88}Ra(n,\gamma)^{227}_{88}Ra \rightarrow ^{227}_{89}Ac \rightarrow ^{227}_{90}Th \rightarrow ^{222}_{88}Ra. \tag{1.2}$$

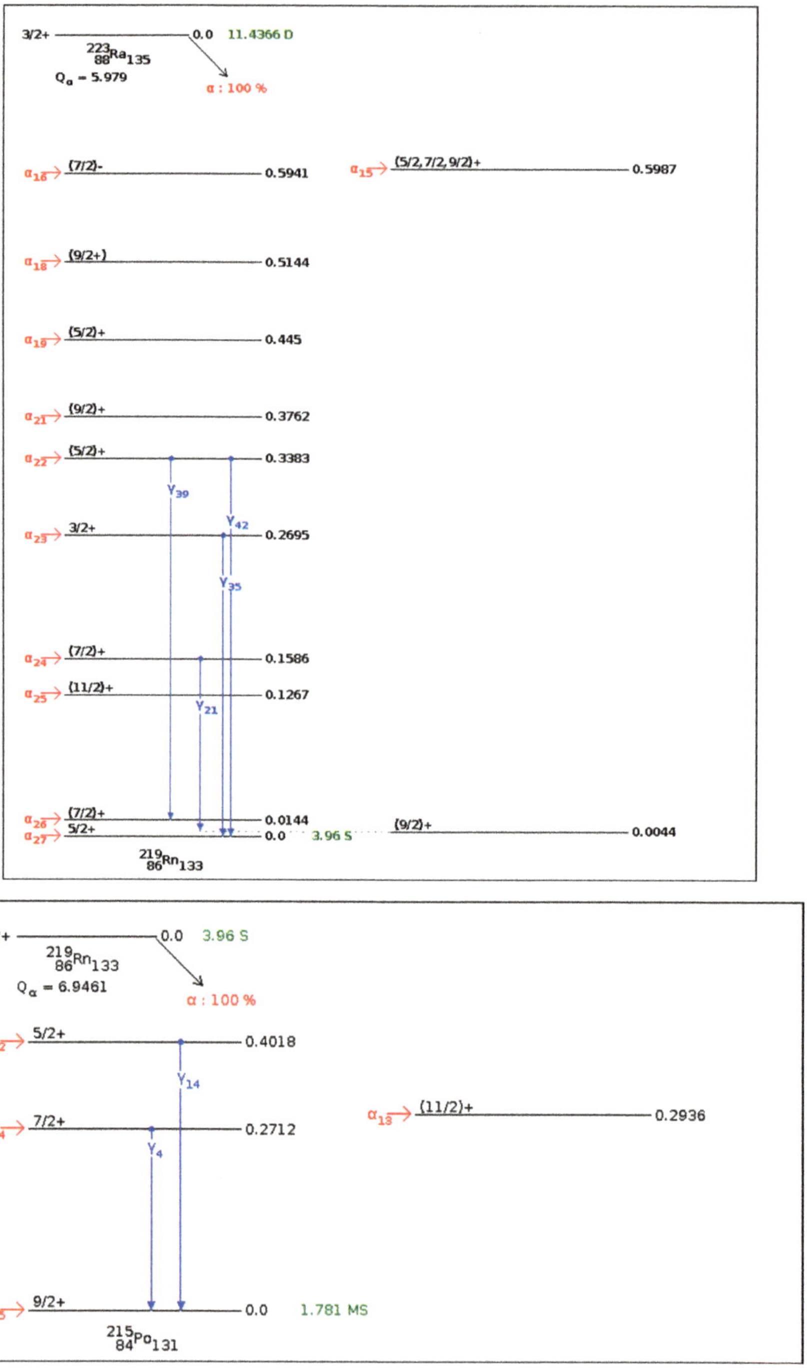

Figure 1.2 Decay scheme for Ra-223 and its daughters (unit: MeV). (Figure from Medical Internal Radiation Dose.)

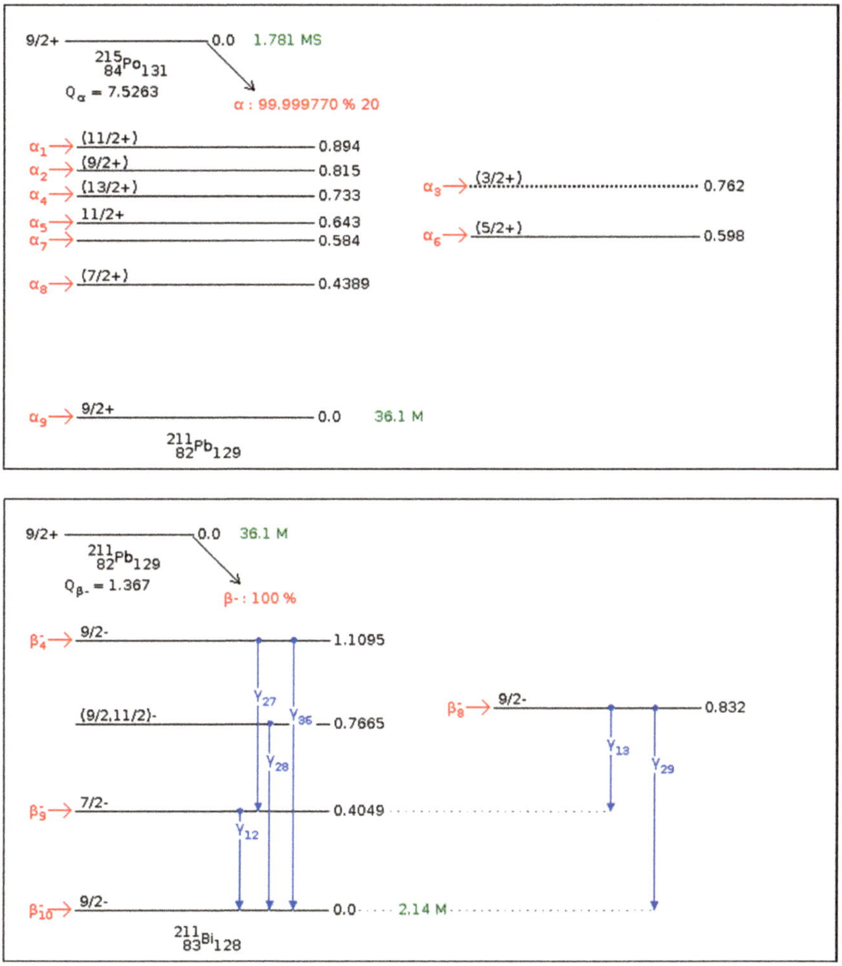

Figure 1.2 (continued)

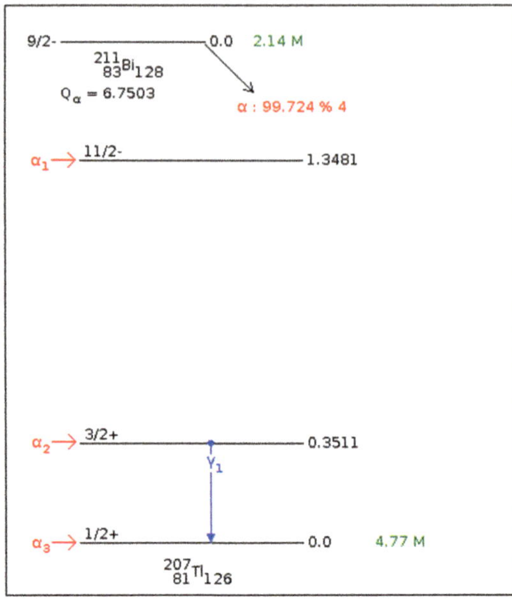

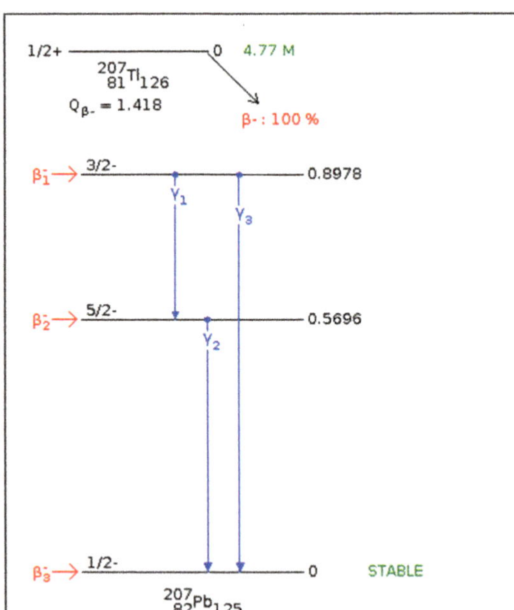

Figure 1.2 (continued)

1.4.3 Precautions

Radiation Safety

Given the low amount of injected material, concerns over caregiver, staff, and general public exposure is negligible. However, care should be taken when handling bodily fluids, especially the feces, as contamination and re-ingestion is the primary concern (FDA-5). For the same reasons, vigilance with regard to spills and potential contamination is required. Even though ^{223}Ra has a short range in tissue, it has a high LET and can result in significant toxicity if ingested or if it comes into contact with the skin.

Shielding

Given the low activities and short range of alpha-particles from radium-223 in tissue, Xofigo does not require significant shielding. However, it is still recommended to exercise ALARA precautions when handling this RPT.

Biological Concerns

The primary route of clearance for Xofigo is fecal excretion (Dauer et al. 2014), about half of which clears within 24–48 hours. Patients are recommended to sit while urinating and flush twice after using the restroom.

The most common potential toxicity is bone marrow suppression (FDA-4). Low blood counts will result in the patient being denied the full course of treatment.

1.5 Yttrium-90 Ibritumomab Tiuxetan

1.5.1 Treatment

Indication

Ibritumomab tiuxetan, commercially known as ZEVALIN® (Acrotech Biopharma, Inc.), is a CD20-directed radiotherapeutic antibody tagged with ^{90}Y. It is used for the treatment of adult patients with relapsed or refractory low-grade or follicular B-cell *non-Hodgkin's lymphoma* (NHL) and previously untreated follicular NHL who achieve a partial or complete response to first-line chemotherapy (FDA-6; Mondello et al. 2016).

Typical Prescription

The recommended prescription for Zevalin is 0.3–0.4 mCi/kg (11.1–14.8 MBq/kg), depending on platelet counts. Total dosages are typically 777–1110 MBq (21–30 mCi), with a maximum of 1184 MBq (32 mCi) (Kim et al. 2010). Pretreatment diagnostic SPECT imaging with (^{111}In)-ibritumomab tuxetan as a surrogate was historically used 1 week before treatment. Indium-111 decays via electron capture with a half-life of 2.8 days and has two imaging photons, one with a 90% branching ratio at 171 keV and another 245 keV with a 94% branching ratio. However, these pretreatment scans are generally not required anymore due to a lack of reported variability between patients' biokinetics of Zevalin (Conti et al. 2005).

Route of Administration

Zevalin is administered intravenously with rituximab (FDA-6).

1.5.2 Decay Scheme

Yttrium-90 is a "pure" β^--emitter; that is, there are only scarce numbers of associated photons in the decay. Yttrium-90 beta particles have a maximum energy of 2.28 MeV and a range in water of about 11 mm, with a physical half-life of 64.1 hours (2.67 days) (FDA-6). The decay scheme is shown in Figure 1.3.

Yttrium-90 may be produced in several ways: (1) eluted from a ^{90}Sr source (half-life of 28.1 years), which is produced as a by-product of ^{235}U fission, or (2) by bombarding stable ^{89}Y with neutrons:

$$^{235}_{92}U(n,f)^{90}_{38}Sr \rightarrow ^{90}_{39}Y \tag{1.3}$$

$$^{89}_{39}Y(n,\gamma)^{90}_{39}Y \tag{1.4}$$

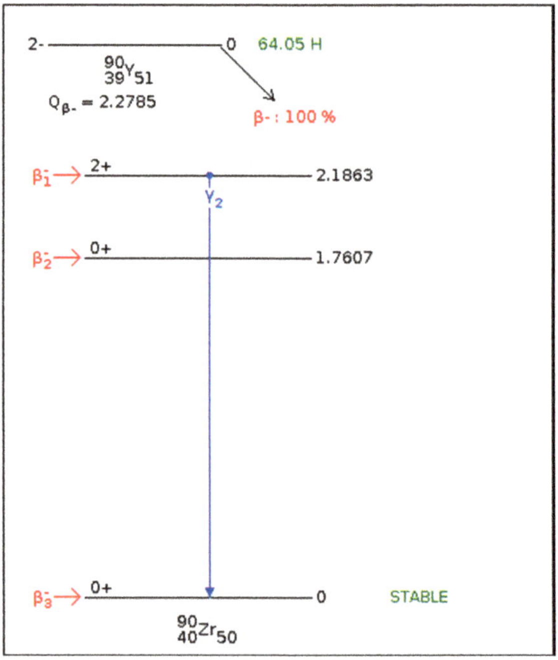

Figure 1.3 Decay scheme for Yttrium-90 (units: MeV). (Figure from Medical Internal Radiation Dose.)

1.5.3 Precautions

Radiation Safety

Even though the range of ^{90}Y beta particles is short in water/tissue, with very few prompt photons, there is still a broad spectrum of bremsstrahlung photons. Both require appropriate shielding, use of PPE, and the practice of ALARA principles when handling Zevalin both during preparation and when administering to the patient (Jødal et al. 2009).

Shielding

About 1.3 cm of Lucite or acrylic works well to attenuate beta emissions. To shield bremsstrahlung emissions, one can add 3–6 mm of lead on the outside of the primary shielding (Jødal et al. 2009).

Biological Concerns

The mean effective half-life of Zevalin in blood is 27 hours (range 14–44 h). The primary clearance mechanism is urinary excretion, and it accounts for the elimination of about 10% of unbound ^{90}Y within the first 12 to 24 hours (Mondello et al. 2016). Common toxicities include neutropenia, leukopenia, thrombocytopenia, anemia, infection, asthenia, musculoskeletal symptoms, and gastrointestinal symptoms. Patients should also not receive live vaccines shortly after receiving Zevalin (Mondello et al. 2016). The major limiting organ for toxicity is the bone marrow. In particular, myelotoxicity may be aggravated by concurrent chemotherapy (Baechler et al. 2010). Studies have been made projecting Zevalin administrations at myeloablative regimens using autologous stem cell support. In these cases, the liver was considered to be the primary organ at risk (Frey et al. 2006; Hobbs et al. 2013).

1.6 Yttrium-90 Microspheres

There are two brands of ^{90}Y microspheres available in the United States today: TheraSphere® (glass microspheres, Boston Scientific Corporation) and SIR-Spheres® (resin microspheres, Sirtex Medical Ltd). Their diameters range from 20 to 60 microns (Bhangoo et al. 2015).

1.6.1 Treatment

Indication

Yttrium-90 microspheres are used to treat primary or metastatic cancer in the liver. Yttrium-90 microsphere radioembolization utilizes the unique physiology of the liver, which has a dual blood supply: the portal vein and hepatic artery. The healthy liver parenchyma receives most of its blood supply from the portal vein, whereas tumors receive nearly all of their blood supply from the hepatic artery. Yttrium-90 microspheres are administered via catheter in interventional radiology through the hepatic artery, embolize within the tumor vasculature, and remain as a permanent implant due to their size. As such, ^{90}Y-microspheres are not a typical RPT as they do not re-localize or clear as a function of time, and there are typically up to 10 million particles that carry the activity as opposed to the Avogadro's number of molecules present in most RPTs. TheraSphere is approved for the treatment of unresectable hepatocellular carcinoma, while SIR-Sphere is approved for unresectable metastatic colorectal cancer in the liver (Sirtex; TheraSphere). TheraSphere is also commonly used to treat other liver diseases under humanitarian device exemption. A fraction of the spheres may exit the liver, return to the vasculature, and embolize in the lung. A *lung shunt fraction* (LSF) is determined using surrogate (planar or SPECT) imaging of ^{99m}Tc-MAA (macro aggregated albumin) surrogate. If the LSF is too high or the lung absorbed dose is too high, ^{90}Y-microspheres may be contra-indicated. Technetium-99m is a metastable excited state of ^{99}Tc which decays by gamma emission with a half-life of 6.0 hours and which emits a 140-keV photon with a 90% branching ratio used for imaging.

Typical Prescription

Prescriptions typically range from 0.3–20 GBq, depending on the brand, tumor size/burden, lung shunting, etc. TheraSphere activity is determined by prescribing absorbed dose to the treated area according to the mass of the liver or target (Bhangoo et al. 2015; Boston Scientific), generally a lobe or segment. SIR-Sphere activity is determined using either a calculation based on *body surface area* (BSA) or other tumor-to-normal uptake ratio modeling, as described by the manufacturer (Sirtex). Activities for both brands of microspheres are adjusted to account for LSF that is determined during a pretreatment ^{99m}Tc-MAA study (Sirtex; Bhangoo et al. 2015). The pre-therapeutic ^{99m}Tc-MAA may also be used in conjunction with SPECT/CT for more patient-specific dosimetry, where the prescription absorbed dose is now calculated to the normal tissue of the targeted lobe or segment. TheraSphere activities come in ready-to-use, patient-specific vials available in increments of 0.5 GBq from 3 GBq to 20 GBq. SIR-Sphere vials come in standard activities (typically 3 GBq), and patient-specific activities must be drawn and prepared in separate vials.

Route of Administration

Both TheraSphere and SIR-Spheres are administered via an intra-arterial delivery through the hepatic artery or more focused arteries that irradiate a specific lobe or segment(s).

1.6.2 Physical Properties

In spite of the (almost) nonexistent prompt photons, post-treatment imaging of ^{90}Y microsphere distribution can be performed using PET or SPECT. There is a branching ratio of 32 parts per million internal pair (positron-electron) production in the decay scheme that allows for PET imaging. Similarly, the large number of electrons create a sufficient number of bremsstrahlung photons that allows for SPECT quantification. Since the microspheres remain as a permanent implant, dosimetry calculations are greatly simplified and depend only on the activity and mass of the target.

The two products are produced differently. SIR-spheres are resin microspheres on which ^{90}Y is attached by ion exchange, then immobilized through precipitation as a phosphate salt. Theraspheres are created as glass beads containing ^{89}Y. The beads are then bombarded with neutrons to create the ^{90}Y as per Equation (1.4).

1.6.3 Precautions

Radiation Safety

As with any RPT, appropriate PPE should be used when preparing for and administering the ^{90}Y microspheres. If spilled, the microspheres can bounce and spread readily on surfaces (Medical Physics Consultants, Inc.).

Shielding

For a beta-emitter, the shielding is best provided with a low-atomic-number material such as acrylic. The low-atomic-number shielding reduces the amount of bremsstrahlung radiation. Lead can also help reduce subsequent bremsstrahlung production.

Biological Concerns

Due to biocompatible material of the microspheres, once microspheres are trapped permanently into the liver, they are not metabolized or excreted and, therefore, no biological clearance is expected. Within 12 days of administration, approximately 95% of the radiation from ^{90}Y has physically decayed (TheraSphere). Potential adverse events include acute pancreatitis, radiation pneumonitis, acute gastritis, acute cholecystitis, and radioembolization-induced liver disease. Most of these adverse events are the result of nontarget uptake of the ^{90}Y microspheres, often fed by extraneous vessels. These extraneous vessels are generally identified and coiled pre-injection during the procedure. Post-treatment imaging with SPECT or PET is highly recommended to verify ^{90}Y microsphere distribution (Sirtex; TheraSphere; Boston Scientific).

1.7 Samarium-153 Lexidronam

Samarium-153 lexidronam, commercially known as QUADRAMET®, is radioactive ^{153}Sm chelated with *ethylenediaminetetramethylenephosphonic* acid (EDTMP).

1.7.1 Treatment

Indication

Quadramet is indicated for pain relief from bone metastases, specifically in those patients with confirmed osteoblastic metastatic bone lesions that enhance on ^{99m}Tc-MDP bone scan (Lantheus Medical Imaging). It has been investigated as a curative for several bone cancers (Loeb et al. 2009; Loeb et al. 2010). The chelator mimics the chemistry of calcium, such that it is taken up in bone- forming areas of the body.

Typical Prescription

The recommended dose Quadramet is 1.0 mCi/kg (Lantheus Medical Imaging). Myeloablative regimens have been given up to 30 mCi/kg (Anderson et al. 2002).

Route of Administration

Quadramet is typically administered intravenously over a period of one minute through a secure in-dwelling catheter. Activity is provided in fixed quantities, and the precise activity prescribed to the patient must be drawn into a syringe. The administration is followed with a saline flush (Lantheus Medical Imaging).

1.7.2 Decay Scheme

Samarium-153 is a β^- emitter with a half-life of 46.3 hours and a maximum energy (Q-value) of 808 keV. It has a 103-keV photon with a branching ratio of 30% that is used for imaging. The decay scheme is shown in Figure 1.4.

Samarium-153 is produced from neutron bombardment of samarium oxide of ^{152}Sm, the stable isotope of samarium:

$$^{152}_{62}Sm(n,\gamma)^{153}_{62}Sm \tag{1.5}$$

1.7.3 Precautions

Radiation Safety

Patients who are candidates for Quadramet may present with incontinence. Please consult with your radiation safety officer and local regulatory guidelines regarding the proper handling and disposal of incontinence pads/diapers, as they will be contaminated with ^{153}Sm radioactivity (Lantheus Medical Imaging).

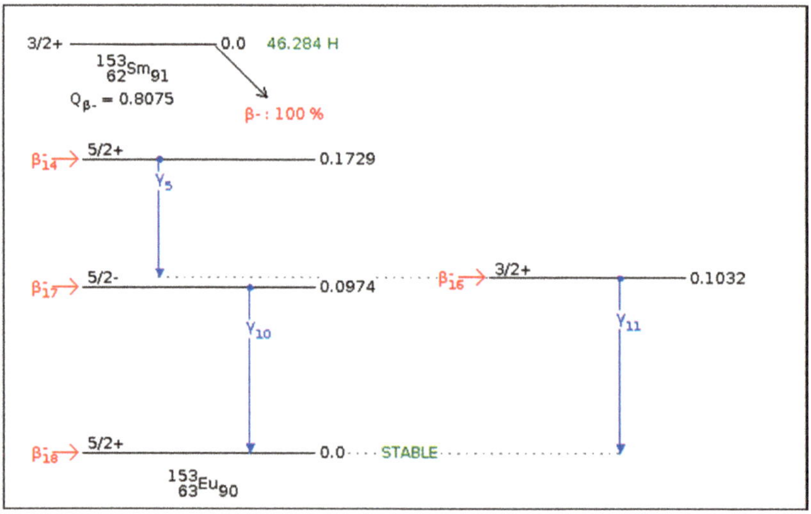

Figure 1.4 Decay scheme for ^{153}Sm (unit: MeV). (Figure from Medical Internal Radiation Dose.)

Shielding

The average and maximum beta particle ranges in water respectively are 0.5 mm and 3.0 mm. The specific gamma-ray emission constant for ^{153}Sm is 0.46 R/mCi-hr at 1 cm (1.24×10^{-5} mSv/MBq-hr at 1 m). For the lead shielding, the half-value thickness of lead (Pb) for ^{153}Sm is approximately 0.10 mm. Quadramet should be stored in a lead-shielded container and frozen until use (Lantheus Medical Imaging).

Biological Concerns

Quadramet is primarily excreted through the urine with approximately 25–50% eliminated through the urine in the first 6 hrs. However, the amount of excreted activity typically decreases with increasing tumor burden (Lantheus Medical Imaging). Quadramet is known to cause bone marrow suppression (Lantheus Medical Imaging) at myelobative regimes, up to 30 mCi/kg. No additional acute toxicities were recorded, although renal toxicity is believed to be of concern (Anderson et al. 2002; Hobbs et al. 2010).

1.8 Lutetium-177 DOTATATE

Lutetium-177 DOTATATE, commercially known as LUTATHERA® (Novartis/ Advanced Accelerator Applications), is a peptide receptor radionuclide therapy (PRRT) (FDA-7).

1.8.1 Treatment

Indication

Lutathera is used to treat either metastatic or inoperable somatostatin receptor-positive *gastroenteropancreatic-neuroendocrine tumors* (NETs) in the foregut, midgut, and hindgut (FDA- 7). Gallium-68-DOTATATE PET can be used to identify NET that are somatostatin receptor-positive, and a positive scan is a requirement for therapy (Anderson et al. 2021). PRRT has been used in Europe for over a decade, but has only recently been approved in the United States.

Typical Prescription

Lutathera is prescribed as 200 mCi (7.4 GBq) dosages in 4 cycles spaced 8 weeks apart (FDA- 7).

Route of Administration

Lutathera is injected intravenously either using a gravity infusion technique or using a programmable infusion pump (Maughan et al. 2021). To reduce the chance of renal toxicity, an amino acid solution containing L-lysine and L-arginine are concurrently administered with the Lutathera infusion to allow for faster renal clearance.

1.8.2 Decay Scheme

Lutetium-177 is primarily a β^- emitter (0.498 MeV max, 0.133 MeV mean) with a physical half-life of 6.65 days (Medical Internal Radiation Dose; FDA-7; Maughan et

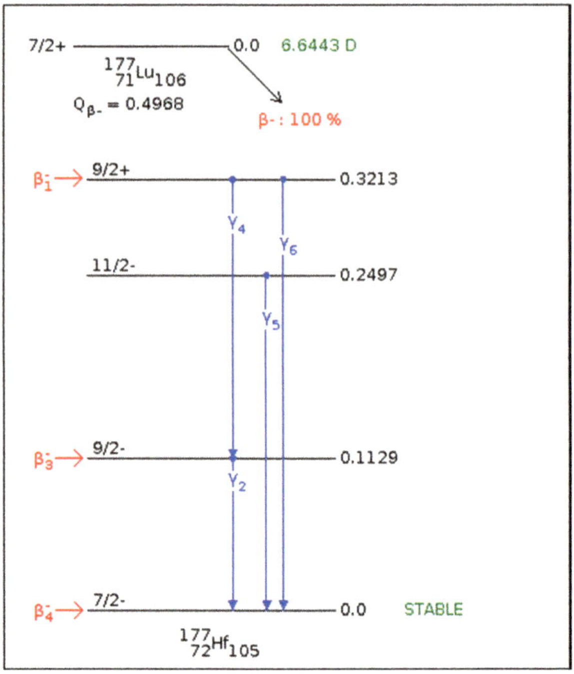

Figure 1.5 Decay scheme for ^{177}Lu (unit: MeV). (Figure from Medical Internal Radiation Dose.)

al. 2021). There are associated gamma emissions that allow for post-treatment imaging via SPECT (see Figure 1.5), particularly the 208-keV emission with an 11% branching ratio.

The companion diagnostic, ^{68}Ga, is a β^+ emitter with a 2.9-MeV maximum positron energy and a 67.7-minute half-life. The relatively short half-life of ^{68}Ga means that although useful as a diagnostic, it cannot be used reliably as a quantitative pre-therapeutic for the pharmacokinetics of the therapeutic.

As with a number of other radioactive isotopes used in medicine, ^{177}Lu is produced from neutron bombardment of the stable isotope of lutetium:

$$^{176}_{71}Lu(n,\gamma)^{177}_{71}Lu. \tag{1.6}$$

There may be a small amount of ^{177m}Lu contaminant present in the produced ^{177}Lu. Studies have shown there to be little dosimetric effect from the contaminant, but regulatory bodies may require inclusion of ^{177m}Lu on the radioactive materials license or careful consideration of waste management with its long half-life (160.4 days).

1.8.3 Precautions

Radiation Safety

Due to long infusion times of the amino acids (3–4 hrs), renal clearance, and the tendency of patients to demonstrate nausea and emesis, the treatment area for Lutathera patients should be appropriately prepared to prevent radiation contamination (Maughan et al. 2021). Patients should also have a dedicated restroom that is restricted to Lutathera only until decontaminated and released to the public/clinical area (Maughan et al. 2021). Be sure to survey the entire area thoroughly when releasing a room/area. Further recommendations on Lutathera treatment are provided by Hope et al. (2019) and Maughan et al. (2021).

Shielding

For maximum energy (497 keV) of emitted beta, the ranges in water and lead are 1.9 mm and 0.3 mm, respectively (Jødal et al. 2009). The gamma constant for ^{177}Lu is 0.028 mrem/hr per mCi at 1 meter (North Carolina Health Physics Society).

Biological Concerns

The radiation from Lutathera will be detectable in the urine for up to 30 days following administration. Within 4 hr of administration, the distribution of Lutathera goes through the kidneys, tumor lesions, liver, spleen, and, occasionally, the pituitary and thyroid glands. Co-administering with amino acids reduces the median absorbed dose to the kidneys by 47% (34% to 59%) and increases the blood clearance of Lutathera by 36% (Das et al. 2019). While studies have shown that radiation exposure to the general public from patients using public transportation for trips (e.g., by plane) after administration may be below safe thresholds, this does not take into account the danger of contamination during the first few days after administration. Patients should not use public forms of transportation for the first 48 hours.

Acute toxicities include nausea and emesis, typically related to the amino acids that are administered concurrently. Other toxicities may include renal toxicity or hematotoxicty, either as reduced blood cell counts or as the potential for myeloplastic syndrome (FDA- 7).

1.9 Lutetium-177 Vipivotide Tetraxetan

1.9.1 Treatment

Indication

Lutetium-177 vipivotide tetraxetan, commercially known as PLUVICTO® (Novartis/Advanced Accelerator Applications), is a targeted radioligand therapy used to treat adult patients with *prostate-specific membrane antigen* (PSMA)-positive *metastatic castration-resistant prostate cancer* (mCRPC) who have been treated with androgen receptor pathway inhibition and taxane-based chemotherapy (CardinalHealth). Before receiving Pluvicto, patients receive a Ga-68 gozetotide (LOCAMETZ) PET scan to help determine eligibility and uptake of the tumors (FDA-1). Common side effects include fatigue, dry mouth, nausea, anemia, decreased appetite, and constipation (FDA-9).

Typical Prescription

Pluvicto is prescribed as 7.4 GBq (200 mCi) in 6 cycles spaced 6 weeks apart (FDA-9).

Route of Administration

Pluvicto is administered intravenously either with a syringe fitted with a syringe shield (with or without syringe pump), gravity infusion-based technique from a vial, or infusion pump-based administration from a vial (FDA-9).

1.9.2 Precautions

Radiation Safety

As with all RPTs, appropriate use of PPE should be utilized when preparing and administering Pluvicto. Patients who are candidates for Pluvicto may present with incontinence. Patients should also have a dedicated restroom that is restricted to Pluvicto only until decontaminated and released to the public/clinical area (Maughan et al. 2021). After treatment administration, be sure to survey the entire area thoroughly when releasing a room/area. Please consult with your radiation safety officer and local regulatory guidelines regarding the proper handling and disposal of incontinence pads/diapers as they will be contaminated with Lu-177 radioactivity.

Biological Concerns

Pluvicto elimination half-life is 41.6 hours and is primarily excreted through the urine (FDA-9). As with Lutathera, patients may be kept in a dedicated room for several hours post-injection while they clear most of their waste into a dedicated bathroom, or they may be released immediately, provided they are given specific instructions and are not using public transportation but will return home quickly.

Organs at risk are similar to those at risk during treatment with Lutathera, but also include the salivary and lacrimal glands, which also express PSMA.

References

Abraham, P. and S. Acharya (2010). "Current and emerging treatment options for Graves' hyperthyroidism." *Ther. Clin. Risk Manag.* 6:29–40. doi:10.2147/tcrm.s5229.

Anderson, P. M., G. A. Wiseman, A. Dispenzieri, C. A. Arndt, L. C. Hartmann, W. A. Smithson, et al. (2002). "High-dose samarium-153 ethylene diamine tetramethylene phosphonate: low toxicity of skeletal irradiation in patients with osteosar-coma and bone metastases." *J. Clin. Oncol.* 20(1):189–96.

Anderson, R. C., E. M. Velez, B. Desai, and H. Jadvar (2021). "Management Impact of 68Ga-DOTATATE PET/CT in Neuroendocrine Tumors." *Nucl. Med. Mol. Imaging* 55(1):31–37. doi:10.1007/s13139-020-00677-0.

Baechler, S.,R. F. Hobbs, H. A. Jacene, F. O. Bochud, R. L. Wahl, and G. Sgouros (2010). "Predicting hematologic toxicity in patients undergoing radioimmunotherapy with 90Y-ibritumomab tiuxetan or 131I-tositumomab." *J. Nucl. Med.* 51(12):1878–84.

Bayer, Inc. (2022). "Product Monograph XOFIGO radium Ra-223 dichloride solution for injection." https://www.bayer.com/sites/default/files/xofigo-pm-en.pdf. Accessed 11/21/2022.

Benabdallah, N., W. Scheve, N. Dunn, et al. (2021). "Practical considerations for quantitative clinical SPECT/CT imaging of alpha particle emitting radioisotopes." *Theranostics* 11(20):9721–37. Published Sep 27, 2021. doi:10.7150/thno.63860.

Bhangoo, M. S., D. R. Karnani, P. N. Hein, et al. (2015). "Radioembolization with Yttrium-90 microspheres for patients with unre-sectable hepatocellular carcinoma." *J. Gastrointest. Oncol.* 6(5):469–78. doi:10.3978/j.issn.2078-6891.2015.056.

Bockisch, A., L. Freudenberg, S. Rosenbaum, and W. Jentzen (2006). "(124)I in PET imaging: impact on quantification, radio-pharmaceutical development and distribution." *Eur. J. Nucl. Med. Mol. Imaging* 33(11):1247–48.

Boston Scientific (2004). "TheraSphere Package Insert." http://www.therasphere.com/physicians-package-insert/package-insert-eu-en.pdf. Accessed 11/28/2022.

CardinalHealth (2022). "FDA-approved radiopharmaceuticals." https://www.cardinalhealth.com/content/dam/corp/web/documents/fact-sheet/cardinal-health-fda-approved-radiopharmaceuticals.pdf. Accessed 11/20/2022.

Conti, P. S., C. White, P. Pieslor, A. Molina, J. Aussie, and P. Foster (2005). "The role of imaging with (111)In-ibritumomab tiuxetan in the ibritumomab tiuxetan (zevalin) regimen: results from a Zevalin Imaging Registry." *J. Nucl. Med.* 46(11):1812–18.

Das, S., T. Al-Toubah, G. El-Haddad, and J. Strosberg (2019). "177Lu-DOTATATE for the treatment of gastroenteropancreatic neu-roendocrine tumors." *Expert Rev. Gastroenterol. Hepatol.* 13(11):1023–31. doi:10.1080/17474124.2019.1685381.

Dauer, L. T., M. J. Williamson, J. Humm, et al. (2014). "Radiation safety considerations for the use of $^{223}RaCl_2$ DE in men with castration-resistant prostate cancer." *Health Phys.* 106(4):494–504. doi:10.1097/HP.0b013e3182a82b37.

FDA-1. "FDA approves Pluvicto for metastatic castration-resistant prostate cancer." https://www.fda.gov/drugs/resources-information-approved-drugs/fda-approves-pluvicto-metastatic-castration-resistant-prostate-cancer." Accessed 2/4/2023.

FDA-2. "Highlights of Prescribing Information: AZEDRA." https://www.accessdata.fda.gov/drugsatfda_docs/label/2018/209607s000lbl.pdf. Accessed 11/21/2022.

FDA-3. "Highlights of Prescribing Information: Sodium Iodide I-131 Capsules Therapeutic for oral use." https://www.accessdata.fda.gov/drugsatfda_docs/label/2018/016517s016lbl.pdf. Accessed on 11/27/2022.

FDA-4. "Highlights of Prescribing Information: XOFIGO." https://www.accessdata.fda.gov/drugsatfda_docs/label/2018/203971s014lbl.pdf. Accessed 11/27/2022.

FDA-5. "Highlights of Prescribing Information: XOFIGO." https://www.accessdata.fda.gov/drugsatfda_docs/label/2019/203971s016lbl.pdf. Accessed 11/21/2022.

FDA-6. "Highlights of Prescribing Information: ZEVALIN." https://www.accessdata.fda.gov/drugsatfda_docs/label/2013/125019s210s213lbl.pdf. Accessed 11/21/2022.

FDA-7. "LUTATHERA (lutetium Lu177 dotatate)." https://www.accessdata.fda.gov/drugsatfda_docs/label/2018/208700s000lbl.pdf. Accessed 11/28/2022.

FDA-8. "Metastron Strontium-89 Chloride injection." https://www.accessdata.fda.gov/drugsatfda_docs/label/2013/020134s012lbl.pdf. Accessed 11/28/2022.

FDA-9. "PLUVICTO (lutetium Lu-177 vipivotide tetraxetan)." https://www.accesdata.fda.gov/drugsatfda_docs/label/2022/215833s000lbl.pdf. Accessed 11/28/2022.

Frey, E., B. He, G. Sgouros, I. Flinn, and R. Wahl (2006). "Estimation of post-therapy marrow dose rate in myeloablative Y-90 ibri-tumomab tiuxetan therapy." *J. Nucl. Med.* 47(Supplement 1):156P.

Henriksen, G., D. R. Fisher, J. C. Roeske, O. S. Bruland, and R. H. Larsen (2003). "Targeting of osseous sites with alpha-emitting 223Ra: comparison with the beta-emitter 89Sr in mice." *J. Nucl. Med.* 44(2):252–59.

Hobbs, R. F., R. L. Wahl, E. C. Frey, Y. Kasamon, H. Song, P. Huang, et al. (2013). "Radiobiologic optimization of combination radi-opharmaceutical therapy applied to myeloablative treatment of non-Hodgkin lymphoma." *J. Nucl. Med.* 54(9):1535–42.

Hobbs, R. F., R. L. Wahl, M. A. Lodge, M. S. Javadi, S. Y. Cho, D. T. Chien, et al. (2009). "124I PET-based 3D-RD dosimetry for a pedi-atric thyroid cancer patient: real-time treatment planning and methodologic comparison." *J. Nucl. Med.* 50(11):1844–47.

Hope, T. A., A. Abbott, K. Colucci, et al. (2019). "NANETS/SNMMI Procedure Standard for Somatostatin Receptor-Based Pep-tide Receptor Radionuclide Therapy with 177Lu-DOTATATE." *J. Nucl. Med.* 60(7):937–43. doi:10.2967/jnumed.118.230607.

IAEA. *Manual for Reactor Produced Radioisotopes.* IAEA-TECDOC-1340. Vienna: International Atomic Energy Association, 2003.

Jentzen, W., R. F. Hobbs, A. Stahl, J. Knust, G. Sgouros, and A. Bockisch (2010). "Pre-therapeutic I-124 PET(/CT) dosimetry confirms low average absorbed doses per administered I-131 activity to the salivary glands in radioiodine therapy of differentiated thyroid cancer." *Eur. J. Nucl. Med. Mol. Imaging* 37(5):884–95.

Jødal, L. (2009). "Beta emitters and radiation protection." *Acta Oncol.* 48(2):308–13. doi:10.1080/02841860802245163.

Kim, Y. C., Y. H. Kim, S. H. Uhm, et al. (2010). "Radiation safety issues in y-90 microsphere selective hepatic radioembolization therapy: possible radiation exposure from the patients." *Nucl. Med. Mol. Imaging* 44(4):252–60. doi:10.1007/s13139-010-0047-7.

Lantheus Medical Imaging, Inc. "Quadramet prescribing information." https://www.accessdata.fda.gov/drugsatfda_docs/label/2017/020570s008lbl.pdf. Accessed 11/28/2022.

Loeb, D. M., E. Garrett-Mayer, R. F. Hobbs, A. R. Prideaux, G. Sgouros, O. Shokek, et al. (2009). "Dose-finding study of (153)Sm-EDTMP in patients with poor-prognosis osteosarcoma." *Cancer* 115(11):2514–22.

Loeb, D. M., R. F. Hobbs, A. Okoli, A. R. Chen, S. Cho, S. Srinivasan, et al. (2010). "Tandem dosing of samarium-153 ethylenedia-mine tetramethylene phosphoric acid with stem cell support for patients with high-risk osteosarcoma." *Cancer* 116(23):5470–78.

Mandel, S. J. and L. Mandel (2003). "Radioactive iodine and the salivary glands." *Thyroid* 13(3):265–71.

Maughan, N. M., H. Kim, Y. Hao, et al. (2021). "Initial experience and lessons learned with implementing Lutetium-177-dotatate radiopharmaceutical therapy in a radiation oncology-based program." *Brachytherapy* 20(1):237–47. doi:10.1016/j.brachy.2020.07.004.

Medical Internal Radiation Dose (MIRD) format from the Evaluated Nuclear Structure Data File (ENSDF). Version available at http://www.nndc.bnl.gov/ensarchivals/. Accessed on 11/27/2022.

Medical Physics Consultants, Inc. (2020). "Radiation Safety for Yttrium-90 Microsphere Therapy." https://www.mpcphysics.com/wp-content/uploads/Y-90-Microspheres-Radiation-Safety-Inservice-Nuclear-Techs-10-6-2020.pdf. Accessed 11/28/2022.

Medscape. "Sodium iodide I-131 (rx)." https://reference.medscape.com/drug/hicon-sodium-iodide-i-131-999924. Accessed on 11/16/2022.

Mondello, P., S. Cuzzocrea, M. Navarra, and M. Mian (2016). "90 Y-ibritumomab tiuxetan: a nearly forgotten opportunity." *Oncotarget* 7(7):7597–609. doi: 10.18632/oncotarget.6531. PMID: 26657116; PMCID: PMC4884941.

Morgan, R., D. Wermuth, E. Molina, and M. Perraillon (2021). "Utilization and Cost of Radium-223 Dichloride (Xofigo®) for Treatment of Metastatic Castration-Resistant Prostate Cancer (mCRPR) in the US Medicare Population." *J. Nucl. Med.* 62(supplement 1):1309.

Musolino, S. (2002). *Radionuclide and Radiation Protection Data Handbook.* Book review. *Health Phys.* 83(1)136. https://ehs.stanford.edu/reference/sr-89-radionuclide-fact-sheet.

National Center for Biotechnology Information. PubChem Compound Summary for CID 5388879, Strontium chloride SR-89. https://pubchem.ncbi.nlm.nih.gov/compound/Strontium-chloride-SR-89. Accessed 11/28/2022.

North Carolina Health Physics Society. "I-131: Nuclide Safety Data Sheet." http://www.hpschapters.org/northcarolina/NSDS/131IPDF.pdf. Accessed 1/18/2023.

North Carolina Health Physics Society. "Lu-177: Nuclide Safety Data Sheet." http://www.hpschapters.org/northcarolina/NSDS/177LuPDF.pdf. Accessed 1/18/2023.

Palot Manzil, F. F. and H. Kaur. *Radioactive Iodine For Thyroid Malignancies.* Treasure Island (FL): StatPearls Publishing, 2022. Available from: https://www.ncbi.nlm.nih.gov/books/NBK580567/.

RSNA (2022). Radioactive Iodine (I-131) Therapy for Hyperthyroidism. Radiological Society of North America, Inc. (RSNA). https://www.radiologyinfo.org/en/info/radioiodine. Accessed on 11/16/2022.

Radswiki, T., D. Bell, Y. Weerakkody, et al. Thyroid scintigraphy (I-123). Reference article, Radiopaedia.org. https://doi.org/10.53347/rID-12707. Accessed on 1/19/2023.

Ravichandran, R., J. Binukumar, and A. A. Saadi (2010). "Estimation of effective half life of clearance of radioactive Iodine (I) in patients treated for hyperthyroidism and carcinoma thyroid." *Indian J. Nucl. Med.* 25(2):49–52. doi:10.4103/0972-3919.72686.

Silberstein, E. B., A. Alavi, H. R. Balon, et al. (2012). "The SNMMI practice guideline for therapy of thyroid disease with 131I 3.0." *J. Nucl. Med.* 53(10):1633–51. doi:10.2967/jnumed.112.105148.
Sirtex (2004). Approved indication, Package Insert (IFU) for SIR-Spheres® User's Manual. Sirtex Medical Limited. https://www.sirtex.com/media/jswdjodw/ifu-005-eu-sir-spheres-microspheres-ifu-for-eu.pdf. Accessed 11/28/2022.
Smith, D. S. and M. G. Stabin (2012). "Exposure rate constants and lead shielding values for over 1,100 radionuclides." *Health Phys.* 102(3):271–91. doi:10.1097/hp.0b013e318235153a.
TheraSphere™ - P200029. https://www.fda.gov/medical-devices/recently-approved-devices/theraspheretm-p200029. Accessed 11/28/2022.
University of Michigan. Facilities & Operations Environment, Health, and Safety. "Environmental, Health, and Safety Guideline: I-131." https://ehs.umich.edu/wp-content/uploads/2016/04/Iodine-131.pdf. Accessed 11/18/2022.
UW Hospital & Clinics. Nuclear Medicine Procedure, I-131 Sodium Iodide for Thyroid Cancer Therapies. https://www.radiology.wisc.edu/wp-content/uploads/2017/12/I-131_Sodium_Iodide_for_Thyroid_Cancer_March_2017.pdf. Accessed on 11/27/2022.

Chapter 2

Nuclear Regulatory Commission Regulations and Medical Event Reporting

Katie Tapp, Ph.D.

Medical Radiation Safety Team
U.S. Nuclear Regulatory Commission
Washington, D.C.

2.1 Introduction

The U.S. Nuclear Regulatory Commission (NRC) is an independent agency created by Congress in 1974 to ensure the safe use of radioactive materials for beneficial civilian purposes while protecting people and the environment. As described in the NRC medical use policy statement[1], the NRC has developed regulations for the use of radioactive material in medicine to ensure adequate radiation safety for workers, the general public, and patients. These regulations apply to all NRC medical use licensees. However, most medical users of radioactive material are located in *Agreement States*. Agreement States are states that the NRC has relinquished its regulatory authority to license and regulate certain radioactive materials, such as those used in medicine, to those states. Agreement States have regulations similar to the NRC for their licensees, however one should always check with their state if they have specific regulatory questions, as there are some differences. Information about Agreement States can be found on the NRC website at https://www.nrc.gov/about-nrc/state-tribal/agreement-states.html.

1. Medical Use of Byproduct Material; Policy Statement, Revision, 65 Fed. Reg. 47654 (August 3, 2000). https://www.nrc.gov/reading-rm/doc-collections/commission/policy/65fr47654.pdf.

2.2 NRC Regulations for Radiopharmaceutical Therapy

For NRC licensees, regulations for the use of radioactive material in medicine can be found in 10 Code of Federal Regulations (CFR) Part 35. To ensure radiation safety for patients, regulations include that physicians who use radioactive material have adequate training and experience, calibration of radiopharmaceutical dosages and sources used in medical treatments, and medical event reporting. To ensure adequate radiation safety for workers and the general public, regulations include requirements for surveys, waste handling and disposal, and patient release. This section will discuss some of these requirements that are specific to radiopharmaceutical therapy.

2.2.1 Training and Experience Requirements

The NRC's regulations have specific minimum training and experience criteria that physicians must meet to become authorized for a given specific type of medical use of radioactive material. Once authorized for medical use, the physician becomes known as an *authorized user* (AU). The purpose of the NRC training and experience criteria is that the NRC expects an AU to be able to independently fulfill all the radiation safety-related duties for the specific medical use for which the physician is authorized. Also per NRC regulations, AUs can be authorized to be radiation safety officers for uses that they are authorized for.

Per regulations, there are four different types of authorizations a physician can receive for radiopharmaceutical therapy. The differences between the types is based on what use the physician desires and their training and experience. The types are:

1. the use of all radiopharmaceuticals for therapy or which a written directive is required (10 CFR 35.390),
2. the oral administration of sodium iodide I-131 requiring a written directive in quantities less than or equal to 1.22 gigabecquerels (10 CFR 35.392),
3. the oral administration of sodium iodide I-131 requiring a written directive in quantities greater than 1.22 gigabecquerels (10 CFR 35.394), and
4. the parenteral administration of unsealed byproduct material requiring a written directive (10 CFR 35.396).

As required by 10 CFR 35.27, the AU is required to provide radiation safety supervision of all staff who receive, possess, use, or transfer radioactive material used in medicine. This supervision must include instruction to such staff on written radiation protection procedures, written directive procedures, regulations contained in 10 CFR 35, and the licensee's license conditions with respect to the use of material. As there is no requirement to have an authorized medical physicist involved in radiopharmaceutical therapies, medical physicists involved in radiopharmaceutical therapy are not required to meet the authorized medical physicist training and experience described in NRC regulations per NRC requirements. However, this does not preclude the States or facilities from having their own minimum training or experience requirements or expectations for medical physicists outside of NRC requirements.

2.2.2 Calibration Requirements

The NRC requirements for determination of dosages used in radiopharmaceutical therapy are contained in 10 CFR 35.63. Requirements state that licensees using unit dosages can determine dosage by direct measurements themselves or by doing decay correction of activity or activity concentration as determined by specific manufacturer or preparer as listed in 10 CFR 35.63. If the dosage is not a unit dosage, then licensees must determine the activity themselves before medical use. If the licensees are measuring the dosage, they must use an appropriately calibrated instrument as required per 10 CFR 35.60, which states that the licensee must calibrate the instrumentation using nationally recognized standards or the manufacturer's instructions. Dose calibration will be discussed further later in this book.

2.2.3 Waste Requirements

Historically, most radiopharmaceutical waste was handled and disposed of following requirements contained in 10 CFR 35.92, which is known as the *decay-in-storage requirement*. This requirement states that a licensee may hold radiological material with a physical half-life of less than or equal to 120 days for decay-in-storage before disposal without regard to its radioactivity if the licensee monitors the material at the surface before disposal and determines that its radioactivity cannot be distinguished from the background radiation level with an appropriate radiation detection survey meter set on its most sensitive scale and with no interposed shielding. The licensee must remove or obliterate all radiation labels, except for radiation labels on materials that are within containers and that will be managed as biomedical waste after they have been released from the licensee.

Today, most radioisotopes used still have a half-life of less than 120 days. However, some of the newer radiopharmaceuticals contain radiological impurities that have half-lives longer than 120 days. When a radiological impurity is known to be present or detected by appropriate survey methods as required by 10 CFR 35.92, then licensees must dispose of the waste material as low-level radioactive waste in accordance with the requirements in 10 CFR Part 20 Subpart K, "Waste Disposal."

2.3 Medical Events for Radiopharmaceutical Therapy

2.3.1 Written Directives

As stated in the NRC medical policy statement, the NRC will minimize intrusion into medical judgments affecting patients and into other areas traditionally considered to be a part of the practice of medicine. In this extent, the NRC will not intrude into the AU's decision to treat a patient or the medical judgment of the AU as to what medical treatment a patient should receive. To ensure radiation safety of patients, the NRC regulations focus on ensuring the patient receives the desired treatment of the AU. Therefore, the NRC regulations focus on a regulatory term known as a *written directive*, which is defined in 10 CFR Part 35 as an AU's written order for the administration of radioactive material or radiation from radioactive material to a specific patient or human research subject. A written directive must be prepared for all radiopharmaceutical therapies, any administration of iodine-131 sodium iodide greater than 1.11 megabecquerels (30 microcuries), and any therapeutic dose of radiation from byproduct materials.

The requirements for written directives are found in 10 CFR 35.40 and 10 CFR 35.2040. All written directives must be dated and signed by an AU before administration, unless the patient's condition was emergent in nature such that a delay to provide a written directive would jeopardize the patient's health. In addition, the written directive must contain the patient or human research subject's name. For radiopharmaceutical therapy, written directives must also include the radioactive drug, dosage, and route of administration. A revision can be made to an existing written directive if the revision is dated and signed by the AU before administration, unless the patient's condition was emergent in nature such that a delay to provide a written directive would jeopardize the patient's health. Licensees must retain a copy of each written directive for a minimum of 3 years.

2.3.2 Medical Event Reporting Criteria for Radiopharmaceutical Therapy

To ensure ongoing radiation safety for patients, the NRC requires licensees to report events when a patient does not receive the treatment the AUs prescribed in their written directive. These events are recorded at the NRC as medical events. Reporting and NRC analysis of medical events helps the NRC and licensees identify root causes of events and ensures that corrective actions are taken to prevent recurrence.

The medical event criteria specifically important to radiopharmaceutical therapy are located below. Note these criteria are modified for simplicity and to focus on those that are typically cited for radiophamaceutical therapy at the time of publication. Please refer to 10 CFR 35.3045 or your Agreement State regulations for a complete and current list of medical event criteria as you evaluate events at your institution.

1. An administration that results in a dose that differs from the dose that would have resulted from the prescribed dosage by more than 0.05 Sv (5 rem) effective dose equivalent, 0.5 Sv (50 rem) to an organ or tissue, or 0.5 Sv (50 rem) shallow dose equivalent to the skin, AND ONE OF THE FOLLOWING:
    - the total dosage delivered differs by a prescribed dosage of 20 percent or more or falls outside the prescribed dose range, or
    - the fractionated dose delivered differs from the prescribed dose for a single fraction, by 50 percent or more.
2. A dose that exceeds 0.05 Sv (5 rem) effective dose equivalent; 0.5 Sv (50 rem) to an organ or tissue, or 0.5 Sv (50 rem) shallow dose equivalent to the skin, AND ONE OF THE FOLLOWING:
    - an administration of a wrong radioactive drug,
    - an administration of a radioactive drug by the wrong route of administration,

- an administration of a dosage to the wrong individual or human research subject, or
- an administration of dosage delivered by the wrong mode of treatment.

3. A dose to the skin or an organ or tissue other than the treatment site that exceeds by:
   - 0.5 Sv (50 rem) or more than expected dose to that site from the procedure if the administration had been given in accordance with the written directive prepared or revised before administration, and
   - 50 percent or more the expected dose to that site from the procedure if the administration had been given in accordance with the written directive prepared or revised before administration.
4. Any event resulting from intervention of a patient or human research subject in which the administration results or will result in unintended permanent functional damage to an organ or a physiological system, as determined by a physician.

2.3.3 Reporting Medical Events

Required Reporting Criteria

The required reporting criteria for NRC licensees are contained in 10 CFR 35.3045 (c)-(g). These specify that licensees should report an event by telephone no later than the next calendar day after discovery of the medical event. NRC licensees should report to the NRC Operations Center at (301) 816–5100. Note, a violation is not associated with all medical events. However, failure to report a medical event within 24 hours of discovery is a violation of NRC requirements. The NRC recognizes the timeframe for reporting is short, but this is necessary for the licensee, industry, and NRC to react if there is a generic issue that may impact other treatments. Therefore, it is recommended to report a suspected event even if the licensee is unsure if the event meets the criteria or not, and then work with the NRC or Agreement State to determine if the incident met the criteria or not. Agreement State licensees should contact their Agreement State to determine how to report an event in their state.

In addition to notifying the NRC or Agreement State, the licensee shall provide notification of the event to the referring physician and also notify the individual who is the subject of the medical event no later than 24 hours after its discovery, unless the referring physician personally informs the licensee either that he or she will inform the patient or that, based on medical judgment, telling the patient would be harmful. The licensee is not required to notify the patient without first consulting the referring physician. If the referring physician or the affected patient cannot be reached within 24 hours, the licensee shall notify the patient as soon as possible thereafter. The licensee may not delay any appropriate medical care for the patient, including any necessary remedial care as a result of the medical event, because of any delay in notification. To meet the requirements of this paragraph, the notification of the patient who is the subject of the medical event may be made instead to that patient's responsible relative or guardian. If a verbal notification is made, the licensee shall inform the patient, or appropriate responsible relative or guardian, that a written description of the event can be obtained from the licensee upon request. The licensee shall provide such a written description if requested. Aside from this notification requirement, nothing in the NRC regulations affect any rights or duties of licensees and physicians in relation to each other or to the patient, or the patient's responsible relative or guardians, affected by the medical event.

In accordance with 10 CFR 35.3045, a more complete written report is required within 15 days after discovery of the medical event. This report must include

- the licensee's name;
- the name of the prescribing physician;
- a brief description of the event;
- why the event occurred;
- the effect, if any, on the individual(s) who received the administration;
- what actions, if any, have been taken or are planned to prevent recurrence; and
- certification that the licensee notified the patient (or the patient's responsible relative or guardian), and if not, why not.

This report may not contain the patient's name or any other information that could lead to identification of the patient. A separate annotated copy of the report shall be provided to the NRC with the name of the patient, patient identifying number, and a copy of the annotated report to the referring physician.

Reporting Best Practices

The information collected on medical events is invaluable in assessing trends or patterns, identifying generic issues or generic concerns, and recognizing any inadequacies or unreliability of specific equipment or procedures. The reported information is critical for initiating a timely and effective understanding as to why the event occurred and identifying any actions necessary to improve the effectiveness of NRC and Agreement State regulatory programs. In order to better support the trending reviews and the NRC's goal of identifying generic issues, the following best practice concepts for reporting should be considered.

When reporting on the root cause, include enough detail that an uninvolved individual would have a full understanding of the medical event, such as the manufacturer or model of the device associated with the event, if applicable; including manufacturer, model, or specifications of supporting equipment associated with the event, such as specific IV pump or gauge size; and reporting specifically what was prescribed as well as what was actually administered. Additionally, include what type of staff was present at the time of the event. With so many new treatment protocols and varying state requirements, do not assume the regulators know the typical staffing for the procedure. Include any relevant information that preceded the event and how the event was identified. Finally, specifically highlight if the event involved a common industry-wide practice or procedure.

When reporting on corrective actions, identify what the short-term and long-term corrective actions are and how they are linked to the event. Include when the corrective actions will be implemented. Finally, indicate when a corrective action may be relevant on a generic level for other procedures, and identify if the corrective action involves a common industry-wide practice or procedure. Finally, when reporting the generic information about the event, it is helpful to provide the information from the manufacturer and all medical and technical information about the event, including any adverse effects that are expected as a result of the event or a statement that no adverse effects are expected.

2.3.4 Dosimetry Requirements

The NRC requires that licensees develop, implement, and maintain written procedures to provide high confidence that each administration is in accordance with the written directive. At a minimum, these procedures must include:

- verification of the identity of the patient or human research subject;
- verification that the administration is in accordance with the treatment plan, if applicable, and the written directive;
- checking both manual and computer-generated dose calculations; and
- determining if a medical event has occurred.

As the medical use of radioactive material is becoming more complex, more is needed in these procedures to ensure high confidence that administrations are given in accordance with the written directive. Most procedures include a number of individuals involved in the treatment process. For example, in radiopharmaceutical therapy, while the authorized user (AU) prescribes the treatment, there are likely other individuals involved in scheduling the procedure, ordering and calibrating the dosage, and delivering the treatment. Therefore, instructions are needed to clearly communicate to all professional team members involved in the treatment process. Complicated processes of this nature require good planning and clear, understandable procedures. To help ensure that all personnel involved in the treatment fully understand instructions in the procedure or written directive, the licensee should instruct all workers to seek guidance with the AU if they do not understand or something seems different. Licensees should also consider verification of written directives by at least one qualified person in addition to the individual who prepared the dosage (e.g., the AU, medical physicist, or nuclear medicine technologist).

As required by 10 CFR 35.41, licensees must determine if a medical event has occurred. This requirement is not specific as to how licensees are expected to assess if a medical event occurred in radiopharmaceutical therapy. Therefore, licensees are not required to perform post-treatment dosimetry to determine if a medical event occurred during a radiopharmaceutical treatment. However, if post-treatment dosimetry does identify a medical event, licensees are required to report it by the next calendar day.

When an incident occurs, licensees may need to perform dosimetry to determine if the medical event criteria effective dose threshold has been met. For many incidents, dosimetry calculations needed for regulatory reasons can be done using dosimetry information contained in the package inserts. However, as procedures become more complex, some incidents may need more detailed dosimetry to determine if a dose threshold was exceeded. This occurred in a recent event where a patient was administered the wrong radiopharmaceutical during an investigational study. The patient was prescribed to receive 0.0405 mCi of Th-227 tagged with a specific antibody-chelator conjugate. However, the Th-227 was incorrectly tagged by the manufacturer with the antigen instead. In this situation, the licensee had to do a dosimetric evaluation, which determined several organs were expected to receive greater than the 0.5 Sv reporting threshold, and the event was reported.

One important point to consider when complex dosimetry is needed, it might take more than 24 hours to complete after the incident is discovered. It is recommended by the NRC to report suspected events even if the dosimetry has not been fully completed. For NRC licensees, reporting a medical event is not a violation on its own, but failure to report a medical event within the required time frame is. By reporting the event early, the licensee can also work with regulators to ensure they do not have questions regarding the dosimetry method being used. If the dosimetric evaluation had determined the event was not reportable, the NRC has a mechanism to retract events from the reporting system.

2.3.5 Root Cause Evaluation of Events

While medical events rarely result in patient harm, the purpose of medical event reporting is to identify the causes of the events in order to correct them, to prevent their recurrence, and to allow the NRC to notify other licensees of the events so they can avoid them. While an incident that occurred might not have caused harm to a patient this time, it is possible that a similar event with a similar cause could result in harm in the future. Therefore, it is important for licensees to conduct a detailed evaluations to determine the root cause of all events to identify ways to minimize reoccurrence. In addition, sharing details of the event with others—including details on the root cause and corrective actions taken to minimize reoccurrence—can help other institutions avoid similar incidents in the future.

To prevent future events, the true root cause of the event needs to be identified and communicated. Evaluations that simplistically identify human factors or training as the cause do not add insights on how to prevent future events. Evaluations should continue to evaluate the incident, looking for causes that might not be obvious at first. Evaluations should include all departments and staff that were involved, especially as treatments become more complex and more departments are becoming involved. When it appears lack of training might be a contributing factor, institutions should look deeper to ensure the staff has the resources and time needed to receive appropriate training. Institutions should also ensure staff had appropriate supervision, especially when a procedure is new to the staff or when a procedure is one they do not do often. When human factors appear to be a contributing factor, the institutions should evaluate if the procedures were adequate to prevent human factors from causing an event. The institution should consider if universal time outs or checklists could prevent a similar type of incident in the future. If the institution identifies a root cause that is related to a standard industry procedure or a federal standard, the institution should report specifics to the regulatory body and industry, if possible, to avoid incidents at other institutions.

Finally, the regulatory body understands some medical events occur due to limitations in the practice of medicine. However, detailed reports of these medical events allow the NRC and the industry to track them and communicate with others, such as researchers and developers.

Many have asked what the NRC does with the medical event information it receives. The NRC evaluates each event to determine if there are any generic implications that all users need to be aware of quickly. Events related to an update in treatment planning software is an example of something the NRC would want to communicate with the industry quickly. In addition, the NRC evaluates all events to determine if there are new trends in types of events or root causes. The more details regarding the root cause contained in reports helps regulators identify if a generic issue is ongoing and communicate that information. When trends are identified, the NRC communicates them to the industry using information notices, which can be found at https://www.nrc.gov/materials/miau/med-use-toolkit/info-notices.html. When an event is related to a device issue, the NRC communicates with manufacturers and the Food and Drug Administration to try to avoid future events. In addition, the NRC and the Advisory Committee of Medical Use of Isotopes make public presentations that describe all reported events in the past year. These presentations describe the details of each event, licensee-identified root causes, and corrective actions to prevent events. These presentations can be found on the NRC medical toolkit website, https://www.nrc.gov/materials/miau/med-use-toolkit.html. Finally, the NRC issues an annual nuclear material events database report which provides details on select significant events and the ten-year trend on total events. These reports can be found at https://nmed.inl.gov/.

2.3.6 Examples of Recent Radiopharmaceutical Therapy Events

At the time of this publication, there were 11 reported radiopharmaceutical therapy medical events in fiscal year 2022 (October 1, 2021, to September 30, 2022) and 9 reported in fiscal year 2021 (October 1, 2020, to September 30, 3021). While these events are a very small fraction of all the radiopharmaceutical therapy procedures performed in the United States, they are important to consider to prevent similar types of events from occurring again. When you look at the following examples, think about if it is possible for a similar event to occur at your facility, and what are some actions you could take to prevent them.

Below are some examples of recent medical events.

Example 1: Standardized doses don't match the written directive dose. Many of the new radiopharmaceutical treatments currently have protocols that use standardized dosages over multiple fractions. However, based on patient factors such as lab results, the AU may need to change the standardized dosages on a case-by-case basis. Over the past two years, three of the reported events occurred when patients received the full standardized dosages for their treatments after the AUs prescribed smaller dosages based on the patients' lab results. All three of these events could have been prevented if the staff administering the dosage had checked the prescription (i.e., written directive) against the vial activity prior to administration. As administrations become more patient-specific, these types of preventable events are expected to become more common. These events highlight the need of administration staff to take a time out before administration for each fraction where they can verify the correct patient and activity prior to administration.

Example 2: Failure to verify vial activity prior to administration. Another event that could have been prevented also involved the administration staff not checking the vial activity versus the prescription (i.e., written directive) prior to administration. In this example, a Xiofigo patient was scheduled to receive 3.47 MBq of Ra-223, but the procedure had to be canceled on the day of treatment due to the patient's low blood pressure. The licensee kept the dosage in the hot lab for decay. Approximately one month later, the patient came back for treatment but was accidentally given the dosage in the original vial, which had since decayed to 0.63 MBq. The patient was able to come back the next day and receive the remaining activity, but this event is another example of the importance of checking both the name and the activity of the desired dosage prior to administration.

Example 3: Protocol schedule not followed. Many of the new radiopharmaceutical treatment protocols include multiple steps or additional treatment administrations in addition to the radiopharmaceuticals, such as alternating radiopharmaceutical fractions with chemotherapy. One recent event occurred when a patient receiving an infusion of Lu-177 informed the AU that chemotherapy had been received the day before, and not after the Lu-177 infusion in accordance with the protocol. The AU immediately stopped the infusion, resulting in an underdose, which is why the NRC was notified. This event demonstrates the importance to check the patient's entire protocol, especially if multiple departments are involved, prior to each administration to ensure that the treatment is going in accordance with the physicians' desires.

Examples 4 and 5: Incorrect setup and administration errors with newer therapies. Over the past two years, the NRC has seen a number of events due to incorrect set up or administration errors as sites begin new radiopharmaceutical therapies and therapies become more complex. In one example, a nurse inadvertently removed a tube occluding clamp and opened the roller clamp on the flush bag line at the beginning of an I-131 Iomab-B treatment. This led to a leaking tube in the infusion system, resulting in the patient receiving only 53% of the prescribed dose. Luckily, no contamination occurred. Following this event, the licensees' corrective action was to ensure the nuclear medicine supervisor and radiopharmacist were trained on the operation of the infusion pump and would be solely in charge of the pump and infusion system for all future patients. They also developed a checklist for setting up the infusion system which the radiopharmacist and nuclear medicine supervisor would go through step-by-step.

Another example of incorrect setup leading to an event happened with a Xiofigo administration. The administration apparatus includes a three-way stopcock to allow for administration of saline and the radium dichloride. In this example, an incorrect cap was used on the unused port of the three-way stopcock. The cap that was used was designed to maintain sterility of the port connection, not to prevent flow through the port, which led to a leak and an underdose to the patient. In another example, a technician had difficulty establishing a patient IV injection site and flow for a Lu-177 Lutathera treatment. After several attempts to establish access, the licensee decided to stop treatment, which resulted in an underdose to the patient. During the licensees' root cause investigation, it was discovered that an incorrect gauge needle was used, contributing to the event.

These examples show the need to ensure all staff is trained on the administration procedures and administration supplies prior to performing administrations. These examples also show the benefits of having checklists developed for radiopharmaceutical procedures.

Chapter 3

Dosimetry in Radiopharmaceutical Therapy: One Physician's Perspective

Chaitanya Divgi, MD
CEO
Divgi Consulting LLC
Meadowbrook, PA

3.1 Introduction

Radiopharmaceutical therapy has, as its underpinning, a comprehensive understanding of radiopharmaceuticals and absorbed dose biodistribution that has, in turn, enabled an understanding of dosimetry. Dosimetry has been used as a guide to determining the appropriate amount of radioactivity for imaging radiopharmaceuticals. It is also being used with increasing frequency to identify organs at risk and determine the possible range of therapeutic radioactivity that may be utilized without unacceptable/irreversible toxicity.

Tumor dosimetry has, however, very rarely been used to clinically guide the amount of administered radiopharmaceutical—this despite the widespread preclinical application of dosimetry to characterize tumor and normal tissue dosimetry to arrive at estimates of therapeutic index. In this author's estimate, there are compelling reasons why this is so, and why this is unlikely to change in the near future. Dosimetry has, however, been invaluable in understanding the range of therapeutic radioactivity in clinical trials, and even more importantly in understanding (and thus protecting) putative organs at risk. The commonly used therapeutic radiopharmaceutical activities are based on, indeed contingent on, an understanding of organ-at-risk dosimetry and, in those instances when clinical imaging of the radiopharmaceutical or its surrogate is feasible, of therapeutic index. These attributes make possible a rational approach to treatment scheduling.

This chapter, with its strong clinical bias, will briefly review the role of radiopharmaceutical dosimetry in selection of the administered activity of approved therapeutic radiopharmaceuticals to illustrate the lack of tumor dosimetry, limiting itself to systemic (oral and intravenous) radiopharmaceutical therapy. We will then summarize the critical role of preclinical dosimetry and speculate on the role of clinical dosimetry in therapeutic radiopharmaceutical development, including the relative lack of an appreciation of the importance of tumor dosimetry in activity determination or treatment scheduling.

3.2 Where We Are Now: Dosimetry in Radiopharmaceutical Therapy

Iodine-131 is arguably the first therapeutic radiopharmaceutical to be used, albeit initially (in the 1940s) in hyperthyroidism, a nonmalignant disorder (Leslie et al. 2003). The group at Memorial Sloan-Kettering Cancer Center, notably Richard Benua and Robert Leeper (Benua et al. 1962) in the 1960s, pioneered the use of dosimetry in patients with thyroid cancer to determine absorbed dose to *organs at risk* (OAR). However, an arbitrary fixed activity was proposed by several groups, eventually becoming the preferred mode of activity determination (Klubo-Gwiezdzinska et al. 2011). The availability of iodine-124, a positron-emitter with a three-day half-life suitable for serial imaging and, thus, dosimetry, enabled tumor visualization and dosimetry (Zanzonico et al. 2015), though it became useful primarily for determining the presence of low iodine-avid (or small volume) disease not visualized with ^{131}I or ^{123}I imaging (Grewal et al. 2016).

The lack of an "imageable" photon precluded imaging in the development of phosphorus-32 (Pandit-Taskar et al. 2004) and strontium-89 (Zenda et al. 2014), radionuclides used to palliate metastatic bone pain. Samarium-153 (which can be assessed with single photon imaging), labeled to a polyphosphonate (EDTMP), was developed along similar lines: activity escalation to *dose-limiting toxicity* (DLT), which was hematopoietic (Serafini et al. 1998).

Radium-223, an alpha emitter initially explored as a metastatic bone pain palliation therapy, was found to increase overall survival in men with bone metastases from *castration-resistant prostate cancer* (CRPC). It was subsequently approved by both U.S. and European regulatory agencies for this indication (Parker et al. 2013). Perhaps not uniquely for therapeutic radiopharmaceutical development, the therapeutic activity and schedule for ^{223}Ra was empirically determined, with the activity escalation trial (Carrasquillo et al. 2013) in the United States being initiated after the agent was already being used in efficacy studies in Europe (Nilsson et al. 2012). The absence of useful photons led to dosimetry not being incorporated in the evaluation of this agent.

The classical phased development route recommended by the FDA—Phase 1 to determine the therapeutic amount, followed by Phases 2 and 3 to determine efficacy and utility—was pioneered in the radioimmunotherapy of lymphoma, with two antibodies, one labeled with yttrium-90 and the other with iodine-131. Unfortunately, neither of these efficacious agents is utilized at this time. Both of these agents target a B-cell receptor, CD20, over-expressed in follicular lymphomas. Yttrium-90 ibritumomab tiuxetan was evaluated in a Phase 1 activity-escalation trial where targeting was demonstrated with indium-111 as the imaging surrogate; dosimetry was not used to determine absorbed dose to bone marrow, the primary organ at risk (Spies 2004). On the other hand, ^{131}I-tositumomab was evaluated in an absorbed dose-based escalation Phase 1 trial, and the whole-body radiation dose (0.75 Gy) was used to determine the variable ^{131}I therapeutic activity (following multiple whole-body measurements of a sub-therapeutic amount of ^{131}I labeled tositumomab) (Wahl 2005). This therapy indeed remains the sole instance of a patient-specific radiopharmaceutical therapy based on normal-organ dosimetry.

Most recently, lutetium-177 labeled peptides/ligands targeting *prostate-specific membrane antigen* (PSMA) have been developed for therapy of metastatic CRPC. Although ^{177}Lu has photons that can be readily imaged using most SPECT devices, dosimetry has not been utilized to determine therapeutic activity. This may have been because imaging demonstrated radiopharmaceutical uptake in the lacrimal and salivary glands, with the therapeutic activity consequently being determined by the extent of toxicity to these organs (Novartis 2022). PET/CT with gallium-68 labeled ligand is routinely used pre-therapy in these patients (NIH); the short half-life precludes meaningful dosimetry (even for these rapidly clearing ligands), and imaging is utilized merely to indicate presence of target on disease.

In summary, therapeutic radiopharmaceutical activity determination has been made without regard to normal organ (or tumor) dosimetry. The rationale for doing so is compelling. Dosimetry as the determinant to activity determination must give way to toxicity-based measurements when the activity calculated based on OAR absorbed dose is so high that only a lower activity is clinically feasible or when the OAR, such as the salivary gland, does not lend itself readily to absorbed dose measurement. (The U.S. National Cancer Institute-based Common Terminology Criteria for Adverse Events (CTCAE) (NIH 2022), is almost universally used for toxicity estimates and classification.) Moreover, radiopharmaceutical dosimetry measurements are contingent, especially for extra-hematopoietic organs at risk, on imaging, which is not feasible for those nuclides that do not emit clinically imageable photons. Finally, conventional absorbed dose-finding studies for systemic cancer therapies are based on CTC parameters of toxicity, and this convention is easy to understand and implement.

Nevertheless, dosimetry has a critical role to play in radiopharmaceutical therapy, as discussed in the following sections.

3.3 The Relevance of Dosimetry in Radiopharmaceutical Therapy

All radiopharmaceutical development is contingent upon dosimetry. Imaging and therapeutic radiopharmaceuticals alike need a robust preclinical assessment that includes biodistribution and targeting studies, usually in a rodent model, that provides dosimetric information adequate to determine the initial radiopharmaceutical activity. When the radiopharmaceutical is being developed as a therapeutic, an appropriate xenograft or other tumor-bearing animal model is essential to establish a therapeutic index—where tumor absorbed dose is determined to be sufficient at putative maximum absorbed dose to organ(s) at risk.

3.3.1 Preclinical Assessments

We will assume for the purposes of this section that an appropriate rodent model exists to assess radiopharmaceutical biodistribution, and that an appropriate tumor-bearing animal model exists to estimate radioactive uptake and retention/clearance in tumors.

Most preclinical radiopharmaceutical dosimetry conducted to establish a starting human activity is carried out in tumor-naïve animals. The radiopharmaceutical is administered to cohorts of animals and its biodistribution is assessed by *in vivo* (animal SPECT or PET) or *ex vivo* ("cut and gamma count") measurements over a period of time that encompasses the biologic clearance of the ligand and the physical half-life of the nuclide. Quantitation of organ residence quantity/time by imaging is reliable with most current animal imaging devices. Advantages include incorporation of the entire organ/body of the same animals over time and a significant reduction in the number of animals required.

Cut-and-count methods are more time-tested. Cohorts of animals are sacrificed at suitable intervals, and their organs and blood are harvested, weighed, and measured (aliquots) in gamma well counters. These methods are, of course, the only methods available for those nuclides that have no imageable photons. Cut-and-count methods require more animals than *in vivo* imaging methods. There is also the concern that since different animals are measured at various time points, differences between the animals may result in variation in measurements. This has largely not been borne out, and there has been good concordance between the two methods (Hoppin et al. 2011). That would, in turn, argue for greater adoption of the imaging method in those instances where the nuclide is imageable, given the reduction in sacrificed animals.

Determination of the therapeutic potential of a radiopharmaceutical needs to be made preclinically in a suitable animal model. Increasing activities of the radiopharmaceutical are administered to cohorts of tumor-bearing animals, and decreases in tumor size are evaluated along with measures of animal well-being (e.g., weight, appetite, and survival). Such studies are indeed valuable in preclinical assessment of therapeutic potential. They need to be accompanied by thorough dosimetric assessments (obtained as in the biodistribution studies mentioned above, in tumor-bearing animals). Dosimetry will provide two unique measurements: the therapeutic index (an estimate of tumor absorbed dose relative to OAR absorbed dose) and, more importantly, the starting activity for the clinical activity escalation (Phase 1A) trial.

Preclinical dosimetry assessments thus determine the starting activity of the imaging/therapeutic radiopharmaceutical, as well as a compelling rationale for development of therapeutic radiopharmaceuticals.

3.3.2 Clinical Dosimetry Assessments

Radiopharmaceuticals are introduced in a clinical trial based on information obtained from preclinical dosimetry, as discussed above.

For nontherapeutic imaging radiopharmaceuticals, this information enables identification of an amount of radioactivity that is generally accepted to not result in "unacceptable" radiation risk to the patient. Predicates abound, simplifying the decision to choose an activity (based on earlier studies using molecules with comparable biodistribution). Nontherapeutic imaging radiopharmaceuticals are frequently used prior to therapeutic studies with the same targeting moiety to establish dosimetry and tumor targeting (and to further buttress the choice of initial therapeutic radioactivity in the concurrent/subsequent activity escalation study). At least the first few (3 to 5) patients would need to have biodistribution studies to verify that the activities chosen from the animal studies do not deliver too high an absorbed dose; in that unlikely event, radioactivity would be appropriately decreased for subsequent patients.

For therapeutic radiopharmaceuticals, the activity would be one that does not deliver to the patient an absorbed dose that is known to injure organs at risk. Most therapeutic radiopharmaceuticals have a nontherapeutic imaging surrogate, either

- lower activity amounts of the therapeutic nuclide (e.g., ^{177}Lu, ^{131}I) or
- an imaging nuclide that has comparable biologic behavior when labeled to the targeting moiety (e.g., ^{111}In for ^{225}Ac, ^{124}I-124 for ^{211}At).

When a radiopharmaceutical enters the first-in-human clinical trial, the criteria that determine radioactivity and sampling times (for pharmacokinetics and imaging for biodistribution and tumor targeting) will have been determined based on careful preclinical dosimetry.

3.3.3 Clinical Trials

The first few (3 to 5) patients studied in the sub/nontherapeutic ("PK") setting in the first-in-human clinical trial will be evaluated primarily to confirm that the preclinical dosimetric assumptions for the amount of administered radioactivity are appropriate. It is important to note that the dosimetric assumptions that are made are appropriate not only for marrow absorbed dose (which can be estimated using serum clearance data) but also for critical normal organs associated with systemic radiopharmaceutical therapy. Such critical organs notably include kidney, liver, and lung, which need serial imaging, planar single photon at the minimum and PET/CT in an ideal setting. In an activity-escalation Phase 1 therapeutic trial, this is usually accomplished in the first cohort of three patients.

It is well recognized that the impetus to build in a therapy protocol in the first-in-human study is compelling. The studies described above may well be incorporated into this schema, ideally by carrying out the PK component prior to administering the therapeutic radiopharmaceutical. While ideal, this can be challenging logistically, since analysis of the PK data needs to be completed before therapy administration to ensure that the activity is not excessive.

With therapeutic nuclides that have imageable emissions (^{177}Lu, ^{131}I), the first cohort can allow imaging and be therapeutic. Ideally, though, analysis of this cohort would need to occur before absorbed dose escalation (or, where feasible, between patients in the cohort) to ensure that preclinical dosimetry assumptions are appropriate.

Therapy escalation in Phase 1 radiopharmaceutical trials is activity-based. However, the initial activity is based on careful dosimetry (preclinical, confirmed in the first clinical cohort); therefore, the activity escalation schema is also an absorbed dose escalation schema.

Phase 1 systemic therapy trials in patients with cancer are "dose-finding," with the goal of assessing safety/toxicity and arriving at a *Recommended Phase 2 Dosage* (RP2D). It is certainly feasible that there is no *"dose" limiting toxicity* (DLT) at the activity predicted to be dose-limiting for a given organ at risk. In those instances where OAR toxicity can be detected in real time (marrow, liver), continued escalation, ideally following the same convention as in the prior escalation steps, is feasible (and perhaps ethically essential when there is evidence of efficacy). Radiopharmaceutical therapy is perhaps unique in that dose can refer to mass of targeting moiety, amount of radionuclide radioactivity, or radiation absorbed dose to normal organs or tumors.

The kidney, however, presents a conundrum since radiation nephropathy is frequently delayed, and there is no consensus on predictive biomarkers for such late toxicity. Careful monitoring of renal function (even when the kidney is not the primary OAR) is thus essential to prepare for this event and also to be better informed for the development of a plan to address this conundrum.

Regardless of the clinical trial design, it is essential that careful dosimetry be carried out in the first few patients in the first-in-human trial. The goal of such dosimetry would be to confirm preclinical absorbed dose assumptions made, and thus to confirm the activity/absorbed dose escalation design.

3.3.4 Limitations

Radiopharmaceutical therapy revolves around normal organ absorbed dose and toxicity. Tumor dosimetry, which is essential for sealed-source radiotherapy, is notably irrelevant for determining the amount of therapeutic radioactivity. However, this is not a unique situation, and it is indeed analogous to other systemic therapies, all of which were developed based on normal organ toxicity (be they chemotherapy, immunobiologic therapy, or inhibitors of enzymatic function). Moreover, sealed-source radiotherapy is also focused on normal organs at risk, which determine not only feasibility and treatment volumes but also fractionation schema and total absorbed dose delivered.

The need to conduct biodistribution/PK studies over time to arrive at meaningful dosimetry estimates is another limitation that is cumbersome for initial patient cohorts. However, this is a sacrifice of the first few patients for the greater good—dosimetry need not be conducted throughout the clinical trial since radiopharmaceutical biodistribution is generally comparable across patients, and normal organ dosimetry results from the initial cohort are generally applicable to all patients for a particular therapeutic radiopharmaceutical.

Obstacles surrounding non-hematopoietic organ dosimetry determination with nuclides that do not emit imageable photons are similarly surmountable. Surrogates with similar physical half-lives, photon/positron energies, and compatible radiochemical properties exist for most therapeutic nuclides.

3.4 Conclusion

Dosimetry is the unsung hero of radiopharmaceutical therapy. It posits therapeutic potential, determines the safe starting activity and schedule, and predicts toxicity. Radiopharmaceutical therapy development would be haphazard and fraught without careful dosimetry. Tumor dosimetry does not play an active role in therapeutic development, but with better positron-emitting nuclides entering clinical investigation, it will further refine therapeutic delivery, particularly with respect to fractionation and dose intensity. The best is yet to come.

References

Benua, R. S., N. R. Cicale, M. Sonenberg, and R. W. Rawson (1962). "The relation of radioiodine dosimetry to results and side effects in the treatment of metastatic thyroid cancer." *Am. J. Roentgenol. Radium Ther. Nucl. Med.* 87:171–82.

Carrasquillo, J. A., J. A. O'Donoghue, N. Pandit-Taskar, J. L. Humm, D. E. Rathkopf, S. F. Slovin, M. J. Williamson, K. Lacuna, A. K. Aksnes, S. M. Larson, H. I. Scher, and M. J. Morris (2013). "Phase I pharmacokinetic and biodistribution study with escalating doses of ^{223}Ra-dichloride in men with castration-resistant metastatic prostate cancer." *Eur. J. Nucl. Med. Mol. Imaging* 40(9):1384–93. doi: 10.1007/s00259-013-2427-6.

Grewal, R. K., A. Ho, and H. Schöder (2016). "Novel Approaches to Thyroid Cancer Treatment and Response Assessment." *Semin. Nucl. Med.* 46(2):109–18. doi:10.1053/j.semnuclmed.2015.10.010.

Hoppin, J., K. D. Orcutt, J. Y. Hesterman, M. D. Silva, D. Cheng, C. Lackas, and M. Rusckowski (2011). "Assessing antibody pharmacokinetics in mice with in vivo imaging." *J. Pharmacol. Exp. Ther.* 337(2):350–58. doi: 10.1124/jpet.110.172916.

Klubo-Gwiezdzinska, J., D. Van Nostrand, F. Atkins, et al. (2011). "Efficacy of dosimetric versus empiric prescribed activity of 131I for therapy of differentiated thyroid cancer." *J. Clin. Endocrinol. Metab.* 96(10):3217–25. doi: 10.1210/jc.2011-0494.

Leslie, W. D., L. Ward, E. A. Salamon, S. Ludwig, R. C. Rowe, and E. A. Cowden (2003). "A randomised comparison of radioiodine doses in Graves' hyperthyroidism." *J. Clin. Endocrinol. Metab.* 88:978–83.

NIH (2022). National Institutes of Health. Cancer Therapy Evaluation Program. https://ctep.cancer.gov/protocoldevelopment/electronic_applications/ctc.htm#ctc_50. Accessed 02 April 2023.

Nilsson, S., P. Strang, A. K. Aksnes, L. Franzèn, P. Olivier, A. Pecking, J. Staffurth, S. Vasanthan, C. Andersson, and Ø. S. Bruland (2012). "A randomized, dose-response, multicenter phase II study of radium-223 chloride for the palliation of painful bone metastases in patients with castration-resistant prostate cancer." *Eur. J. Cancer* 48(5):678–86. doi: 10.1016/j.ejca.2011.12.023.

Novartis (2022). Highlights of Prescribing Information. https://www.novartis.com/us-en/sites/novartis_us/files/pluvicto.pdf. Accessed 02 April 2023.

Pandit-Taskar, N., M. Batraki, and C. R. Divgi (2004). "Radiopharmaceutical therapy for palliation of bone pain from osseous metastases." *J. Nucl. Med.* 45(8):1358–65. PMID: 15299062.

Parker, C., S. Nilsson, D. Heinrich, S. I. Helle, J. M. O'Sullivan, S. D. Fosså, A. Chodacki, P. Wiechno, J. Logue, M. Seke, A. Widmark, D. C. Johannessen, P. Hoskin, D. Bottomley, N. D. James, A. Solberg, I. Syndikus, J. Kliment, S. Wedel, S. Boehmer, M. Dall'Oglio, L. Franzén, R. Coleman, N. J. Vogelzang, C. G. O'Bryan-Tear, K. Staudacher, J. Garcia-Vargas, M. Shan, Ø. S.Bruland, O. Sartor; and ALSYMPCA Investigators (2013). "Alpha emitter radium-223 and survival in metastatic prostate cancer." *N. Engl. J. Med.* 369(3):213–23. doi: 10.1056/NEJMoa1213755.

Serafini, A. N., S. J. Houston, I. Resche, D. P. Quick, F. M. Grund, P. J. Ell, A. Bertrand, F. R. Ahmann, E. Orihuela, R. H. Reid, R. A. Lerski, B. D. Collier, J. H. McKillop, G. L. Purnell, A. P. Pecking, F. D. Thomas, and K. A. Harrison (1998). "Palliation of pain associated with metastatic bone cancer using samarium-153 lexidronam: a double-blind placebo-controlled clinical trial." J. Clin. Oncol. 16(4):1574–81. doi: 10.1200/JCO.1998.16.4.1574.

Spies, S. M. (2004). "Imaging and dosing in radioimmunotherapy with yttrium 90 ibritumomab tiuxetan (Zevalin)." *Semin. Nucl. Med.* 34(1 Suppl 1):10-13. doi: 10.1053/j.semnuclmed.2003.11.004.

Wahl, R. L. (2005). "Tositumomab and (131)I therapy in non-Hodgkin's lymphoma." *J. Nucl. Med.* 46 Suppl 1:128S–40S. PMID: 15653661.

Zanzonico, P., J. A. Carrasquillo, N. Pandit-Taskar, J. A. O'Donoghue, J. L. Humm, P. Smith-Jones, S. Ruan, C. Divgi, A. M. Scott, N. E. Kemeny, Y. Fong, D. Wong, D. Scheinberg, G. Ritter, A. Jungbluth, L. J. Old, and S. M. Larson (2015). "PET-based compartmental modeling of (124)I-A33 antibody: quantitative characterization of patient-specific tumor targeting in colorectal cancer." *Eur. J. Nucl. Med. Mol. Imaging*. 42(11):1700–06. doi: 10.1007/s00259-015-3061-2.

Zenda, S, Y. Nakagami, M. Toshima, S. Arahira, M. Kawashima, Y. Matsumoto, H. Kinoshita, M. Satake, and T. Akimoto (2014). "Strontium-89 (Sr-89) chloride in the treatment of various cancer patients with multiple bone metastases." *Int. J. Clin. Oncol.* 19(4):739–43. doi: 10.1007/s10147-013-0597-7.

Chapter 4

Dose Calibrator QA, Traceability, and Uncertainty

Denis E. Bergeron, Ph.D.[1] and Brian E. Zimmerman, Ph.D.[1]

[1]Radiation Physics Division
National Institute of Standards and Technology
Gaithersburg, MD

4.1 Introduction

Radiopharmaceutical dosages are assayed in terms of activity (SI unit becquerel, Bq). An accurate assay of the activity to be administered helps to assure that the patient receives the minimum absorbed dose required for a quality diagnostic image or for response to therapy. Administered activity is the key initial input for radiopharmaceutical dosimetry in the *medical internal radiation dose* (MIRD) schema (Bolch et al. 2009) and increasing focus on absolute quantitation in medical imaging has brought increased importance to precision in activity assays. Reduced uncertainties in SPECT and PET images allow medical practitioners to measure smaller changes over time (e.g., disease remission or progression) and support comparisons of images acquired at different sites with different scanners. For imaging and therapeutic radiopharmaceuticals alike, most activity assays are achieved with radionuclide calibrators (often referred to as *dose calibrators*).

The term *radionuclide calibrator* refers to a well-type reentrant pressurized ionization chamber with attendant electronics to display a reading directly in activity units. For a detailed discussion of the principles of operation, the reader is referred to Schrader (1997) and Schrader (2007). In the present contexts, a very brief overview is sufficient. The x-rays, γ-rays, and β particles emitted by a decaying radioactive source

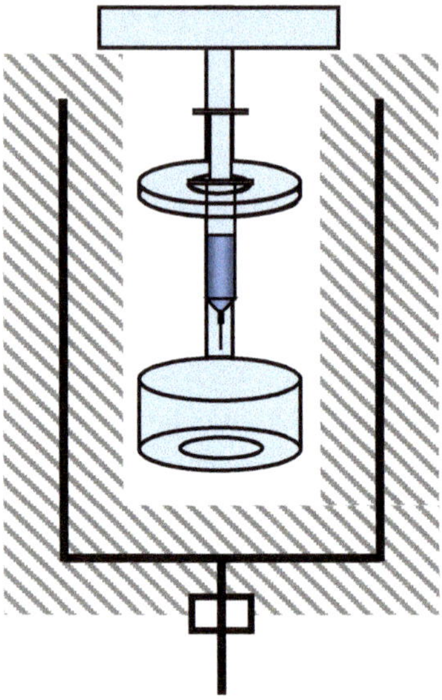

Figure 4.1 Sketch of a reentrant well-type ionization chamber with an electrode configuration (solid black) designed to minimize sensitivity to source displacement in both the vertical and radial axes. A syringe is shown suspended in a "dipper" that places the source near the center of the well. (Adapted from Weiss 1973.)

in the well of an ionization chamber interact with the fill gas, producing ions and electrons that create an electrical current as they move toward electrodes. This current is the instrument response, which can be measured with an electrometer and converted according to a selected calibration setting to give source activity. The ionization chambers in commercially available radionuclide calibrators are constructed with aluminum alloy or steel walls. Variations in wall thickness, subject to manufacturing tolerances, can result in variation in instrument response, especially for radionuclides emitting low-energy photons or β particles. Variations in the pressure of the fill gas may also result in slight variations in instrument response and are sometimes used by manufacturers to match an individual chamber's response to an exemplar. Historically, imperfect seals in some pressurized chambers have resulted in slow gas leaks, leading to gradually decreasing instrument response over time. Source positioning in the well is typically achieved with a "dipper" that places the source near the center of the well, where variations in response due to sample height (or, by proxy, volume) are minimal. In some cases, syringes suspended from a hanger may suffer from decreased geometric efficiency as the solution height approaches the top opening of the chamber (Figure 4.1).

Radionuclide calibrators have served as the main instruments for activity assays in clinical and preclinical settings for many decades now, so many documentary standards and "best practices" guides for radionuclide calibrators have been published (e.g., ANSI 2004; NPL 2006; AAPM 2012). Recently, stakeholder communities have raised concerns that many of these documents require updates. The AAPM has answered this call by forming a new Task Group on Traceability and Standardization for Activity Measurements in Radionuclide Therapy. Similarly, IEC and ISO have convened a joint working group to update relevant documentary standards (https://www.iec.ch/ordsf?p=103:14:613407998629659::::FSP_ORG_ID,FSP_LANG_ID:27382,25).

A common theme in all of these documents, both published and forthcoming, is that reliable routine activity assays must be supported by appropriate quality assurance practices.

4.2 Quality Assurance

Routine quality tests for radionuclide calibrators are described in instrument manuals and in documentary standards. The overview of critical daily and routine tests given here focuses on practical considerations for efficiently establishing and maintaining a reliable measurement system.

Any radionuclide calibrator quality assurance program necessarily assumes proper installation according to the instrument manual (AAPM 212). Appropriate environmental conditions, as described in the manual, should be established and characterized. Some daily tests require a reference data set with established control limits. Daily background tests (see Section 4.2.2), for example, require a reference value to determine whether today's reading deviates from expectations. During installation (or after any instrument repair or change in environmental conditions), a set of background readings should be taken and recorded; the mean of these readings can be adopted as the reference value. The standard deviation of these readings can be used to set control limits; setting the limits at three times the standard deviation means that about 99% of future background readings should fall within the control limits.

A similar approach to reference data should be taken for the daily constancy test (see Section 4.2.3). An excellent graphical way to monitor long-term trends is the *control chart* (ISO 2013). Figure 4.2 gives an example control chart used for daily background checks of a radionuclide calibrator, including several points corresponding to unexpectedly high background readings that would require prompt investigation. The control chart is also very useful as a visual tool for spotting longer-term trends that may be overlooked by simply comparing a daily reading to a reference value.

Some key routine quality assurance tests are described in the remainder of this section. Other tests are recommended in instrument manuals and documentary standards, but they are not covered here. These include repeatability checks to assess the dispersion of readings acquired under controlled conditions and reproducibility checks intended principally to assess variance due to source placement.

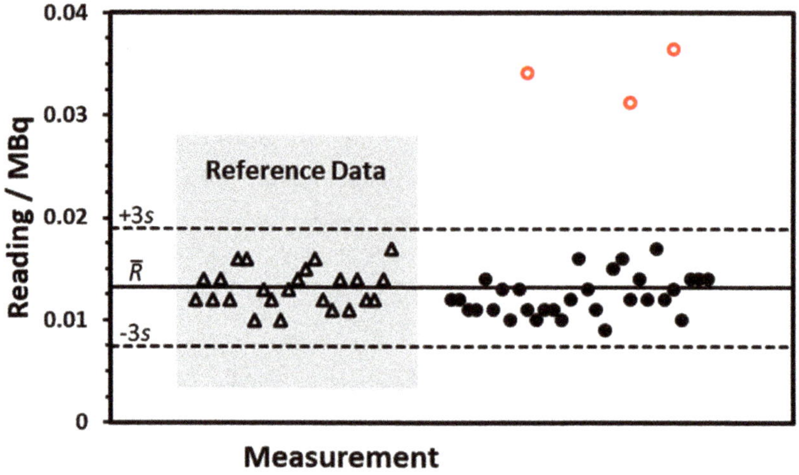

Figure 4.2 A control chart established for evaluating daily background measurements. In the gray box are the reference measurements acquired during commissioning of the radionuclide calibrator. The mean of these readings, $\bar{R}$, is adopted as the reference background and the standard deviation, *s*, is used to calculate response limits. For a system under statistical control, it is expected that ≈99% of readings (for a normal distribution) should fall within ±3*s* of $\bar{R}$. The red points illustrate readings that clearly indicate the system is not in statistical control. Such readings would prompt an investigation to find the source of the elevated background.

4.2.1 Instrument and Data Checks

Daily test protocols for radionuclide calibrators are described in instrument manuals. These tests include a check of the system zero adjustment, high voltage, and built-in data.

The zero adjustment concerns the signal measured by the electrometer when its input is short-circuited and checks the proper functioning of the electronic amplification and measuring circuits. A larger-than-expected drift (see specific instrument manual) during zero adjustment indicates instrument malfunction, requiring repair.

The high voltage check concerns the bias at the terminals of the ionization chamber. Proper function corresponds to bias on the plateau region of the chamber's current-voltage curve, resulting in almost complete collection of the positive and negative ions generated in the gas by a radioactive source. The voltage reading observed in this test should be recorded and compared to the nominal value specified by the manufacturer. The use of a control chart will help to identify any long-term drift. Fluctuation outside of the tolerances stated in the manual or significant drift (especially downward) indicates malfunction, requiring repair.

For some commercial radionuclide calibrators, a software routine checks the built-in nuclide-specific data. Failure may indicate a software problem, and action consistent with the instrument manual should be taken.

4.2.2 Background

The background is the radionuclide calibrator reading observed with no radioactive source or contamination in the well chamber or in its immediate surroundings. Background measurements must be made with the source holder inside the ionization chamber. As part of a daily check, a background reading should be recorded and compared to the reference value.

A higher-than-expected background reading usually indicates contamination or an unexpected source nearby. In the case of contamination, it is usually possible to isolate the contaminated equipment (most often this is the source holder or protective well liner) and decontaminate. If the elevated background reading is attributed to a nearby source, then the preferred solution is to remove the source from the vicinity. In the event that source removal is not possible, then background subtractions can be applied to activity measurements. Ideally, this involves bracketing a source measurement with background measurements before and after; the average background reading can then be subtracted from the source reading. On some radionuclide calibrators, it is possible to implement an automatic background subtraction, but this option should be used only with caution since an elevated background caused by a nearby source may decrease over time as the source decays.

4.2.3 Accuracy and Constancy

The daily constancy test is intended to check the stability of a radionuclide calibrator's response over a relatively long period of time. It can identify electrometer failure or slow loss of ionization chamber pressure. The test is performed with a long-lived check source of sufficient activity to yield a response at least 100 times larger than background and suited to repeated use for an extended period of time. As part of a daily check, a check source reading should be recorded and compared to the decay-corrected reference value for that source at that calibration setting.

A higher-than-expected reading for the check source may indicate elevated background (though this should have been identified during the daily background check), contamination of the check source, or an operator error (wrong calibration setting, wrong holder, wrong decay correction, etc.). An other-than-expected reading may indicate compromised integrity of the check source, failing electronics, loss of ionization chamber pressure, or operator error. A failure of the daily constancy check should prompt an investigation and remeasurement. In the event the problem cannot be resolved, repair or recalibration may be necessary.

The accuracy test is intended to confirm that the radionuclide calibrator gives the correct reading for a traceably calibrated reference source when measured at the appropriate calibration setting. If the constancy test is performed at the appropriate calibration setting and with a traceably calibrated reference source, then a dedicated accuracy test is unnecessary. Otherwise, accuracy tests are typically performed annually.

4.2.4 Linearity

The linearity check is intended to confirm that the radionuclide calibrator gives a predictable linear response over the full range required for routine assays. This test is usually performed with the decaying source method. A source of a short-lived radionuclide with high initial activity is measured repeatedly as the source decays, thus sampling the different electrometer ranges. The decay-corrected response is expected to be constant over time if the radionuclide calibrator response is linear. The decaying source method requires a source of a radionuclide with a well-known half-life and with very low radioimpurities. In the nuclear medicine setting, ^{18}F (as FDG) and ^{99m}Tc are commonly used.

An alternative approach is to use a set of radioactive sources with well-known activity ratios to sample the relevant electrometer ranges. Suitable commercial products for this purpose are available.

4.3 Traceable Calibrations

4.3.1 Overview of Activity Standards for Radiopharmaceuticals

The National Metrology Institute for the United States is the National Institute of Standards and Technology (NIST). NIST is responsible for the development and dissemination of activity standards. National standards for activity are based on measurements with *primary methods*, as defined by the International Vocabulary of Metrology (VIM; JCGM/WG2 2012). Put simply, a primary measurement method is one used to obtain a measurement result without reference to a standard certified for a quantity of the same kind. For radioactivity, most primary measurement methods today are based on liquid scintillation counting (Collé 2009).

NIST standards for radionuclides currently used in Food and Drug Administration (FDA)-approved radiopharmaceuticals are summarized in Table 4.1. In addition to these, NIST has recently developed standards for several radionuclides with emerging applications in radiotherapy: ^{212}Pb, ^{224}Ra, and ^{227}Th (Bergeron et al. 2022a; Napoli et al. 2020).

4.3.2 Establishing Traceability to Standards

The key to metrological traceability to a standard is the "documented unbroken chain of calibrations, each contributing to the measurement uncertainty" (JCGM/WG2 2012). Traceability is thus a hierarchical concept with uncertainty increasing at each level as the number of calibrations separating a measurement from a primary standard increases. Figure 4.3 depicts this hierarchy with the International System of Units (SI) at the top of a pyramid. National Metrology Institutes (NMIs) such as NIST realize the SI by measurements with primary method. The International Bureau of Weights and Measures (BIPM) facilitates "Key Comparison" exercises to assist NMIs in establishing equivalence of standards (Karam 2007). NMIs can also establish equivalence of measurement standards by submitting calibrated sources for measurement on the International Reference System (SIR) (Ratel 2007), or by hosting the Transfer Instrument of the International Reference System (SIRTI) (Michotte et al. 2013).

As discussed by Zimmerman and Judge (2007), the roles played by and relationships among Secondary Standards Laboratories, radiopharmacies, and clinical end-users can vary from country to country and radionuclide to radionuclide. In cases

Table 4.1 A listing of radionuclides used in FDA-approved radiopharmaceuticals for which the National Institute of Standards and Technology (NIST) has established activity standards

Radionuclide	Reference
F-18	Fitzgerald et al. 2014
Cu-64	Bergeron et al. 2018
Ga-67	Michotte et al. 2021
Rb-82	Hoppes et al. 1987
Sr-89	Coursey et al. 1998
Y-90	Coursey et al. 1993
Tc-99m	Michotte et al. 2010
In-111	Zimmerman and Cessna 2018
I-123	Bergeron et al. 2014
I-125	Zimmerman and Cessna 2000
I-131	Zimmerman and Bergeron 2016
Xe-133	Zimmerman and Cessna 2000
Lu-177	Bergeron and Cessna 2018
Tl-201	Coursey et al. 1990
Ra-223	Zimmerman et al. 2015

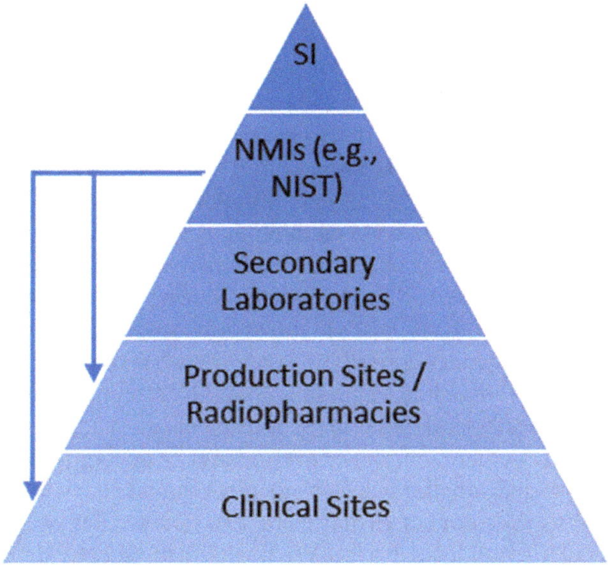

Figure 4.3 Traceability of radioactivity measurement is a hierarchical concept. National Metrology Institutes (NMIs) like NIST establish traceability to the International System of Units (SI) and work with secondary laboratories, production sites, radiopharmacies, and clinical sites to enable traceable calibrations. More commonly, clinical sites establish traceability to NIST through a chain of calibrations.

where NMIs interact directly with production sites, radiopharmacies, or even clinical sites, there are fewer links (calibrations) in the traceability chain, so uncertainties may be smaller.

For clinical sites, traceability is established by calibrating local instrumentation with a source with activity that is traceable to a national standard. Procedures and special considerations for achieving traceable calibration with a radionuclide calibrator are discussed in Section 4.3.3.

Radionuclide calibrator manufacturers recommend calibration settings for some clinically important radionuclides. Such recommendations may be found in tables in instrument manuals or even as nuclide-specific "buttons" on the control unit. Before using recommended settings of this type, users should confirm that the indicated source geometry (i.e., specific container and volume) is consistent with the source to be measured. Often, the manufacturer-recommended setting is valid for a NIST-traceable flame-sealed 5 ml glass ampoule. The expected radionuclide calibrator response for a plastic syringe in the hanging position may be significantly different (see Section 4.3.3). Moreover, local shielding and backscatter conditions can vary significantly from site to site, affecting the appropriate calibration setting. Users should always verify calibration settings using a traceably calibrated source on their instrument under the measurement conditions that will be encountered during clinical assays.

Similarly, radionuclide calibrator settings for various radionuclides can be found in the literature. Here again, differences in source geometry and local conditions must be considered. In addition, many calibrations are based on gamma-ray spectrometry measurements, potentially with no traceability to primary standards. In such cases, uncertainties associated with the nuclear decay data (gamma-ray emission intensities) and spectrometer calibration (Lépy et al. 2015) are often underestimated. Sometimes authors provide enough description of their measurement conditions and raw data to make it possible to account for updates in relevant decay data or differences in source geometry (see, e.g., Beattie et al. 2014 and Bergeron et al. 2019), but such cases seem to be the exception, not the rule. Even for calibrator settings published by NMIs with clear documentation of the activity calibration and uncertainty assessment, users should verify calibration settings locally using a traceably calibrated source under the measurement conditions that will be encountered during clinical assays.

Suppliers of drug products will provide a traceable activity with each shipped source and may also give recommendations on procedures and settings for checking the activity with a radionuclide calibrator. Since local conditions may vary, users should verify calibration settings on their instrument under the measurement conditions that will be encountered during clinical assays. Any large discrepancy between local measurements and supplier recommendations should be investigated and, if confirmed, reported to the supplier of the calibrated source. For drug products, users should always defer to guidance from the radiopharmacy/supplier.

4.3.3 Practical Calibration

With a traceably calibrated source, the correct calibration setting for a radionuclide calibrator can be determined by "dialing in" or by calibration curve (Zimmerman and Cessna 2000). The *dialing-in method* simply requires the user to adjust the calibration

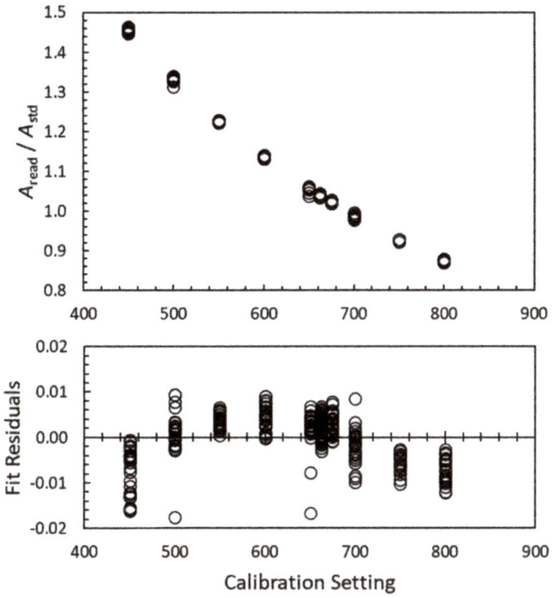

Figure 4.4 Data acquired with a Capintec CRC-15R radionuclide calibrator with a calibrated ^{212}Pb source. The standard activity is read with a calibration setting of 690. The residuals of a fit to the data are used to calculate an uncertainty component that is combined with the uncertainty on the standard activity and the uncertainty from the decay correction to give the combined standard uncertainty on the activity. These data were reported in (Bergeron et al. 2022a).

setting to give the correct (decay-corrected) activity for the calibrated source. The uncertainty on the calibrated activity should be combined in quadrature with the uncertainty associated with the instrument reading, the decay correction, and possibly other factors (see Section 4.4).

The *calibration curve method* can be useful if the exact activity of the source is not known at the time of calibration. In the clinical setting, this might occur if the source material has been transferred to another container and corrections for activity loss cannot immediately—but will ultimately—be made. To build the calibration curve, readings should be taken at a series of calibration settings, ideally centered on the setting that gives the approximate expected activity and covering a wide range (e.g., readings are ±50% of the approximate expected activity). After the data are collected and all readings are decay-corrected to a common reference time, they may be fit to a curve (see Figure 4.4[1]). The uncertainty on the calibrated activity should be combined in quadrature with the uncertainty estimated from the fit residuals.

1. Certain commercial equipment, instruments, or materials like those identified in the caption to Figure 4.4 are made to foster understanding. Such identifications do not imply recommendation by the National Institute of Standards and Technology, nor do they imply that the materials or equipment identified are necessarily the best available for the purpose.

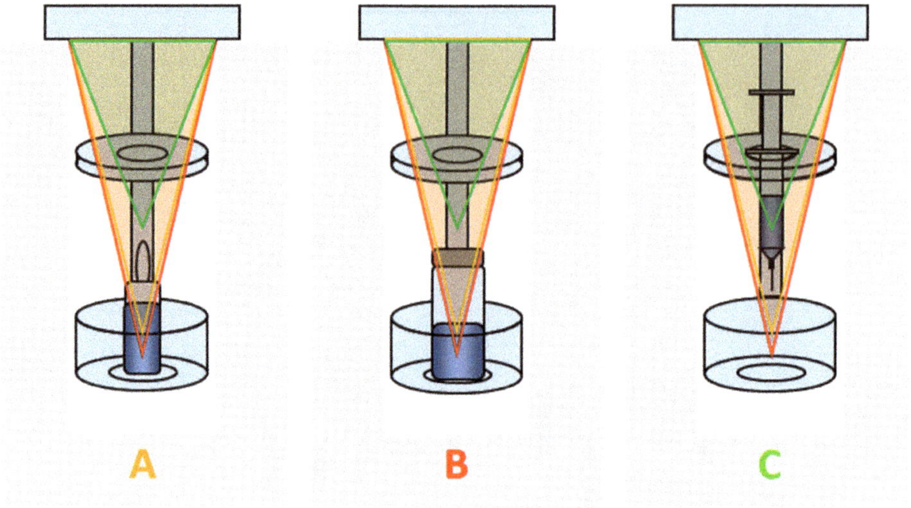

Figure 4.5 Radionuclide calibrator response can be sensitive to source geometry. In this graphic, a flame-sealed ampoule (a) contains the same volume as a dose vial (b). A hanging syringe (c) positions the source much higher in the well-type ionization chamber (illustrated in gray), reducing geometric efficiency.

Often, the traceably calibrated source will be provided in a geometry (i.e., container and volume) that differs from the most convenient geometry for clinical calibrations. For example, ^{177}Lu sources are provided in dose vials, but practitioners often prefer to make radionuclide calibrator measurements in syringes. As illustrated in Figure 4.5, radionuclide calibrator response may be very sensitive to changes in source geometry. The solution in the wider vial has a shorter column height and photons have, on average, more solution to traverse before interacting with the ionization chamber gas volume. Increased attenuation in the solution volume may result in decreased calibrator response. In addition, the change in solution height may change the efficiency of the response slightly as the number of photons incident on the sensitive volume of the chamber changes. Finally, the difference in height affects the geometric efficiency of the radionuclide calibrator. This is best illustrated by comparing the triangles in (a), (b), and (c). The green triangle originates from the approximate center of the syringe source shown in (c). The yellow and red triangles originating from the ampoule and vial are much more acute. The triangles illustrate how the solid angle subtended by the ionization chamber opening increases with source height, allowing more photons to escape without interacting with the ionization chamber gas volume. Similar height effects result when sample volume increases. Finally, in cases like ^{177}Lu or ^{90}Y where bremsstrahlung contributes significantly to radionuclide calibrator response, differences in the container composition or wall thickness must be considered.

Users can determine the appropriate calibration setting for a given source geometry by quantitatively transferring a calibrated source to the preferred geometry. A simple procedure would be:

1. ***Measure the activity of a calibrated source in a geometry*** (e.g., a 10 ml dose vial containing 2 ml of solution) for which a local radionuclide calibrator setting has been confirmed.

2. ***Transfer the entire active solution to the new geometry*** (e.g., a 3 ml syringe containing 2 ml of solution). Measure the emptied container on the radionuclide calibrator to confirm that the amount of residual active solution is negligible (the reading should be < 1% of the reading for the full source).

3. ***Measure the new container in the configuration to be adopted for future assays.*** Adjust the calibration setting until the correct activity (determined in the preceding steps) is displayed.

Similar procedures can be developed for dilutions or partial transfers by mass or by volume. Confidence in residual measurements can be improved by establishing volume corrections. This is most easily achieved by adding a nonradioactive solution to an initially small volume of a radioactive solution, assuring homogeneous mixing, and measuring the radionuclide calibrator response at numerous volumes. The total activity is the same for all points, so any change in the radionuclide calibrator reading implies a volume dependence of the radionuclide calibrator response that may require correction.

Calibrations can be more complicated for some emerging therapeutic agents based on decay chain radionuclides. There can be a period of time during which the ingrowth of progeny complicates the expected radionuclide calibrator response, and the usual decay corrections with a single decay constant do not apply. The interested reader is referred to Collins et al. (2019) and Bergeron et al. (2022b), but for here it is sufficient to point out that for these radionuclides it is even more important to carefully follow guidance from the radiopharmacy/supplier.

4.4 Uncertainty Evaluation for Dose Calibrator Measurements

The term *uncertainty* is often used interchangeably with *error*. However, the term error refers to the difference between a measured value and a true value. When error is systematic, it is referred to as *bias*. Identified sources of bias can be corrected; the corrections carry uncertainty. As stated in the GUM (JCGM/WG 1 2008),

> *"...when all of the known or suspected components of error have been evaluated and the appropriate corrections have been applied, there still remains an uncertainty about the correctness of the stated result, that is, a doubt about how well the result of the measurement represents the value of the quantity being measured."*

Uncertainty, then, is a measure of how well a value is known. Without some expression of uncertainty, measurement results cannot be meaningfully interpreted or compared.

For radionuclide calibrator measurements, complications associated with decay corrections can provide a salient illustration of the difference between bias and uncertainty. Some radiopharmaceuticals are distributed in "unit dose" form, with a stated activity calibrated for a predetermined reference time when patient administration is planned. When the unit is administered before or after the reference time, a bias is introduced. The bias is corrected by applying the usual decay correction:

$$A_{admin} = A_{ref}e^{-\lambda(t_{admin}-t_{ref})} \tag{4.1}$$

where the administered activity A_{admin} is greater than or less than the unit dose activity A_{ref}, depending on the difference between the actual administration time, t_{admin}, and the reference time for the unit dose activity, t_{ref}. Each variable in the correction equation carries an associated uncertainty. The uncertainty on the decay constant, λ, will typically be very small, but the uncertainty on the reference activity and on the administration and reference times must be combined to calculate the uncertainty on the administered activity.

As an example, Figure 4.6a illustrates the uncertainty on a ^{18}F activity assay due to the uncertainty on the decay constant. The figure shows that a 1 h decay correction introduces an uncertainty on the activity of < 0.005%; for a 4 h decay correction, it would be 0.02%. These uncertainties are not expected to be clinically significant. On the other hand, Figure 4.6b shows the bias on the administered activity resulting from a difference between the reference time and the administration time. This bias can, of course, be corrected using Equation (4.1), but it is important to account for the uncertainty on the correction arising from the uncertainty on the time difference ($t_{admin} - t_{ref}$). If, for example, we are correcting for a 6.5% bias due to administration of a unit dose 10 min before the reference time, but there is a 3 min uncertainty on the difference between the administration time and the reference time, then the correction itself introduces a 1.9% uncertainty on the administered activity.

For a routine activity assay with a radionuclide calibrator, the administered activity (A_{admin}) at t_{admin} might be calculated according to a measurement equation like the following:

$$A_{admin} = \left(A_f \times C_{f,decay} \times C_{f,vol}\right) - \left(A_r \times C_{r,decay} \times C_{r,vol}\right) \tag{4.2}$$

where A_f is the radionuclide calibrator reading for a full syringe prior to administration, and A_r is the radionuclide calibrator reading of the residual activity left in the syringe after administration. $C_{f,decay}$ is the appropriate decay correction (see Equation (4.1)) to t_{admin} and $C_{f,vol}$ is a previously determined volume correction, accounting for a difference between the volume in the syringe being assayed and the volume for which the calibration setting was determined. The decay and volume corrections appropriate to the assay of the residual material ($C_{r,decay}$ and $C_{r,vol}$) must also be applied.

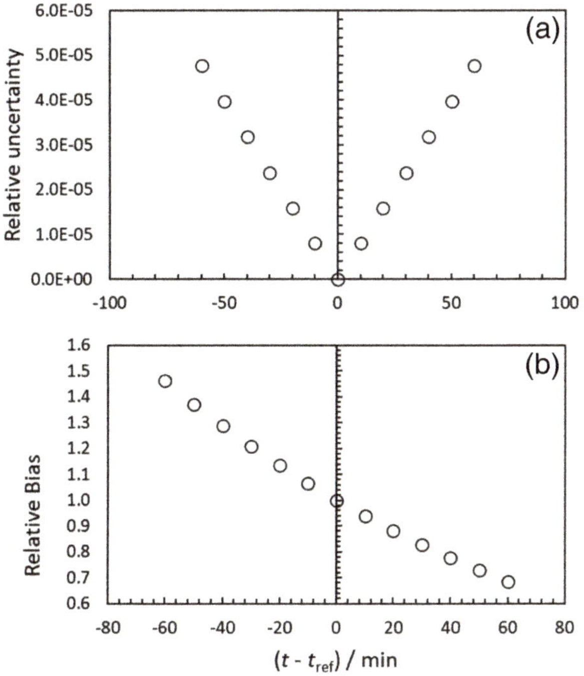

Figure 4.6 (a) The relative uncertainty on a ^{18}F activity assay resulting from the uncertainty on the decay constant ($T_{1/2}$ = 1.82890(23) h; DDEP 2022) plotted as a function of time relative to the reference time. (b) The relative bias on a ^{18}F activity assay plotted as a function of time relative to the reference time; this bias is corrected by applying the decay correction, as in Equation (4.1).

The overall uncertainty in A_f can be estimated by combining the standard deviation of repeated activity readings from the radionuclide calibrator, $s_{A(f)}$, estimated from the standard deviation on repeat readings or "by eye" according to the range or observed readings over a short period, and $u_{setting}$, the uncertainty (in activity units) from the calibration setting used[1] that will, in most cases, be the largest contributor to the combined uncertainty, as follows:

$$\frac{u_{A_f}}{A_f} = \sqrt{\left(\frac{s_{A(f)}}{A_f}\right)^2 + \left(\frac{u_{setting}}{A_f}\right)^2} \tag{4.3}$$

1. Recall that the radionuclide calibrator actually measures current. Thus, $u_{A,f'}$ is an estimate of the uncertainty on the current reading. To get A_f in activity units, a factor based on the calibration setting is applied; $u_{setting}$ expresses the uncertainty associated with the calibration, including the uncertainty on the activity of the calibrated source used to determine the correct calibration setting.

We can next consider the corrected activity reading, A'_f, which is the ($A_f \times C_{f,decay} \times C_{f,vol}$) term in Equation (4.2). Its relative uncertainty can be calculated according to

$$\frac{u_{A(f')}}{A'_f} = \sqrt{\left(\frac{s_{A(f)}}{A_f}\right)^2 + \left(\frac{u_{setting}}{A_f}\right)^2 + \left(\frac{u_{f,decay}}{C_{f,decay}}\right)^2 + \left(\frac{u_{f,vol}}{C_{f,vol}}\right)^2} \tag{4.4}$$

where $u_{f,decay}$ is the uncertainty on the decay correction, including any uncertainty due to timing (see above) and $u_{f,vol}$ is the uncertainty on the volume correction. Equation (4.4) shows that for terms that are multiplied in the measurement equation, the relative uncertainties are added in quadrature.

The second term in Equation (4.2) is the correction for the residual activity left in the syringe after administration. The radionuclide calibrator reading for the residual activity, A_r, is corrected for decay and volume as A_f to calculate A'_r and its uncertainty, $u_{A(r')}$, is estimated in the same manner. Finally, the combined standard uncertainty on A_{admin} in Equation (4.2) is estimated by combining the uncertainties from A'_f and A'_r. When terms are added or subtracted, their absolute uncertainties are added in quadrature, as

$$u_{A(admin)} = \sqrt{u^2_{A(f')} + u^2_{A(r')}}\,. \tag{4.5}$$

So, a very large relative standard uncertainty on the residual will contribute minimally to the combined relative standard uncertainty on A_{admin}, assuming the residual activity itself is relatively small.

In some cases, the measurement equation may include more corrections than are depicted in the example shown in Equation (4.2). The principles for combining relative and absolute uncertainties are general and can be applied.

Occasionally, it may be more convenient to propagate uncertainties through a complex set of calculations by varying the input parameters according to their uncertainties (JCGM/WG 1 2008; JCGM/WG 1 2020). This may be carried out analytically or using Monte Carlo methods.

Finally, the uncertainty estimators discussed in this section are assumed to be "unexpanded." That is, they correspond to one standard deviation. The combined standard uncertainty can be multiplied by a coverage factor (expressed as "k") to give an expanded uncertainty with increased statistical probability that the "true" value of the quantity being measured falls within the range described. Ignoring statistical caveats, in general use, a coverage factor of $k = 2$ corresponds approximately to the 95% confidence band, while $k = 3$ corresponds to 99% confidence.

Acknowledgments

We are grateful to NIST colleagues Ron Collé, Ryan Fitzgerald, and Bert Coursey for critical insights and valuable discussions of this material.

References

American Association of Physicists in Medicine (AAPM) (2012). "The Selection, Use, Calibration, and Quality Assurance of Radionuclide Calibrators Used in Nuclear Medicine." College Park: AAPM, Report of AAPM Task Group 181.

American National Standards Institute (ANSI) (2004). "Calibration and Usage of 'Dose Calibrator' Ionization Chambers for the Assay of Radionuclides." National Committee on Radiation Instrumentation, N42, Institute of Electrical and Electronic Engineers, Inc., New York.

Beattie, B. J., K. S. Pentlow, J. O'Donoghue, and J. L. Humm (2014). "A recommendation for revised dose calibrator measurement procedures for ^{89}Zr and ^{124}I." *PLoS ONE* 9:e106868.

Bergeron, D.E., J. T. Cessna, D. B. Golas, R. K. Young, and B. E. Zimmerman (2014). "Dose calibrator manufacturer-dependent bias in assays of ^{123}I." *Appl. Radiat. Isot.* 90:79–83.

Bergeron, D. E. and J. T. Cessna (2018). "An update on 'dose calibrator' settings for nuclides used in nuclear medicine." *Nucl. Med. Comm.* 39:500–4.

Bergeron, D. E., J. T. Cessna, R. Fitzgerald, L. Pibida, andB. E. Zimmerman (2018). "Standardization of ^{64}Cu activity." *Appl. Radiat. Isot.* 139:266–73.

Bergeron, D. E., J. T. Cessna, R. Fitzgerald, L. Pibida, and B. E. Zimmerman (2019). "Standardization of I-124 by three liquid scintillation-based methods." *Appl. Radiat. Isot.* 154:108849.

Bergeron, D. E., J. T. Cessna, R. P. Fitzgerald, L. Laureano-Pérez, L. Pibida, and B. E. Zimmerman (2022a). "Primary standardization of ^{212}Pb activity by liquid scintillation counting." *Appl. Radiat. Isot.* 190:110473.

Bergeron, D. E., K. Kossert, S. M. Collins, and A. J. Fenwick (2022). "Realization and dissemination of activity standards for medically important alpha-emitting radionuclides." *Appl. Radiat. Isot.* 184:110161.

Bolch, W. E., K. F. Eckerman, G. Sgouros, and S. R. Thomas (2009). "MIRD Pamphlet No. 21: A generalized schema for radiopharmaceutical dosimetry—standardization of nomenclature." *J. Nucl. Med.* 50:477–84.

Collé, R. (2009). "Radionuclidic standardization by primary methods: An overview." *J. Radioanal. Nucl. Chem.* 280:265–73.

Collins, S. M., J. D. Keightley, P. Ivanov, A. Arinc, S. M. Jerome, A. J. Fenwick, and A. K. Pearce (2019). "The potential radio-immunotherapeutic α-emitter 227^{Th} – part I: standardization via primary liquid scintillation techniques and decay progeny ingrowth measurements." *Appl. Radiat. Isot.* 145:240–50.

Coursey, B. M., D. D. Hoppes, A. T. Hirschfeld, S. M. Judge, D. H. Woods, M. J. Woods, E. Funck, H. Schrader, and A. G. Tuck (1990). "Standardization and decay scheme of ^{201}Tl." *Appl. Radiat. Isot.* 41:289–91.

Coursey, B. M., J. M. Calhoun, and J. T. Cessna (1993). "Radioassays of yttrium-90 used in nuclear medicine." *Nucl. Med. Biol.* 20:693–99.

Coursey, B. M., F. J. Schima, D. B. Golas, O. T. Palabrica, A. Suzuki, and M. A. Dell (1998). "Measurement standards for strontium-89 for use in bone palliation." *Appl. Radiat. Isot.* 49:335–44.

Decay Data Evaluation Project (DDEP). http://www.lnhb.fr/nuclear-data/nuclear-data-table/, accessed November 2022.

Fitzgerald, R., B. E. Zimmerman, D. E. Bergeron, J. C. Cessna, L. Pibida, and D. S. Moreira (2014). "A new NIST primary standardization of ^{18}F." *Appl. Radiat. Isot.* 85:77–84.

Hoppes, D. D., B. M. Coursey, F. J. Schima, and D. Yang (1987). "National Bureau of Standards Decay-Scheme Measurements of Strontium-82-Rubidium-82." *Appl. Radiat. Isot.* 38:195.

International Organization for Standardization (ISO) (2013). "Control charts—Part 2: Shewhart control charts." Geneva: ISO, 7870-2:2013(E).

Joint Committee for Guides in Metrology Working Group 2 (JCGM/WG 2) (2012). *International vocabulary of metrology—Basic and general concepts and associated terms (VIM)*, 3rd ed. Sevres: BIPM, JCGM 200-2012.

Joint Committee for Guides in Metrology Working Group 1 (JCGM/WG 1) (2008). *Evaluation of measurement data—Guide to the expression of uncertainty in measurement.* Sevres: BIPM, JCGM 100-2008.

Joint Committee for Guides in Metrology Working Group 1 (JCGM/WG 1) (2020). *Guide to the expression of uncertainty in measurement—Part 6: Developing and using measurement models.* Sevres: BIPM, JCGM GUM-6:2020.

Karam, L. (2007). "Application of the CIPM MRA to radionuclide metrology." *Metrologia* 44:S1-S6.

Lépy, M. C., A. Pearce, and O. Sima (2015). "Uncertainties in gamma-ray spectrometry." *Metrologia* 52:S123.

Michotte, C. and R. Fitzgerald (2010). "Activity measurements of the radionuclide ^{99m}Tc for the NIST, USA in the ongoing comparison BIPM.RI(II)-K4.Tc-99m." *Metrologia* 47:06027.

Michotte, C., M. Nonis, C. Bobin, T. Altzizoglou, and G. Sibbens (2013). "The SIRTI: a new tool developed at the BIPM for comparing activity measurements of short-lived radionuclides world-wide." Rapport BIPM-2013/02.

Michotte, C., S. Courte, M. Nonis, R. Coulon, S. Judge, G. Ratel, R. Fitzgerald, K. Kossert, and O. Nähle (2021). "Update of the BIPM comparison BIPM.RI(II)-K1.Ga-67 of activity measurements of the radionuclide ^{67}Ga to include the 2010 result of the PTB (Germany) and the 2010 result of the NIST (United States)." *Metrologia* 58:06017.

Napoli, E., J. T. Cessna, R. Fitzgerald, L. Pibida, R. Collé, L. Laureano-Pérez, B. E. Zimmerman, and D. E. Bergeron (2020). "Primary standardization of ^{224}Ra activity by liquid scintillation counting." *Appl. Radiat. Isot.* 155:108933.

National Physical Laboratory (NPL) (2006). "Protocol for Establishing and Maintaining the Calibration of Medical Radionuclide Calibrators and Their Quality Control." NPL, Measurement Good Practice Guide No. 93.

Ratel, G. (2007). "The Systéme International de Référence and its application in key comparisons." *Metrologia* 44:S7–16.

Schrader, H. (1997). *BIPM Monographie-4: Activity Measurements with Ionization Chambers.* Bureau International Des Poids et Mesures, Sevres, France.

Schrader, H. (2007). "Ionization chambers." *Metrologia* 44:S53.

Weiss, H. M. (1973). "4π γ-ionization chamber measurements." *Nucl. Instrum. Meth.* 112:291–97.

Zimmerman, B. E. and J. T. Cessna (2000). "Experimental determinations of commercial 'dose calibrator' settings for nuclides used in nuclear medicine." *Appl. Radiat. Isot.* 52:615–19.

Zimmerman, B. E. and S. Judge (2007). "Traceability in nuclear medicine." *Metrologia* 44:S127–32.

Zimmerman, B. E., D. E. Bergeron, J. T. Cessna, R. Fitzgerald, and L. Pibida (2015). "Revision of the NIST standard for ^{223}Ra: New measurements and review of 2008 data." *J. Res. NIST* 120:37–57.

Zimmerman, B. E. and D. E. Bergeron (2016). "(Mis)use of ^{133}Ba as a calibration surrogate for ^{131}I in clinical activity calibrators." *Appl. Radiat. Isot.* 109:250–53.

Chapter 5

Patient Release Criteria and Instructions

Wesley E. Bolch[1], Ph.D., PE, FAAPM
and Adam Kesner[2], Ph.D.,DABR

[1]Director, Advanced Laboratory for Radiation Dosimetry Studies
J. Crayton Pruitt Family Department of Biomedical Engineering
University of Florida
Gainesville, FL

[2]Department of Medical Physics
Memorial Sloan Kettering Cancer Center
New York, NY

5.1 Introduction

Important radiological safety aspects in the use of radiopharmaceuticals for cancer therapy are the conditions and regulations under which the patient may be released from the hospital following treatment. Historically, the U.S. Nuclear Regulatory Commission (NRC) and its Agreement States required patients receiving radiopharmaceutical therapy to remain hospitalized until the retained activity in the patient was less than 1.11 GBq (30 mCi) or that the dose rate at 1 m from the patient was less than 0.05 mSv/h. However, the NRC amended its regulations concerning radiopharmaceutical therapy patients in a revision to 10 CFR 35.75 effective May of 1997. These revised regulations allowed for the release from medical confinement of the patient if the *total effective dose equivalent* (or TEDE)[1] to individuals exposed to the released patient would not likely to exceed 5 mSv. This new provision allows greater flexibility in patient management. Studies by Schneider and McGuire (1996) had indicated that this replacement of the traditional activity and dose rate-based release criteria with a new dose-based criterion would result in shorter hospitalization stays, reduced health care costs, and possibly yield a positive impact on the psychological well-being of both patients and family member and/or caregivers. Guidance to licenses on determining when patients may be released based upon these new criteria were provided in NRC Regulatory Guide 8.39 dated April 1997 (NRC 1997). That regulatory guide was recently updated (Revision 1) in April 2020 (NRC 2020) to include:

1. more details regarding instructions to patients,
2. consideration of actions to be taken upon the death of the patient, and
3. further guidance to the breastfeeding mother to reduce radiation doses to the infant or child.

Details of Revision 1 to Regulatory Guide 8.39 are given in this chapter. It is noted, however, that the NRC has developed a Revision 2 to this same regulatory guidance document, but this document, as of this writing, has not yet approved by the Commission (NRC 2021).

5.2 Nuclear Regulatory Guide 8.39 Revision 1–April 2020

5.2.1 Introduction

The NRC is the governing body which establishes rules and regulations for safe use of radioactive materials for beneficial civilian purposes while protecting people and the environment. The NRC regulates uses of nuclear materials, such as in nuclear medicine, through licensing, inspection, and enforcement of its requirements. In some states, called *Agreement States*, nuclear materials are regulated by the state agencies, who commit to implementing rules which are at least as strict as the NRC. Thus, the NRC guidance represents the minimally acceptable oversight. All NRC regulatory guidance documents, and the code of federal regulations, are freely available for download.

In 1997, the NRC promulgated Title 10 of the Code of Federal Regulations (10 CFR) 35.75—*Release of individuals containing unsealed byproduct material or implants*

1. The Total Effective Dose Equivalent (or TEDE) includes both external and internal exposures, whereas in this Regulatory Guide, the presumption is that the exposure is from external gamma-rays and x-rays only. The Effective Dose Equivalent is the predecessor to the Effective Dose given in ICRP Publication 103. The EDE was originally introduced in ICRP Publication 26.

containing byproduct material". The NRC developed this regulation as a revision of 10 CFR Part 20 in 1991, which itself had revised dose limits to members of the general public in 10 CFR 20.1301. These new regulations, however, did not address exposures from the release of patients. While doses should be maintained *as low as reasonably achievable* (ALARA), the NRC determined that a dose limit of 5 mSv would provide adequate protection from exposure to patients treated with radiopharmaceuticals (or brachytherapy sources). This "Patient Release Rule" allows a licensee to authorize the release of a patient from its control if the TEDE to any other individual (e.g., medical staff, family member, caregiver), from exposure to the released patient, is not likely to exceed 5 mSv. In addition, 10 CFR 35.75 requires that a licensee provide the released individual, or the patient's family or other caregivers, with appropriate instructions, including written instructions, on recommended actions to maintain doses to other individuals ALARA if the TEDE to any other individual is likely to exceed 1 mSv.

5.2.2 Operational Equations

In Regulatory Guide 8.39 (Rev 1), the following expression is used to compute the cumulative exposure (or TEDE) to an individual (medical staff, family member, caregiver) exposed to the *radiopharmaceutical therapy* (RPT) patient (NRC 2020). This expression is taken from NCRP Report No. 37, which was published in 1970 (NCRP 1970):

$$D(t) = \frac{3.46\,\Gamma Q_0 T_p E\left(1 - e^{\frac{-(\ln 2)t}{T_p}}\right)}{r^2} \tag{5.1}$$

where D_t = the cumulative exposure at time t [in the traditional units of the roentgen (R)],

34.6 = a factor that is the product of the hours-to-day conversion (24 hours/day) and the inverse of the logarithm of 2,

Γ = the specific gamma-ray constant for a point radionuclide source (in units of R/h per mCi at 1 cm distance),

Q_0 = the initial activity of the point source in mCi at the time of release,

T_p = the physical half-life of the radionuclide in days,

E = the occupancy factor (fraction of time the individual is within distance r of the released RPT patient), and

t = the distance from the point source of interest in cm, and is the exposure time in days.

Several assumptions are made by the NRC in the application of Equation (5.1) to compute the amount of activity in the RPT patient that would permit hospital release.

- The dose to the exposed individual is a point estimate of dose at the individual's distance from the RPT patient and, thus, there is no allowance for body shielding of incident gamma-rays or x-rays of the exposed person.

- The RPT patient is similarly considered a point source of the radionuclide and thus there is no consideration of body self-shielding of gamma rays or x-rays emerging from the patient's source organs and tissues.
- The parameter D_t is not the absorbed dose to the exposed individual, but instead the air exposure at the individual's location (distance r from the RPT patient) in the traditional units of the roentgen (R). It is further assumed that this quantity in units of R is equivalent to a TEDE of 10 mSv (or 1 rem in the traditional units).
- A conservative estimate of the dose to the exposed individual is taken to be its value following total physical decay of the radionuclide. As such, the bracketed term in the numerator of Equation (5.1) is set at unity (as $t \rightarrow \infty$).
- Values of both the physical half-lives and specific gamma ray constants may be found in Appendix A of Regulatory Guide 8.39 (Rev 1).
- Default values of retained radiopharmaceutical activities at which patients may be released are computed using only physical radionuclide decay, and thus there is no allowance for biological uptake and elimination of the radionuclide in the source organs. Modifications are thus needed to Equation (5.1) to accommodate these additional considerations (addressed below).
- For radionuclides with physical half-lives greater than 1 day, and for the cases in which biological elimination is not considered, the assumption is that the individual who is likely to receive the highest dose from exposure to the RPT patient would receive a dose of 25% of the dose to full physical decay. This is effectively applying an occupancy factor of 0.25 to the exposure scenario.
- For radionuclides with physical half-lives less than or equal to 1 day, an occupancy factor of between 0.75 and 1.0 may be more appropriate (the value of 1.0 is the default value).

In summary, the operational expression of Equation (5.1) is thus revised to Equations (5.2) and (5.3) ($T_p \leq 1$ day):

$$D(\infty) = \frac{3.46\ \Gamma Q_0\, T_p\, (0.25)}{r^2} \tag{5.2}$$

$$D(\infty) = \frac{3.46\ \Gamma Q_0\, T_p\, (1.0)}{r^2} \tag{5.3}$$

These expressions consider the exposure to be from external photons only (gamma rays and x-rays emerging from the RPT patient) and do not include any dose contribution from inhaled or ingested radionuclides. Furthermore, they are not applicable to breastfeeding infants or children who continue to breastfeed (see Section 5.2.4.2).

5.2.3 Release Criteria

Regulatory Guide 8.39 (Rev 1) provides the licensee three options for releasing the RPT patient from their hospital or facility. These include (1) release based on the administered activity, (2) release based on a measurement of external dose rate, and (3) release based upon a patient-specific dose calculation.

5.2.3.1 Release of Patients Based on Administered Activity

Equations (5.2) and (5.3) may be used to compute the value of Q_0 by setting the value of $D(\infty)$ to 5 mSv and solving for this initial activity. If the administered activity to the patient is not greater than Q_0, the patient may be released and no record of instructions provided to the patient are required. The one exception is if the patient is breast-feeding an infant or child (see below). If, however, the administered activity exceeds the value of Q_0, the licensee may release the patient when the retained whole-body activity has decayed to a value at or below Q_0. In this case, 10 CFR 35.75(c) and 10 CFR 35.2075(a) require the licensee to maintain a record of the basis for authorizing the patient release as it is based upon *retained* activity and not *administered* activity.

Values of are provided in Table 1 (column 1) of Regulatory Guide 8.29 (Rev 1) in units of both GBq and mCi. In this chapter, we recreate these values in our **Table 5.1** using physical half-lives and specific gamma ray constants provided in the Regulatory Guide in its Appendix A. Of the 24 radionuclides listed in the Regulatory Guide, 10 are of current interest in radiopharmaceutical therapy, and these are considered on the left side of Table 5.1. As noted in the Regulatory Guide and in Equation (5.2) and Equation (5.3), an occupancy factor of 0.25 at 1 meter is assumed for those radionuclides with physical half-lives greater than 1 day; otherwise, the occupancy factor is set to unity. As presented in the operational Equation (5.1), there is no consideration of body shielding of gamma or x-rays in either the RPT patients or the exposed individual. Also, these limits are not applicable to pure beta-emitters such as ^{89}Sr and ^{90}Y. As shown, the Regulatory Guide does not include information for alpha-emitters currently used in clinical practice or in clinical trials. It further does not consider some important therapeutic beta-emitters—^{166}Ho and ^{177}Lu—or the Auger-emitters ^{193m}Pt and ^{195m}Pt.

In Table 5.1, equations (5.1) and (5.2) are further evaluated on the right-side columns of the table using different input parameters. For the physical half-lives, ICRP Publication 107 was consulted (ICRP 2008). Also, recently recalculated values of the specific gamma ray constant are taken from Peplow (2020), which were computed using decay schemes (photon energies and yields) from ICRP Publication 107 (ICRP 2008), and effective dose per fluence coefficients (for the reference adult) in the antero-posterior (AP) irradiation orientation taken from ICRP Publication 116 (ICRP 2010). These values thus allow for computation of retained activity for release (in GBq) using equation (5.2) and (5.3) for all listed therapeutic radionuclides.

As the Peplow (2020) dose coefficients were computed using the ICRP Publication 110 adult reference computational phantoms, body shielding of the exposed individual is now considered (as opposed to a point estimate of exposure or TEDE). In general, with the exception of ^{188}Re and ^{153}Sm, values of Γ given in Peplow (2020) are lower than those assumed in Regulatory Guide 8.39 (Rev 1), and thus the values of Q_0 are proportionally higher. In the most extreme case, the value of Q_0 given in the Regulatory Guide is 1.49 GBq for ^{103}Pd, while it is 7.42 GBq using these revised values of T_p and Γ.

Table 5.1 Retained activity (in GBq) and externally measured dose rates (mSv/h) for authorizing patient release. Columns on the left side of the table recreate approved values given in Regulatory Guide 8.39 Revision 1. Values of physical half-lives and specific gamma-ray dose constants are taken from those in Appendix A of this regulatory guide. On the right side of this table are evaluations of equations (5.2) and (5.3) using physical half-lives from ICRP Publication 107 (ICRP 2008), and specific gamma-ray dose constants from Peplow (2020).

Radionuclide	Tp (Appendix A) (days)	Occupancy Factor	Γ (Appendix A) (R/h) per mCi at 1 cm	Γ (Appendix A) (mSv/h) per MBq at 1 m	Retained Activity for Release (GBq)	Dose Rate at 1 m for Release (mSv/h)	Tp (ICRP 107)	Γ (Peplow 2020) (mSv/h) per MBq at 1 m	Retained Activity for Release (GBq)	Dose Rate at 1 m for Release (mSv/h)
Alpha-Emitters										
At-211		0.25					10.00	9.123E-05	0.63	0.058
Bi-212		1.00					0.04	1.479E-04	23.23	3.437
Bi-213		1.00					0.03	2.369E-05	192.71	4.564
Ra-223		0.25					11.43	4.552E-05	1.11	0.051
Ac-225		0.25					10.00	3.080E-05	1.88	0.058
Th-227		0.25					16.86	6.576E-05	0.52	0.034
Beta-Emitters										
I-124		0.25					4.18	1.471E-04	0.94	0.138
I-131	8.04	0.25	2.20	5.95E-05	**1.21**	**0.072**	8.02	5.483E-05	1.31	0.072
Y-90	These two radionuclides are pure beta-emitters						These two radionuclides are pure beta-emitters			
Sr-89										
Ho-166		0.25					1.12	4.174E-06	124.02	0.518
Lu-177		0.25					6.65	5.176E-06	16.80	0.087
Re-186	3.777	0.25	0.20	5.41E-06	**28.31**	**0.153**	3.72	3.183E-06	48.84	0.155
Re-188	0.708	1.00	0.26	7.03E-06	**29.05**	**0.204**	0.71	8.490E-06	24.02	0.204
Sm-153	1.946	0.25	0.43	1.15E-05	**25.86**	**0.297**	1.94	1.187E-05	25.13	0.298
Auger-Emitters										
In-111	2.83	0.25	3.21	8.68E-05	**2.35**	**0.204**	2.80	6.149E-05	3.35	0.206
I-123	0.55	1.00	1.61	4.35E-05	**6.04**	**0.263**	0.55	2.845E-05	9.19	0.261
I-125	60.14	0.25	1.42	3.84E-05	**0.25**	**0.010**	59.40	1.254E-05	0.78	0.010
Pt-193m		0.25					4.33	1.957E-06	68.21	0.133
Pt-195m		0.25					4.02	1.239E-05	11.61	0.144
Sn-117m	13.61	0.25	1.48	4.00E-05	**1.06**	**0.042**	13.76	2.536E-05	1.66	0.042
Pd-103	16.69	0.25	0.86	2.32E-05	**1.49**	**0.035**	16.99	4.582E-06	7.42	0.034

Table 5.2 Derivation of specific gamma-ray dose constants for alpha-emitting radionuclides to include contributions by decay chain progeny. Branching ratios are taken from ICRP Publication 107 (ICRP 2008).

Radionuclide (Daughters)	Branch Ratio	Γ (Peplow 2020) (mSv/h) per mBq at 1 m	Radionuclide (Daughters)	Branch Ratio	Γ (Peplow 2020) (mSv/h) per mBq at 1 m	Radionuclide (Daughters)	Branch Ratio	Γ (Peplow 2020) (mSv/h) per mBq at 1 m
At-211		5.906E-06	**Bi-212**		1.344E-05	**Bi-213**		1.818E-05
(Po-211)	0.582	1.098E-06	(Po-212)	0.641	8.222E-06	(Po-213)	0.979	5.031E-09
(Bi-207)	0.418	2.026E-04	(Tl-208)	0.359	3.596E-04	(Tl-209)	0.021	2.632E-04
						(Pb-209)	1.00	0.000E+00
	Total:	**9.123E-05**		**Total:**	**1.479E-04**			
							Total:	**2.369E-05**
Ra-223		2.107E-05	**Ac-225**		2.796E-06	**Th-227**		2.024E-05
(Rn-219)	1.00	8.446E-06	(Fr-221)	1.00	4.278E-06	(Ra-223)	1.00	2.107E-05
(Po-215)	1.00	2.516E-08	(At-217)	1.00	3.539E-08	(Rn-219)	1.00	8.446E-06
(Pb-211)	1.00	8.828E-06	(Bi-213)	1.00	1.818E-05	(Po-215)	1.00	2.516E-08
(Bi-211)	1.00	6.837E-06	(Po-213)	0.979	5.031E-09	(Pb-211)	1.00	8.828E-06
(Tl-207)	0.997	3.084E-07	(Tl-209)	0.021	2.632E-04	(Bi-211)	1.00	6.837E-06
(Po-211)	0.003	1.098E-06	(Pb-209)	1.00	0.000E+00	(Tl-207)	0.997	3.084E-07
						(Po-211)	0.003	1.098E-06
	Total:	**4.552E-05**		**Total:**	**3.080E-05**			
							Total:	**6.576E-05**

For the alpha-emitters shown in Table 5.1, one must give consideration to the external dose contribution of the decay progeny. In **Table 5.2**, these cumulative values of the specific gamma ray dose constants are computed as linear sums of the values for each daughter radionuclide weighted by the branching fraction. These total values of Γ are then carried forward in the upper right portion of Table 5.1. For the alpha-emitters of Table 5.1, values of Q_0 are shown to range from a low of 0.52 GBq for ^{227}Th to a high of 193 GBq for ^{213}Bi.

Let us consider the case of Ra-223 chloride (Xofigo) treatment of patients with late-stage *castration-resistant prostate cancer* (CRPC). The FDA package insert indicates an administration of 50 kBq per kg of body weight given at 4-week intervals for 6 injections. For a 70-kg individual, the administered activity per injection would be (50 kBq/kg) × (70 kg) × (MBq / 1000 kBq) or 3.5 MBq, which is substantially below our derived value of Q_0 of 1.11 GBq or 1110 MBq. We also acknowledge that even with our revised values of Γ to include body shielding by the exposed individual, there is yet no accounting of body self-shielding of the RPT patient, which would be substantial given the lower energies of x-rays that dominate the decay spectra of these therapeutic alpha-emitting radionuclides.

5.2.3.2 Release of Patients Based on Measured Dose Rate

The second option for patient release entails a measurement of external dose rate. When administered activities exceed values of Q_0 given in Table 5.1, licensees may still release their RPT patients when measured dose rates at 1 m from the patient surface are no greater than values given in the adjacent columns of Table 5.1. As with values of Q_0, two values of dose rates for patient release are given—one using values of T_p and Γ taken from Appendix A of the Regulatory Guide 8.39 (Rev 1) and one from ICRP Publication 107 and Peplow (2020), respectively. The dose rate permitting patient release for Ra-223 is shown to be 0.051 mSv/h or 51 mSv/h at 1 meter distance. As reported in Dauer et al. (2014), measured dose rates from patients administered Ra-223 dichloride at their institution were typically less than 2 mSv/h at contact. More importantly, they averaged 0.02 mSv/h at 1 meter immediately following administration (less than the 1-meter dose-rate limit of 0.051 mSv/h). As such, these patients were able to be released following their therapy administrations and subsequent imaging session as needed for dosimetric analysis.

5.2.3.3 Release of Patients Based on Patient-Specific Dose Calculations

The third option available to the license is to further consider patient-specific parameters. With this method, in accordance with 10 CFR 30.75(a), the licensee must compute the maximum most likely dose to the exposed individual on a case-by-case basis (e.g., different exposure scenarios). If the calculated maximum likely dose to the exposed individual is no greater than 5 mSv, the licensee may release the RPT patient. Patient-specific parameters may include:

1. considering biological uptake and elimination of the radiopharmaceutical,
2. occupancy factors less than 0.25 at 1 meter,
3. tissue shielding within the patient, and

4. tissue shielding within the exposed individual.

This information may be obtained by reviewing patient data and in-person conversations with patients to determine their travel and housing arrangements. In these cases, 10 CFR 35.2075(a) requires the licensee to maintain a record of the basis for authorizing the patient release. Further guidance and example cases for patient-specific considerations of maximum exposures to individuals in contact with the released RPT patient are given in Appendix B of Regulatory Guide 8.39 (Rev 1). Example cases are also provided in the NUREG 1556 guidance document series (https://www.nrc.gov/reading-rm/doc-collections/nuregs/staff/sr1556/index).

When considering both physical and biological elimination of retained radiopharmaceutical activity in the patient, values of the effective half-life of the radiopharmaceutical may be taken into account to demonstrate compliance with dose limits given in 10 CFR 35.75. The effective half-life is defined as:

$$T_{eff} = \frac{T_b T_p}{T_b + T_p} \tag{5.4}$$

where T_b is the biological half-life of the radiopharmaceutical and T_p is the physical half-life of the radionuclide. In an example of ^{131}I therapy, the radiopharmaceutical (sodium iodide) can be modeled using two physiological compartments: (1) extrathyroidal iodide (e.g., existing outside the thyroid) and (2) thyroidal iodide following uptake by the thyroid gland. The effective half-lives for the extrathyroidal and thyroidal fractions (F_1 and F_2, respectively) can be computed as:

$$T_{eff_1} = \frac{T_{b_1} T_p}{T_{b_1} + T_p} \text{ and } T_{eff_2} = \frac{T_{b_2} T_p}{T_{b_2} + T_p} \tag{5.5}$$

However, simple exponential excretion models do not account for (1) the time over which ^{131}I is absorbed from the stomach to blood and (2) the period over which ^{131}I is within the contents of the urinary bladder. Failure to account for these factors could result in an underestimate of the dose to another individual exposed to the released RPT patient. Therefore, Regulatory Guide 8.39 (Rev 1) makes a conservative approximation to account for these factors by assuming that, during the first 8 hours following administration, approximately 80% of the administered ^{131}I is removed from the body at a rate determined by the physical half-life of ^{131}I.

An expression to compute the dose from a patient administered ^{131}I may thus have three components.

- ***The first component is the dose for the first 8 hours following administration.*** This component comes directly from Equation (5.1) using the physical half-life and an activity reduction factor of 0.80.
- ***The second component is the dose from the extrathyroidal component from 8 hours to total decay.*** In this component, the first exponential term represents the activity at t = 8 hours based on physical decay. The second exponential

term represents the activity from t = 8 hours to total decay based upon the effective half-life of the extrathyroidal component.

- ***The third component then is the dose from the thyroidal component for 8 hours to total decay.*** This is computed in the same manner as the second component. The full expression for this exposure scenario would be:

$$D(\infty) = \frac{34.6\,\Gamma Q_0}{r^2}\left\{T_p E_1(0.8)\left(1-e^{\frac{-(\ln 2)(0.33d)}{T_p}}\right) + T_{eff1}F_1E_2e^{\frac{-(\ln 2)(0.33d)}{T_p}} + T_{eff2}F_2E_2e^{\frac{-(\ln 2)(0.33d)}{T_p}}\right\} \tag{5.6}$$

where F_1 and F_2 are the uptake fractions for the extrathyroidal and thyroidal compartments, respectively, and E_1 and E_2 are the occupancy factors for the first 8 hours and from 8 hours to total decay, respectively. For use of ^{131}I therapy for thyroid ablation, acceptable values given in Appendix B of Regulatory Guide 8.39 (Rev 1) are F_1 = 0.20 and F_2 = 0.80, with T_{eff_1} = 0.32 days and T_{eff_2} = 5.2 days. For ^{131}I treatment of thyroid remnants following surgical removal of the thyroid, acceptable values given in Appendix B of the Regulatory Guide are F_1 = 0.95 and F_2 = 0.05 with T_{eff_1} = 0.32 days and T_{eff_2} = 7.3 days.

For an administration of 7400 MBq of ^{131}I to a thyroid cancer patient, Equation (5.6) yields a maximum dose to an exposed individual at 1 meter distance of 4.53 mSv, and thus the patient could be released under 10 CFR 35.75, assuming that the above assumptions were justified for that given individual patient, and the patient was given release instructions.

5.2.4 Instructions to Patients

5.2.4.1 Activities and Dose Rates Requiring Instructions

In accordance with 10 CFR 35.75(b), licensees must provide instructions to the released patients, including written instructions on how to maintain doses to other individuals ALARA following their release. These instructions are a requirement if the maximum TEDE to exposed individuals exceeds 1 mSv. Values of activity at or above which patients must be given these instructions, or dose rates measured at 1 meter distance at or above which patients must be given these instructions, can be determined as 1/5th of the values given in Table 5.1. These values thus reflect the change in dose limit from 5 mSv (release criterion) to 1 mSv (instruction criterion). If the patient release was based upon a patient-specific dose calculation, the dose limit of 1 mSv to the maximal exposed person is still the criterion above which instructions to the released patient must be given.

5.2.4.2 *Additional Instructions for Breast-Feeding After Release*

When the released patient is female and anticipates breastfeeding her infant or child following hospital release, 10 CFR 35.75(b) requires that the licensee give the patient further instructions regarding guidance on the interruption or discontinuation of breastfeeding to ensure that dose to the infant or child does not exceed 1 mSv. Table 3 of Regulatory Guide 8.39 (Rev 1) addresses an array of radiopharmaceuticals and provides (1) values of Q_0 above which instructions to the released patient are required, (2) values of Q_0 above which records of instruction assignment are required, and (3) recommended durations of interruption (hours, days, or complete cessation) of breastfeeding.

As an example, Table 3 of the Regulatory Guide 8.39 (Rev 1) states that for the administration of ^{123}I[NaI] to a female patient, instructions on breastfeeding precautions must be given for administered activities above 20 MBq, and records must be maintained of this notification for administered activities above 100 MBq. The recommended duration of interruption of breastfeeding is 3 days.

5.2.4.3 *Instruction Content*

Written instructions to the released patient should give consideration to the patient's destination upon the release and the ability of the patient or caregiver to understand and follow the release instructions. The licensee should thoroughly ascertain the patient's post-treatment destination and the method by which the patient plans to travel to that destination. This information will allow best estimates of the likely cumulative radiation doses to other members of the public (e.g., family members and caregivers) and, therefore, the appropriate protective measures that will ensure these doses are ALARA and that dose limits will likely be met.

These instructions should always be tailored to the specific patient situation and the type and amount of radiopharmaceutical administered. Instructions should also be provided on how long the precautions should be followed. As a guideline, the licensee may consider using several (3 to 5) effective half-lives of the radiopharmaceutical. Example elements of these instructions might include the following:

Concerns over radionuclide contamination

- Wash hands frequently and bathe daily.
- Wash laundry separately from others.
- Use dedicated or disposable kitchen utensils, and do not share them with others.
- Use a dedicated sole-use bathroom, if possible. Always sit on the toilet. Flush the toilet twice after each use.
- Use disposable gloves and flushable wipes when cleaning.
- Discard trash separately, and hold it to allow for radioactive decay.

Concerns over radiation exposure

- Sleep alone in a separate bedroom.

- Abstain from any intimate contact.
- Avoid preparing or sharing food with others.
- Avoid using public transportation.
- Maintain good hydration, as directed by a physician.
- Minimize the amount of time in close proximity to other people, especially children and pregnant women.

The licensee should instruct family members and caregivers to notify the treating medical facility of any medical emergency or if the patient dies. The licensee should also tell the patient how to clean up an area contaminated with body fluids, such as urine or vomit.

5.2.4.4 Death of a Patient Following Radiopharmaceutical Administration

If the licensee learns that a patient has died shortly after administration of a therapeutic radiopharmaceutical, the treating authorized user and the *radiation safety office* (RSO) should be notified immediately. The RSO or the authorized user should perform an assessment of the type and amount of retained activity, based on patient records.

If the death occurs in a hospital, access to the room occupied by the deceased should be controlled until the room has been surveyed and decontaminated as necessary. A specified form of patient identifier (e.g., bracelet, body tag) should be attached with relevant information to the radioactive body. A body bag may need to be used to contain the leakage of radioactive material. To minimize external radiation, the body may need to be retained in a secured area. Radiation safety procedures to be applied in practice for handling the body should be determined in close consultation with the RSO and an authorized user at the facility where the therapy was administered. If a patient is subsequently admitted to a different hospital or facility after release and then passes away, the local RSO should be involved. Further instructions are given in Section 2.4 of Regulatory Guide 8.38 (Rev 1).

5.2.5 Recordkeeping

5.2.5.1 Records of Patient Release

The NRC has no requirement for recordkeeping on the release of patients who were released in accordance with administered activities below values of Q_0 given in Table 5.1. However, if the release of the patient is based on a dose calculation that considered retained activity, an occupancy factor of less than 0.25 at 1 meter, the effective half-life, or shielding by tissue, 10 CFR 35.2075(a) requires the licensee to maintain a record of the basis for authorizing the patient's release. This record should include the patient's identifier in a way that ensures that confidential patient information is not traceable or attributable to any specific patient, the radiopharmaceutical administered, the administered activity, and the date of the administration. In addition, depending on the basis for authorizing the release of patients, records should include the following information:

- ***For Immediate Release of a Patient Based on a Patient-Specific Calculation.*** The record should include the equations used, including the patient-specific factors and the bases used in computing the dose to the person exposed to the patient and the resultant calculated dose. The patient-specific factors may include the effective half-life and uptake fraction for each component of the biokinetic model, the time that the physical half-life was assumed to apply to retention, and the occupancy factors. The record should include the basis for the selection of each of these values.
- ***For Immediate Release of a Patient Based on a Measured Dose Rate.*** The record should include the results of the measurement, the specific survey instrument used, and the name of the individual performing the survey.
- ***For Delayed Release of a Patient Based on a Radioactive Decay Calculation.*** The record should include the time of the administration, the date and time of release, and the results of the decay calculation.
- ***For Delayed Release of a Patient Based on a Measured Dose Rate.*** The record should include the results of the survey meter measurement, the specific survey instrument used, and the name of the individual who performed the survey.

5.2.5.2 Records of Instructions for Breastfeeding Patients

If a patient's failure to interrupt or discontinue breastfeeding could result in a dose to the infant or child in excess of 5 mSv, 10 CFR 35.2075(b) requires a record that the licensee gave these patient instructions. The record should include the patient's identifier, the radiopharmaceutical administered, the administered dosage, the date of the administration, and whether instructions were provided to the patient who could be breastfeeding an infant or child. The patient's identifier should be prepared in a way that would ensure that confidential information is not traceable or attributable to a specific patient.

5.3 NCRP Report No. 155 – December 2006

5.3.1 Introduction

Following the original release of NRC Regulatory Guide 8.39 in April 1997, further guidance was needed within the medical health physics community for the assessment of likely doses from exposure to released RPT patients under specific exposure scenarios. In June of 2000, a series of generalized algorithms for determining time of release and duration of post-release radiation precautions was published by Zanzonico et al. (2000). These concepts were then further expanded by the NCRP in the release of NCRP Report No. 155 (NCRP 2006)—the first major update to Report No. 37 (NCRP 1970). In that report, the prior work of Zanzonico and colleagues was carried forward and expanded to include the develop of a calculational tool for patient release criteria written in Microsoft Excel. Appendix A of NCRP Report No. 155 provides extensive guidance on the operational equations and needed parameter values to

Table 5.3 Occupancy factors, index distances, and effective dose limits given in Appendix A of NCRP Report No. 155 (NCRP 2006)

Group / Activity	Occupancy Factor	Index Distance (m)	Effective Dose Limits (mSv)
Members of the patient's family			
Non-sleeping partner/adult	0.25	1.0	5
Nonpregnant sleeping partner	0.33	0.3	5
Pregnant sleeping partner	0.33	0.3	1
Pregnant women/children	0.25	1.0	1/1
Child held by patient	0.20	0.3	1
Public / Coworker	0.33	1.0	1

consider a variety of realistic exposure scenarios. For each exposure group, recommended values of occupancy factors, index exposure distances, and effective dose limits are given. **Table 5.3** summarizes these values.

5.3.2 Operational Equations

The following material is taken from Appendix A of NCRP Report No. 155. The time-dependent effective dose rate at an index r_j distance from a patient containing radioactivity may be approximated using the following equations taken from Zanzonico et al. (2000):

$$\dot{E}(r_j,t) = \dot{K}_a(r_j,t)(E/K_a) = \sum_{i=1}^{n} \dot{K}_a(r_j,0)_i (E/K_a) e^{\frac{-(\ln 2)t}{T_{e_i}}} \tag{5.7}$$

where $\dot{E}(r_j,t)$ = the effective dose rate (mSv/h) at an index distance r_j (meters) from the patient at time t post-administration (days),

$\dot{K}_a(r_j,t)$ = the air kerma rate (Gy/h) at an index distance r_j (m) from the patient at time t (days) post-administration,

E/K_a = the effective dose per unit air kerma coefficient (Sv/Gy),

$\dot{K}_a(r_j,0)$ = the zero-time intercept of the exponential component i of the time-dependent air kerma rate measured at an index distance r_j (m) from the patient,

T_{e_i} = the effective half-life (days) of the non-decay corrected total-body activity for compartment i of a multi-exponential function, and

n = the number of exponential compartments required to describe the time-dependent total-body activity.

Implicit in Equation (5.7) is the overestimation by a factor of approximately 2 when using the air kerma rate, that is actually measured in practice and applied throughout the analysis that follows. Also implicit in Equation (5.7) is the assumption that the time-dependent total-body activity follows a multi-exponential function, such as:

$$A(t) = A(0)\sum_{i=1}^{n} F_i \, e^{\frac{-(\ln 2)t}{T_{e_i}}} \tag{5.8}$$

where $A(t)$ is the total-body activity (Bq) at time t (days) post-administration, and F_i is the zero-time intercept of exponential component i of the total-body activity expressed as a fraction of administered activity such that $\sum_i F_i = 1$. The foregoing activities are decay corrected (i.e., activity is corrected for radioactive decay) back to the time of administration. In principle, each of the exponential components represents a metabolic compartment, and its effective half-life is related to the rate of clearance of activity from that compartment.

In addition, implicit in these equations is the assumption that activity is being continuously eliminated from the patient's body from the time of administration. Of course, no activity is biologically eliminated from the body before the patient's first void or defecation post-therapy. Within these first several hours post-therapy, the only means of elimination of activity is physical decay. Various studies have accounted for this fact by appending an effective dose contribution term with an effective half-life equal to the radionuclide's physical half-life, as is done in Equation (5.6) earlier. The duration of time during which an effective half-life is equal to the radionuclide's physical half-life is taken as 8 hours for orally administered radiopharmaceuticals such as [^{131}I]NaI, but only 3 hours for intravenously administered radiopharmaceuticals such as ^{131}I-labeled anti-B1 monoclonal antibodies.

The values of the parameters F_i and T_{e_i} of the multi-exponential function in Equation (5.8) may be obtained from sources in the pertinent scientific literature or, preferably, determined empirically for individual patients by performing serial measurements of total-body activity following a pre-therapy tracer administration of the therapeutic radiopharmaceutical and mono- or multi-exponential curve-fitting of the resulting time-activity data. Following the determination of the parameters F_i and T_{e_i}, Equation (5.7) may be rewritten as:

$$\dot{E}(r_j, t) = \dot{K}_a(r_j, 0)(E / K_a)\sum_{i=1}^{n} F_i \, e^{\frac{-(\ln 2)t}{T_{e_i}}} \tag{5.9}$$

Thus, the effective dose at an index distance r_j from a radioactive patient from time t_1 to time t_2 post-administration is given by the following expression:

$$E(r_j)_{t_1 \to t_2} = (24) T_{r_j} \dot{K}_a(r_j, 0)(E / K_a)\sum_{i=1}^{n} \int_{t_1}^{t_2} F_i \, e^{\frac{-(\ln 2)t}{T_{e_i}}} \, dt \tag{5.10}$$

where $E(r_j)_{t_1 \to t_2}$ = the effective dose (mSv) at an index distance r_j from the patient from time t_1 to time t_2 post-administration (days),

T_{r_j} = the occupancy factor for an individual at a distance r_j (m) from the radioactive patient, and

24 = the conversion of time in days to time in hours.

Equation (5.10) may be further generalized to yield the effective dose from a radioactive patient from time t_1 to time t_2 ($E_{t_1 \to t_2}$) post-administration, including all possible occupancy factors, as:

$$E(r_j)_{t_1 \to t_2} = (24)\sum_{j=1}^{m} T_{r_j} \dot{K}_a(r_j, 0)(E / K_a) \sum_{i=1}^{n} \int_{t_1}^{t_2} F_i \, e^{\frac{-(\ln 2)t}{T_{e_i}}} dt \qquad (5.11)$$

By substituting $t_1 = 0$ and $t_2 = \infty$ in Equation (5.11) and evaluating the resulting definite integral, the effective dose from the exposure to the radioactive patient over all time can be obtained as:

$$E = (24)\sum_{j=1}^{m} T_{r_j} \dot{K}_a(r_j, 0)(E / K_a)(1.44)\sum_{i=1}^{n} F_i \, T_{e_i} \qquad (5.12a)$$

$$E = (24)\tilde{a}\sum_{j=1}^{m} T_{r_j} \dot{K}_a(r_j, 0)(E / K_a) \qquad (5.12b)$$

where $\tilde{a} = \tilde{A} / A_0$ = the time-integrated activity coefficient (TIAC—formally termed the residence time τ) for total-body activity

$\tilde{A}$ = the time-integrated activity (cumulative decays) for total-body activity, and

A_0 = the administered activity to the patient. Here, $\tilde{a} = 1.44\sum_i F_i \, T_{e_i}$.

When the patient is to be released from medical confinement, the effective dose for any person associated with the patient arising from the air-kerma rate of such a patient at the release time ($t_{release}$) shall not exceed the applicable dose limit. The time of release [$(t_{release})_{limit}$] then becomes the controlling factor in determining the effective dose at time of release [$E(t_{release})_{limit}$] that is then compared to the effective dose limit (E_{limit}). Thus, by substituting $t_1 = (t_{release})_{limit}$ and $t_2 = \infty$, Equation (5.11) becomes:

$$E_{limit} = (34.6)\sum_{j=1}^{m} T_{r_j} \dot{K}_a(r_j, 0)(E / K_a)\sum_{i=1}^{n} F_i \, T_{e_i} e^{\frac{-(\ln 2)(t_{release})_{limit}}{T_{e_i}}} \qquad (5.13)$$

For the whole period after release, where 34.6 = 1.44 × 24, a factor for calculating the time-integrated activity coefficient (hours) from the fractional zero-time intercepts F_i and the effective half-lives T_{e_i}.

An important aspect of Equation (5.13) is that it cannot be solved analytically for the needed parameter $(t_{release})_{limit}$ for a multi-exponential total-body time-activity function. However, the time post-administration for release of the patient from medical confinement $t_1 = (t_{release})_{limit}$ and the duration of various radiation precautions may be determined iteratively from Equation (5.13).

If the effective dose E at an index distance of 1 m from the radiopharmaceutical therapy patient as given by Equation (5.13) does not exceed the effective dose limit of 5 mSv, the patient may be released after administration of the agent following the restrictions on a first void and time restrictions as discussed previously. If, however, E at an index distance of 1 m from the radiopharmaceutical therapy patient given by Equation (5.13) does exceed the effective dose limit of 5 mSv, possible values of the time post-administration of release of the patient from medial confinement $[(t_{release})_{5\ \mathrm{mSv}}]$ may be substituted into the right side of Equation (5.13). Beginning with a time of 1 day, and substituting time values in 1-day increments, the smallest time value (1-day, 2-day, 3-day, ...) that yields an effective dose by Equation (5.13) equal to or less than the dose limit is the time post-administration of release of the patient from medical confinement $[(t_{release})_{limit}]$.

Similarly, the duration in terms of times post-administration of the post-release radiation precautions (such as not working, avoiding pregnant women and children, limiting holding of children, and sleeping partners not sleeping together) may be deterred with Equation (5.13) by substitutions of dose limits, index distances, and occupancy factors that are most applicable to these respective groups, as well as measured zero-time air kerma rates. Increments shorter than 1 day may be used for this calculation. Depending on the groups of interest which, in turn, are dependent upon the circumstances of the individual patient, the pertinent values of the effective-dose limits, the index distances, and occupancy factors will vary, as well the exact form of Equation (5.13). As noted in NCRP Report No. 155, it is useful to calculate these times using a computerized spreadsheet program. One is provided upon request from the NCRP, as well as being demonstrated below.

For a therapeutic radiopharmaceutical for which the time-dependent total-body activity follows a mono-exponential function [setting n=1 in Equation (5.7)], Equation (5.13) can be reduced to the following:

$$E_{limit} = (34.6)\sum_{j=1}^{m} T_{r_j} \dot{K}_a(r_j,0)(E/K_a)T_e e^{\frac{-(\ln 2)(t_{release})_{limit}}{T_e}} . \tag{5.14}$$

Mono-exponential total-body time-activity functions apply, at least approximately, to many therapeutic radiopharmaceuticals. In contrast to Equation (5.13), Equation (5.14) can be solved explicitly for $(t_{release})_{limit}$, the time of release from medical confinement following radiopharmaceutical therapy treatment.

It must be recognized that equations (5.13) and (5.14) offer a more realistic estimate of the effective dose to individuals exposed to release radiopharmaceutical therapy patients than the conservative base equations given in NRC Regulatory Guide 8.39 (eee Equation (5.1)). The NCRP Report 155 methodology thus provides a rigorous framework for performing patient-specific calculations of effective dose to exposed individuals allowed under option 3 of that guidance document. Three aspects of the NCRP methodology are notable:

1. There is explicit consideration of biological elimination of the radiopharmaceutical through the use of effective half-lives.
2. There is explicit consideration of body self-shielding of emitted gamma-rays and x-rays through the use of patient-specific air kerma measurement.

3. There is explicit consideration of body shielding of the exposed individual through the use of air kerma-to-effective dose coefficients, which are derived Monte Carlo simulations in whole-body computational reference phantoms as described in ICRP Publication 116 (ICRP 2010).

5.3.3 Example Case Study: Patient with Metastatic Thyroid Cancer

As a working example of the NCRP Report 155 methodology for computing patient release conditions, we here provide Example 1 for metastatic thyroid cancer given in Appendix B of that NCRP report. For this particular example, the NCRP Excel spreadsheet screens are shown in Figure 5.1.

We consider a post-thyroidectomy and post-radioiodine ablation patient with metastatic thyroid cancer who has been treated with an administered activity of 6.475 GBq (175 mCi) of ^{131}I as sodium iodide. Based on kinetic measurement of a pre-therapy administration of 75 MBq (2 mCi) of [^{131}I]NaI tracer, this patient was found to exhibit a bi-phasic (bi-exponential) retention function (i.e., total body time-activity function), where 95% of the administered activity was eliminated with an effective half-life of 0.32 days, and the remaining 5% was eliminated with an effective half-life of 5.8 days. The ^{131}I time-integrated activity coefficient (or "residence time") is computed to be 0.86 days or 20.5 hours. The relevant expression for the TIAC is given as:

$$\tilde{a} = \frac{\tilde{A}}{A_0} = 1.44 \sum_{i=1}^{2} \frac{A_i}{A_0} T_{e_i} = 1.44 \sum_{i=1}^{2} F_i \, T_{e_i} \tag{5.15}$$

where T_{e_i} is the effective half-life for compartment i and F_i is the fraction of administered activity at time $t = 0$ for compartment i of the time-activity function of the whole-body. The TIAC $\tilde{a}$ has units of time. We further note that immediately following administration of the 74 MBq (2 mCi) tracer, exposure rates were measured with a survey meter yielding 0.26 mR/h and 3 mR/h at distances of 1 m and 0.3 m, respectively.

Accordingly, as shown in Figure 5.1, all of the column 1 values in the spreadsheet are below 5 mSv (500 mrem), so this patient could be released on the same day the treatment was administered. However, the patient should avoid holding small children for > 10 minutes per day for 21 days post-treatment, and only if the patient avoided the children altogether during the first day after treatment when the patient was the most radioactive and, therefore, producing the highest exposure rate. This restriction is necessitated by the fact that the values in the "1-day" sub-column in column 5 do not decrease to below 1 mSv unit after day 21 post-treatment. As indicated in the other sub-columns in column 6, this "child-avoidance" period is shortened somewhat if the patient avoided children altogether for longer than 1 day immediately post-treatment. Conversely, if the patient did not avoid children during that first day post-treatment, the subsequent "child-avoidance period" is much longer, as the values in the 0-day sub-column in column 6 do not decrease to < 1 mSv even up to day 40 post-treatment.

Because all of the column 4 values after the first day post-treatment are < 1 mSv, the patient can return to work on day 2 post-treatment. Likewise, the general radiation

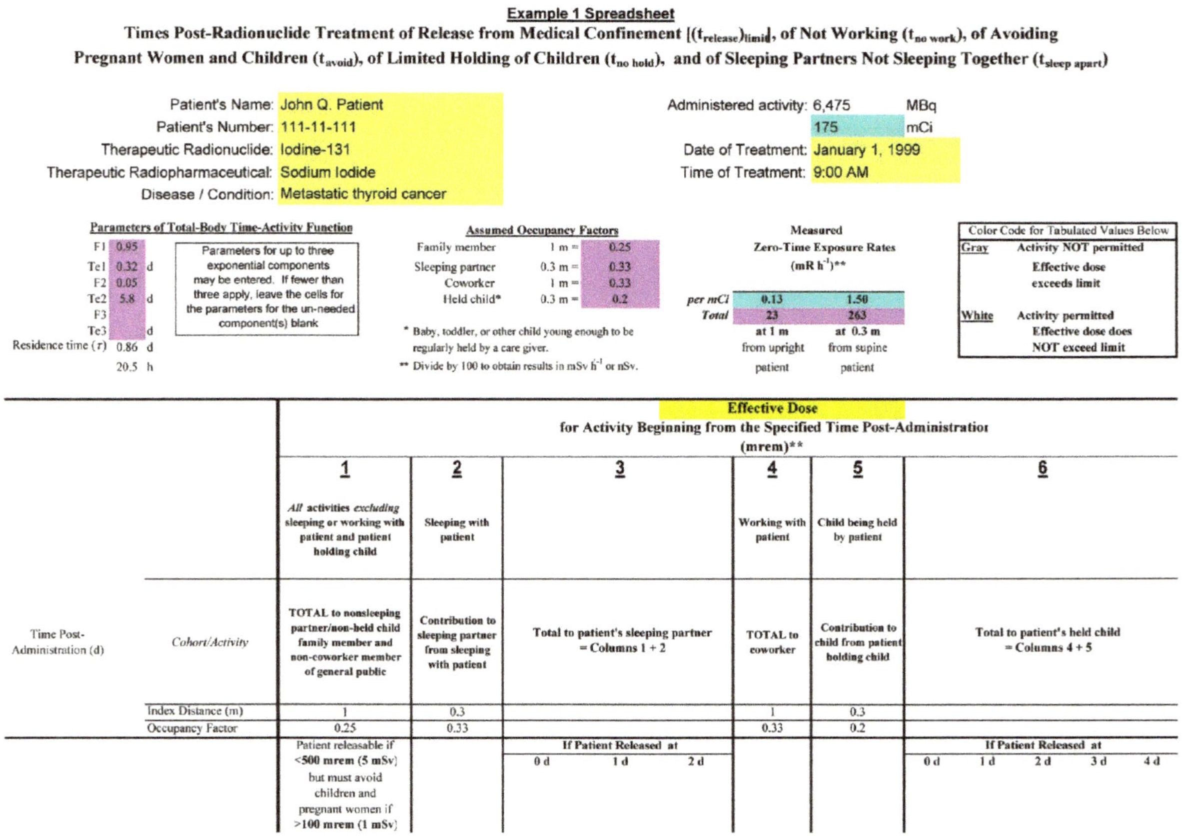

Example 1 Spreadsheet

Times Post-Radionuclide Treatment of Release from Medical Confinement $[(t_{release})_{limit}]$, of Not Working ($t_{no\ work}$), of Avoiding Pregnant Women and Children (t_{avoid}), of Limited Holding of Children ($t_{no\ hold}$), and of Sleeping Partners Not Sleeping Together ($t_{sleep\ apart}$)

Patient's Name:	John Q. Patient	Administered activity:	6,475	MBq
Patient's Number:	111-11-111		175	mCi
Therapeutic Radionuclide:	Iodine-131	Date of Treatment:	January 1, 1999	
Therapeutic Radiopharmaceutical:	Sodium Iodide	Time of Treatment:	9:00 AM	
Disease / Condition:	Metastatic thyroid cancer			

Parameters of Total-Body Time-Activity Function

F1	0.95	
Te1	0.32	d
F2	0.05	
Te2	5.8	d
F3		
Te3		d
Residence time (τ)	0.86	d
	20.5	h

Parameters for up to three exponential components may be entered. If fewer than three apply, leave the cells for the parameters for the un-needed component(s) blank

Assumed Occupancy Factors

Family member	1 m =	0.25
Sleeping partner	0.3 m =	0.33
Coworker	1 m =	0.33
Held child*	0.3 m =	0.2

* Baby, toddler, or other child young enough to be regularly held by a care giver.

** Divide by 100 to obtain results in mSv h^{-1} or mSv.

Measured Zero-Time Exposure Rates ($mR\ h^{-1}$)**

	at 1 m from upright patient	at 0.3 m from supine patient
per mCi	0.13	1.50
Total	23	263

Color Code for Tabulated Values Below	
Gray	Activity NOT permitted Effective dose exceeds limit
White	Activity permitted Effective dose does NOT exceed limit

Effective Dose for Activity Beginning from the Specified Time Post-Administration (mrem)**

Time Post-Administration (d)		1	2	3	4	5	6
		All activities *excluding* sleeping or working with patient and patient holding child	Sleeping with patient		Working with patient	Child being held by patient	
	Cohort/Activity	TOTAL to nonsleeping partner/non-held child family member and non-coworker member of general public	Contribution to sleeping partner from sleeping with patient	Total to patient's sleeping partner = Columns 1 + 2	TOTAL to coworker	Contribution to child from patient holding child	Total to patient's held child = Columns 4 + 5
	Index Distance (m)	1	0.3		1	0.3	
	Occupancy Factor	0.25	0.33		0.33	0.2	
		Patient releasable if <500 mrem (5 mSv) but must avoid children and pregnant women if >100 mrem (1 mSv)		If Patient Released at 0 d 1 d 2 d			If Patient Released at 0 d 1 d 2 d 3 d 4 d

Figure 5.1 NCRP Excel spreadsheet for the case of a metastatic thyroid cancer patient (taken from Appendix B of NCRP Report 155).

0	119	1,778	1,897			156	1,078	1,196				
1	58	875	993	754		77	530	649	589			
2	46	696	814	666	585	61	422	540	480	468		
3	41	608	727	597	524	54	369	487	427	415		
4	36	538	657	536	470	47	326	445	385	373	367	
5	32	478	596	482	423	42	290	408	348	336	330	325
6	28	424	542	434	380	37	257	375	315	303	297	293
7	25	376	495	392	343	33	228	347	286	274	269	264
8	22	334	452	355	309	29	202	321	261	249	243	238
9	20	296	415	321	280	26	180	298	238	226	220	215
10	18	263	381	292	253	23	159	278	218	206	200	195
11	16	233	352	265	230	21	141	260	200	188	182	177
12	14	207	326	242	209	18	125	244	184	172	166	161
13	12	184	302	221	191	16	111	230	170	158	152	147
14	11	163	282	203	175	14	99	217	157	145	139	135
15	10	145	263	187	160	13	88	206	146	134	128	124
16	9	128	247	172	147	11	78	196	136	124	118	114
17	8	114	232	159	136	10	69	188	127	115	110	105
18	7	101	220	148	126	9	61	180	120	108	102	97
19	6	90	208	138	117	8	54	173	113	101	95	90
20	5	80	198	129	109	7	48	167	107	95	89	84
21	5	71	189	121	102	6	43	161	101	89	83	79
22	4	63	181	114	96	6	38	157	96	84	79	74
23	4	56	174	108	90	5	34	152	92	80	74	70
24	3	49	168	102	85	4	30	148	88	76	70	66
25	3	44	162	97	81	4	27	145	85	73	67	62
26	3	39	157	93	77	3	24	142	82	70	64	59
27	2	34	153	89	74	3	21	139	79	67	61	57
28	2	31	149	85	70	3	19	137	77	65	59	54
29	2	27	146	82	68	2	16	135	75	63	57	52
30	2	24	143	80	65	2	15	133	73	61	55	51
31	1	21	140	77	63	2	13	132	71	59	53	49
32	1	19	138	75	61	2	11	130	70	58	52	47
33	1	17	135	73	60	1	10	129	69	57	51	46
34	1	15	133	72	58	1	9	128	67	55	50	45
35	1	13	132	70	57	1	8	127	66	54	49	44
36	1	12	130	69	56	1	7	126	65	54	48	43
37	1	10	129	68	55	1	6	125	65	53	47	42
38	1	9	128	67	54	1	6	124	64	52	46	42
39	1	8	127	66	46	1	5	124	63	51	46	41
40	0	7	126	58	46	1	4	123	63	51	45	40

Figure 5.1 (continued)

precautions at home need to be observed only for the first day post-treatment subject to the other dose restrictions. On the other hand, the patient should not sleep in the same bed with a sleeping partner until day 7 after treatment, as the values in the 0-day sub-column in column 3 do not decrease to < 5 mSv until day 7 post-treatment. However, if the patient's sleeping partner is pregnant, her dose limit is lowered to 1 mSv and the patient should, therefore, not sleep in the same bed until day 24 after treatment, and only if he avoided his sleeping partner altogether during the first day after administration of the radiopharmaceutical therapy, as the values in the 1-day sub-column of column 9 do not decrease to < 1 mSv until day 24 post-treatment.

Example take-home instruction sheets for this particular example are given over the next three pages in Figure 5.2.

Example 1: Instruction Sheets

Radiation Safety Precautions for Radiopharmaceutical Therapy Patients

Note: Please carefully read and follow the instructions in this document.

If you or your health care provider have any questions or concerns regarding the radionuclide therapy you have received, please contact:

Dr. J. Smith
Nuclear Medicine Attending Physician

at (111) 222-3333
Telephone Number

or 0000
Emergency or Pager Telephone Number

Patient John Q. Patient, Medical Record Number 111-11-111, received a therapeutic dose of 6.475 MBq (175 mCi) of Iodine-131 Sodium Iodide at General Hospital on January 1, 2004 at 9:00 am and *should* observe the following radiation safety precautions at home as follows.

- ☐ Avoid close contact [less than 1 meter (3 feet) away from] with pregnant women and children until 1 day after the administration of the radionuclide therapy.
- ☐ Do not hold or embrace children for more than 10 minutes a day until 21 days after the day your radionuclide therapy was administered. *This requires that you avoid contact with these children altogether during your* first day *after the day your radionuclide therapy was administered.*
- ☐ Unless you work alone (for example, driving a truck), do not return to work until 1 day after the day your radionuclide therapy was administered.
- ☐ Do not sleep in the same bed with your sleeping partner until 7 days after the day your radionuclide therapy was administered.
- ☐ However, if your sleeping partner is pregnant, do not sleep in the same bed with your sleeping partner until 24 days after the day your radionuclide therapy was administered. This requires that you avoid contact with your pregnant sleeping partner *altogether* during your first day after the day your radionuclide therapy was administered.

Figure 5.2 Example instruction sheets provided in the clinical example in Appendix B of NCRP 155.

Patient: Patient, John Q.

MRN: 111-11-111

In addition, the following precautions *should* be observed until 1 day after the administration of your radiopharmaceutical therapy.

- To the extent that is *reasonable*, generally try to remain as far away from individuals around you as possible.
- After using the toilet, flush twice and, as usual, wash your hands. If possible, use paper towels to dry your hands and dispose the paper toweling in the trash.
- You *should* otherwise observe good personal hygiene and may shower, bathe, shave, etc. as you normally would, rinsing the shower stall, tub or sink thoroughly after use.
- Wipe up any spills of urine, saliva and/or mucus with tissues or a small amount of disposable (*i.e.*, flushable) paper toweling, and dispose of the tissue or toweling down the toilet.
- Use nondisposable plates, bowls, spoons, knives, forks and cups. If possible, you *should* wash plates, bowls, spoons, knives, forks and cups which you use, using a separate sponge or wash cloth from that used by the rest of your household. Rinse the sink thoroughly after use, wipe the fixtures with paper towels, and dispose of the paper toweling in the trash.
- If you use a dishwasher, wash your plates, bowls, spoons, knives, forks and cups separately from those of the rest of your household.
- Use the same set of plates, bowls, spoons, knives and forks for 1 day after your radionuclide therapy.
- Store and launder your soiled/used clothing and bed linens separately from those of the rest of your household, running the rinse cycle two times at the completion of machine laundering.
- Do not share food or drinks with anyone.
- After using the telephone, wipe the receiver (especially the mouth piece) with paper towels, and dispose the paper toweling in the trash.

Figure 5.2 (continued)

Signature Section (example release form)

I have read this form and all of my questions have been answered. By signing below, I acknowledge that I have read and accept all of the information above.

____

Signature of patient or personal representative

____

Print name of patient or personal representative

____

Date

____

Relation of personal representative to patient

Figure 5.2 (continued)

5.4 Advanced Methods of Patient-Specific Dose Calculations

Apart from guidance on patient release criteria given in NRC Regulatory Guide 8.39 and in NCRP Report No. 155, more advance methods have been proposed in the literature through the direction use of 3D whole-body computational phantoms and Monte Carlo radiation transport simulations.

In these studies, pairs of computational phantoms are applied—one representing the radioactive patient and one representing the family member (or other members of the public). In these studies, a biokinetic model is invoked to identify source organs in the patient phantom (or at least body regions of activity) that can be time-dependent in nature. In Han et al. (2013), the study team used members of the UF/NCI library of adult and pediatric reference phantoms in radiation transport studies of patient-to-family-member exposures.

The sets of exposure scenarios explored in this study are shown Figure 5.3. These included:

- face-to-face exposures between an adult male or female patient and pediatric or adult family members with all combinations computed at different separation distances,

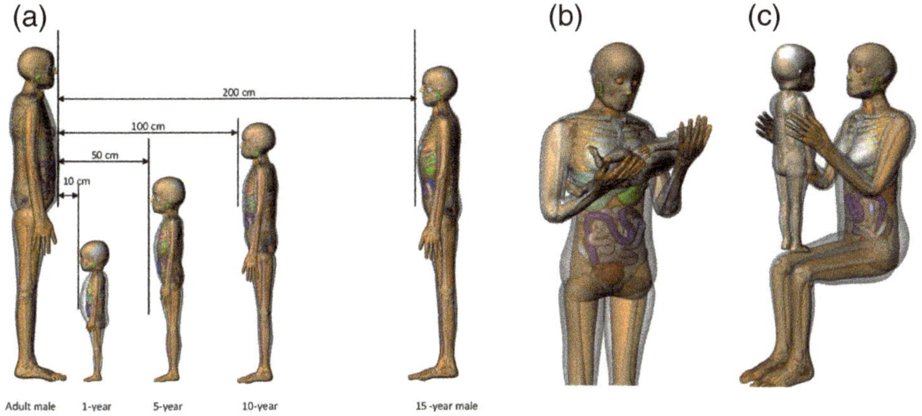

Figure 5.3 Exposure scenarios explored within the study of Han et al. (2013). (a) Lateral views of the adult male patient phantom facing the 1-year, 5-year, 10-year, and 15-year male family member phantoms at distances of 10, 50, 100, and 200 cm, respectively. The distance of 75 cm is now shown for clarify. Further views of the adult female patient phantom holding (b) a newborn phantom at her chest and (c) the 1-year phantom on her lap.

- the adult female patient holding her newborn child (male or female), and
- the adult female patient holding her 1-year-old (male or female) in her lap.

In the traditional MIRD schema of internal dosimetry, all identifications of source organs and all assignments of target organs are within the radiopharmaceutical therapy patient, and thus:

$$D(r_T) = \sum_S \tilde{A}_S \times S(r_T \leftarrow r_S) \tag{5.16}$$

where $D(r_T)$ = the absorbed dose to the target organ of the patient,
$\tilde{A}_S$ = the measured (or estimated) time-integrated activity within each source organ of the patient, and
$S(r_T \leftarrow r_S)$ = the radionuclide S values defined as the absorbed dose to the target organ r_T per radioactive decay in the source organ r_S.

In Han et al. (2013), the authors extended the MIRD schema to consider both the radioactive patient (PT) and the exposed family member (FM):

$$D(r_{FM,T}) = \tilde{A}_{PT} \times S(r_{FM,T} \leftarrow r_{PT}) \tag{5.17}$$

where $D(r_{FM,T})$ = the absorbed dose to a given target organ in the family member phantom,
$\tilde{A}_{PT}$ = the whole-body time-integrated activity within the patient phantom, and
$S(r_{FM,T} \leftarrow r_{PT})$ = the patient-to-family member radionuclide S values where the source region is the whole-body activity of

the patient and the target region is a given tissue T within the family member phantom.

By applying a weighted sum of the organ doses in the family member phantoms (using ICRP tissue weighting factors), one obtains the effective dose to the family member computational model:

$$E(r_{FM}) = \tilde{A}_{PT} \times \sum_{T} w_T \times S(r_{FM,T} \leftarrow r_{PT}) \tag{5.18}$$

In this study, the authors considered two ^{131}I source distributions within the adult patient phantoms: *hyperthyroidism* (HT) and *differentiated thyroid cancer* (DTC). For the HT patient phantom, the source organ was considered to be the thyroid, as this gland has been demonstrated to uptake 80% or more of available circulating radioiodine in this disease. A thyroid source distribution is also applicable to approximate radioiodine uptake in post-operative thyroid gland remnants in patients undergoing ^{131}I remnant ablation after thyroidectomy or in patients with residual iodine-avid metastatic lymph node involvement in the neck.

For the DTC phantom, it was assumed that the modeled patient had undergone total or near-total thyroidectomy typically with next ^{131}I uptake of $< 5\%$ of the administered activity. Metastases are then most frequently found in lungs (49%), skeleton (25%), or both (15%). Thus, in order to be conservative, the authors considered source regions for simulating the DTC patient to include all sites of significant iodine localization to include the lungs, thyroid bed (simulating the gland remnants or neck lymph nodes), salivary glands, stomach (contents and walls), small intestine (contents and walls), colon (contents and walls), liver, kidneys, and urinary bladder (contents and walls). Skeletal metastases were not considered as they may occur fairly randomly within any given patient. Source decay sites were sampled uniformly across all declared source regions and, thus, an implicit assumption was uniform activity concentration of ^{131}I in all source tissues of the DTC patient phantoms.

Results of the study by Han et al. (2013) are shown in Figures 5.4 through 5.6. Figures 5.4 and 5.5 plot the effective dose to the family member per unit time-integrated activity in the adult patient phantom—male or female for ^{131}I treatment of hyperthyroidism or differentiated thyroid cancer, respectively. In both Figures 5.4 and 5.5, the horizontal line gives the equivalent estimate of this quantity by assuming a point source and dose point as given in Option 1 of NRC Regulatory Guide 4.39. Values of patient-to-family member S values and corresponding effective doses to the newborn and one-year-old child held by the adult female HT or DTC patient are shown in Figure 5.6. It is further noted that while this study made the assumption of a time-independent source distribution of ^{131}I within the patient models, the methodology can be readily extended to consider time-dependent activity distributions and independent source organs within the patient phantom.

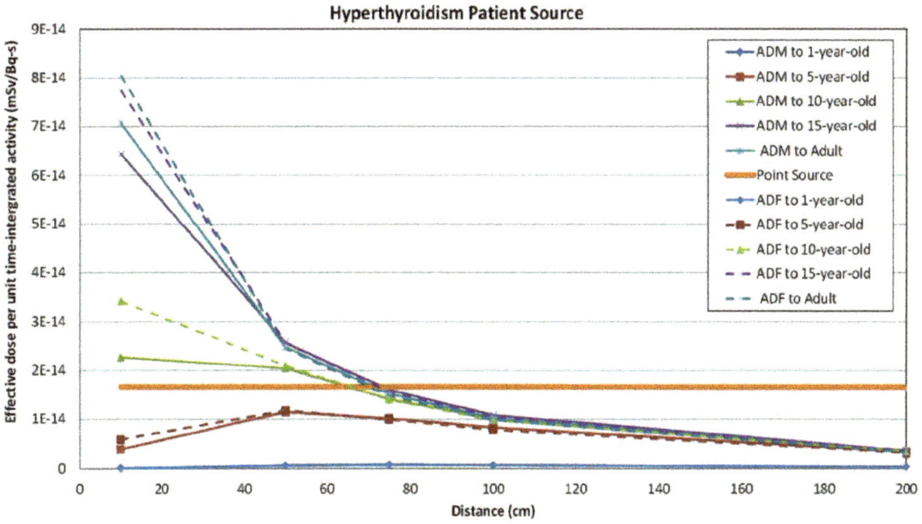

Figure 5.4 Effective dose per time-integrated activity (mSv/Bq-s) to family member phantoms from the adult male and adult female hyperthyroid (HT) patient phantoms as a function of separation distance.

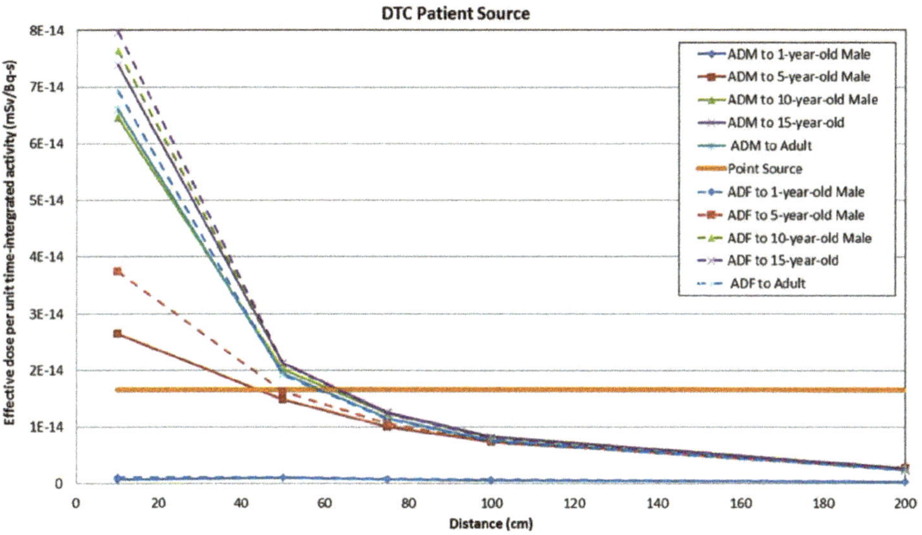

Figure 5.5 Effective dose per time-integrated activity (mSv/Bq-s) to family member phantoms from the adult male and adult female differentiated thyroid cancer (DCT) patient phantoms as a function of separation distance.

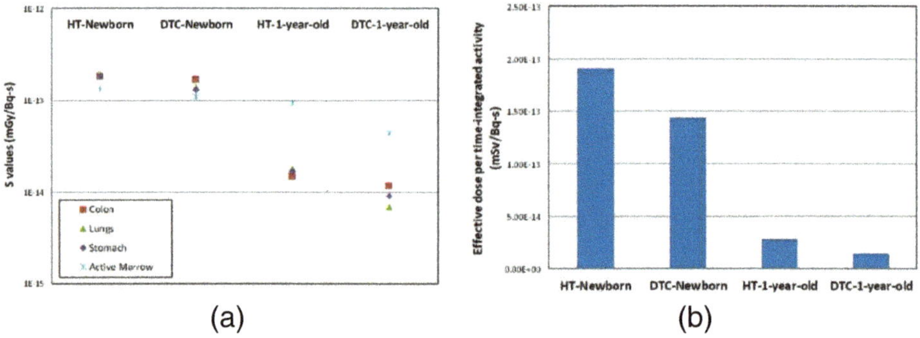

Figure 5.6 (a) S values (mGy/Bq-s) to colon, lungs, stomach, and active bone marrow and (b) effective dose per time-integrated activity (mSv/Bq-s) to the newborn and one-year-old child from exposure to the adult female patient phantoms treated for either hyperthyroidism (HT) or differentiated thyroid cancer (DTC). Exposure orientations correspond to those in Figure 5.3.

5.5 Summary

Current federal regulations given in sections of 10 CFR Part 35 permit the licensee to authorize the release of any radiopharmaceutical therapy patient from its control if the effective dose (or, more specifically, the total effective dose equivalent) to any other individual from exposure to the release patient is not likely to exceed 5 mSv (10 CFR 35.75). Furthermore, the licensee is required to provide the released patient (or the patient's parent or guardian) with written instructions on actions recommended to maintain doses to other individuals (such as family members, coworkers, and members of the general public) as low as reasonably achievable (ALARA) if the effective dose (or specifically the TEDE) to these other persons is likely to exceed 1 mSv. If the dose to a breastfeeding infant or child could exceed 1 mSv if the patient does not interrupt breastfeeding, the instructions shall also include (1) guidance on the interruption or discontinuation of breastfeeding and (2) information on the consequences of failure to follow this guidance (10 CFR 35.75(b)). The licensee is further required to maintain a record of the basis for authorizing the release of the radioactive patient for 3 years after the date of release (10 CFR 35.75(c) and 10 CFR 35.2075(a)), and is also required to maintain a record of instructions provided to the breastfeeding female patient for the same period of 3 years (10 CFR 35.75(d) and 34.2075(b)).

This chapter has reviewed several methods for making decisions on the time of patient release and the calculation of effective dose to exposed members of the public or family members. These include three options under NRC Regulatory Guide 8.39 to include point-source-based methods of administered activity, retained total-body activity, or measured dose rate at 1 meter distance. Further guidance—including exposures to sleeping partners and the holding of infants—is provided in NCRP Report No. 155. Finally, studies in the literature can provide more sophisticated techniques and dose coefficients that stem from the use of 3D anatomical models of both the radiopharmaceutical therapy patient and the exposed family member.

References

Dauer, L. T., M. J. Williamson, J. Humm, J. O'Donoghue, R. Ghani, R. Awadallah, J. Carrasquillo, N. Pandit-Taskar, A. K. Aksnes, C. Biggin, V. Reinton, M. Morris, and J. St. Germain (2014). "Radiation safety considerations for the use of Ra-223 dichloride in men with castration-resistant prostate cancer." *Health Phys.* 106:494–504.

Han, E. Y., C. Lee, L. McGuire, T. L. Brown, and W. E. Bolch (2013). "Organ S values and effective doses for family members exposed to adult patients following I-131 treatment: a Monte Carlo simulation study." *Med. Phys.* 40:083901.

ICRP (2008). ICRP Publication 107: Nuclear decay data for dosimetric calculations. *Ann. ICRP* 38:1–26.

ICRP (2010). ICRP Publication 116: Conversion coefficients for radiological protection quantities for external radiation exposures. *Ann. ICRP* 40:1–257.

NCRP. NCRP Report No. 37: *Precautions in the Management of Patients Who Have Received Therapeutic Amounts of Radionuclides.* Bethesda, MD: National Council on Radiation Protection and Measurements, 1970.

NCRP. NCRP Report No. 155: *Management of Radionuclide Therapy Patients.* Bethesda, MD: National Council of Radiation Protection and Measurement, 2006.

NRC. Regulatory Guide 8.39: *Release of Patients Administered Radioactive Materials.* Washington, D.C.: U.S. Nuclear Regulatory Commission, 1997.

NRC Regulatory Guide 8.39 (Revision 1): *Release of Patients Administered Radioactive Materials.* Washington, D.C.: Nuclear Regulatory Commission, 2020.

NRC Regulatory Guide 8.39 (Proposed Revision 2): *Release of Patients Administered Radioactive Materials.* Washington, D.C.: Nuclear Regulatory Commission, 2021.

Peplow, D. E. (2020). "Specific gamma-ray dose constants with current emission data. *Health Phys.* 118:402–16.

Schneider, S. and S. A. McGuire. *Regulatory Analysis on Criteria for the Release of Patients Administered Radioactive Material.* Washington, D.C.: U.S. Nuclear Regulatory Commission, 1996.

Zanzonico, P. B., J. A. Siegel, and J. St. Germain (2000). "A generalized algorithm for determining the time of release and the duration of post-release radiation precautions following radionuclide therapy." *Health Phys.* 78:648–59.

Chapter 6

Radiation Safety in Radiopharmaceutical Therapy

Jessica Clements, MS, DABR, FAAPM

Imaging Physicist and Radiation Safety Officer
Department of Radiology
The University of Vermont Medical Center
Burlington, VT

6.1 Introduction

This chapter aims to provide a basic framework for radiation safety requirements related to *radiopharmaceutical therapies* (RPTs) and describes activities that will involve the facility's *Radiation Safety Officer* (RSO). The use of radioactive materials in the United States is regulated at the federal level by the Nuclear Regulatory Commission (NRC). Many states are classified as Agreement States with the NRC, meaning they have entered into agreements that grant the State authority to license and inspect facilities located within state borders, except for federal jurisdiction land located in states. This chapter will focus on the NRC requirements in effect at the time of writing. It is imperative for licensees to be aware of state, county, city, and other government agencies that may also regulate the medical use of radioactive materials within their jurisdiction.

6.2 New Program Preparation

6.2.1 Radioactive Materials License Review

Once a decision has been made to perform a new RPT, one of the first steps is to review the *radioactive materials license* (RML). The facility must be approved for the proposed radiopharmaceutical, which is likely to be categorized as 10 CFR 35.300 Unsealed Byproduct Material Written Directive Required or 10 CFR 35.1000 Other medical uses of byproduct material or radiation from byproduct material (Table 6.1). If the RPT is not included on the radioactive materials license, an amendment is required to add it. The licensing agency will provide a list of the required information to include with the amendment request, which may include the radionuclide, activity possession limit, intended use, proposed users, operating procedures, and emergency procedures. Manufacturers may provide licensing toolkits to assist with amendments. The time to process an amendment will vary and could take several months to complete.

6.2.1.1 Activity Possession Limit

Careful consideration is required when selecting the maximum activity possession limit for the radioactive materials license. The value chosen represents the maximum activity at the facility at one time. It includes delivered doses pending administration, ordered doses that have been received but the administration was canceled, unused portions of administered doses, and waste. Maximum possession limits are typically grouped on an RML for all of category 35.300 together, and include all instances of I-131, Ra-223, and Lu-177. The 35.1000 uses are individually specified for each 35.1000 item, with some examples shown in Table 6.2.

6.2.1.2 Authorized Users

Authorized users (AU) are physicians who meet the training and experience requirements of the respective category defined by 10 CFR Part 35. A newly hired or trained physician must provide training and experience credentials to the licensee so an amendment to the RML can be submitted to add the AU to the license. AUs must be approved on a facility RML prior to practicing at a facility, even if already approved on another RML. Requirements for each authorization category are listed in 10 CFR 35 Subparts D through K with specific guidance for 35.1000 categories found in NRC Licensing Guidance accessed at in the NRC Medical Use Tool Kit located at https://nrc.gov/materials/miau/med-use-toolkit/emerg-licensed-med-tech.html#reg. The NRC maintains a list of medical specialty boards whose certification process has been recognized to meet NRC training and education requirements. The NRC defines an *Authorized Medical Physicist* (AMP) for therapies using modalities such as remote afterloading units and stereotactic radiosurgery units. The role of an AMP has not been defined by the NRC in radiopharmaceutical therapy. The American Association of Physicists in Medicine defined a *Qualified Medical Physicist* (QMP) for each of

Table 6.1 Radioactive materials license categories and therapeutic radionuclide examples

10 CRF Part 35	Description	Examples
Subpart E 300	Use of unsealed byproduct material for which a written directive is required	• NaI I-131 • Ra-223 dichloride • Lu-177 pharmaceuticals
Subpart K 1000	Other medical uses of byproduct material or radiation from byproduct material (Emerging Medical Technologies)	• Y-90 microspheres

Table 6.2 Sample possession limits for radioactive materials

Nuclide	Form	Possession Limit
A. Any byproduct material permitted by 10 CFR 35.300	A. Any	A. 1.2 Ci total
B. Radioactive material permitted by 10 CFR 35.1000 as specified below:	A. Any	A. 1.2 Ci total
1. Yttrium-90	1. Microspheres (BWXT Medical Ltd. Model TheraSphere)	1. Total not to exceed 111 GBq (3 Ci). Each source vial not to exceed 20 GBq (540 mCi).
2. Yttrium-90	2. Microspheres (Sirtex Medical Model SIR-Spheres)	2. Total not to exceed 111 GBq (3 Ci). Each source vial not to exceed 11 GBq (296 mCi).

the subfields of medical physics in Professional Policy PS 7-A (AAPM 2018). Additionally, the AAPM recommends that individual patient dosimetry for therapies involving radiopharmaceuticals must be performed by a QMP (Clements et al. 2018).

6.2.1.3 Treatment Location

A suitable treatment location must be identified. Often, the treatment location will be in an existing clinical space in nuclear medicine or radiation oncology as opposed to an area of new construction. When identifying a treatment location, there are several considerations. The treatment location must be authorized for the administration of radioactive materials. Authorized locations are defined in the facility's radioactive materials license, and an amendment may be required if a new location is to be added. The space used during administration should have enough room for an uptake chair and ideally will be located near a patient bathroom that may be used exclusively by the patient during and immediately following the treatment. Ideally, the treatment space should be enclosed with controlled access. Doses to members of the public and personnel must not exceed regulatory effective dose limits, and exposure to personnel must be *as low as reasonably achievable* (ALARA).

Table 6.3 Regulatory effective dose limits

Occupational Limits for Adults Defined in 10 CFR 20.1201			
Location	**Annual Limit (rems)**	**Annual Limit (Sv)**	**Measurement Details**
Total Effective Dose Equivalent (TEDE)	5	0.05	When the external exposure is determined by measurement with an external personal monitoring device, the deep dose equivalent (DDE) is used in place of the total effective dose equivalent. DDE is estimated at a tissue depth of 1 cm.
Lens Dose Equivalent (LDE)	15	0.15	LDE is estimated at a tissue depth of 0.3 cm.
Shallow Dose Equivalent (SDE)	50	0.5	To the skin of the whole body or to the skin of any extremity. Extremity exposure is measured with a ring dosimeter. SDE is estimated at a tissue depth of 0.007 cm averaged over an area of 1 cm^2.
Public Exposure Limits as Defined by 10 CFR 20.1301			
Total Effective Dose Equivalent (TEDE)	0.1	1	Licensees must demonstrate compliance by measurement or calculation. This dose limit applies to licensed operation that does not include: • contributions from background radiation • any administration the individual has received • exposure from individuals administered radioactive material and released • voluntary participation in medical research programs • from disposal of radioactive material into sanitary sewerage
Unrestricted area dose	2 mrem in any one hour	0.02 mSv in any one hour	If the licensee permits members of the public to have access to controlled areas, the limits for members of the public continue to apply to those individuals.

Table 6.3 lists regulatory effective dose limits. Additional shielding beyond basic building materials may be required. To estimate exposure rates surrounding therapeutic administration locations, consider the exposure rate constant of the radionuclide, half-value layer, and potential workload. Technical factors are provided in Table 6.4 for therapeutic radiopharmaceuticals.

6.2.2 Protocol Development

Treatment teams are typically multidisciplinary. Early in the program development process, it's ideal to invite all stakeholders to develop a written protocol which may include and is not limited to:

- referrals and scheduling,
- determination of whether treatment is indicated,
- contraindications,

Table 6.4 Technical factors used to determine if shielding may be needed in theranostic treatment areas

Radio-pharmaceutical	Radio-isotope	Emission Type	Exposure Rate Constant (R cm^2 / mCi h)	Half-Life (days)	HVL (mm Pb)	Administration Guidance from Package Inserts
Sodium iodide	I-131	Beta Gamma	2.2[1]	8.04[2]	2.74[1]	Hyperthyroidism: 148–370 MBq (4–10 mCi) Thyroid carcinoma: 3700–5550 MBq (100–150 mCi)
Azedra	I-131	Beta Gamma	2.2[1]	8.04[2]	2.74[1]	<62.5 kg 296 MBq/kg (8 mCi/kg) >62.5 kg 18,500 MBq (500 mCi) for up to 2 cycles
Xofigo	Ra-223	Alpha Beta Gamma	0.77[1]	11.43[3]	0.69[1]	50 kBq/kg body weight (1.35 μCi/kg) at 4-week intervals for 6 administrations
Lutathera	Lu-177	Beta Gamma	0.181[1]	6.7[4]	0.542[1]	7.4 GBq (200 mCi) at 8- week intervals for 4 administrations
Pluvicto	Lu-177	Beta Gamma	0.181[1]	6.7[4]	0.542[1]	7.4 GBq (200 mCi) at 6-week intervals for 6 administrations
Quadramet	Sm-153	Beta Gamma	0.481[1] 0.46[5]	1.946[2] 1.93[5]	0.0876[1] 0.1[5]	37 MBq/kg (1 mCi/kg)
Zevalin	Y-90	Beta	N/A[2]	2.67[2]		14.8 MBq/kg (0.4 mCi/kg) with a maximum of 1184 MBq (32 mCi)

[1]Smith et al. 2012; [2]NRC 2020; [3]ACMUI 2012; [4]NRC 2022; [5]FDA 2017

- patient education,
- pre-treatment steps, including premedication, cessation of medication, lab tests, or imaging,
- treatment planning,
- written directive,
- administration area preparation, including signs, absorbent paper,
- patient preparation,
- required supplies and equipment,
- radiopharmaceutical administration,
- radiation safety surveys,
- patient release, including release criteria and instructions to the patient, and
- emergency procedures, including spills and adverse events.

6.2.3 Training

All personnel involved in the protocol must be appropriately trained in advance of the first treatment. Their responsibilities should be clearly described in the facility's written protocols. It is ideal to conduct one or more dry runs in advance of the first treatment to practice and refine procedures. The vendor may provide specific training about delivery devices or other elements of the administration protocol.

6.2.4 Instrumentation

The hot lab and imaging instrumentation must be calibrated and be able to detect the theranostic agents utilized. See Chapter 4 for a discussion of dose calibrator quality assurance, traceability, and uncertainty. Instruments used to conduct area surveys and removable contamination (wipe) surveys after administration must be able to detect microcurie quantities of contamination. A calibrated ion chamber survey instrument is ideal for measuring the exposure rate from the patient after administration to ensure that release criteria are met. See Chapter 5 for more details on release criteria and take-home instructions. See Chapters 7–10 for imaging system calibrations and requirements.

6.2.5 Contamination Surveys and Removable Contamination

Contamination surveys are typically performed after a patient treatment, spill, or contamination event. Wipe tests are used to quantify amounts of removable contamination. Wipe samples are counted in liquid scintillation counters, sodium iodide crystal well counters, or solid state well counters. The expected efficiency for the type and energy of the sample being measured must be established before starting a new therapy. The NRC Regulatory Guide 1556 Vol. 9 Rev. 3 includes model procedures for contamination surveys and suggested trigger levels in unrestricted areas per 100 cm^2 as: 200 dpm for I-131 and Ra-223, 100 dpm for other alpha emitters, and 1000 dpm for beta-gamma emitters.

6.3 Handling Radioactive Material

Every facility must develop a radiation safety program defined by formal policies and procedures that describe safe practices for handling and using radioactive materials. These describe practices for managing public and occupational exposure, operating and emergency procedures, accurate measurement of radiopharmaceuticals, required personal protective equipment, shipping and receiving radioactive material, surveys, recordkeeping, and waste management. Policies and procedures are reviewed on a regular basis, typically annually, but should also be reviewed when beginning a new procedure. This section lists considerations for new radiopharmaceuticals that may apply to existing policies and procedures in the radiation safety program.

6.3.1 Receiving Radioactive Material

The requirements for receiving packages are found in 10 CFR 20.1906 Procedures for receiving and opening packages. Received packages must be surveyed for contamina-

Table 6.5 Materials used to shield various radiation emission types

Emission	Material
Alpha	No extra shielding
Beta	Plastic
Gamma	Lead or Tungsten
Positron	Lead or Tungsten

tion as soon as practical after receipt of the package, but not later than three hours after the package is received if received during normal business hours. If received after working hours, the package must be surveyed within three hours after opening the next business day. Therapeutic agents are often delivered by courier instead of standard radiopharmacy deliveries. The packages may be dropped at a loading dock or receiving department instead of direct delivery to the nuclear medicine department. A workflow may need to be developed to ensure compliant receipt of radioactive material by an alternate route.

6.3.2 Specialized Shields

Specialized equipment—such as syringe shields, dose carriers, infusion pump shields, and waste storage containers—may be required depending on the radiopharmaceutical emission (Table 6.5). Special attention is required for mixed emission sources. While the syringe or vial material containing alpha emitters is adequate, the presence of beta and gamma emissions may require the addition of a shield. In the case of mixed beta and gamma emissions, the order of shielding material matters. Beta emitters should be shielded first by plastic and followed by lead or tungsten to attenuate the gamma emissions. If the betas encounter the high-Z material first, bremstrahlung x-rays could be produced.

6.3.3 Shipping Packages

Hazmat employers must train and test hazmat employees according to 49 CFR 172.704. Hazmat employees may have tasks such as loading, unloading, or handling hazmat or preparing hazmat for transportation. Training must be completed within 90 days of becoming a hazmat employee. The new hazmat employee may perform hazmat duties in the first 90 days prior to completing training if they work under the direct supervision of an existing hazmat employee. Recurrent training must be completed at least every three years and includes general awareness, function-specific awareness, safety, security awareness, and in-depth security training if a security plan is required.

6.4 Waste Management

6.4.1 Facility Waste

Every program will generate radioactive waste and must have a radiation protection program in place for waste management and disposal. Licensed materials must be disposed of in accordance with NRC requirements. Prior to disposal, waste must be secured in appropriately shielded storage containers.

6.5.1.1 Decay-In-Storage

The NRC requirements for decay-in-storage are contained in 10 CFR 35.92. A licensee may hold material with a physical half-life of less than or equal to 120 days for decay before disposal in the regular waste stream. Waste is typically held for 10 half-lives, until it cannot be distinguished from background radiation level with an appropriate survey instrument. A survey must be made and documented, and all visible radiation labels must be removed or obliterated before entering the standard waste stream.

6.4.1.2 Commercial Disposal

Waste with a half-life greater than 120 days that is ineligible for decay-in-storage may be disposed of with a commercial broker that is licensed to manage waste. The applicable regulations are described in 10 CFR 30.41(b), 10 CFR 20.2006, and 10 CFR 20.2008. It may also be possible to request an amendment to a facility radioactive materials license to request decay-in-storage for longer-lived radiopharmaceutical waste.

6.4.1.3 Release Into Sanitary Sewerage

Excreta from patients that have received administration of radioactive material are exempt from limits for release of licensed material into sanitary sewerage. For general discharge into sanitary sewerage, conditions described in 10 CFR 20.2003 must be met. Agencies other than the one issuing a facility radioactive materials license may have state regulations and permitting requirements for release of material into the sewer. Model procedures for release into the sewer are included in Appendix W of Regulatory Guide 1556 Vol. 9 Rev. 3 (NRC 2019b).

6.4.1.4 Impurities

Some radiopharmaceuticals may contain accompanying long-lived radionuclides as a result of the production methods used. Table 6.6 lists common theranostic agents and details about potential impurities present in each. The half-lives of long-lived contaminants may exceed the standard NRC limit of a half-life less than 120 days for decay-in-storage. Each licensee must develop procedures for detecting impurities and potentially dispose of waste containing them with an authorized licensee.

6.4.2 Managing Outpatient Treatment Waste

NRC Regulatory Guide 8.39 (NRC 2020) Revision 1 details current guidance for the Release of Patients Administered Radioactive Material. It is suggested that the licensee provide release instructions that may include measures necessary to limit the transfer of radioactive contamination to others. Licensees may need to evaluate the need for outpatients to dispose of potentially contaminated trash from their residences. Holding trash for decay may be required to avoid a refusal or return of household waste from a landfill with detection procedures in place. At the time of writing, a

Table 6.6 Potential impurities found in common theranostic agents

Radionuclide	Impurity	Impurity Half-Life	Presence	NRC Guidance
Ga-68	Ge-68	270.8 days	Due to breakthrough.	ML19106A376 (NRC 2019a)
Lu-177	Lu-177m	160.4 days	Lu-177 produced by the direct method of neutron irradiation of Lu-176 targets (carrier added).	ML22318A150 (NRC 2022)
Ra-223	Ac-227 Th-227	22 years 18.7 d	According to the manufacturer, break-through tests are performed on each individual batch produced.	ML12320A450 (NRC 2012)
Y-90	Y-88 Y-91 Eu-154 Eu-152 Co-57 Co-60	106.6 days 58.5 days 8.8 years 13.6 years 270.9 days 5.27 years	May be present in all forms of Y-90 microspheres.	ML063470020 (NRC 2007)

draft Revision 2 of Regulatory Guide 8.39 (NRC 2021) issued in 2021 was under consideration of the NRC Advisory Committee on the Medical Uses of Isotopes.

6.6 Conclusion

Radiation safety programs require review and possible changes in advance of starting a new theranostic program. This chapter provides several considerations for radiation safety personnel to consider and includes regulatory references and guidance. Management of outpatient waste is currently being discussed among professional societies and regulators.

References

AAPM (2018). American Association of Physicists in Medicine Professional Policy PS &-A Definition of a Qualified Medical Physicist. Published August 2, 2018. Accessed at https://www.aapm.org/org/policies/details.asp?id=2551&type=PS.

ACMUI (2012). Advisory Committee on the Medical Uses of Isotopes. Final Radium-223 Chloride Report, November 20, 2012. Accessed at https://www.nrc.gov/docs/ML1232/ML12326A568.pdf.

Clements, J. B., C. T. Baird, S. F. de Boer, L. A. Fairobent, T. Fisher, J. H. Goodwin, D. A.Gress, J. L. Johnson, K. L. Kolsky, G. S. Mageras, R. M. Marsh, M. C. Martin, B. Parker, D. C. Pavord, M. C. Schell, J. A. Seibert, D. M. Stevens, R. B. Tarver, C. G. Waite-Jones, and N. Wingreen (2018). "AAPM medical physics practice guideline 10.a.: Scope of practice for clinical medical physics." *J. Appl. Clin. Med. Phys.* 19:11–25. https://doi.org/10.1002/acm2.12469.

FDA (2017). Quadramet Prescribing Information. Issued September 2017. Accessed at https://www.accessdata.fda.gov/drugsatfda_docs/label/2017/020570s008lbl.pdf.

NRC (2007). U. S. Nuclear Regulatory Commission. Information Notice 2007-10: Yttrium-90 Theraspheres and Sirspheres Impurities. Issued March 15, 2007. Accessed at https://www.nrc.gov/reading-rm/doc-collections/gen-comm/info-notices/2007/in200710.pdf.

NRC (2012). U. S. Nuclear Regulatory Commission. Radium-223 Dichloride: Bayer Responses to NRC Questions. Issued November 8, 2012. Accessed at https://www.nrc.gov/docs/ML1232/ML12320A450.pdf.

NRC (2019a). U. S. Nuclear Regulatory Commission. Germanium-68/Gallium-68 Pharmaceutical Grade Generators Licensing Guide. Issued July 2019. Accessed at https://www.nrc.gov/docs/ML1910/ML19106A367.pdf.

NRC (2019b). U. S. Nuclear Regulatory Commission. Consolidated Guidance About Materials Licenses NUREG-1556, Volume 9, Rev. 3. Issued September 2019. Accessed at https://www.nrc.gov/docs/ML1925/ML19256C219.pdf.

NRC (2020). U. S. Nuclear Regulatory Commission. Regulatory Guide 8.39 Revision 1. Accessed at https://www.nrc.gov/docs/ML1923/ML19232A081.pdf.

NRC (2021). U. S. Nuclear Regulatory Commission. Draft Regulatory Guide DG-8061 (Proposed Revision 2 to Regulatory Guide 8.39) issued 2021. Accessed at https://www.nrc.gov/docs/ML2132/ML21326A168.pdf.

NRC (2022). U. S. Nuclear Regulatory Commission. Revision of 2018 Memorandum to the Regions, "Licensing of Lutetium-177." Issued December 8, 2022. Accessed at https://www.nrc.gov/docs/ML2231/ML22318A150.pdf.

Smith, D. S. and M. G. Stabin (2012). "Exposure rate constants and lead shielding values for over 1,100 radionuclides." *Health Phys.* 102(3):271–91. doi: 10.1097/hp.0b013e318235153a. PMID: 22420019.

Chapter 7

Concepts and Requirements for SPECT/CT Quantification

Eric C. Frey, Ph.D., FSNMMI

Professor, Division of Radiological Physics
Russell H. Morgan Department of Radiology and Radiological Science
Baltimore, MD
and Founder and CEO, Rapid, LLC

7.1 Introduction

Accurate estimates of activities in objects of interest are the cornerstone of accurate dosimetry. Many of the radionuclides used in therapeutic applications are single photon emitters and can thus be imaged using single-photon imaging techniques. Many early studies used planar imaging and quantification techniques because early SPECT imaging was not quantitative. However, developments in instrumentation—especially the development and clinical adoption of SPECT/CT scanners, tomographic reconstruction, and methods of compensating for image-degrading factors—have resulted in the clinical availability of quantitative SPECT images. Many studies have demonstrated the superiority of SPECT or hybrid SPECT/planar methods over purely planar methods for dosimetry applications. However, obtaining quantitative SPECT images is not as automatic as it is on PET systems; obtaining SPECT images that provide activity estimates suitable for dosimetry requires paying attention to the selection of instrumentation, acquisition parameters, reconstruction and compensation methods and parameters, and quantification techniques. This chapter discusses the various concepts that must be understood to make decisions necessary to obtain quantitative SPECT. It also provides illustrations of the achievable quantitative accuracies and precisions for several radionuclides relevant to targeted radionuclide therapy. More detailed and quantitative discussions of some of the issues can be found elsewhere (Dewaraja et al. 2012; Agency 2014). These documents are available for free download from the Society of Nuclear Medicine and Molecular Imaging and the International Atomic Energy Agency (IAEA), respectively.

7.2 SPECT/CT Instrumentation: Key Concepts and Components

The most common clinical SPECT/CT systems are based on combining two large-field-of-view scintillation cameras and multi-slice diagnostic CT systems (see Figure 7.1). The images are acquired sequentially in time with the order of acquisition depending largely on preference.

7.2.1 Scintillation Cameras

Figure 7.2 shows the components of a scintillation camera. Gamma rays are emitted isotropically and are incident on the camera from all directions. The gamma camera generates a digitized projection image of activity emitted from the object. The major components and their role in determining image quality for quantitative applications are discussed below. More general, comprehensive, and quantitative descriptions of

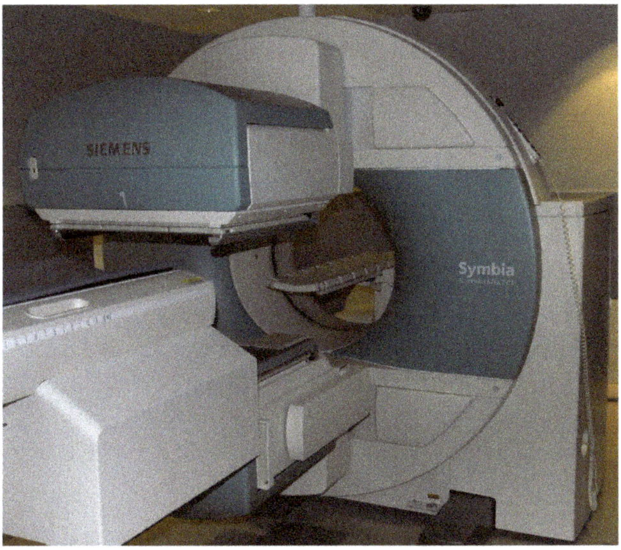

Figure 7.1 Typical clinical SPECT/CT system. The systems are typically comprised of two large-field-of-view (FOV) scintillation cameras (axial FOV ~40 cm and transaxial FOV ~50–60 cm) and a 4–16 slice diagnostic CT scanner.

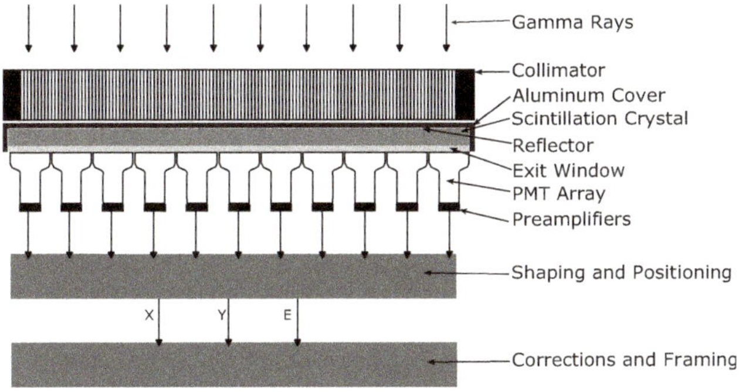

Figure 7.2 Schematic cross section of gamma camera.

gamma cameras can be found (Cherry et al. 2012; Bailey and IAEA 2014). The latter reference is available free from the International Atomic Energy Agency.

7.2.2 Collimators

7.2.2.1 Resolution-Sensitivity Trade-Off

The collimator acts as a mechanical lens, creating a relationship between positions in the 2D projection image and directions in space. Collimators are the key determinant of image resolution and noise properties: the *full-width at half-maximum* (FWHM) spatial resolution of the collimator is approximately proportional to the diameter of the collimator holes and inversely proportional to the hole length (for sources far from the collimator). The collimator sensitivity (typically specified in counts per activity per time) is proportional to the square of the hole diameter and inversely proportional to the hole length. The FWHM spatial resolution is also approximately proportional to the distance from the collimator face, resulting in complex spatial variation of the resolution in SPECT images.

One result of the above properties is that there is a trade-off between spatial resolution and noise, with the sensitivity inversely proportional to the square of the resolution. For quantitative applications, the collimator resolution largely determines the magnitude of *partial volume effects* (PVEs), i.e., the blurring of activity between adjacent organs.

Different resolutions and sensitivities are required for different imaging tasks. For example, quantifying uptake in large organs like the liver requires less resolution than quantifying activity in smaller objects like tumors. Similarly, imaging using a small *administered-activity* (AA) for therapy planning requires collimators with greater sensitivity (or longer imaging times) than imaging after a therapeutic administration using a large AA. As a result, classes of collimators with different resolution-sensitivity trade-offs are often available: high sensitivity, all or general purpose (AP or GP), high-resolution (HR), and ultra-high-resolution (UHR) collimators, in order of decreasing FWHM, are often available from manufacturers. The exact relationship between the collimator resolution and resolution class is manufacturer dependent, and not all manufacturers offer all the above classes.

7.2.2.2 Septal Penetration and Scatter Properties

Another important property of collimators is their ability to block photons at oblique angles. Detected photons that scatter in or penetrate through the septa result in long tails in the *point response function* (PRF), reduced contrast in projection images, and complicate compensation for the collimator response (see below for details). Thus, using appropriate collimators to reduce septal penetration and scatter is important.

The amount of septal penetration and scatter is determined by the composition and thickness of the septa (the walls between the collimator holes), the aspect ratio (diameter over length) of the collimator holes (thicker septa and smaller aspect ratios increase the path length through the septa, thus reducing penetration), and the energy of incident photons. Collimators designed for different ranges of photon energies are available from manufacturers, typically extra-high-energy (EHE, >450 keV), high-energy (HE, for photon energies ~300–450 keV), medium-energy collimators (ME, for ~160–300 keV), and low-energy collimators (LE, <~160 keV). Collimators designed for higher-energy photons have lower sensitivities at the same resolution due to the use of thicker septa. Not all combinations of energy range and resolution properties are available. For example, HR and UHR collimators are typically available only for LE designs, and ME and HE collimators are typically available only in the GP resolution class.

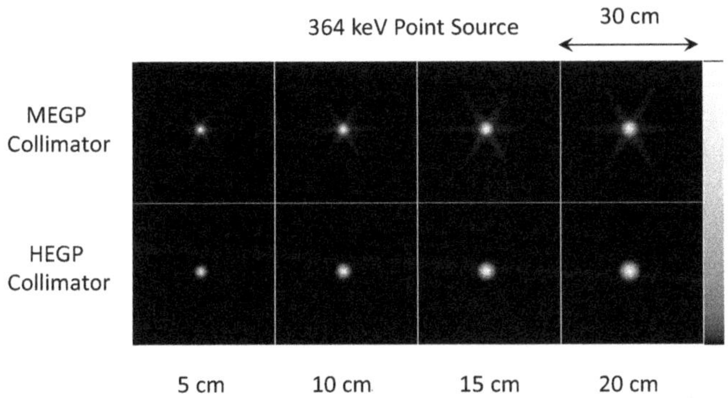

Figure 7.3 Simulation of image of a 364-keV point source using two different collimators at four distances from the face of the collimator. The images are shown on a logarithmic gray scale to emphasize low-intensity features.

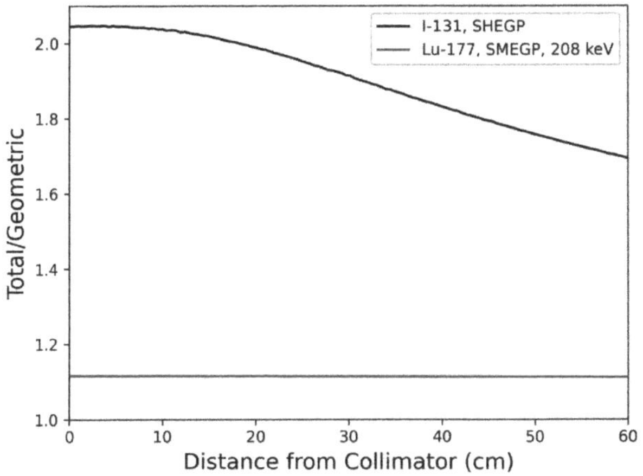

Figure 7.4 Plot of the total-to-geometric count ratio as a function of distance for ^{131}I using a Siemens HEGP collimator and ^{177}Lu using a Siemens MEGP collimator.

Choosing an appropriate design based on the photon energy spectrum produced by the radionuclide is an important task in specifying an imaging protocol. Figure 7.3 is a simulated image of a 364-keV photon from ^{131}I at four distances from the face of the collimator using HEGP and MEGP collimators. Note the presence of the star-shaped artifacts due to septal penetration and scatter for the MEGP collimator. Also note that the resolution, as indicated by the size of the central feature, increases with increasing distance, demonstrating the distance-dependence of the spatial resolution.

For quantitative SPECT, accounting for the spatial dependence of the sensitivity is important. The geometric component of the collimator sensitivity (i.e., the photons passing through the collimator holes without interacting) is independent of distance. However, the amount of septal penetration and scatter does depend on distance. Figure 7.4 shows a plot of the ratio of the sum of all photon counts to geometric photon counts (*total to geometric ratio*, TGR) as a function of distance to the collimator face for a ^{131}I and ^{177}Lu image using HEGP and MEGP collimators, respectively. Since the geometric sensitivity is constant with distance, total sensitivity is proportional to the TGR. For ^{177}Lu and the MEGP collimator, the total sensitivity is constant with distance, while there is substantial variation for ^{131}I and the HEGP collimator. Accounting for the effects of septal penetration on the PRF and the spatial variation in sensitivity is important in obtaining quantitative SPECT images. This is especially true when the sensitivity varies with distance. These data demonstrate the importance of selecting appropriate collimators for the radionuclide of interest and the fundamental challenge of imaging radionuclides that emit high-energy photons. This will be discussed in greater depth below.

7.2.2.3 Collimator Geometry

A final property of collimators is their geometry. Parallel hole collimators predominate for quantitative SPECT in therapeutic applications due to the need to estimate the activity distribution in multiple organs of interest, requiring a large *field-of-view* (FOV). There has also been less effort in developing quantitative reconstruction methods for other collimator geometries. For some applications, converging beam collimators can provide improved sensitivity-resolution trade-offs at the cost of a reduced field of view and greater complexity of the reconstruction algorithms. Diverging beam collimators are rarely used, but they provide larger fields of view at the cost of a poorer resolution-sensitivity trade-off and are seldom used in current nuclear medicine practice.

7.2.3 Scintillation Crystal, Photodetector Array, and Detector Electronics

The scintillation crystal converts the energy of incoming gamma rays to an ensemble of visible light photons emitted isotropically. Most current cameras use large single crystals of NaI(Tl). This material is hygroscopic, so must be hermetically sealed. The aluminum cover serves to reflect and possibly shape the distribution of scintillation photons. The scintillation photons pass through a glass exit window and are incident on an array of photodetectors that convert the scintillation photons into an electric signal where the total electrical charge is proportional to the number of scintillation photons incident on the photodetector. Commercial general-purpose SPECT systems typically use photomultiplier tubes as photodetectors.

The output of the photodetectors is a set of electrical signals in time. These signals are converted to a voltage signal where the peak voltage is proportional to the charge, shaped in time to improve count rate performance and the reliability of digitization, and digitized to produce a digital representation of the voltage. The output of the detector electronics in modern digital cameras is, for each incident gamma ray, a collection of digital values. These digital values are proportional to the intensity of scintillation light incident on the PMT. These values are used to estimate the 2D position of interaction of the gamma ray in the plane of the scintillation crystal and the total energy of the gamma ray deposited in the crystal. The digitized gamma ray energy estimate is used to discriminate between scattered and unscattered photons by rejecting photons outside a specified range of energies around the gamma ray's emission energy. The digitized position estimate is used to select the pixel in a 2D image that is incremented, indicating the detection of a photon in the corresponding position in the projection image.

The accuracy of the position estimates is improved via the use of energy, spatial, and uniformity correction algorithms, all of which require calibration procedures. The precision of the estimates of energy and position are ultimately limited by quantum noise due to the random number of scintillation photons produced from a given gamma ray and subsequently converted to the digital representation. The precision of the position and energy estimates determines the intrinsic spatial and energy resolutions of the camera, respectively. The number of scintillation photons produced is proportional to the gamma ray energy and follows a Poisson distribution. Thus, to first order, both the FWHM intrinsic spatial and energy resolutions vary inversely as the square root of energy. Typical spatial and energy resolutions of current scintillation cameras at 140 keV range from 3.2–4 mm and 9% to 10%, respectively.

The intrinsic sensitivity of a crystal is the fraction of a well-collimated beam of photons that is absorbed by the crystal. It is thus a function of the density, composition, and thickness of the crystal and the energy of the incident photons. Typical clinical SPECT systems are designed for imaging the 140-keV photons from ^{99m}Tc and have a crystal thickness of 0.9525 cm. Figure 7.5 shows the relationship between intrinsic sensitivity and photon energy for four crystal thicknesses. Note that the sensitivity approaches unity for 140-keV photons, but decreases rapidly for high-energy photons. For example, for the 364-keV photon from ^{131}I, the sensitivity is less than 40% for a crystal 0.9525 cm thick and approximately 70% for a crystal 2.54 cm thick. The trade-off for the thicker crystal is an approximately linear degradation in intrinsic

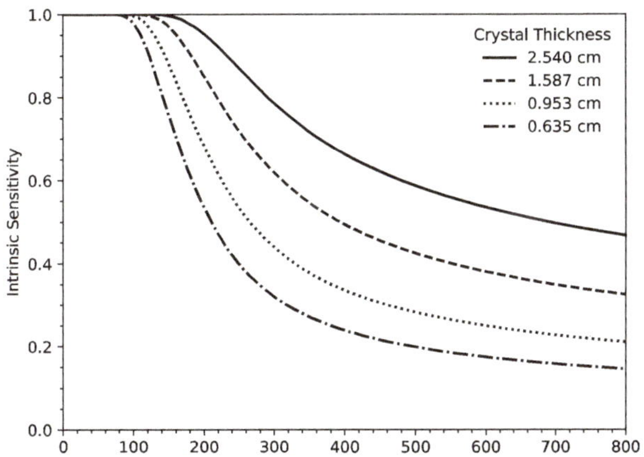

Figure 7.5 Plot of the intrinsic sensitivity as a function of photon energy for four thicknesses of NaI.

spatial resolution. However, as will be discussed in more detail below, the system resolution in SPECT is dominated by the collimator resolution. As a result, crystal thicknesses of 1.59 cm have been and are commercially available and have some advantages for imaging of high-energy photon emitters. Note that for post-therapy imaging where the count rate is high, the benefits of increased intrinsic sensitivity are relatively less important.

The selection of energy windows and window widths can have an important effect on the quantitative reliability. Narrower energy windows, e.g., 15%, are often used for scanners that have higher resolutions, such as 9% at 140 keV, for improved scatter rejection. However, narrow windows result in greater variation in the sensitivity with drifts in energy calibration of the scanner. As a result, the author tends to prefer wider (20%) energy windows, assuming suitable scatter compensation is used. Even wider windows are often needed when imaging x-ray photopeaks due to the presence of multiple emissions and the generally poorer fractional energy resolution at lower energies.

7.2.3.1 SPECT Spatial Resolution

As indicated above, the spatial resolution in SPECT is a function of the collimator and intrinsic spatial resolution. A good approximation is that the system FWHM resolution is equal to the square root of the sum of the squares of the FWHM collimator and intrinsic resolutions. Figure 7.6 shows a plot of the FWHM resolution for the collimator alone and including the intrinsic resolution for an LEHR collimator. The *collimator-detector response* (CDR), i.e., the projection of a point source, is well-approximated by a Gaussian function for collimators where septal penetration and scatter are minor effects.

The 3D point response in SPECT is complicated by the fact that the distance from a position in the image to the collimator face varies, in general, with projection view. Figure 7.7 shows the reconstruction using filtered backprojection of projections from point sources in air distributed along the horizontal axis. A circular orbit was simulated. Note that the reconstructed resolution is symmetric at the center of rotation and is relatively constant in the radial direction. In the tangential direction, the resolution improves toward the right of the image because the projection in the view from the right is so close to the source. The situation is more complicated for the body contouring orbits typically used, but the principle is the same, and the resulting 3D resolution of the SPECT image is spatially varying and asymmetric.

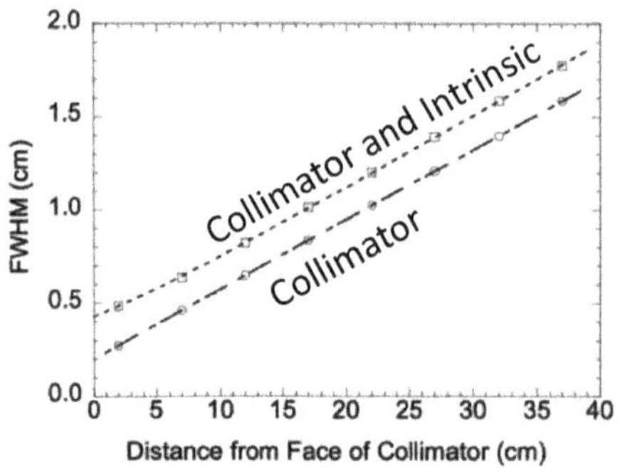

Figure 7.6 Plot of FWHM resolution for an LEHR collimator as a function of distance from the face of the collimator with and without the intrinsic resolution.

Figure 7.7 Left: image of point source phantom in air. Right: reconstructed image of the phantom at left simulating noise-free acquisition with a 360° circular orbit and an LEHR collimator.

7.2.3.2 Count Rate Losses

Due to the finite decay time of the fluorescence processes that produce scintillation photons and time widths of electrical pulses, signals from photons that are close together in time can pile up. One result of this is count-rate-dependent losses of photon counts. In other words, the measured count rate is less than the true count rate. This can be significant in post-therapy imaging where count rates > 100 kcounts per second are possible. Some gamma cameras include counting modes designed to perform better at high count rates, typically with some compromise of other imaging properties, such as energy resolution. There has been little published in the literature on the use of these modes for quantitative imaging.

For scintillation cameras, photon positions and energies are estimated from the entire photomultiplier tube array (or a large fraction of it). In addition, pulses from photons with energies outside the energy windows used for image acquisition contribute to count-rate losses. It is thus reasonable to compensate the entire projection image with a single correction factor, but the factor should be estimated based on the count rate for the full energy spectrum.

There are two idealized models for counting system count rate losses: *paralyzable* and *non-paralyzable*. In both cases, the system is characterized by the resolving (or dead) time, ρ. This is the length of time after arrival of a photon that a second photon would affect the processing. In paralyzable systems, photons arriving during the resolving time of a previous photon extend the processing time by an amount equal to the full resolving time; in nonparalyzable systems photons arriving during the resolving time are ignored and do not reset the processing time counter. Scintillation cameras are comprised of cascaded systems that are characterized separately by these models. The analog parts of the processing chain (e.g., fluorescence decay and charge integration) tend to behave like paralyzable systems, and digital parts (e.g., analog to

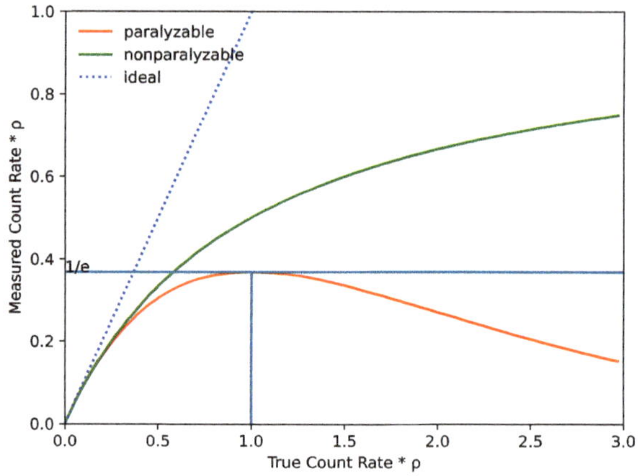

Figure 7.8 Plot of the measured count rate vs the input count rate for ideal paralyzable and non-paralyzable systems having a resolving time of ρ.

digital conversion and image binning) tend to behave like non-paralyzable systems. Typically, paralyzable models are used to compensate for resolving time losses.

In paralyzable systems, the count rate has a maximum measured count rate equal to $1/e\rho$ for a true count rate of $1/\rho$. For a nonparalyzable system, the measured count rate asymptotically approaches a maximum of $1/\rho$. Figure 7.8 illustrates the behavior of these two systems in comparison with an ideal system that has no count-rate losses.

Paralyzable systems are described by the equation

$$N = ne^{-n\rho}, \tag{7.1}$$

where N is the measured count rate and n is the true count rate. Given the resolving time and the measured count rate, it is possible to solve for the true rate numerically or using the Lambert W function, W:

$$n = \frac{-W(-N\rho)}{\rho}. \tag{7.2}$$

This gives two solutions, one on either side of the maximum. Thus, other information must be used to determine which solution is appropriate.

7.2.3.3 Direct Conversion Solid State Gamma Cameras and Novel SPECT Systems

Solid state gamma cameras have begun to be produced commercially. These systems use cadmium-zinc-telluride (CZT) detectors that are directly pixelated. Gamma ray energy is directly converted to a charge signal, resulting in improved energy resolution. This has the potential to provide improved rejection of scattered photons and better ability to distinguish photopeaks (such as those from characteristic x-rays) that are close together in energy. However, impurities and differences in hole and electron mobility, combined with variations in depth of interaction, produce a low-energy tail on the energy response function in contrast to the Gaussian shape typical of scintillation cameras. This tailing effect somewhat reduces the promise of the improved energy resolution. The intrinsic resolution of these systems is determined by the pixel size, which depends on spacing of metal electrical contacts on the detector. The pixel size for CZT detector modules used in clinical systems has been 2.5 mm. Because of their novelty and limitations in their ability to image high-energy photons because of charge sharing between neighboring pixels, there has been relatively less use of these for quantitative SPECT for dosimetry applications. It is expected this will change in the future as therapy becomes a larger part of the nuclear medicine procedure mix.

A significant advantage of direct conversion detectors is their small size. This enables the creation of novel detector geometries, such as the Spectrum Dynamics Veriton and GE's Starguide systems. These systems have multiple small FOV detectors that can be independently scanned, resulting in high sensitivity. There have been few studies of quantitative use of these systems with therapeutic radionuclides, so they are not discussed in this chapter.

7.3 Image Quality for Quantitative SPECT Imaging

Rigorous objective assessment of image quality must be performed with reference to the task. For dosimetry, this means that the quality of image acquisition or processing methods or parameters should be evaluated in terms of their effect on the accuracy and precision of absorbed dose estimates. Images that are optimal for absorbed dose estimation are often not optimal for defect detection or other classification tasks. Similarly, images that are optimal for estimating tumor or organ absorbed dose are likely not optimal for estimating the absorbed dose-volume histogram. Since absorbed dose is estimated from time-activity curves, and assuming that the activities at the different measurements are not statistically correlated, a necessary condition for obtaining optimal absorbed dose estimates is that the activity estimates at each timepoint are optimal. Thus, selection of parameters and methods can be done in terms of optimizing activity.

Indirect physical measures of image quality—including resolution, contrast, and noise power—indirectly affect the accuracy and precision of dose estimates. However, a complete optimization requires going beyond these measures.

Reliability of an absorbed dose estimate depends on both accuracy and precision. Accuracy is often assessed by bias, i.e., the deviation of the value estimated from imaging from the true value. Precision is often measured by the variance, i.e., the sum of the square of the differences of the measured values from the mean. There is often a trade-off between accuracy and precision in selecting imaging or processing parameters. For example, in collimators there is a trade-off between resolution, which affects accuracy for small objects, and sensitivity, which affects quantum noise. For dosimetry, both are typically desirable: accuracy is important so that absorbed dose values for a given patient can be compared to those assessed at other sites or standardized dose limits; precision is important since larger imprecision typically necessitates larger safety margins when setting absorbed dose values. Thus, it is reasonable to weight accuracy and precision equally when optimizing for dosimetry applications. This can be done by using the mean-squared error, equal to the bias squared plus the variance or the average of the square difference between the true and measured values.

It is often difficult to apply these metrics to patients, where the true activity in an organ or tumor is impossible to measure, and obtaining repeat measurements is difficult or practically impossible. As a result, simulations are often used. These can be the use of simulated objects mimicking the clinical imaging problem, such as phantoms or numerical simulations. In all cases, it is important to remember that both accuracy and precision depend on anatomy and uptake distribution, which are random variables, in the context of considering the *a priori* reliability of an absorbed dose estimate. That is, the accuracy and precision will vary among patients.

7.4 Radionuclides for Radiopharmaceutical Therapy

Table 7.1 gives a list of radionuclides that can be imaged with SPECT that are relevant for radiopharmaceutical therapy. A number of relevant properties for imaging are given. Some of these radionuclides, e.g., ^{123}I, are shown because they are useful as imaging surrogates for treatment planning. The energies of photons used in imaging are shown. The downscatter column lists the energy and abundance of high-energy photons emitted during the decay (possibly by daughters) that could contribute downscatter to the image. The energy of those photons should be considered in selecting a collimator. The beta energy and range are given for comparison to the image resolution to help decide on the relevance of using a local dose deposition model for beta particle dosimetry. Collimators that have been or are typically used are shown. In some cases, two collimators are listed. In these cases, the selection of a collimator depends on whether the reconstruction method compensates for septal penetration and scatter. For example, for ^{123}I and ^{67}Cu, a low-energy, high-resolution collimator could be used if compensation for downscatter and septal penetration and scatter is included in the reconstruction in order to allow better quantification of small structures. Some more discussion of this is given below in the section on the CDR as an image-degrading factor.

7.5 Image-Degrading Factors

In an ideal SPECT imaging system, photons would not interact in the body (no object attenuation or scatter), but would be detected only if traveling in the direction perpendicular to the camera. In addition, the septa would be opaque to photons, patients would not move, and there would be no quantum noise. In this case, the activity distribution could be exactly reconstructed from a set of parallel projections acquired over a circular orbit of 180° around the patient using the filtered backprojection algorithm. The factors above are thus referred to as *image-degrading* factors since using filtered

Table 7.1 Properties of radionuclides with applications in radiopharmaceutical therapy

RN	Half-life	Decay Type	Radio-active Daugh-ters	Imageable Daughters	Imageable Photons	Down-scatter	Beta Energy	Max Beta Range	Collimator
In-111	2.87 d	E.C.	No	None	171, 245 keV	None			ME
I-131	8.02 d	β^-	No	None	364 keV	637 keV, 7%	606 keV	2 mm	HE
I-123	13.23 h	E.C.	No	None	159 keV	524 keV, 1.4%			LE or ME
Sm-153	46.5 h	β^-	No	None	103 keV	None	805 keV	3.3 mm	LE
Lu-177	6.65 d	β^-	No	None	112, 208 keV	321 keV, 0.2%	498 keV	1.6 mm	ME
Cu-67	61.8 h	β^-	No	None	92 keV (X), 185 keV	300 keV, 0.8%	575 keV	2.1 mm	LE or ME
Y-90[1]	64.1 h	β^-	No	None	bremsstrah-lung	2.28 MeV brems-strahlung	2.28 MeV	11 mm	HE or ME
Ra-223	11.4 d	α	Yes	Bi-211 (351 keV, 12%)	84 keV (X), 269 keV	832 keV, 3.5%	-		ME
Ac-225[2]	10.0 d	α	Yes	Fr-221 (218 keV) Bi-213 (440 keV)	None	1.57 MeV, 2.1%	-		HE
Th-227	18.8 d	α	Yes	Ra-223	236 keV	832 keV, 3.5%	-		ME or HE
At-211	7.21 h	α	Yes	None	78 keV (X)	1064 keV 31%	-		LE, ME or HE
Pb-203	51.9 h	EC	No	None	279 keV	401 keV, 3.3 %	-		ME or HE
Pb-212[3]	10.6 h	β^-	Yes	None	238 keV	2.6 MeV, 35.6%	-		HE
Ho-166	26.8 h	β^-	No	None	80.6 keV	None	1.8 MeV	8.7 mm	LE

[1]Low-abundance positron mode that can be imaged with PET.
[2]Emits only low-abundance x-rays.
[3]Daughters emit α-particles.

backprojection to reconstruct images affected by these effects results in degraded quantitative accuracy and precision. The following is a discussion of these image-degrading effects and their effect on SPECT images.

7.5.1 Attenuation

Attenuation refers to the loss of photons in the object being imaged compared to the case where there were no interactions of the photons with the object. This is due to photoelectric absorption or Compton or coherent scatter of the photons. Because the body is comprised largely of low-atomic-number elements, Compton interactions dominate for the photon energies used in nuclear medicine and especially for therapeutic radionuclides. Attenuation can be described by the Beer-Lambert Law:

$$I(\tau) = I(0)e^{-\mu\tau}, \tag{7.3}$$

where μ is the narrow-beam attenuation coefficient, τ is the thickness traversed, and $I(\tau)$ is the intensity (or number of photons) present after traversing a thickness τ. For nonuniform attenuators, $\mu\tau$ is replaced by the line integral through the attenuation distribution. Attenuation results in a position-dependent reduction in the number of detected photons compared to imaging the activity distribution in air. Reconstruction without taking this into account results in a spatially varying reduction in the estimated activity. Figure 7.9 shows a phantom comprised of a uniformly emitting and attenuating ellipse containing two circular hot-spots. Projections were simulated modeling attenuation and reconstructed with filtered backprojection (i.e., without attenuation compensation). The intensity in the reconstructed image is reduced toward the center of the object. Accurate estimation of activity in the object requires attenuation compensation.

The narrow beam attenuation coefficient describes the situation where all photons that interact are not detected. This is not the case in real SPECT systems, as some scattered photons are detected. Thus, attenuation compensation tends to amplify the effects of scatter, and scatter compensation is essential when compensating for attenuation to obtain quantitative images. Using a broad beam attenuation coefficient is not a realistic way to accomplish this, as it depends on the details of the geometry and activity distribution and does not compensate for the spatial distribution of scattered photons.

7.5.2 Photon Scatter

Scatter refers to photons that interact with the object via a Compton or coherent scatter interaction and contribute to the measured projections. The cross section for coher-

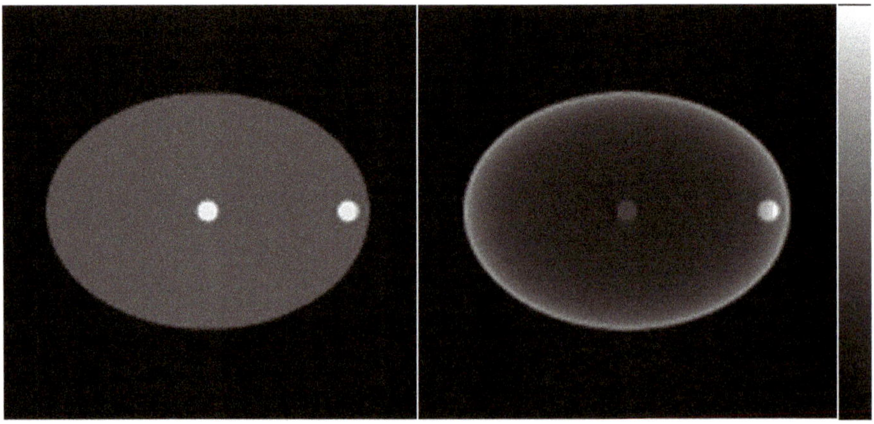

Figure 7.9 Left: phantom comprised of uniform elliptical attenuator/emitter and two circular hot spots. Right: image of phantom reconstructed using filtered backprojection (no attenuation compensation) from projections that include the effects of attenuation.

ent scatter increases with proton number and decreases with energy and is generally less important for typical SPECT energies. However, the fact that photons do not lose energy results in a non-negligible contribution to the distribution of scattered photons, and especially for 80–90 keV x-ray photons that can be useful in imaging therapeutic radionuclides. Compton scatter results a loss of photon energy that increases with larger scatter angles. However, due to the finite energy resolution of gamma cameras and the resulting use of energy windows with finite widths, scattered photons can be detected and contribute to the measured projections, as illustrated in Figure 7.10.

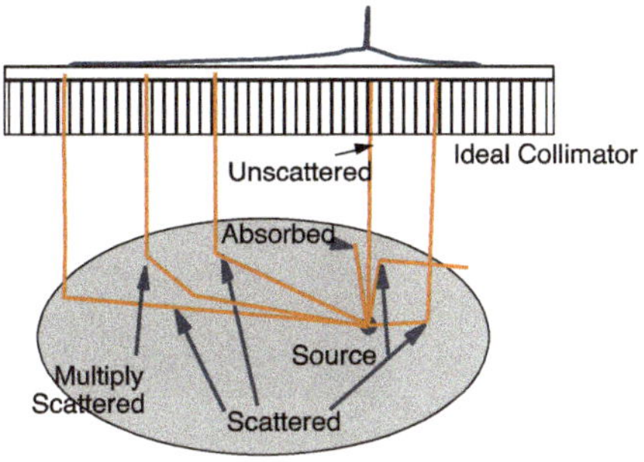

Figure 7.10 Single and multiply scattered photons can pass through the collimator and be detected.

Detectors with improved energy resolution enable the use of narrower energy windows and can reduce the contribution of scatter. However, current CdZnTe detectors have significant low-energy tails resulting from depth-of-interaction effects and the difference in hole and electron mobilities. Thus, using energy windows based on the *full-width at half-maximum* (FWHM) energy resolution will result in reduced sensitivity to unscattered photons. As a result, the full benefit of the improved FWHM is typically not realized in practice.

For therapeutic radionuclides, the energy spectrum often contains high-energy photons. Even if these are low in abundance, they can scatter in the body or camera and contribute to lower-energy photopeak windows.

The distribution of scattered photons is object-dependent and spatially varying. This is illustrated in Figures 7.11 and 7.12. Figure 7.11 shows a transaxial profile through the projection of a ^{99m}Tc line source in a water-filled elliptical cylinder. The scatter response varies in both magnitude and shape as a function of source position and object shape. Figure 7.12 shows the *scatter-to-primary ratio* (SPR) for ^{99m}Tc and ^{201}Tl sources as a function of source depth. The simulation was for a point source in a water-filled cylindrical tank. The ^{201}Tl was imaged using a 30%-wide energy window positioned over the Hg x-ray energies, while ^{99m}Tc was imaged using a 20%-wide energy window centered at 140 keV. This demonstrates that, for this phantom, the relative magnitude of the scatter as described by the SPR increases approximately linearly with depth. Also, it demonstrates that the SPR is different for different radionuclides and energy windows.

Figure 7.13 shows the quantitative impact of scatter on SPECT images. Note that without scatter, attenuation compensation was quite effective in restoring the original activity distribution (the texture in the images is due to noise in the MC simulation). However, when scattered photons were present in the projection data and attenuation compensation without scatter compensation was applied, the reconstructed image had a spurious increase in activity that was most pronounced at the center of the phantom.

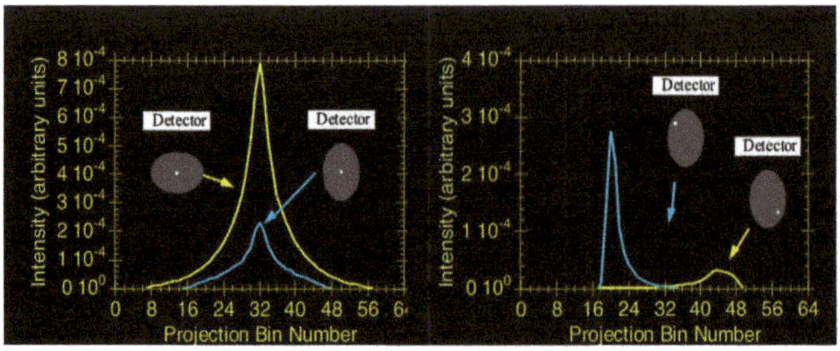

Figure 7.11 Transaxial profiles through the scattered photon projections of a line source in an elliptical cylinder at the positions and orientations indicated.

Figure 7.12 Scatter-to-primary ratio for ^{99m}Tc and ^{201}Tl sources in a cylindrical tank as a function of source depth.

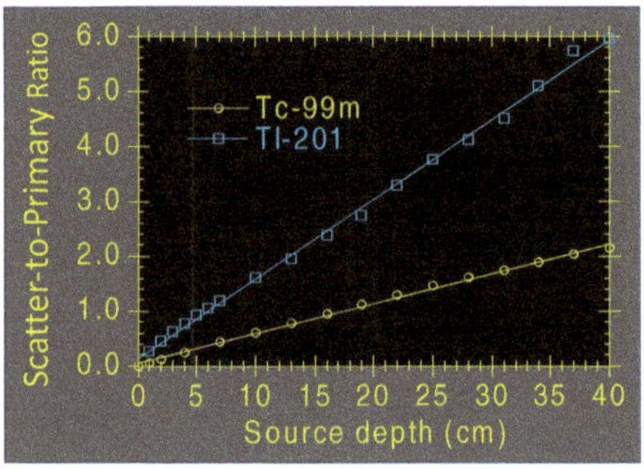

7.5.3 Collimator Detector Response

The *collimator detector response* (CDR) describes the spatial resolution of gamma camera, and *CDR function* (CDRF) is the image of a point source in air. Both the collimator and the detector contribute to the CDR.

For an ideal parallel-hole collimator, only photons perpendicular to the collimator face would pass through the collimator and be detected. However, a vanishingly small number of photons are perpendicular to the collimator face. Thus, holes with finite lengths and diameters are used to physically constrain the direction of photons that can be detected. Similarly, ideal septa (the walls between the collimator holes) would be opaque to the incident gamma rays and only photons that passed through the holes (referred to as *geometrically collimated photons*) would contribute to the image. However, the septal thickness needed to accomplish this would be unacceptably high. As a result, septa have a finite thickness, and some fraction of incident photons can penetrate through them and be detected. In addition, the septa are not perfect absorbers, so photons can scatter from septa and contribute to the detected image. These effects are illustrated in Figure 7.14 and give rise to the septal-penetration, septal-scatter, and geometric components of the collimator response. The septal-scatter and septal-penetration components are especially important for higher photon energies.

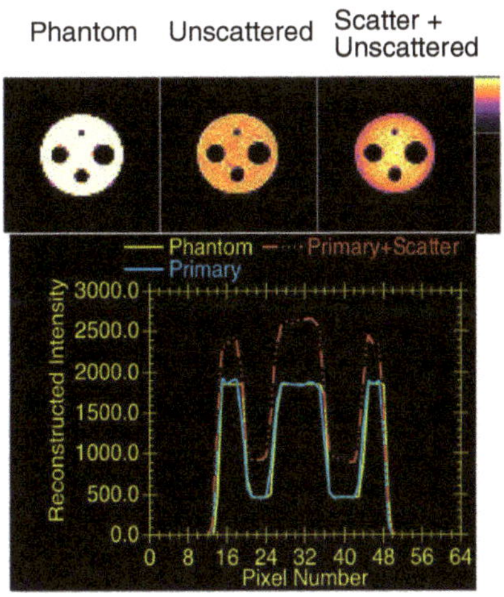

Figure 7.13 Top row: phantom with 4 warm rods and reconstructed images of MC simulated projection data from the phantom modeling ^{99m}Tc and a 20% energy resolution from data without and with scattered photons. The simulation modeled an ideal collimator (perfect spatial resolution), and the attenuator was a uniform water-filled cylinder. The images were reconstructed with compensation for attenuation using the narrow-beam attenuation coefficient. The color scale is windowed to demonstrate the effects of scatter in the background. Bottom row: profiles through the phantoms in the top row.

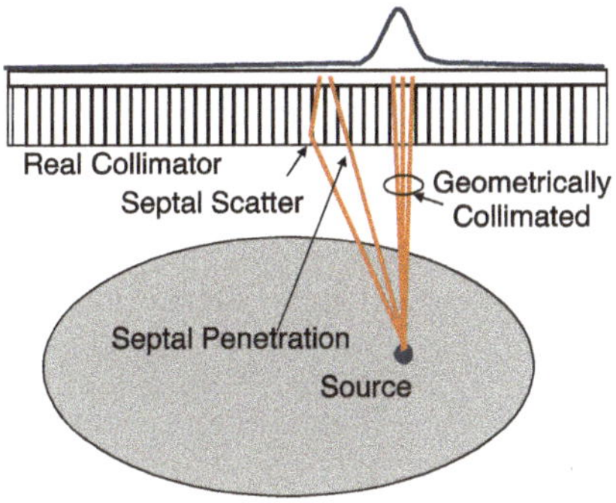

Figure 7.14 Illustration of the kinds of photons contributing to the CDR.

As discussed above, the choice of an appropriate collimator is essential in achieving reliable quantitative SPECT.

The CDR degrades the image in a number of ways. As show above, the CDR is not a point source and results in reduced resolution in the image. The width of the CDR is a measure of resolution and increases (indicating poorer spatial resolution) approximately linearly with distance from the face of the collimator. In SPECT, the net result is spatially varying and asymmetric resolution. This results in partial volume effects, as will be discussed below.

For high-energy photons compared to the design energy of the collimator, the septal penetration and scatter components can produce long tails that add a slowly varying background to the image, degrading contrast in the projections. Septal penetration through the top and bottom edges of the septa reduces the effective hole-length of the collimator and thus reduces image resolution. The amount of septal penetration and scatter that contributes to the image tends to decrease with distance from the collimator. Finally, photons from outside the field of view can, via septal penetration or scatter, contribute counts to the image. The amount of septal penetration and scatter depends on the photon energy and collimator design, while the resolution—determined primarily by the geometric component—depends primarily on the collimator design. The CDR degrades images in two ways:

1. It introduces spatially varying loss of resolution that manifests as partial-volume effects.
2. When septal penetration and scatter are important, it introduces spatially varying sensitivity.

The resolution largely depends on the collimator used and distance between the objects of interest and the collimator.

Figures 7.15–7.17 demonstrate these effects for ^{177}Lu using a MEGP collimator and ^{67}Cu using both LEHR and MEGP collimators. These data were generated using realistic Monte Carlo simulations of a Siemens Symbia gamma camera. Energy windows with full widths of 20% were used in all cases. A point source in air was simulated so that only unscattered photons are incident on the collimator. Many photons were simulated to generate virtually noise-free images. Pixel sizes of 0.33 and 0.479 cm were simulated for ^{177}Lu and ^{67}Cu.

The shapes of the CDRFs are very similar for the MEGP collimators for both radionuclides and all the energy windows. The images in Figure 7.15 show CDRFs for the two energy windows of ^{177}Lu for a MEGP collimator. The CDRFs for ^{67}Cu are shown in Figure 7.16 for both MEGP and LEHR collimators in both linear and logarithmic gray scales. The latter is shown to highlight the tails of the CDRF, which arise from septal penetration and scatter for the LEHR collimator and the 185-keV

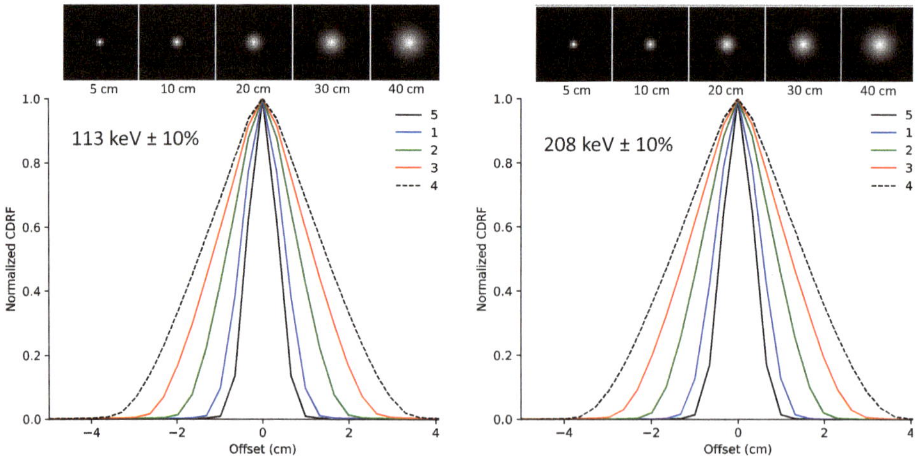

Figure 7.15 CDRFs of ^{177}Lu in 113-keV and 208-keV energy windows. These are for a Siemens MEGP collimator and a pixel size of 3.3 mm. The top row shows images of the CDRF (projection of a point source) for 5 different source-to-collimator distances. The bottom row shows horizontal profiles through the center of the CDRF.

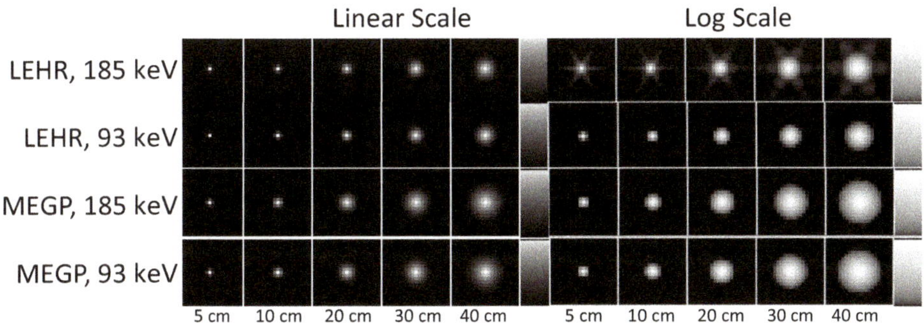

Figure 7.16 Images of the CDRF for a ^{67}Cu source imaged using Siemens LEHR and MEGP collimators in 93-keV and 185-keV energy windows that are 20% wide. The CDRs are shown for the 5 indicated distances to the face of the collimator and are displayed on linear and logarithmic gray scales (the latter to enhance the tails). A pixel size of 0.479 cm was used.

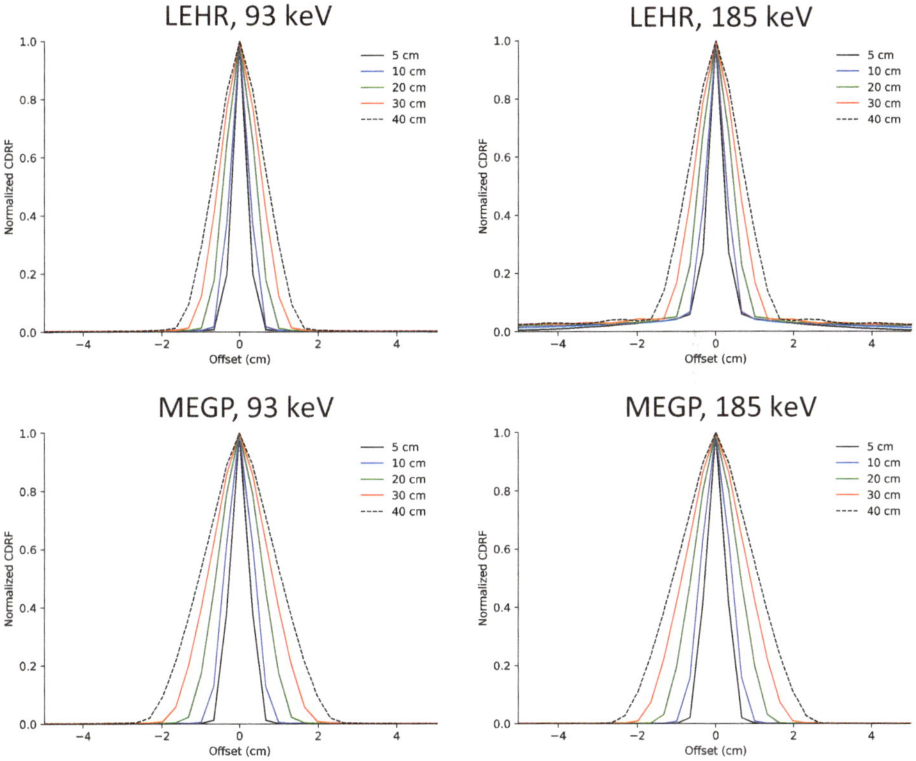

Figure 7.17 Horizontal profiles through the center of the CDRs in Figure 7.16. The profiles are normalized so that the height is 1 to allow comparing the shapes of the CDRFs.

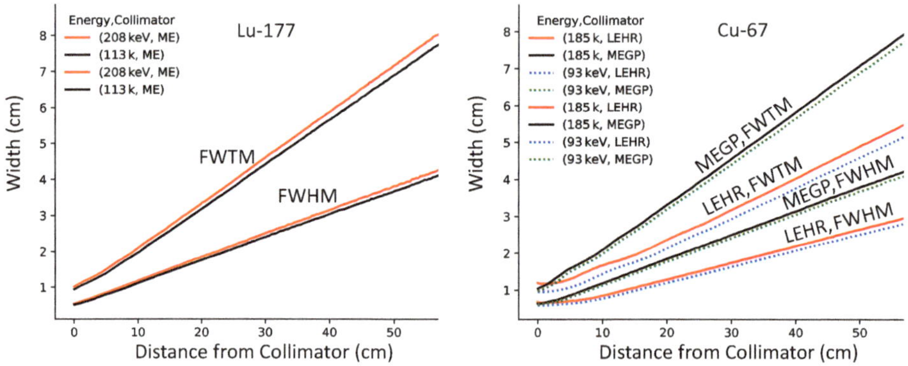

Figure 7.18 Plot of the full widths at half (FWHM) and tenth (FWTM) maximum of the CDRFs for the indicated radionuclides, collimators, and energy windows as a function of source distance. In both cases the solid lines are for the higher-energy window and the dotted lines for the lower-energy one.

energy window. Figures 7.15 and 7.17 show profiles through the CDRFs. Note that the CDRFs for the MEGP collimator are very similar in shape for both radionuclides and energy windows. For the LEHR collimator, the profiles and the star artifact in the CDRF for 185 keV shown on the logarithmic gray scale demonstrate the presence of septal penetration and scatter components. Also note the increase in the width of the CDRFs as a function of distance and that the widths are smaller for the LEHR collimator than the MEGP collimator. The LEHR collimator thus potentially provides higher resolution at the cost of increased septal penetration and scatter.

A quantitative comparison of the CDRFs can be seen in Figure 7.18, where the full width of the CDR at half (FWHM) and tenth (FWTM) maximum are shown. Note that these are along the horizontal direction in the CDRF, and the FWTM data thus shows a direction that is maximally affected by septal penetration and scatter. From these graphs we see that there is slightly better resolution (smaller FWTM and FWHM) for the lower-energy window. This is due to edge septal penetration that effectively decreases the length of the collimator holes. for high-energy photons Also, the FWHM and FWTM curves are very similar for the MEGP collimator for both radionuclides. The difference in widths as a function of energy for ^{67}Cu and the LEHR collimator is small for the lower-energy window and somewhat larger for the 185-keV window.

As discussed above, a second important effect is the spatial variance of the sensitivity. The ratio of the sum of the total (including geometric and septal penetration and scatter components) and geometric CDRFs, the TGR, provides a measure of the excess photons due to septal penetration and scatter. Since the sum of the geometric component is essentially independent of source distance, the TGR as a function of source distance provides a measure of the spatial variation of the total sensitivity.

The TGR is plotted for the radionuclides, collimators, and energy windows discussed above in Figure 7.19. For the MEGP collimator, the low-energy window has

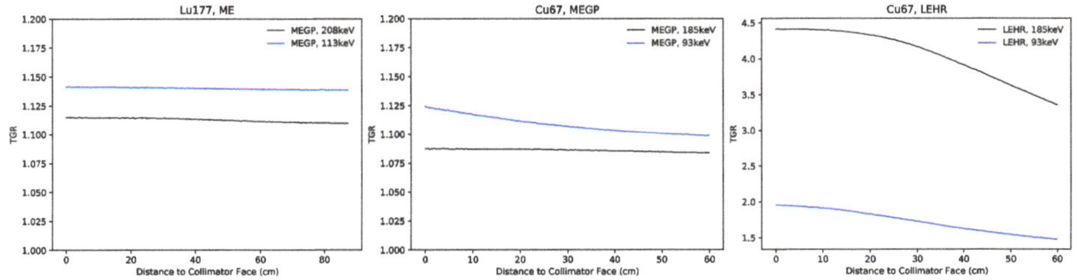

Figure 7.19 Plots of the total-to-geometric count ratio (TGR) for the indicated radionuclides, collimators, and energy windows as a function of source distance. Note that the plot for ^{67}Cu and the LEHR collimator is on a different vertical scale than the other two plots.

an average TGR that is higher by ~2% than the upper window, resulting from down-scatter in the collimator of the high-energy photons. For this collimator, the TGRs for ^{177}Lu are ~2% higher than ^{67}Cu due to the higher energies of its photons. For Lu-177, the spatial variation of the TGR is negligible. For ^{67}Cu and the low-energy window, the TGR varies by about 3% over the 60-cm range in distances shown. As a result, for both radionuclides compensating for the full CDRF is likely not essential for the MEGP collimator.

However, for ^{67}Cu and the LEHR collimator, the TGRs are larger than for the MEGP collimator by a factor of 1.5 for the 93-keV window and 4 for the 185-keV window. In addition, there is a much larger variation in the TGR over distance. Without correction, for the full CDR, these would likely result in significant differences and spatial variation in quantitative accuracy for the two energy windows.

7.5.4 Partial Volume Effects

In SPECT, *partial volume effects* (PVEs) refer to the contribution, due to the limited spatial resolution, of activities in neighboring regions to each other. This effect is illustrated in Figure 7.20. This shows a phantom having a warm background and cool

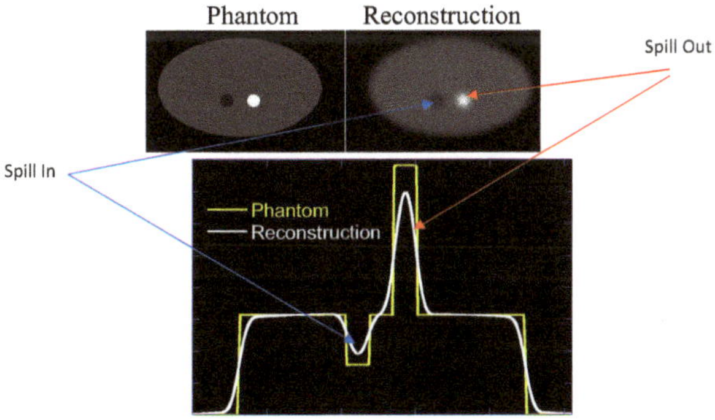

Figure 7.20 Illustration of partial volume effects due to limited resolution in the reconstructed image.

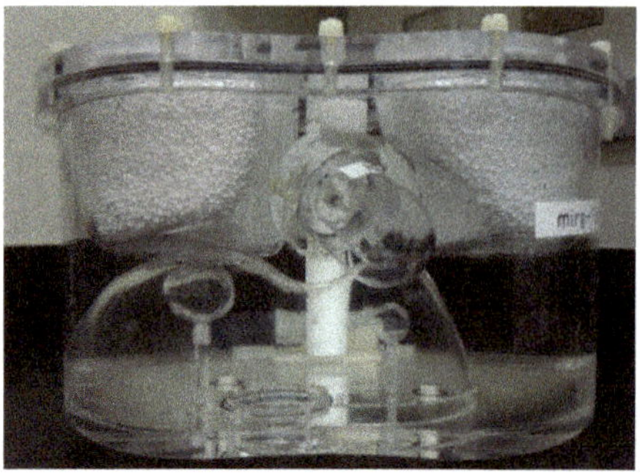

Figure 7.21 Photograph of torso phantom showing the location of the various compartments.

and hot circular objects. The image resulting from a simulated SPECT acquisition modeling the CDR and reconstructed using filtered backprojection is shown. The reconstructed image is a blurred version of the original object. The profiles illustrate the quantitative effects. The activity from the hot circle spills out into the surrounding background, and activity from the warm background spills into the cold circle. If a region of interest was defined over the circles corresponding to the true geometry, the activity inside the hot circle would be underestimated and the activity inside the cold circle would be overestimated. The amount of over- or under-estimation depends on the relative activities of the target and background, the shape of the target, and scanner resolution.

To show the magnitude of PVEs on image quantification, consider the following experiment using ^{131}I chloride in an appropriately buffered water solution using a Data Spectrum Anthropomorphic Torso Phantom with cardiac and two spherical inserts (see Figure 7.21). The phantom also contains lungs filled with a mixture of polystyrene beads and water (specific gravity about 0.3) and a liver compartment. No activity was placed in the lung compartment. The activities and volumes in the other compartments are shown in Table 7.2. The liver and background compartments had the same activity concentration. The images were reconstructed with compensation for attenuation, scatter, and the geometric collimator detector response. Thus, residual error is due largely to partial volume effects. After reconstruction, VOIs based on a CT scan were defined and approximate the true shapes of the compartments. The total activities were estimated in these VOIs and compared with the activities placed in the compartments. The errors were –5, –21, and –32% for the heart, large sphere, and small sphere, respectively. The small error for the heart demonstrates the generally good accuracy of the reconstruction, and the large errors for the spheres demonstrate the impact of PVEs on accuracy for small objects. As expected, the largest error is for the smallest sphere.

Table 7.2 Volumes and activities in phantom shown in Figure 7.21.

	Heart Chamber	Myo-cardium	Large Sphere	Small Sphere	Background
Volume (ml)	59.7	115.3	17.5 (r =1.61 cm)	5.7 (r =1.11 cm)	9580
Activity (mCi)	0.562	0.471	0.136	0.044	8.15
Activity Concentration (mCi/μl)	9.38	4.08	7.77	7.72	0.851

7.5.5 Quantum Noise

The pixel values in projection data in SPECT are random variables due to the random nature of radioactive decay and interactions in the body and camera system. To good approximation, the pixel values in the raw data are statistically independent (random variations in one pixel do not affect the random variations in other pixels) and described by the Poisson distribution. The Poisson distribution gives the distribution of count values that would be obtained in a pixel if the image was acquired many times under the exact same conditions. The formula for the Poisson distribution is:

$$P(x) = \frac{m^x e^{-m}}{x!}, \tag{7.4}$$

where x is the number of counts in the voxel and m is the mean number of counts. The mean number of counts is a positive real number, and the number of counts (measured) is a non-negative integer. Note that the Poisson distribution depends only on a single parameter, the mean number of counts. For SPECT, the mean number of counts is equal to the mean count rate times the acquisition duration. Figure 7.22 shows a plot of the Poisson distribution for four different values of the mean counts, which in this example are integers. The variance, σ^2, of the Poisson distribution is equal to the mean:

$$\sigma^2 = m. \tag{7.5}$$

The *coefficient of variation* (COV), the standard deviation divided by the mean, is thus equal to:

$$COV = \frac{\sigma}{m} = 1/\sqrt{m}. \tag{7.6}$$

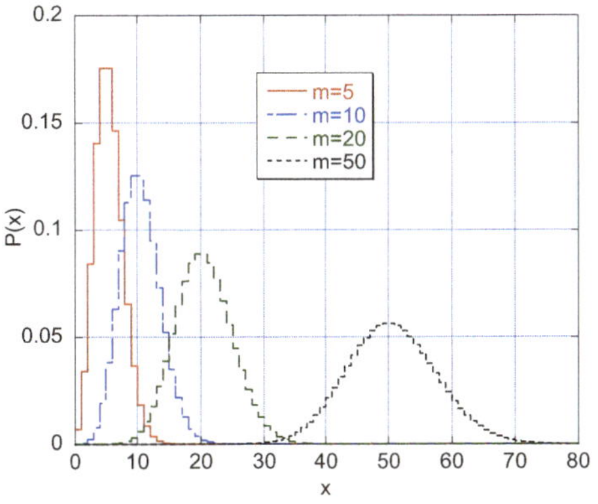

Figure 7.22 Graph of the Poisson distribution for 4 values of the mean number of counts.

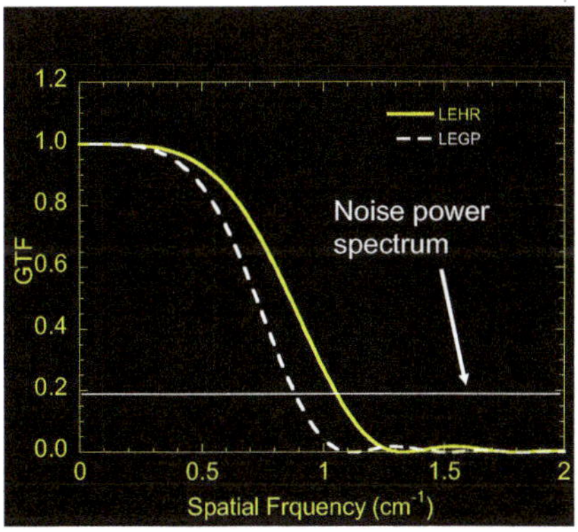

Figure 7.23 Comparison of noise power spectrum and geometric transfer function for LEHR and LEGP collimators.

Thus, the relative noise in the pixel decreases with the square root of the number of counts, and the noise in the projections thus decreases with increasing count rate and acquisition duration.

Considering the noise in the frequency domain can be useful. In the frequency domain, the *noise power spectrum* (NPS) shows the amount of noise as a function of spatial frequency. In raw projection images the noise is independent, and is thus white noise, i.e., it has equal power at all spatial frequencies. Figure 7.23 shows a comparison of the NPS to the *geometric transfer function* (GTF)—the magnitude of the Fourier transform of the geometric component of the CDR—for LEHR and LEGP collimators at one distance. Note that the NPS is constant at all frequencies; the value of this constant is a function of the COV and would thus decrease for higher and increase for lower count data. The GTFs go to zero, reflecting a permanent loss of information at those frequencies. The quantum noise acts on the signal, in some sense, after the blurring due to the collimator. Thus, the only information in the projections at high frequencies is due to noise.

As will be discussed below, SPECT reconstruction amplifies the high-frequency content of the images. For example, the FBP algorithm uses a ramp filter, with frequency increasing linearly with spatial frequency, in the reconstruction. Thus, reconstruction tends to amplify high frequencies. As seen from above, the high frequencies will contain only noise, and the reconstructed images will tend to exhibit high-frequency noise. It is common to use a low-pass filter (e.g., Butterworth or Gaussian) to reduce high-frequency image noise. This is shown in the example in Figure 7.24. Note that while the low-pass filter reduces noise, it also sacrifices resolution and can increase partial volume effects, resulting in a loss of quantitative accuracy for small objects.

Image processing, including tomographic reconstruction, applied to the projections results in correlated noise. Figure 7.25 shows a projection with a uniform mean and simulated Poisson noise without and with filtering with a Gaussian filter having

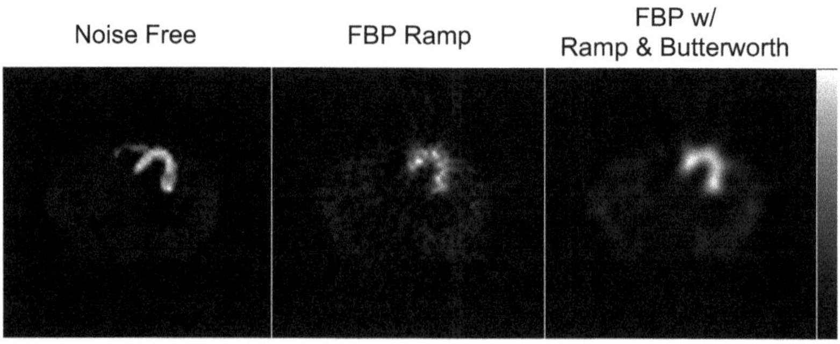

Figure 7.24 Demonstration of the effects of Poisson noise on image quality. The ramp filter alone results in peppery noise due to amplification of high-frequency noise by the filter. Applying a Butterworth low-pass filter reduces noise, but results in loss of image resolution.

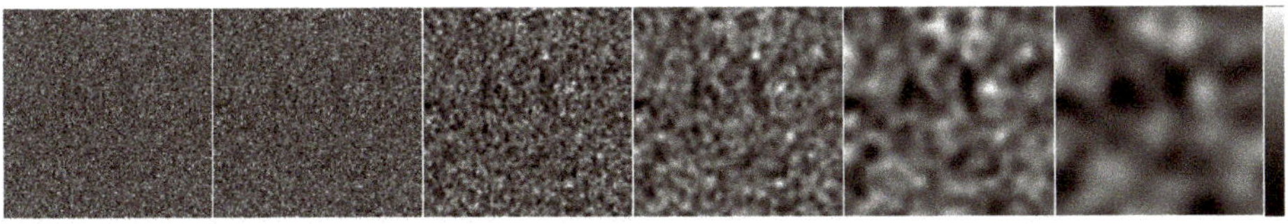

Figure 7.25 An image of a uniform object with a pixel value of 100 corrupted by Poisson noise without processing (unfiltered) and after filtering with a Gaussian filter having FWHMs of 1, 2, 4, 8, and 16 pixels.

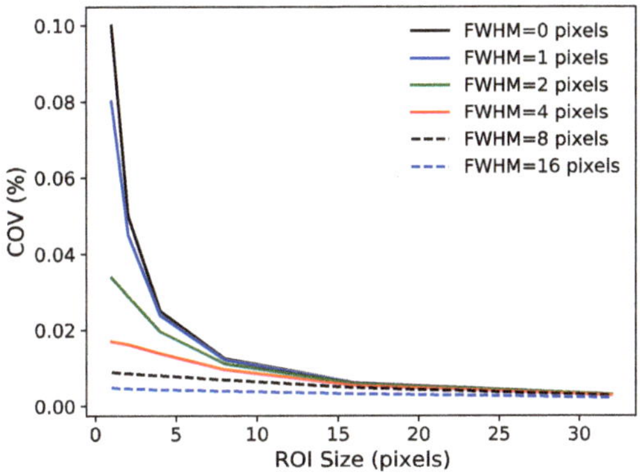

Figure 7.26 Plot of COV over 100,000 Poisson noise realizations of the average values in an ROI with the indicated size applied at the center of a 256x256 image with a mean pixel value of 100 counts filtered with Gaussian filters with the indicated FWHMs. The central 128x128 region of a single noise realization of these images is shown in Figure 7.25.

different widths. The unfiltered noise has a very peppery texture because the pixel values are uncorrelated. However, filtering introduces correlations that appear as blobs.

Quantum noise impacts the precisions of activity estimates. For example, in the unfiltered case in Figure 7.25, the COV in each pixel is 10%. Using a 2×2 VOI includes four times as many counts, so the COV of the total counts or count density in the VOI is 5%; for a 16×16 VOI the COV is less than 1%. This is important as images that are too noisy for visual interpretation may be perfectly acceptable for estimating the activity in an organ, assuming that the organ VOI can be obtained independently, e.g., from a registered anatomical image. Filtering also reduces noise, but it increases PVEs. Thus, for organ or tumor activity quantification, no post-filtering is typically optimal. However, when using voxel dosimetry to estimate 3D dose metrics like a *dose-volume histogram* (DVH), low-pass filtering can help improve DVH reliability.

The following numerical experiment illustrates these effects. A uniform 256×256 image with a pixel value of 100 counts was created, and 100,000 Poisson distributed noise realizations were created. These were each filtered with a Gaussian having the FWHMs listed above, and square ROIs with sizes of 1, 2, 4, 8, 16, and 32 pixels were created and applied to the center of each image to calculate the mean counts in each ROI. The COV was computed over the noise realizations for each ROI size and filter width. The COVs are plotted for each filter width as a function of ROI size in Figure 7.26. Notice that the COV decreases as the size of the ROI increases and with increasing width of the Gaussian filter. Since the original image was flat, the Gaussian filtering does not introduce partial volume effects that would degrade the accuracy.

7.5.6 Count Rate Losses

As discussed above, gamma cameras behave as approximately paralyzable counting systems. In general, the count rate will vary with projection view. Thus, the use of a single post-reconstruction correction factor is not, in principle, sufficient to accurately correct for the loses. To demonstrate the effects, consider the following numerical experiment.

A phantom consisting of a water-filled cylinder with an elliptical cross section having diameters of 22 and 32 cm was simulated with a relative background activity of 1. Three cylindrical rods with diameters of 4 cm were positioned as shown in Figure 7.27 and had 11 times the activity of the background. The phantom was digitized with a voxel size of 0.2 cm. Projections were simulated analytically at 120 projections over 360 degrees and then collapsed to give 0.4 cm projection pixels. Count rate losses were simulated using a resolving time of 3 μs, corresponding to a maximum count rate of 123 kcps for true count rate of 333 kcps. This is a somewhat larger deadtime than most modern cameras, which have maximum count rates of 180–300 kcps. The projections were scaled so that the projection view with the maximum counts was equal to 0.99 times the maximum count rate to avoid problems with the dual-valued nature of the count rate loss curve. A paralyzable deadtime model was applied independently for each projection view. The count rates averaged over all projections were used to calculate a single correction factor, and the count-rate-loss-degraded projections were multiplied by this factor to provide approximate compensation. The original projections, projections degraded by count rate losses, and the degraded projections multiplied by the single correction factor were separately reconstructed using OS-EM with 10 iterations of 24 subsets and attenuation compensation.

The reconstructed images are shown in Figure 7.27. The mean activity in each of the four regions (background and three cylindrical ROIs) was calculated and compared to the true value. The results are shown in Table 7.3. Without count rate losses, the errors were less than 2.3%, indicating the efficacy of attenuation compensation. Without count rate loss correction, the ROI activities were underestimated by 23% to 28%. The count rate loss fraction for the mean count rate was 0.759 and was used to

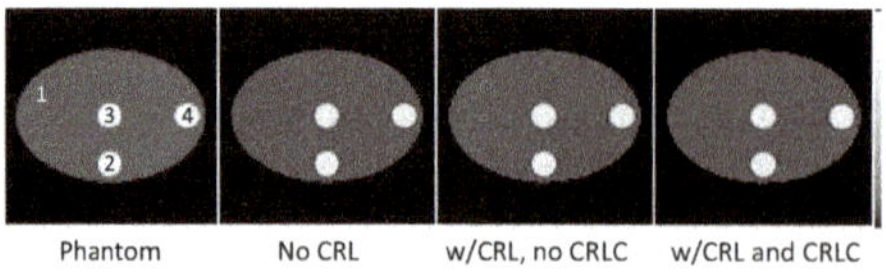

Figure 7.27 Phantom and images reconstructed from data without count rate losses (CRLs) and with CRLs with and without CRL compensation (CRLC).

Table 7.3 Percent errors in activity estimates

	ROI 1 (%)	ROI 2 (%)	ROI 3 (%)	ROI 4 (%)
No CRL	1.28	−1.87	−2.29	−1.62
w/CRL, no CRLC	−23.36	−27.86	−26.23	−26.91
w/CRL, w/CRLC	0.99	−4.98	−2.80	−3.69

correct the measured projections. The residual errors were different for the different cylinders and ranged from 1% for the background to almost 5% for one of the cylinders.

7.5.7 Patient Motion

The projection operation that is inverted by SPECT reconstruction assumes that the activity distribution is same at each of the different projection views. Since projection views in SPECT are acquired sequentially, typically using two gamma cameras, patient motion—whether gross motion such as the patient shifting position during the acquisition, involuntary motion like heart beating or breathing, or redistribution of the activity in the organs over time—can result in errors and artifacts.

Respiratory and cardiac motions happen quickly compared to the scale of SPECT acquisitions. Thus, organs and tumors in the thorax will typically be blurred. Methods for monitoring motion externally and based on the images have been developed. Methods for respiratory or cardiac gating are not widely available for SPECT.

Voluntary motion by patients is another important problem. This is especially important for long acquisitions. Careful attention to patient comfort (e.g., do not try to image patients with arms over their heads, the use of pillows and cushions to improve comfort, and reducing acquisition times as much as practical) are important ways to prevent these problems. Software to compensate for involuntary motion between projections is provided on some vendor workstations. However, the accuracy of these methods and their effects on activity quantification have not been studied.

The use of SPECT/CT systems is appealing in dosimetry applications as the CT images provide both anatomical information for use in defining VOIs and for compensating for attenuation. Patient movement between the SPECT and CT scans can result in misregistration that affects the accuracy of both the VOIs defined using the CT image and attenuation compensation. Registration of CT and SPECT reconstructions (without attenuation compensation) can be used to check for motion between these two acquisitions. Rigid and nonrigid registration methods can used. In general, the registration should be better than the expected resolution of the SPECT images.

Another confounding factor is that CT images are typically acquired very quickly, e.g., in much less than the time required for a single breath. They are thus relatively unaffected by respiratory motion. SPECT acquisitions typically are averaged over one or more breaths at each projection. Thus, the SPECT images will be degraded by respiratory motion blurring. Use of CT images reflecting the anatomy at one phase of motion for attenuation compensation can result in errors in attenuation compensation. A common manifestation of this problem is in the vicinity of the diaphragm: parts of the liver can appear to be inside the lungs in the CT image, resulting in overestimates of activity in that top part of the liver. One solution to this is to exclude that portion of the liver from the liver VOI used to estimate activity. Then, the activity concentration for the portion not affected can be used in conjunction with the liver volume to estimate the total liver activity.

7.6 SPECT Image Reconstruction

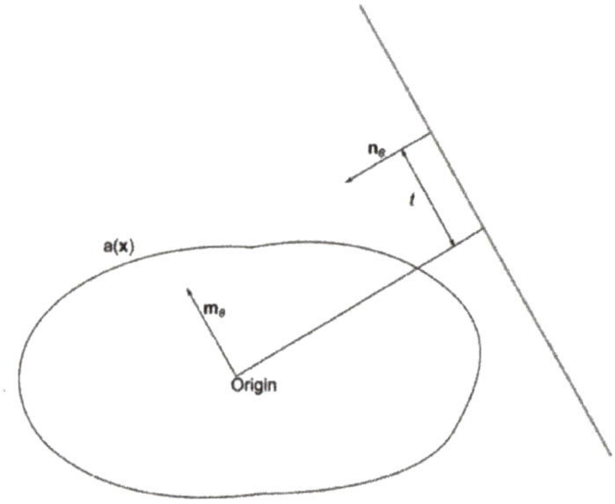

Figure 7.28 Illustration of the SPECT reconstruction problem and geometry.

The purpose of tomographic image reconstruction is to provide an estimate of the 3D activity distribution based on measurements of projections at a set of views around the patient. Historically, SPECT reconstruction has used analytic methods that provide formulas based on solutions to an inverse problem that approximates the SPECT image formation process. However, the approximations, discussed in detail above, result in degradations of SPECT images. These degradations are especially important for dosimetry applications as they result in substantial errors in estimates of organ activities. This section provides a background on SPECT reconstruction methods with an eye toward using them to compensate for the image-degrading factors mentioned above.

7.6.1 Analytic Reconstruction

The reconstruction method used in clinical scanners until the late 1990s, and still used for a variety of applications today, is the *filtered backprojection* (FBP) method. This method is based on solving analytically the problem illustrated in Figure 7.28 in 2D. The goal is to recover the activity distribution, a($\boldsymbol{x}$), where $\boldsymbol{x}$ is a vector describing the position in object space, from the projections $p(t,\boldsymbol{\theta})$ where t is the position in the projection and $\boldsymbol{\theta}$ is the projection angle, as shown in Figure 7.28. The projections are defined as line integrals through object space:

$$p(t,\theta) = \iint a(\boldsymbol{x})\delta(\boldsymbol{m}_\theta \bullet \boldsymbol{x} - t)d\boldsymbol{x} , \tag{7.7}$$

where $\boldsymbol{m}_\theta$ is a unit vector parallel to the projection direction and $\delta(\boldsymbol{x})$ is the Dirac delta function. In this expression, the delta function selects positions, $\boldsymbol{x}$, that lie on the line integral through the object that intersects the detector at a distance t from the origin. These are the projections that would be measured if there was no attenuation or scatter, the collimator-detector system counted only photons perpendicular to the face of the detector, and there was no quantum noise.

This problem can be solved analytically to obtain an estimate of the activity distribution, $\tilde{a}(\boldsymbol{x})$, and gives the following formula for image reconstruction:

$$\tilde{a}(\boldsymbol{x}) = \int_{-\infty}^{\infty} \int_0^{\pi} \left[h(t) * p(t,\theta)\right] \delta(\boldsymbol{x} \bullet \boldsymbol{m}_\theta - t)\, dt d\theta , \tag{7.8}$$

where * denotes convolution and

$$h(t) = \int \left| f \right| e^{2\pi i f t} df \tag{7.9}$$

is a ramp filter. The term in parentheses in Equation (7.8) represents the ramp-filtered projections. The ramp filtering can be applied in the frequency domain or Equation (7.9) can be truncated, since the projections are measured at discrete positions and used to derive the form of a spatial domain convolution kernel. As shown above, application of FPB results in degraded images due to a failure to accurately model the image formation process in SPECT. Analytic solutions for the full SPECT problem have not been found. Instead, iterative methods have been developed that allow handling these effects and result in substantially improved quantification.

7.6.2 Iterative Reconstruction Methods

As noted above, the projections in SPECT are measured at a finite number of discrete projection bins. In addition, images are displayed and manipulated in digital form, i.e., on a finite-sized grid of voxels with uniform spacing. (It should be recognized that this is an approximation: patients are continuous objects. More theoretically correct formulations treat the problem as a continuous to discrete transformation and seek inverses to that problem.) As a result, the SPECT reconstruction problem is often formulated in 3D in digital form as:

$$p_i = \sum_j C_{ij} a_j . \tag{7.10}$$

Here, $\{p_i; \in [1, LMN]\}$ are the projections where i is over the product of the number of projections, L, the number of bins in the axial direction, M, and the number of bins in the transaxial direction, N; $\{a_j; j \in [1, NNM]\}$ (assuming the reconstructed image has the same number of voxels in both orthogonal axes in the transaxial plane and in the axial direction) is the activity distribution; and $C_{ij}; i \in [1, LMN], j \in [1, NNM]$ is the projection matrix. The elements of the projection matrix are the probability that a photon emitted in voxel j is counted in projection bin i. In matrix form, this can be written as:

$$\boldsymbol{p} = \boldsymbol{Ca}, \tag{7.11}$$

where $\boldsymbol{p}$ and $\boldsymbol{a}$ are the projection and activity vectors and $\boldsymbol{C}$ is the projection matrix.

It might first appear that the Equation (7.11) could be inverted using matrix methods. However, the projection matrix is very large, i.e., LNM×NNM. If L, M, and N are all 128, the matrix has 4.4×10^{12} elements. The matrix is different for each patient as it includes the attenuation and scatter inside the body. Thus, it is practically impossible to use direct inversion. An alternative is to use iterative methods.

The iterative methods that have found their way into clinical use are statistical iterative reconstruction methods. These are based on posing the inverse problem as an estimation problem and considering the statistics of the measured data. The general form of this is to pose the problem as a *maximum a posteriori* (MAP) estimation problem. In this, we seek the estimate of the activity distribution, $\boldsymbol{a}$, that maximizes the probability of the activity distribution given the measured projections, $P(\boldsymbol{a}|\boldsymbol{x})$. Using Bayes theorem this can be written as:

$$P(\boldsymbol{a}|\boldsymbol{p}) = \frac{P(\boldsymbol{p}|\boldsymbol{a})P(\boldsymbol{a})}{\int P(\boldsymbol{p}|\boldsymbol{a}')P(\boldsymbol{a}')d\boldsymbol{a}'}, \tag{7.12}$$

where $P(\boldsymbol{p}|\boldsymbol{a})$ is the probability of the projections given the activity distribution, i.e., the statistical likelihood, and $P(\boldsymbol{a})$ is the prior probability of the activity distribution. The integral in the denominator is a normalizing term and does not affect the optimization problem.

There has been a great deal of work on MAP estimation for emission-computed tomography, and some MAP methods are available commercially for PET scanners. However, these are not yet available for SPECT. These methods use prior functions that express something that is known about the activity distribution, such as its smoothness. However, in general, little is known that can be written as an analytical expression and implemented in an optimization framework. As a result, it is common to assume that all activity distributions are equally likely. In this case, the problem reduces to finding the activity distribution that results in the highest likelihood, $P(\boldsymbol{p}|\boldsymbol{a})$. Formally this can be written as:

$$\underset{\boldsymbol{a}}{\arg\max}\, P(\boldsymbol{p}|\boldsymbol{a}). \tag{7.13}$$

There are several appealing properties of Maximum Likelihood solutions to estimation problems in general that are beyond the scope of this chapter.

The values in the projection bins, i.e., the $\{p_i\}$, are independently Poisson distributed. Thus, the likelihood can be expressed as the product of the univariate Poisson likelihood function given in Equation (7.4). In that equation, the mean, m, becomes $\Sigma_j C_{ij} a_j$ and $x = p_i$. The resulting equation can then be optimized to find the $\{a_i\}$ that maximize the multivariate Poisson likelihood (ML). Since the likelihood is a set of products, this is done more easily by maximizing the log of the likelihood, which is a sum. This optimization problem has been solved using the iterative *expectation maximization* (EM) algorithm. The update equation for the ML-EM algorithm for this Poisson likelihood problem gives the estimate of the activity in voxel i after $n+1$ iterations, $\tilde{a}_i^{n+1}$, in terms of the estimate after n iterations, $\tilde{a}_i^n$:

$$\tilde{a}_i^{n+1} = \tilde{a}_i^n \frac{1}{\Sigma_{j'} C_{ij'}} \Sigma_j C_{ij} \frac{p_j}{\Sigma_k C_{kj} a_k^n}, \tag{7.14}$$

In this, $\Sigma_k C_{kj} a_k^n$ is the projection of the current estimate of the activity distribution and $\frac{p_j}{\Sigma_k C_{kj} a_k^n}$ is the ratio of the measured projections and the projection of the current estimate, which can be thought of as a multiplicative error term. The term $\Sigma_j C_{ij} \frac{p_j}{\Sigma_k C_{kj} a_k^n}$ represents the backprojection, using the transpose projection matrix, of the multiplicative error term. This, in effect, spreads the multiplicative error in a projection bin over the activity distribution in proportion to the probability that it contributed to that projection bin. The final term, $\frac{1}{\sum_{j'} C_{ij'}}$, is a normalization term; the term in the denominator is the backprojection of 1s. This thus means that there will be no update (update factor is 1) if all the error terms are 1, i.e., the projection of the current estimate equals the measured projections. Inherent in the update equation is that the voxel values in $\boldsymbol{a}$ are non-negative if the initial estimate of $\boldsymbol{a}$ is non-negative. This is practically appealing since physical activity values cannot be negative.

There are, however, several practical obstacles. First, as already mentioned, the projection matrix is very large and is patient-dependent. One interesting observation based on Equation (7.14), however, is that the update formula only requires the multiplication of $\boldsymbol{C}$ by a vector. This multiplication is referred to as a *projection operator*; the multiplication of the transpose of $\boldsymbol{C}$ by a vector is known as a *backprojection operator*. Thus, the update formula can be implemented by coding backprojection and projection algorithms that model the image formation process. How this is done for the various image-degrading factors will be discussed in the next section.

A second practical problem is that the ML-EM algorithm converges very slowly, requiring hundreds of iterations to achieve a useful solution, especially for quantitative applications. This has been addressed using block iterative algorithms such as the ordered subsets algorithm. The idea of the *ordered subsets* (OS) algorithm is to divide the projections into n subsets each containing L/n projections (n must be chosen to be a factor of L). The update equation in Equation (7.14) can then be applied to this smaller problem that can be computed in less time by a factor of L/n. On the subsequent updates, different subsets of projections are used, and so on. (The composition and order of the subsets is important in achieving good results. In general, it has been found that projections in a subset should be maximally separated by projection angle, and that the projections in adjacent subsets should be as far apart from each other as possible.) The result is that there are n updates in each full iteration, resulting in a speed-up by a factor of n. For example, for $L = 128$ and $n = 32$ subsets, there are 4 projections per subset, and the speedup is by a factor of 32.

When OS are applied to the ML-EM algorithm, the result is conventionally called OS-EM. Algorithms other than the OS-EM have been used, including the *conjugate gradients* method, which can result in even further improvements in convergence speed. However, this comes at the cost of greater algorithmic complexity and a need to specifically implement the non-negativity constraint that is intrinsic in OS-EM. In general, subsets smaller than 4 projections should not be used. Also, for very low count data, reducing the number of subsets (increasing the number of views in a subset) will provide results with less bias and better noise properties.

The key to using iterative algorithms to compensate for image-degrading effects is modeling the degrading effects in the projector and backprojector algorithms. It has been noted empirically that as the complexity of the effects modeled increases, the convergence slows down. That is, it requires fewer iterations to achieve useful images with attenuation compensation alone than when scatter or detector response compensation are modeled. In addition, it has also been observed empirically that including effects that have long ranges, i.e., the CDRF and scatter, in the backprojector results in visually reduced noise amplitude in the reconstructed images.

These observations have led to the use of so-called *dual-matrix algorithms* where a different projection matrix (or algorithm modeling different effects) is used in backprojection than in the projection. For example, attenuation, the full CDR (including septal penetration and scatter), and scatter are modeled in the projector. This improves accuracy as, in the absence of noise and if the models of these effects are perfect, the projection of the estimated true activity distribution would equal the measured activity distribution. Thus, the error terms would be all ones, and the backprojection would result in no update. However, to improve convergence speed while retaining the desirable noise properties, attenuation and the geometric CDR are modeled in the backprojector. That is, object scatter and septal penetration and scatter are omitted in the backprojection.

A final practical consideration is the number of iterations required to obtain a reliable image. It has been observed empirically that noise in the reconstructed image increases with iteration. On the other hand, continued iteration results in better image resolution and reduced partial volume effects. While several theoretical methods for selecting the optimal iteration number and the use of MAP methods with priors that enforce smoothness have been proposed, a practical solution that has been widely adopted is to truncate the iteration after some empirically determined number of iterations. When visual image quality is important, i.e., for a defect detection task, applying a post-reconstruction low-pass filter, such as a Gaussian or Butterworth filter, has been shown to improve image quality. However, for object quantification this increases partial volume effects and is generally not needed nor desirable. For VOI quantification, using a larger number of iterations gives the most reliable estimates considering the combination of bias (which is reduced by more iterations) and noise (which tends to increase at the voxel level with iteration) due to the averaging effect demonstrated above of applying a VOI to the reconstructed image.

7.7 Compensating for Image-Degrading Effects

The current gold-standard method of correcting for image-degrading effects is, as discussed in the previous section, the use of iterative reconstruction methods. The key is to implement models of the image-degrading effects in the projector and (for some effects) backprojector algorithms. The following is an overview of methods used to implement these methods in SPECT and a discussion of the efficacy of the resulting compensation.

7.7.1 Attenuation Compensation

Attenuation compensation requires modeling the attenuation of photons as they pass through the body. Analytically, the attenuated projection, $p_\mu(t,\theta)$, can be written using the geometry and notation of Figure 7.29 as:

$$p_\mu(t,\theta) = \iint a(\boldsymbol{x})\delta(\boldsymbol{m}_0 \bullet \boldsymbol{x} - t)e^{-\int_0^\infty \mu(\boldsymbol{x}+l\mathbf{n}_\theta)dl}d\boldsymbol{x}, \tag{7.15}$$

Figure 7.29 Geometry and notation used to write the expression for the attenuated projection: **x** is the position in space; θ is the projection angle, $\mathbf{n}_\theta$ is a unit vector originating from the center of rotation (Origin) that is perpendicular to the detector (and thus parallel to the projection direction); $\mathbf{m}_\theta$ is a unit vector perpendicular to the projection direction, *t* is the coordinate in the projection from the projected position of the center of rotation into the projection; and ***l*** is the distance along the projection starting at the point **x**.

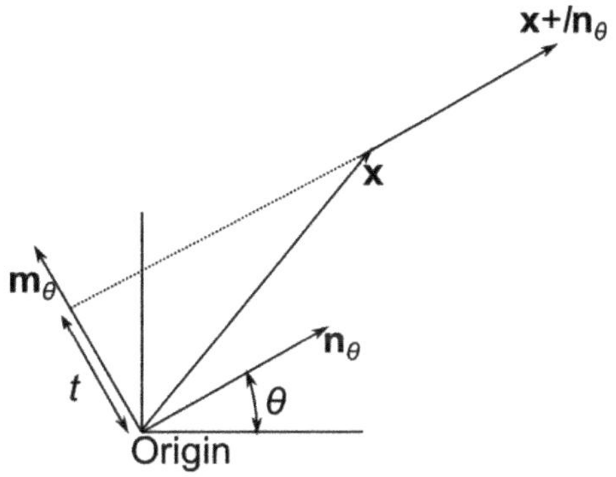

where $\mu(x)$ is the attenuation map, i.e., the attenuation coefficient at each point.

Attenuation modeling can be implemented in a discrete setting using a rotation-based projector/backprojector. This is not necessarily the most efficient computationally for attenuation compensation alone, but it is very efficient when combined with CDR compensation, as will be described below. The idea is to rotate the sampling grid for the attenuation map and activity image so that rows are parallel to the detector, as shown in Figure 7.30. In the rotated grid, moving along columns and computing the sum along the projection rays of the attenuation coefficients allows progressively calculating the attenuation factor for each pixel. These can be applied to the activities in the rotated activity image. The attenuated activity values are then summed along the columns to calculate the value in each projection bin. A backprojector can be implemented by performing the operation in reverse: the values in the projection bin to be backprojected are spread along the columns of a matrix. The attenuation factors in the rotated attenuation map are then applied to each voxel.

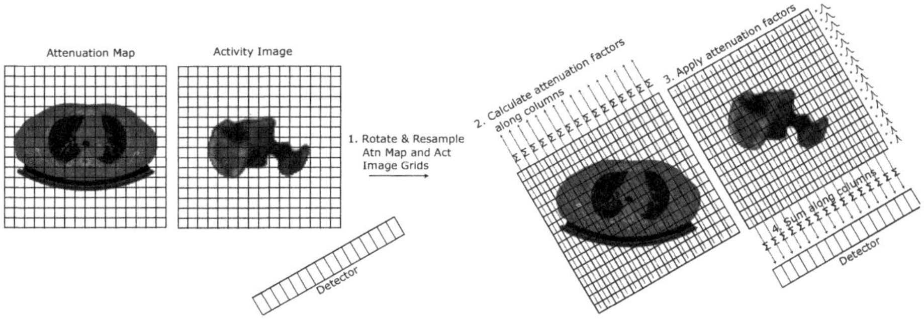

Figure 7.30 Illustration of the implementation of attenuation modeling using a rotation-based projector.

(Note that the attenuation factors should be applied and not their inverse because the backprojector is the transpose of the projector and not the inverse.) Then, the activity image is rotated backwards (by the negative of the angle used in the projection) and added to the backprojected images from other views.

The key issue in the efficacy of reconstruction-based attenuation compensation is the accuracy of the attenuation map. In current practice, attenuation maps are obtained from patient CT scans. These can be from a separate scanner, but most commonly are acquired using a SPECT/CT system that incorporates a CT scanning mechanism. The major issues in the accuracy of the attenuation maps are how well the CT scan is registered to the SPECT image, and the accuracy of the attenuation coefficients in the attenuation map created from the CT image.

Registration of CT to the SPECT is an important issue, and evaluating this should be part of quality control. The use of an external CT scanner requires registration of the CT images to the SPECT prior to reconstruction. This can be complicated if the pose of the patient (e.g., arms at sides or above the head, position of legs, etc.) is different in the two scans. The use of SPECT/CT systems generally vastly simplifies this. (Part of a regular quality control program is to evaluate the registration of the SPECT and CT images, typically using a special phantom.)

For current SPECT/CT systems, the scans are not acquired simultaneously, so the patient position should be maintained between the two scans; if there is motion, registration may be required. Another important issue that can cause problems is respiratory motion. CT scans are acquired very rapidly, in times very short compared to the respiratory cycle. They thus acquire images of the lungs at essentially one point during the respiratory cycle. The duration of a SPECT projection acquisition is, on the other hand, comparable to or longer than that of the respiratory cycle. Thus, SPECT images are blurred by respiratory motion. Using a CT image acquired at one point in the respiratory cycle to generate an attenuation map for a reconstruction of projections blurred by respiratory motion can produce artifacts; these are especially evident in the diaphragm region. Methods for respiratory gating of SPECT or CT acquisitions to account for these effects have been developed, but they are not commercially available or widely implemented in SPECT at this time.

Creating attenuation maps from CT images also requires care. CT images are acquired using polychromatic beams, but the attenuation map should be the attenuation coefficients for the photopeak energy that is being projected. The output of polychromatic CT scans is usually an image in *Hounsfield units* (HU), where –1000 is defined as air, 0 is water, and 1000 is cortical bone. For attenuation compensation, voxel values in the attenuation map should be in units of inverse length at the energy of the photopeak under consideration. Conversion from CT number to attenuation map usually involves piecewise linear scaling, assuming each voxel in the object is comprised of some mix of water and air, or water and bone. The object is considered to be comprised of various densities of water (i.e., combinations of air and water) for CT numbers ≤ 0, a mixture of water and bone for CT numbers between 0 and 1000, and various densities of cortical bone for CT numbers ≥ 1000:

$$\mu(h) = \begin{cases} \frac{1000+h}{1000}\mu_{water} \text{ for } h \le 0 \\ \mu_{water} + \frac{h}{h_{bone}}(\mu_{bone} - \mu_{water}) \text{for } 0 < h < h_{bone} \\ \frac{h}{h_{bone}}\mu_{bone} \text{ for } h > h_{bone} \end{cases} \tag{7.16}$$

where μ_{bone} and μ_{water} are the narrow-beam attenuation coefficients of water and bone, respectively, at the photopeak energy, and h_{bone} is the number of Hounsfield units corresponding to the bone density used for μ_{bone}. For cases where multiple photopeaks are acquired in the same or separate projections, the most theoretically correct treatment is to perform the attenuated projection operation separately for each energy window. However, it is often sufficient to use a single attenuation map that is based on a weighted average of the attenuation coefficients at the relevant photopeaks.

Assuming good registration and accurate values of attenuation coefficients in the attenuation map, attenuation compensation using iterative reconstruction is quite effective. A sample of its efficacy is shown in combination with scatter compensation below. Attenuation compensation does result in spatially varying noise correlations, as will also be demonstrated below.

7.7.2 Scatter Compensation

In iterative reconstruction, scatter can be compensated for using either a scatter estimate obtained outside the reconstruction or by modeling scatter during the iterative process. The former case is most common, and typically involves estimating the scatter using data acquired in other energy windows. The most common method is the *triple-energy-window* (TEW) method. In this method, scatter images are acquired just above and below the photopeak energy window, as illustrated in Figure 7.31. The scatter contribution to the photopeak energy window is then approximated as the area of a trapezoid where the heights are the average counts in the two scatter windows and the length is the width of the photopeak window. This can be written as:

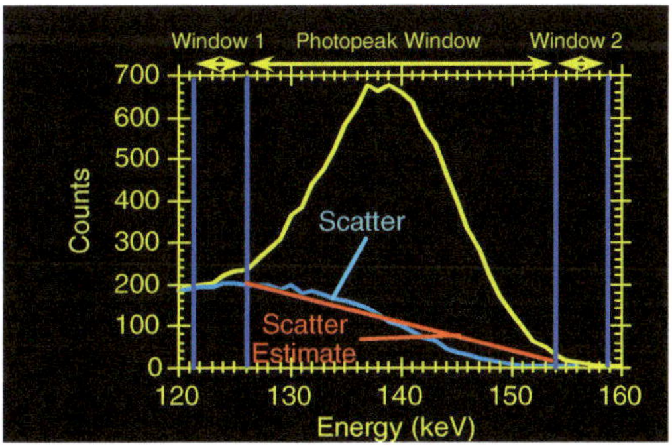

Figure 7.31 Illustration of the energy windows used in TEW scatter compensation.

$$S = \frac{1}{2}\left(\frac{C_1}{W_1} + \frac{C_2}{W_2}\right)W_p, \tag{7.17}$$

where S is the scatter estimate, C_1 and C_2 are the counts in the two scatter windows, p indicates the photopeak window, and W_i is the width of window i (i = 1,2,p). This is applied on a voxel-by-voxel basis in the projection images. Theoretically and practically, the best way to implement this in an iterative reconstruction method is to add the scatter estimate in each projection pixel to the value of the projection calculated from the current estimate prior to taking the ratio with the measured projections. Thus, Equation (7.14) becomes:

$$\tilde{a}_i^{n+1} = \tilde{a}_i^n \frac{1}{\Sigma_{j'} C_{ij'}} \Sigma_j C_{ij} \frac{p_j}{\Sigma_k C_{kj} a_k^n + S_k}, \tag{7.18}$$

where S_k is the scatter estimate in the kth projection bin.

There are theoretical and practical limitations of TEW. First, getting a good estimate of the height of the scatter spectrum at the edges of the photopeak window is a compromise between using narrow windows to get an accurate estimate, since the energy spectrum varies with energy, and using a wide window to reduce noise and get a precise estimate. Low-pass filters can be applied to the scatter estimate to reduce the impact of noise, but this results in bias in the scatter estimate. An example of the problem of noise in the projections is shown in Figure 7.32. This shows a SPECT projection from a single view in the photopeak window and two scatter windows. Note that these are shown on independent gray scales as the scatter windows have fewer counts than the photopeak window.

Using wide scatter windows results in somewhat biased estimates of the scatter. Figure 7.33 shows low-noise Monte Carlo simulations of the scatter in ^{99m}Tc energy windows as a function of energy window width compared to the TEW scatter and an estimate from an analytic scatter model, the *Effective Source Scatter Estimation*

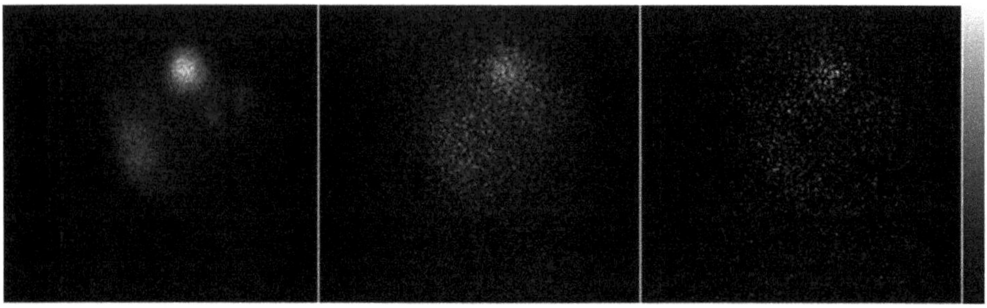

Figure 7.32 Projection from a patient administered a ^{177}Lu-based radiopharmaceutical. The images are, left to right, from the 208-keV photopeak using a 20% wide window, an abutting 12% wide lower window, and an abutting 9% wide upper window. The images are shown on independent gray scales.

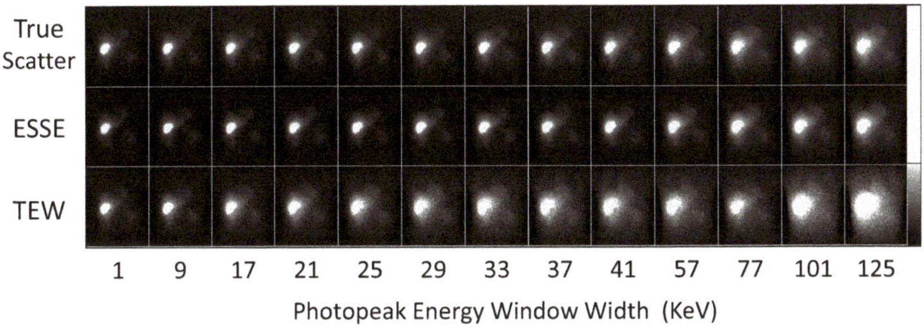

Figure 7.33 Monte Carlo simulated scatter projections (True Scatter) compared to those from the Effective Source Scatter Estimation (ESSE) scatter model and the TEW method for various photopeak energy window widths. All windows are centered on the 140-keV photopeak energy of ^{99m}Tc. The simulation is of the distribution of a myocardial perfusion imaging agent using the XCAT anthropomorphic torso phantom.

Method (ESSE). The TEW estimate tends to be more blurred, which can result in some edge enhancement and suboptimal quantification. TEW is also difficult in cases where there are multiple photopeaks close together, e.g., in the cases where x-ray photons are imaged.

An alternative to independent scatter estimates such as TEW is *model-based* scatter estimation. With model-based estimation, the scatter is modeled in the projection steps of the iterative algorithm either using Monte Carlo or analytic photon transport methods or by a scatter model such as ESSE. Analytic calculation of the scatter distribution is very challenging for SPECT because high-order scatter has significant contribution to measured scatter in the projections. Fast Monte Carlo methods have been developed and are very appealing. Figure 7.33 shows a comparison of the estimate from ESSE compared to a standard (not fast) Monte Carlo code. The ESSE estimate has the advantages in terms of noise and accuracy but takes more computational resources than TEW.

The primary factor affecting efficacy of scatter compensation is the accuracy of modeling the scatter contribution to the projections. If the accuracy is good, then scatter compensation can be quite effective since it adds information to the projection, in contrast to the CDR where high frequency information is lost. To demonstrate this, consider the results of a numerical experiment shown in Figure 7.34. Projections were generated using Monte Carlo simulations and modeled an ideal collimator-detector (perfect spatial resolution). Separate projections were generated from unscattered and scattered photons. The unscattered photons only and the scattered plus unscattered photons were reconstructed with no compensation, with attenuation compensation, and with scatter compensation using the true scattered projections as the scatter estimate:

- Without compensation, there is a substantial and spatially varying underestimate of the activity concentration.

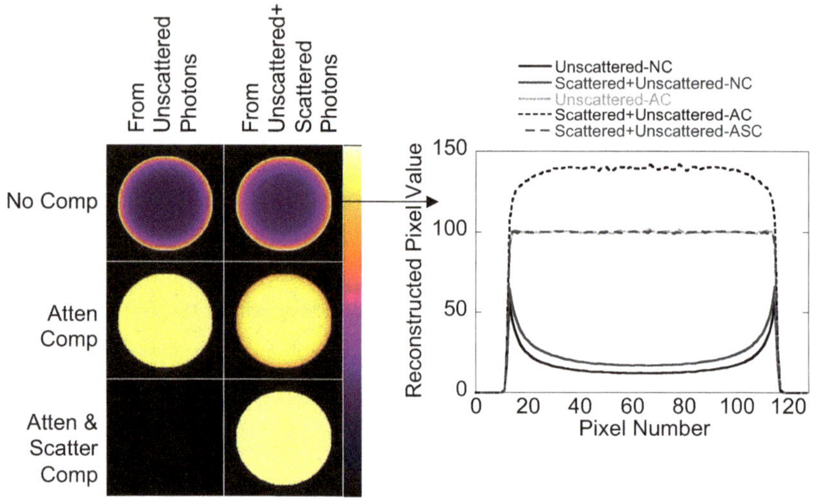

Figure 7.34 Demonstration of efficacy of attenuation and scatter compensation.

- With attenuation compensation alone, there is a substantial and spatially varying overestimate of the activity distribution.
- However, with compensation for the same effects included in the projections, the activity concentration is quite accurately estimated.

7.7.3 Collimator-Detector Response Compensation

Compensation for the CDR requires modeling the effects of the spatially varying resolution in the iterative reconstruction process. One computationally efficient way to do this is to take advantage of the fact that, ignoring the effects of the collimator hole pattern, the CDR is spatially invariant in planes parallel to the face of the collimator. A rotation-based projector/backprojector pair can thus be implemented as illustrated in Figure 7.35. The CDR used can either be the geometric CDR (i.e., modeling only photons that pass through the collimator holes) or can include the effect of septal penetration and scatter. The geometric CDR can be calculated analytically given the collimator hole geometry. The full CDR must be either measured as a function of distance or calculated using, for example, Monte Carlo photon transport methods.

Because the CDR results in a loss of information, the compensation is not complete. This is illustrated in the numerical experiment below with results shown in Figure 7.36. This was a simulation of point sources at the four positions indicated. They were placed on a background with 2% of the intensity of the sources. Projections were simulated using a projection bin size of 0.479 cm and modeling Siemens LEHR and MEGP collimators at 120 projection views over 360°. The CDRFs of the collimators were computed for ^{67}Cu and acquisition in 93 keV and 185 keV energy windows with widths of 20% and were used in the simulation. Attenuation and scatter in the

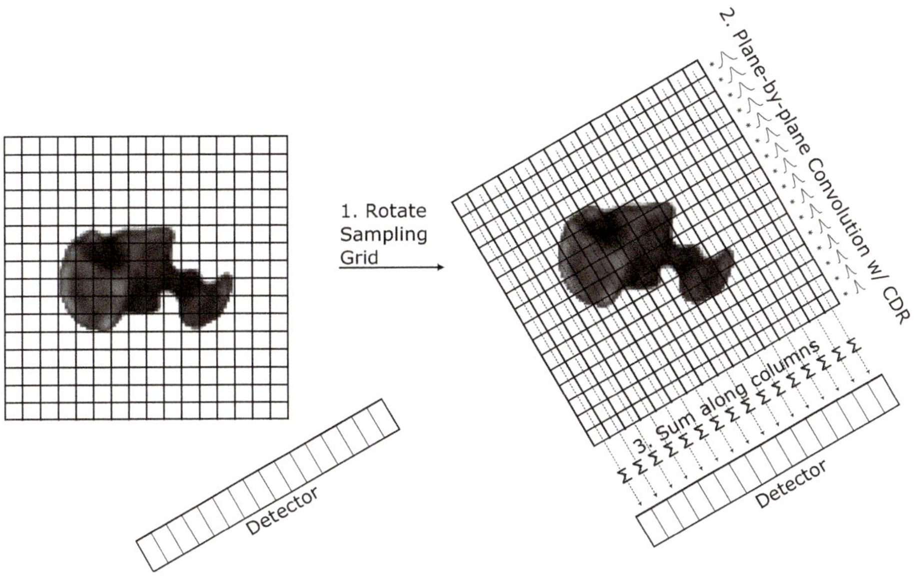

Figure 7.35 Illustration showing the use of a rotation-based projector to model the distance-dependent CDR. The activity image is rotated and resampled so that the rows are parallel to the face of the collimator. The planes parallel to the collimator are then convolved with the appropriate (distance-dependent) CDR. Summing along columns then gives the projection image.

object were not modeled to show purely the effects of the CDR. These were reconstructed using 10 and 50 iterations (24 subsets) without and with CDR compensation, respectively. A smaller number of iterations was used without CDR compensation to due to its faster convergence. The first row of Figure 7.36 shows these reconstructed images. Note the substantially improved resolution for the LEHR collimator compared to the MEGP. The resolution improves with CDR compensation, but the phantom is not perfectly recovered and, in fact, the resolution is quite spatially varying.

To study the properties of the point sources in more detail, phantoms were created separately for each point source plus the background and the background alone. Projections were simulated and reconstructed the same way. The 3D *modulation transfer function* (MTF), i.e., the frequency response of the reconstruction, was computed as follows. The reconstruction of the background was first subtracted from the point source plus background reconstruction, yielding the point response. The magnitude of the 3D discrete Fourier transform was computed and is displayed in Figure 7.36 with frequencies arranged so that zero frequency is in the center. If CDR compensation perfectly recovered the point source, then the MTF would be 1 over the whole space. Figure 7.37 shows profiles through one of the MTFs to provide a more quantitative assessment of the shape.

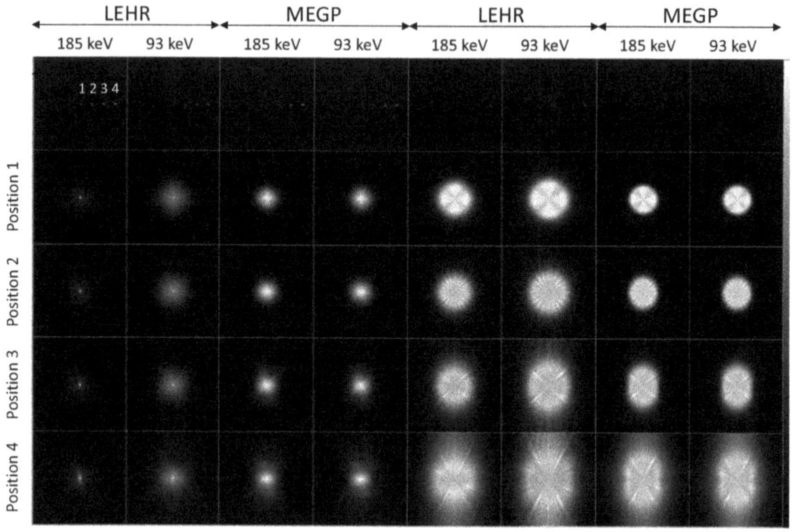

Figure 7.36 Results of numerical experiment to demonstrate the efficacy of CDR compensation. The top row shows the phantom with all four point-source positions reconstructed from a simulation of ^{67}Cu imaging using the two indicated Siemens collimators and energy window centers (width was 20%). The bottom four rows show the MTFs for the point source at the indicated position. In the MTFs, zero frequency is at the center of the image.

Comparing the images without and with CDR compensation shows a number of things. First, for the LEHR collimator—and especially in the 185-keV energy window—the low-frequency components are large (substantially > 1), resulting in a bright spot in the center of the image. This is due to the long low-frequency tails in the CDR of ^{67}Cu at this energy and collimator as discussed above. With the MEGP collimator, this feature is not present. However, the MTF for the LEHR collimator is wider, indicating a wider frequency response and thus better resolution. With CDR compensation, the MTFs are wider and are much flatter than without compensation. The high-intensity, low-frequency feature is no longer present. The MTFs are similar for both energies for the MEGP collimator and slightly narrower (indicating a loss of high-frequency information) for the 185-keV window compared to the 93-keV window for the LEHR collimator. This latter effect is likely due to the slightly larger FWHM resolution for the 185-keV photons noted above. Of importance, in all cases there are some frequencies that are not recovered. Thus, some features will be lost and not recovered. The quantitative effects of this will be discussed below.

An important effect on image quality of CDR compensation is the introduction of ringing effects, shown in Figure 7.38. These artifacts can be explained by the shape of the MTFs shown in Figure 7.36 and the profiles in Figure 7.37: without CDR compensation there is amplification of the low frequencies from septal penetration and scatter, and a slow roll-off of high frequencies resembling a Gaussian filter. With

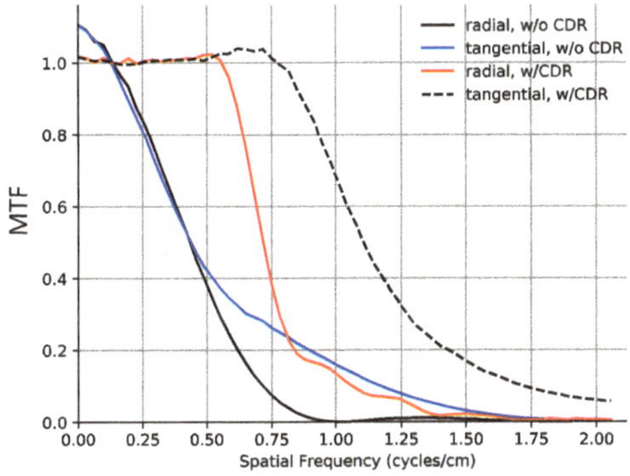

Figure 7.37 Horizontal and vertical profiles through the MTFs in Figure 7.36 for the MEGP collimator, the 93-keV energy window, and source position 3 with and without CDR compensation. The horizontal and vertical directions correspond to radial and tangential directions, respectively, for these points source positions.

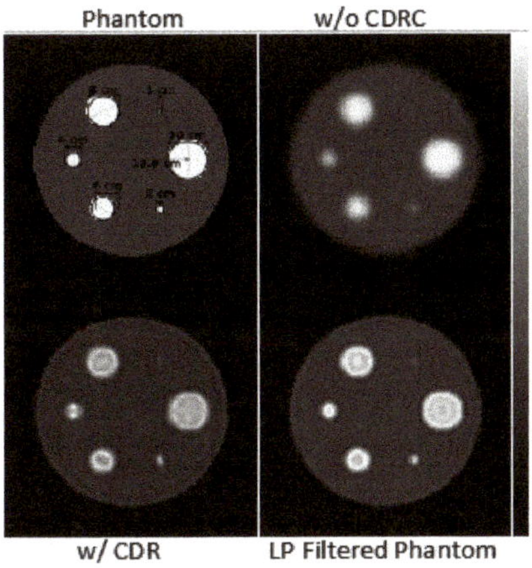

Figure 7.38 Central slice through phantom and reconstructed images without and with CDR compensation (CDRC) compared to the phantom filtered with a Butterworth low-pass filter.

CDR compensation, the zero-frequency amplification is greatly reduced, and the roll-off at high frequencies is much sharper, resembling a high-order Butterworth filter. The effect of this on the reconstructed images is shown in Figure 7.38. Note that the image without CDR compensation is blurry, as with a Gaussian filter. The reconstruction with CDR compensation shows some ringing. A phantom filtered with a Butterworth filter (cut-off frequency of 1 cycle/cm and order 8) is also shown. The filter was only chosen to demonstrate the ringing produced by low-pass filtering with a sharper cutoff and not to produce the exact ringing pattern produced by the spatially varying and asymmetric resolution in the image reconstructed with CDR compensation.

7.7.4 Partial Volume Effects

As shown above, the CDR results in permanent loss of image resolution. This results in both spill-in and spill-out PVEs. Scatter can also result in PVEs, but with accurate scatter compensation these PVEs are smaller than those due to the CDR, so this section will concentrate on the effects of the CDR.

The quantitative effects PVEs are often assessed in terms of the *recovery coefficient* (RC) for a volume of interest:

$$RC = \frac{A_{\text{measured}}}{A_{\text{true}}} \tag{7.19}$$

where A_{true} and A_{measured} are the true and measured activities in the VOI. Generally, the RC is interpreted as a measure of bias, and so the measured activity should be the mean of multiple measurements or from low-noise data. A value of RC=1 represents good activity quantification.

The RC depends on the resolution and size, shape, and activity concentration relative to surrounding (background) structures. To demonstrate these effects, consider the phantom shown in the inset in Figure 7.39 and associated RCs for a range of sphere-to-background contrasts. The phantom contained spheres with sizes ranging from 1 to 4 cm in diameter, each at 5 transaxial positions in the phantom. As demonstrated above, the resolutions at these positions are different, but the same for the same transaxial position in different axial slices. The simulation was repeated for sphere-to-background activity concentration ratios ranging from 1 to 20. Data from both circular and object-contouring orbits were generated using both LEHR and MEGP collimators. The simulation modeled only the geometric component of the CDR to emphasize resolution effects. The activities in the VOIs were estimated using the true (voxelized) spherical VOIs used to generate the phantoms so partial voxel effects (where the geometric object contributes to part of a voxel) were not present. The error bars at each point are the ranges of the RCs observed for that sphere size and contrast resulting from the different positions. Note that there is substantial size and contrast dependence and generally smaller, though still discernible, dependence on transaxial position, as demonstrated by the size of the error bars. As expected, since the MEGP has poorer resolution, the RCs are farther from 1 for the MEGP collimator than for the LEHR; this is especially evident for the smaller spheres.

As shown in Figures 7.36 and 7.38, CDR compensation can improve image resolution. Figure 7.40 shows that improvement in resolution reduces PVEs. The data in Figure 7.40 was generated from the same projections as Figure 7.39, but the images were reconstructed with geometric CDR compensation. There was substantial improvement in RCs. However, the variation in RC with source position is generally larger after CDR compensation.

Several different *PVE compensation* (PVC) methods have been proposed. Perhaps the most common is to use a VOI larger than the physical size of the object to reduce spill-out effects. This works well if the activity concentration in surrounding regions is low compared to that in object. However, if that is not the case, then the larger VOI will increase spill-in effects. This method is also somewhat subjective.

A second commonly used method is to measure RCs using a set of spheres with different diameters. Activities measured in objects can then be divided by the RC of a sphere having the same volume. However, as demonstrated in Figure 7.39, the RC depends not just on volume, but on background activity and, to a lesser extent, position. The background dependence can be reduced by compensating for spill-in of activity from the background. Another problem is that there is significant shape dependence. For example, a sphere and an ellipsoid having the same volume can have very different RCs depending on the eccentricity of the ellipsoid. RCs could be measured for a range of object shapes, but this can quickly become very complicated.

A method that implicitly accounts for shape is to estimate both the spill-in and spill-out PVEs using the exact object geometry. This is essentially what the *geometric transfer matrix* (GTM) method does. In this method, the spill-out and spill-in effects are quantified by the GTM, G, where the matrix elements are $G_{ij}, i, j \in [1, \ldots, N]$,

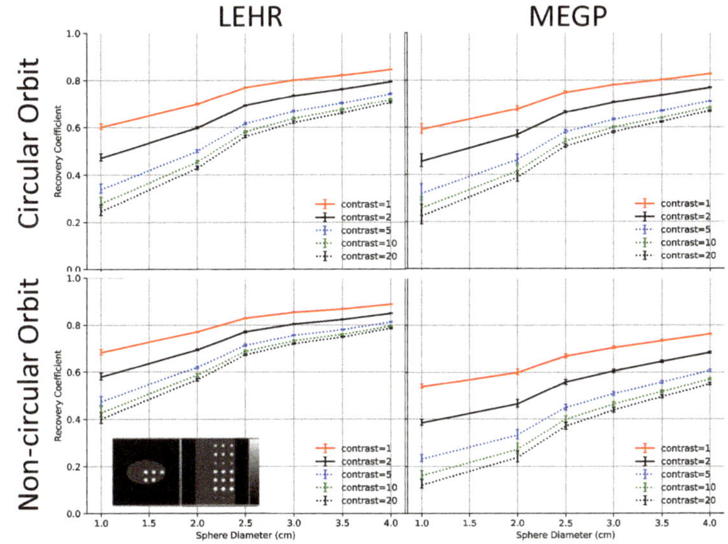

Figure 7.39 Plot of the recovery coefficient as a function of sphere diameter and contrast relative to the background for a simulation of ideal LEHR and MEGP collimators (no septal penetration and scatter). Slices through the phantom perpendicular to and along the axis of rotation are shown in the image inset at lower left. Spheres at 5 different positions in the transaxial direction and 7 different positions parallel to the axis of rotation were simulated. The error bars indicate the mean recovery coefficient over the 4 positions and error bars indicate the range of values seen for each sphere.

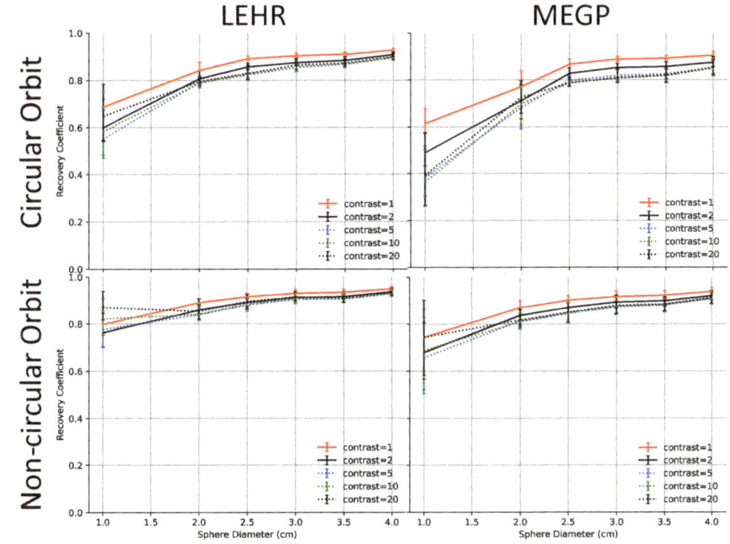

Figure 7.40 Plot of the recovery coefficient as a function of sphere diameter and contrast relative to the background for a simulation of ideal LEHR and MEGP collimators. This is like Figure 7.39, but the projection included septal penetration and scattering, and the reconstruction included full CDR compensation.

where N is the number of VOIs. The element G_{ij} represents the spill-out of activity from region i into region j (or the spill-in of activity from region j into region i. These matrix elements can be calculated by:

$$G_{ij} = V_j \bullet (R * V_i)\,, \tag{7.20}$$

where V_i is the VOI image for VOI i, that is the image that has all voxels inside VOI i set to 1 and voxels outside set to zero, and R is the 3D point response of the imaging system. The dot product of V_j and the blurred V_i represents the sum of counts in the blurred image in V_j. The PV-corrected activities can be calculated from the measured activities using:

$$\mathrm{A}^{PVC} = \mathrm{G}^{\mathrm{inv}}\mathrm{A}\,, \tag{7.21}$$

where A^{PVC} is the vector containing VOI i activities after compensation, A is the vector of activities estimated from the image, and $\mathrm{G}^{\mathrm{inv}}$ is the matrix inverse of the GTM. This method is not available commercially, but can be implemented easily, provided an estimate of the resolution kernel, R, is available.

One shortcoming of the GTM method as applied to SPECT is that it assumes that the resolution is spatially varying. An alternative for SPECT is to replace convolution by a single resolution kernel with sequential application of projection and reconstruction operators:

$$G_{ij} = V_j \bullet (\mathcal{R}\mathcal{P}Vi)\,, \tag{7.22}$$

where $\mathcal{R}$ is the reconstruction operator and $\mathcal{P}$ is the projection operator that models the CDR. This thus requires one extra projection and reconstruction operation for each VOI. In principle, the projection should model all physical effects, and the reconstruction operator should model the same effects as in the reconstruction to which PVC will be applied. Thus, for iterative reconstruction methods that can be quite time-consuming, this can add substantial computational complexity. In addition, for nonlinear algorithms such as OS-EM, this equation is not rigorously correct, and perturbation methods should be used, necessitating additional reconstructions.

A final set of methods incorporates PVC into image reconstruction. This can be done, for example, by penalizing deviations from the mean activity concentration in VOIs or by enforcing that all the activity concentrations in a VOI are constant. There has been relatively little study of these methods as applied to SPECT activity quantification.

The efficacy of PVC depends largely on how accurate the assumptions are and the method used. Some results will be shown below for perturbation-based GTM PVC (pGTM) to indicate the potential improvements that can be obtained.

7.7.5 Quantum Noise

Quantum noise, once present in the data, cannot really be compensated for in the same sense as the other physical effects. In iterative reconstruction, it is considered in designing the statistical reconstruction algorithm. For example, the ML-EM method explicitly models the effects of noise in the cost function that is optimized in the iterative process. This provides, in theory, some useful properties from estimation theory.

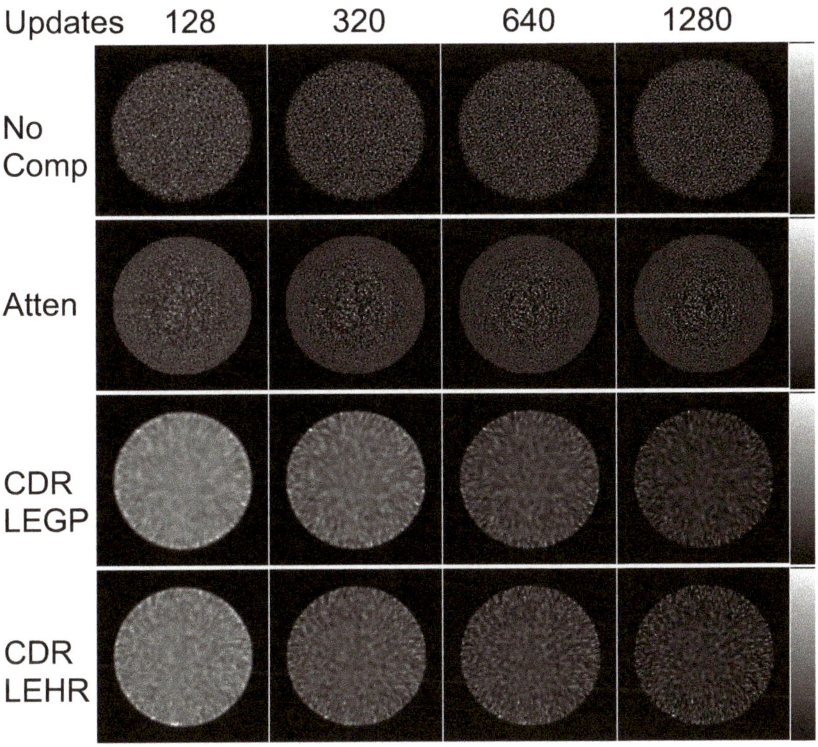

Figure 7.41 Example of noise correlations in reconstructed images resulting from compensation for various image-degrading factors. The phantom in all cases was a uniform elliptical cylinder. Projections were simulated including no effects, attenuation, and the geometric CDR for LEGP and LEHR collimators. The projections were reconstructed with compensation for the same effects modeled in the projections using OSEM with 32 subsets per iteration and from 4 to 40 iterations. The number of updates is the product of the number of iterations and the number of subsets per iteration.

However, the resulting activity estimates will still be imprecise; that is, there will be variation in activity estimates around the mean if measurements are repeated.

As discussed above, image processing, including SPECT reconstruction, produces correlated noise in reconstructed images even though the data in the projections are statistically independent. These correlations produce textured noise in the reconstructed images. Figure 7.41 shows the effect of compensation for the different physical image-degrading factors and number of updates (iteration times number of subsets). Note that the noise texture is spatially varying and depends on the effects compensated for.

From Figure 7.41 it is evident that noise will be a substantial problem for small objects. This will be demonstrated in the section illustrating the achievable accuracy and precision. However, for large objects, spatial averaging that occurs when applying a VOI reduces the impact of noise. This is demonstrated in the Figures 7.42 and 7.43, where sample images and average and standard deviations of percent errors are shown for simulation of acquisition durations per bed position ranging from 1.5 to 30 minutes. The data were simulated using the XCAT phantom and a validated Monte Carlo simulation code. The activity distributions were based on clinical ^{111}In ibritumomab tiuxetan (Zevalin) SPECT/CT studies and an AA of 185 MBq. Low-noise projections were generated, scaled to model the different acquisition durations, and Poisson noise was simulated. Images were reconstructed using OS-EM with compensation for atten-

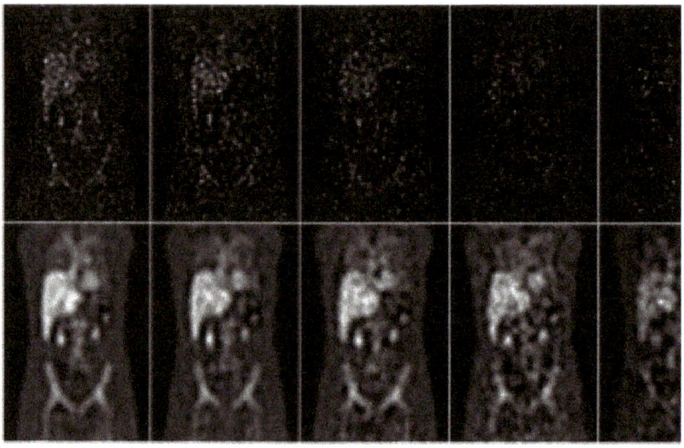

Figure 7.42 Coronal images of reconstructed simulated images of ^{111}In ibritumomab tiuxetan (Zevalin) corresponding, from left to right, to total acquisition durations from all projections of 30, 15, 9, 3 and 1.5 minutes per FOV. The top row is unfiltered, and the bottom row is after application of a Butterworth filter to improve visual image quality

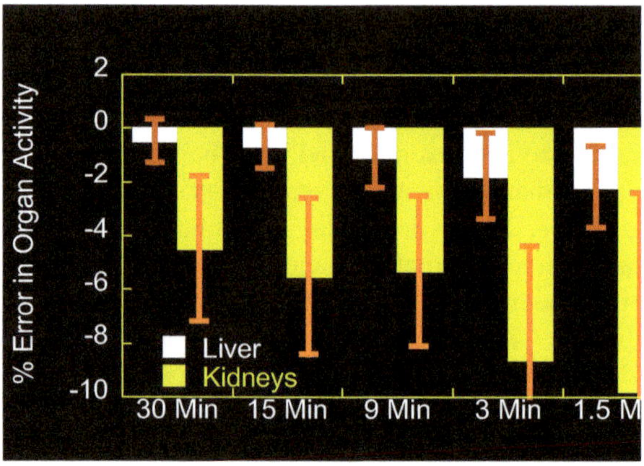

Figure 7.43 Average percent error and standard deviations of percent errors (error bars) in kidney and liver activities over 50 noise realizations for the indicated total acquisition durations per bed position.

uation, scatter, and the full CDR using 30 iterations and 24 subsets per iteration. The errors in activity estimates increased with decreasing acquisition duration, but even very short acquisition times gave rise to modest increases in error, especially in comparison to the marked increase in visual noise in the images.

7.7.6 Count Rate Losses

As discussed above, the finite time required to process signals from detected photons in the gamma camera can result in lost counts, and the fraction of counts lost depends on the count rate. Compensating for these losses requires knowledge of the relationship between true and measured count rates. While gamma cameras are not exactly paralyzable systems, this theoretical framework can be used to model and compensate for count rate losses to good approximation. A key requirement is measurement of the resolving time, τ. Since the paralyzable model does not exactly describe the behavior of the gamma camera, it is desirable to obtain measurements over the range of expected count rates and fit the resulting data to the model rather than using simpler methods, such as a two-source method.

A relatively simple and reliable way to estimate ρ is to acquire images of a decaying source over enough half-lives to span the count rates of interest. Since the full energy spectrum determines the count rate losses, using a phantom that approximates the scatter distribution expected from patients is desirable. A spherical insert inside an elliptical or circular cylinder phantom with no background activity provides a balance between ease in filling the phantom and generating a representative scatter spectrum. Image data should be acquired in the photopeak windows and additional energy windows that span the energy range supported by the gamma camera.

Let $\{N(t_i),\ i=1,\ldots m\}$ be the set of measured count rates over the entire energy spectrum acquired at m different timepoints. The true count rate at time t, $n(t)$, is given by:

$$n(t) = SA_0 e^{-\frac{\ln 2}{\tau}t}, \tag{7.23}$$

where S is the sensitivity, i.e., the detected counts, excluding the count rate losses, per unit time, A_0 is the activity of the source, and τ is the half-life of the radionuclide. Using Equation (7.1), the formula giving the measured count rate of a paralyzable system in terms of the true rate, we have:

$$N(t) = SA_0 e^{-\frac{\ln 2}{\tau}t} \exp\left(-\rho SA_0 e^{-\frac{\ln 2}{\tau}t}\right). \tag{7.24}$$

This equation can be fit to the measured count data to estimate the product SA_0 and ρ using, for example, least squares curve fitting methods.

The results of the above calibration can be used to correct the acquired data as follows. The SPECT data should be acquired for the same energy windows as used for the calibration above. This includes the photopeak window and auxiliary windows that, together, span the entire useful energy spectrum of the camera. For each projection, the measured count rate in the full spectrum is obtained by summing the counts in the photopeak and auxiliary windows. The resolving time estimate from the experiment and fitting procedure is used to calculate the true count rate from the measured count rate and the resolving time using Equation (7.2). The ratio of the measured and true resolving time represents the fraction of counts lost. The photopeak energy window is then divided by the inverse of this ratio to obtain an estimate of the count-rate-loss-corrected projections. This rescaling is repeated for all energy windows used in the reconstruction (i.e., scatter windows and additional photopeak windows) using the same ratio as for the photopeak window. The resulting data can then be reconstructed using the desired compensations. Ideally, these correction factors should be applied inside the iterative reconstruction process to preserve the Poisson nature of the projection data. However, this capability may not be available, and it is likely to have only a modest impact on the reconstructions, especially if the correction factors are similar for the different projection views. If it is not possible to scale the individual projections, averaging the correction factors over all the projection reviews and applying this average factor to the reconstructed images is also an alternative that, though not theoretically rigorous, can be used.

7.8 Achievable Quantitative Accuracy and Precision

The previous sections have discussed SPECT imaging and the factors and compensations needed to achieve quantitative SPECT. The following section provides examples of achievable quantitative accuracy for a variety of radionuclides.

Evaluating quantitative methods requires consideration of both accuracy and precision. The accuracy is usually measured in terms of the bias, the deviation of the mean of a set of measurements from the true value. The precision is often measured using the *coefficient of variation* (COV), the standard deviation divided by the mean. Assessing these on patient data is practically impossible as true activities can seldom be obtained, and it is usually not ethically or practically possible to perform repeat measurements. Animal studies are expensive, and there are likely to be substantial errors in measurement of organ activities even if animals are sacrificed and organs are harvested and counted.

One alternative that is often used is physical phantom studies. These have the advantage of exactly modeling the imaging system. However, even with anthropomorphic phantoms, physical phantoms are at best an approximation of human anatomy: organ shapes are often unrealistic, and activities are surrounded by walls with finite thickness that can affect spill-out and spill-in PVEs. A final difficulty is that it is expensive and time-consuming to obtain the multiple repeat measurements needed to assess precision. One solution for the final problem is to acquire high-count (low-noise data) and then to treat the resulting projections as the mean counts and use those to simulate multiple sets of noisy data using a Poisson distributed random number generator.

A final alternative is the use of simulation studies. In these, realistic human anatomies can be modeled. Phantoms such as the XCAT, Zubal, or ICRP phantoms are available that have been created from patient imaging data. The XCAT and similar phantoms go beyond this by fitting continuous functions to organ surfaces so that models can be transformed to model motion or variations in organ shape, etc. One limitation is that the activity distribution inside organs is usually assumed to be uniform. Once a digitized activity distribution has been created, there are several *Monte Carlo simulation* (MCS) codes that can accurately model most aspects of the image formation process. The GATE code from the OpenGATE collaboration is highly realistic and can be used to create detailed models of the gamma camera and collimators. It tends to have a somewhat steep learning curve, and simulating noise-free projections can be computationally demanding, especially for digitized phantoms. The SIMIND code models the collimator and crystal realistically. It uses simple models for the materials behind the crystal and does not model the camera shielding. However, it is less computationally demanding and has a menu-driven application for setting simulation parameters. None of these codes models count-rate loss effects in current implementations.

Since MCS codes often use variance reduction techniques to improve estimation of the mean projections, the noise is typically not Poisson distributed. Thus, many photons are typically simulated to generate low-noise projections. Simulating organs and objects of interest separately and scaling and summing them allows efficiently modeling variations in activity distribution over time. The low-noise projections can be used directly to assess bias, and multiple noise realizations can be generated to study precision. Noisy projections are generated by scaling the low-noise projections to represent appropriate acquisition time and administered activities so that they represent mean projections. Then, multiple Poisson noise realizations can then be generated using Poisson-distributed pseudo random number generator codes.

A full and rigorous evaluation of accuracy and precision likely involves physical phantom and simulation studies. The physical phantom studies are essential to validate simulation codes and estimate accuracy for real scanners; the simulations allow studying accuracy for realistic objects shapes, studying the variations in accuracy due to variations in object shape, spatial proximity, and organ uptake. These studies can also be useful in optimizing acquisition and data processing parameters in terms of quantitative reliability. The following sections show examples of evaluation of quantitative SPECT to illustrate the accuracies and precisions that are achievable.

7.8.1 Example 1: ^{131}I Physical Phantom Study

Iodine-131 is perhaps the oldest therapeutic radionuclide and is still widely used for thyroid cancer and hypothyroidism therapy. Iodine-131 is a beta emitter and emits a 364-keV gamma ray that is used for imaging. There are some low-abundance, high-energy photons (637 keV 7.39%, and 722 keV 1.8%) that, after scattering in the body, collimator, crystal, and camera, can contribute counts in the photopeak window. We have developed a model-based reconstruction method that includes modeling of attenuation, object scatter, downscatter of the high-energy photons, and the CDR using the OS-EM algorithm (Song, Du et al. 2011). The method was initially evaluated using a phantom study in the context of ^{131}I tositumomab (Bexxar) treatment planning. The phantom is pictured in Figure 7.44, and the activities and volumes in the various compartments are shown in Table 7.4. Data were acquired on a Phillips Precedence

Table 7.4 Activities and volumes in the various compartments of the phantom in Figure 7.44

	Blood Pool	Myocardium	Large Sphere	Small Sphere	Background
Volume (ml)	59.7	115.3	17.5 (r = 1.61 cm)	5.7 (r = 1.1 cm)	9580
Activity (mCi)	0.562	0.471	0.136	0.044	8.15

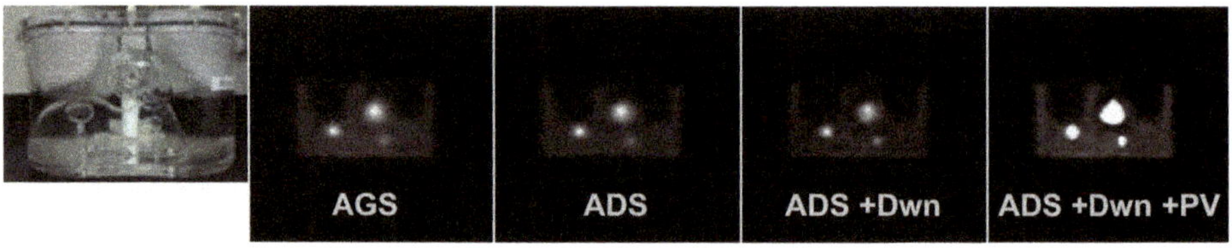

Figure 7.44 Leftmost: Data Spectrum Anthropomorphic Torso phantom with cardiac and sphere inserts used in the evaluation. The images are vertical slices through images reconstructed with compensation for various combinations of attenuation (A), scatter (S), the geometric CDR (G), the full CDR (D), downscatter from high-energy photons (Dwn), and partial volume effects.

SPECT/CT system using a 20% energy window centered at 364 keV and an HEGP collimator. The images were reconstructed with various combinations of compensations for the image-degrading effects discussed above; sample images after 30 iterations (24 subsets per iteration) are shown in Figure 7.44. The system sensitivity was measured using a planar image of a syringe containing a known ^{131}I activity. VOIs were defined manually using the registered CT images, and the activities in the different compartments were calculated and compared to the true activities placed in the compartments. Though different activity concentrations were used in the heart and blood pool, the image resolution was not good enough to resolve the myocardium, so only the total activity in the heart (myocardial wall plus blood pool) was evaluated. PVC was performed in the reconstruction by constraining the activity of all the VOIs (but not the background) to be constant.

The errors in the activity estimates are shown in Table 7.5. Since the data is relatively low in noise, the error is likely largely bias. Without compensating for septal penetration and scatter (i.e., G compensation) and downscatter, the activity in all compartments was substantially overestimated. The combination of all compensations provided the smallest errors for the two spheres, but downscatter compensation did not provide the best results for the heart. One possible reason is that the PVC method used modeled the activity in the entire heart as uniform, but the activity concentration in the myocardial wall compartment was lower in the blood pool. Using a single average activity concentration for the entire region could lead to overestimation of spill-

Table 7.5 Percent errors of activity estimates for the Anthropomorphic Torso Phantom

Compensations	Heart	Large Sphere	Small Sphere
AGS	–15.21	–26.12	–32.72
ADS	13.85	–10.48	–19.33
ADS + Dwn	3.03	–14.25	–25.19
ADS + Dwn + PV	5.56	–8.16	–12.27

The percent error was calculated as % error = (estimate-true)/true.

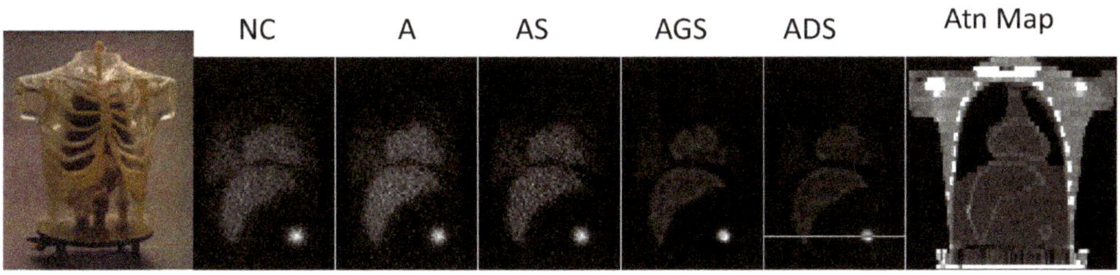

Figure 7.45 Leftmost: picture of RSD Torso phantom used in the experiment. The images are coronal slices through reconstructed images obtained using no compensations (NC) and with combinations of attenuation (A) model-based scatter (S), geometric CDR (G), and the full CDR (D) compensations. A corresponding slice through the attenuation map is also shown.

out effects, resulting in overestimation of the total activity. This data also demonstrates that accuracies, even with PVC, will be poorer for small objects than for larger ones.

7.8.2 Example 2: ^{111}In Physical Phantom Study

The second example is for ^{111}In, a radionuclide that can be an imaging surrogate for radionuclides such as ^{90}Y. In this case, the activity distribution modeled that of ^{111}In ibritumomab tiuxetan (He, Du et al. 2005; He and Frey 2006; He, Wahl et al. 2008; He, Du et al. 2009; He, Wahl et al. 2009). An RSD Torso Phantom (Figure 7.45) was used, and the compartments were filled with the following activities: heart, 27.66 MBq; lungs, 27.32 MBq; liver, 49.16 MBq, background, 20.28 MBq (activity concentration ratios 19:5:20:1). Two spherical inserts with inside diameters of 22 and 34 mm were placed in the phantom and had activities of 2.26 and 1.25 MBq (sphere-to-background ratios of 156:1 and 17:1, respectively). The phantom was imaged on a GE Discovery VH (2.54-mm thick crystal) SPECT/CT system with an MEGP collimator. Projections were acquired at 120 views over 360 degrees, 0.442-cm voxels, and two summed 14%-wide energy windows centered over the two ^{111}In photopeaks. An acquisition time of 45 s per view was used to obtain low-noise projections. The

Table 7.6 Percent errors in activity estimates for ^{111}In physical phantom study

Compartment	NC (%)	A (%)	AS (%)	ADS (%)	ADS+PV (%)
Heart	–77.60	24.63	–11.76	–3.72	–2.11
Lungs	–62.78	31.39	–0.96	4.23	6.45
Liver	–74.38	29.22	–7.47	2.71	4.14
20.6 cc sphere	–78.88	–14.85	–29.81	–3.36	–1.97
5.6 cc sphere	–88.24	–51.53	–56.75	–21.55	–11.94

The headings above the columns indicate the compensations used. The same notation is used as in Figure 7.45 with the addition of PV for partial volume compensation.

projections were reconstructed with various combinations of compensations (see Figure 7.44). Up to 30 iterations with 24 subsets per iteration were used, a number of iterations that produced convergence in the activity estimates. VOIs were defined manually using the CT image. Post-reconstruction PVC was applied using a perturbation-based GTM method (pGTM). The sensitivity was measured using a syringe filled with a known activity and a planar image acquired using the same energy windows and used to convert reconstructed voxel values to activities. The activities estimated from the images were compared to the true activities, and the percent differences were computed.

Table 7.6 shows the percent errors for the different compartments and compensation methods. For compartments corresponding to normal organs, the best results were obtained with attenuation, scatter, and full CDR compensation (ADS). The addition of PVC resulted in substantial improvement for the smallest sphere, modest improvement for the large sphere and heart, and a small increase in error for the lungs and liver. In general, except for the smallest sphere, the errors were less than 5% with ADS compensation. These results were better than with ^{131}I, likely due to a combination of better collimator resolution and more accurate modeling of the image-degrading effects.

7.8.3 Example 3: ^{111}In Anthropomorphic Phantom Simulation—Organ Activity Estimation

A Monte Carlo simulation study was performed to assess the accuracy in an anatomically realistic setting and to assess precision of the organ activity estimates (He, Du et al. 2005). The NCAT phantom was used, and organ activities were based on those measured 24 hours post injection in a clinical ^{111}In ibritumomab tiuxetan study. Activity in major blood vessels was simulated based on blood sampling. The same acquisition parameters were used as for the physical phantom experiment. A validated Monte Carlo simulation was used to generate low-noise projections for each organ. The projections were scaled to represent an acquisition duration of 30 minutes per bed position and an injected activity of 185 MBq of ^{111}In. Fifty Poisson noise realizations were generated using a pseudorandom number generator. Figure 7.46 shows the phan-

Table 7.7 Percent bias ± COV for the various organs and tissues

Organ	A (%)	AS (%)	AGS (%)	ADS (%)	ADS + PVC (%)
Heart	33.75 ±0.52	25.58 ±4.24	−1.55 ±0.40	−1.21 ±0.42	2.31 ±0.43
Lungs	17.50 ±0.60	5.42 ±3.39	20.63 ±0.77	20.88 ±0.82	11.47 ±0.86
Liver	40.32 ±0.93	30.94 ±3.92	−2.10 ±0.43	−2.21 ±0.45	−1.76 ±0.43
Kidneys	33.32 ±0.29	21.40 ±5.39	−5.04 ±1.38	−4.73 ±1.46	2.52 ±1.32
Spleen	16.61 ±0.27	7.90 ±4.55	−3.09 ±1.15	−1.26 ±1.19	4.33 ±1.09
Marrow	5.25 ±0.53	−0.94 ±2.36	−7.94 ±0.65	−5.46 ±0.69	3.40 ±0.69
Blood Vessels	−23.05 ±0.64	−28.60 ±1.47	−20.25 ±0.97	−15.22 ±1.06	3.61 ±0.90

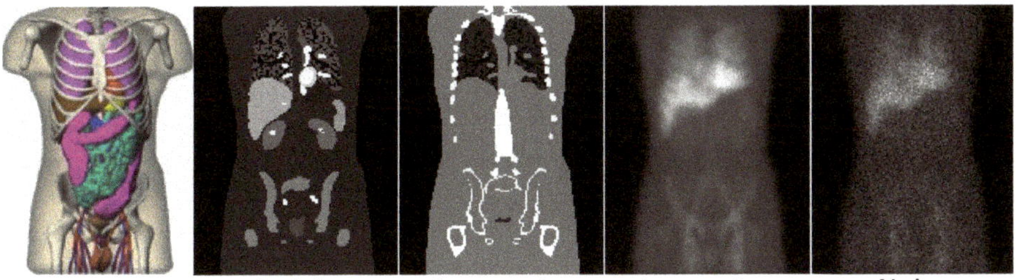

Figure 7.46 Phantom, coronal slices through the activity and attenuation distributions, and low-noise and noisy projections from the ^{111}In simulation study.

tom, a slice through the activity and attenuation distributions, and low-noise and noisy projections. The noisy projections were reconstructed using the same set of compensations used in the physical phantom experiment. The activities in the organs were compared to the true simulated values, and percent bias and COV over the 50 noise realizations were computed.

The results of the comparison are shown in Table 7.7 for each of the organs. Note that the errors are all less than 5% when all compensations were applied except for the lungs. The larger error for the lungs is likely due to incomplete compensation for partial volume effects resulting from the low activity in the lungs, the nonuniform activity distribution resulting from the airways, and proximity of the liver and blood vessels resulting in large spill-in effects. The errors are the same order of magnitude as in the phantom study. The differences are likely a combination of more accurate VOIs and attenuation map, and less uncertainty in the activities placed in the phantom and calibration syringe in the simulation study. The COVs are all smaller than the biases, demonstrating the effect of averaging over large VOIs in reducing the effects of quantum noise.

Table 7.8 Tumor sizes, activities, and tumor/background ratios inside NCAT phantom

	T1	T2	T3	T4	T5	T6	T7	T8	T9
Diameter (cm)	3.09	0.88	2.21	0.88	3.09	2.65	1.33	1.77	2.21
Volume (ml)	15.32	0.36	5.56	0.36	15.32	9.98	1.33	2.77	5.56
Activity (MBq)	1.42	0.08	3.36	0.33	5.00	1.20	0.25	0.39	3.72
Tumor/Bkg Ratio	4.40	11.01	5.21	12.0	18.83	6.93	10.75	8.05	10.52
Organ	Lungs	Lungs	Spleen	Spleen	Bckgnd	Bckgnd	Bckgnd	Heart	Liver

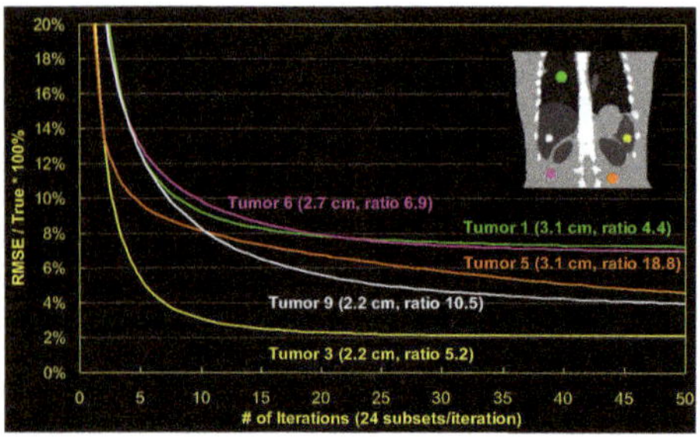

Figure 7.47 Plot of % RMSE vs. iteration for spheres larger than 2.2 cm in diameter.

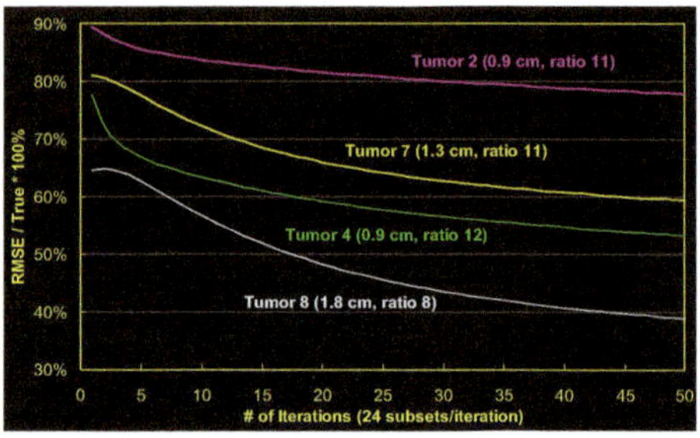

Figure 7.48 Plot of % RMSE vs iteration for spheres with diameters ≤ 1.8 cm.

7.8.4 Example 4: ^{111}In Anthropomorphic Phantom Simulation—Tumor Activity Estimation

To demonstrate the effects of object size and activity concentration relative to the background, we used the same simulation data as above but added projections of simulated spheres. The sizes and tumor/background ratios are shown in Table 7.8. Fifty noise realizations were generated, and the images were reconstructed with compensation for ADS as above. The percent bias, COV, and root mean square error (RMSE) were computed as functions of iteration up to 50 iterations (24 subsets per iteration).

The results are shown in Figure 7.47 for the selected spheres with diameters ≥ 2.2 cm. For these spheres, which are larger than the FWHM resolution, the achievable RMSE errors were less than 10%. Figure 7.48 shows similar data for spheres smaller than 2 cm. In this case, the errors are much larger. For both the large and small spheres, even though image noise increased with iteration, the errors in the activity estimates continued to decrease up to 50 iterations. The data also demonstrate that accuracy depends on the size, location, and target-to-background ratio in a complicated way. For the smaller spheres, the RMSE was substantially larger and convergence was slow. This indicates that it may not be possible to obtain reliable activity estimates for objects smaller than the resolution.

To investigate the efficacy of PVC, we implemented a reconstruction-based method, OS-VOI-SV, that enforced constant activity concentration inside VOIs dur-

Table 7.9 Accuracy and precision in tumor activity estimates

	Diameter (cm)	OS-EM (%)	OS-VOI-SV (%)
Tumor 1	3.09	−9.2 ±12.3	−6.7 ±2.8
Tumor 2	0.88	−63.4 ±60.0	−47.0 ±21.7
Tumor 3	2.21	−5.8 ±12.0	−1.8 ±1.4
Tumor 4	0.88	−58.6 ±47.0	−7.6 ±5.7
Tumor 5	3.09	−7.9 ±8.5	−4.3 ±2.0
Tumor 6	2.65	−11.8 ±13.6	−1.8 ±3.7
Tumor 7	1.33	−60.2 ±53.9	−68.0 ±13.0
Tumor 8	1.77	−48.6 ±36.7	−51.2 ±15.0
Tumor 9	2.21	−9.1 ±9.4	−6.1 ±1.4

during reconstruction. The VOIs provided were at ½ the voxel size of the reconstruction, and the method modeled the effects of partial VOI contributions to reconstruction voxels. Explicit VOIs were provided for the same seven organs as above, but the body background was represented with independent voxels. Up to 50 iterations were used. Percent bias and COVs were calculated for each tumor with and without PVC.

The results of this PVC are shown in Table 7.9. For Tumors 2, 7, and 8, all with diameters less than 2 cm, the biases and COVs, even after PVC, were quite large. For tumors larger than 2 cm, the errors were all less than 7% after PVC. These results indicate the potential of PVC for small objects, but challenges remain for objects less than the size of the image resolution.

7.8.5 Example 5: ^{111}In Voxel-Level Dosimetry

Voxel dosimetry involves applying voxel kernels or Monte Carlo radiation transport methods to activity images to generate dose-rate images. Integration over time of the dose-rate images at the voxel level gives 3D dose distributions. *Dose volume histograms* (DVHs) can then be generated and used to calculate various metrics of efficacy or toxicity such as *equivalent uniform dose* (EUD) or *tumor control probability* (TCP). Evaluating the impact of accuracy and precision of quantitative SPECT images on DVHs is thus an important area of research. The use of simulation techniques is one way to approach this problem.

This section presents results of a simulation study to investigate optimization of reconstruction parameters—specifically the number of iterations and post-reconstruction filters—on the accuracy of cumulative DVHs (CDVHs) (Cheng, Hobbs et al. 2013). Simulated images using the methods for the ^{111}In simulation study were used including the same phantom, activity distribution, and acquisition parameters except that 128 projection views were simulated instead of 120. The simulated activity concentration inside the liver and kidney cortex was uniform. Noise was added to low-noise projections to produce noisy ones. Images were reconstructed using OS-EM with ADS compensation, 24 subsets, and up to 50 iterations.

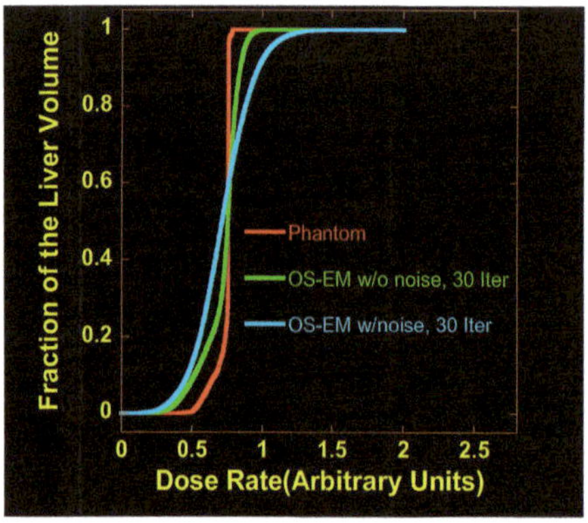

Figure 7.49 Plot of CDRVH for liver imaged 1-hour post-injection for the phantom and reconstructions without and without quantum noise.

The CDVH is corrupted by both noise and partial volume effects. Figure 7.49 illustrates this by showing a plot of the cumulative dose-rate volume histogram (CDRVH) for liver from the phantom and images reconstructed without and with noise. The image without noise deviates from the phantom image due to partial volume effects, i.e., blurring at the edges of the liver. When noise is added, the deviations are larger due to some voxels having higher and lower values than the mean due to the quantum noise. As seen above, noise increases with iteration while partial volume effects decrease. Thus, there should be an optimal iteration number. Adding a post-reconstruction filter provides another degree of freedom for optimizing the CDRVH. Figure 7.50 shows plots of the root normalized mean square error as a function of iteration for no filtering and five different Gaussian post-reconstruction filter widths. The curves are shown for the liver at 1 hour and the kidneys at 144 hours post administration. The optimal parameters are different for different timepoints and organs. Table 7.10 shows the optimal parameters for the five different timepoints for the liver, kidneys as a whole, and the renal cortices alone. When an FWHM is not listed, it means that the case of no filter was optimal. In general, the liver—which is larger and thus is more sensitive to noise than PVEs—requires a smaller number of iterations and the use of a

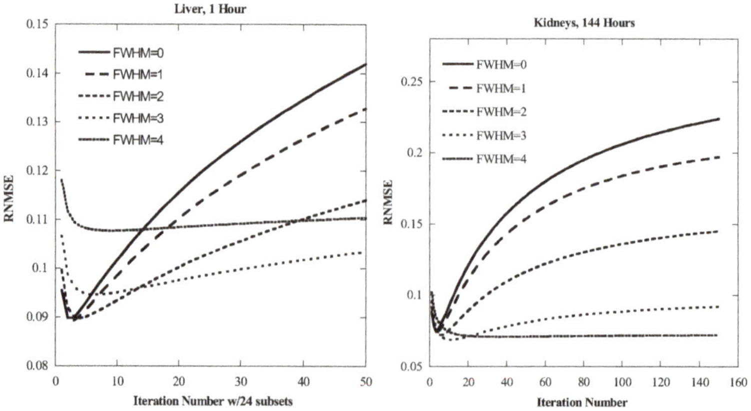

Figure 7.50 Plot of RNMSE of CDRVH for liver at 1 hour and kidneys at 144 hours as a function of iteration number for various Gaussian post-filter FWHM values.

Table 7.10 Optimal values of iteration number and Gaussian Filter FWHM for different organs and timepoints

	Liver	Kidneys	Renal Cortices
1 hour	Iteration 3, FWHM 1.5	Iteration 35	Iteration 53
5 hours	Iteration 3, FWHM 1.5	Iteration 34	Iteration 50
24 hours	Iteration 3, FWHM 1.5	Iteration 26	Iteration 36
72 hours	Iteration 2, FWHM 0.5	Iteration 22, FWHM 1.5	Iteration 24, FWHM 1.5
144 hours	Iteration 2 FWHM 1.0	Iteration 18, FWHM 3.5	Iteration 32, FWHM 5

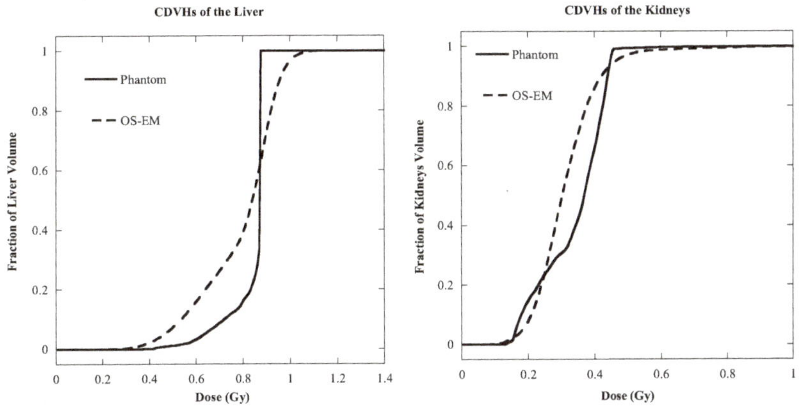

Figure 7.51 Cumulative dose-volume histograms (CDVHs) for liver and Kidneys using optimal parameters for each organ and timepoint.

filter. The renal cortices—which are small and thus require high resolution—need more iterations and no filter except for the last timepoint.

Figure 7.51 shows the results of using the optimal parameters for the liver and kidneys CDVHs. There is significant deviation from the true histograms. This is especially true near the low- and high-volume fraction portions of the curve, and it is these regions of the curve that are often important for predicting toxicity and efficacy. These data demonstrate the challenge of using 3D dose distributions for treatment planning in radiopharmaceutical therapy.

7.9 Imaging Alpha Emitters: Ra-223 Organ Activity Estimation

Alpha emitters are a new and promising tool for cancer therapy. Their high linear energy transfer results in a large number of double-strand breaks and, thus, high lethality. The ability to image their distribution directly could enable treatment plan-

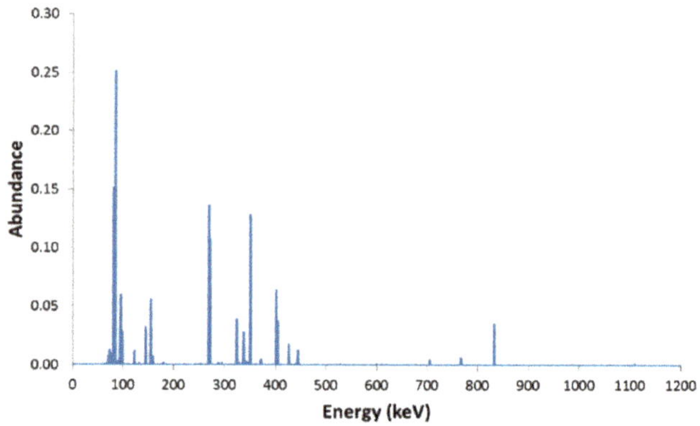

Figure 7.52 Photon emission energy spectrum from Ra-223.

ning using absorbed dose. However, due to their high lethality, therapeutic activities are often low. In addition, alpha emitters tend to have complicated energy spectra with multiple photons, including high-energy photons, that are difficult to collimate and shield. This section presents phantom and simulation results demonstrating the feasibility of using SPECT to image the distribution of alpha emitters.

Radium-223 has a complicated energy spectrum (see Figure 7.52).The NCAT phantom was used as in the studies above to model human anatomy. We used three energy windows (75.6–92.4 keV, 138.6–169.4 keV, and 242.1–295.9 keV) to acquire projection data. A multi-energy range method was developed to model the energy dependence of the attenuation, scatter, and CDR. The method simulates scatter and primary photons separately. Five energy ranges are used to model the primary photons, with separate scaling factors for the attenuation map and CDRFs for each range. The effective source scatter estimation (ESSE) method was used to model the scatter in each of the three energy ranges, and different attenuation map scaling factors and CDRFs were used for each range. The model was implemented in the projection step of an OSEM reconstruction method. The backprojection modeled attenuation and the geometric CDR, but not the full CDR or scatter, to improve convergence and reduce computation time.

Figure 7.53 Picture of the phantom and spherical insert.

A circular cylinder phantom 22 cm in diameter by 20 cm long with a 3.52-cm diameter spherical insert at the center phantom was used (Figure 7.53). A total

Table 7.11 Organ activities in MBq at the four simulated timepoints in the ^{223}Ra simulation study

Organ	4 h	24 h	72 h	144 h
Small intestine	3.34	0.18	0.00	0.00
Upper large intestine	0.00	3.75	1.67	0.50
Lower large intestine	0.00	1.08	1.15	0.62
Red marrow	3.15	2.57	1.56	0.74
Cortical bones	1.93	1.57	0.96	0.46
Remainder	0.33	0.09	0.03	0.03

Table 7.12 Percent errors in organ activities for the ^{223}Ra simulation study

Organ	4 h (%)	24 h (%)	72 h (%)	144 h (%)
Small intestine	−3.8 ±0.6%	*	*	*
Upper large intestine	*	9.7 ±0.6%	8.4 ±0.8%	6.5 ±1.7%
Lower large intestine	*	11.4 ±1.7%	12.8 ±1.7%	12.6 ±1.8%
Bones	−7.3 ±0.5%	−7.0 ±0.7%	−9.0 ±0.7%	−12.9 ±1.0%

Percent errors are the average of the estimate minus the true activity divided by the true activity.
The uncertainties shown are standard deviations calculated over 50 noise realizations.
* indicates that the mean organ activity concentration was very close to that of the body background and the percentage errors were not calculated.

of 195.9 kBq of ^{223}Ra dichloride was added to water inside the sphere, giving an activity concentration of 8.611 kBq/ml. A solution with an activity concentration of 0.315 kBq/ml was used to fill the background. Projections were acquired on a Siemens Symbia using an MEGP collimator, 120 views over 360°, a projection bin size of 0.48 cm, and a duration of 45 s per view. The images were reconstructed using 20 iterations of OS-EM (12 subsets per iteration). A VOI was defined manually over the sphere, and the sum of voxel values inside the sphere were converted to activity using the sensitivity measured from a planar image of a syringe with known activity. The pGTM partial volume compensation method was applied to compensate for PVEs. After 20 iterations (24 subsets per iteration) the error in the total activity in the sphere was 22% without partial volume compensation and 7% with PVC.

A simulation study was performed to further evaluate the method for realistic activity concentrations and object shapes. The NCAT phantom was used, and activities were set based on values found in the literature (Table 7.11). Projections were simulated using the SIMIND MCS code and scaled to represent an acquisition duration of 30 s per view per FOV. Fifty statistically independent Poisson-distributed noise realizations were generated. A sample set of projections is shown in Figure 7.54.

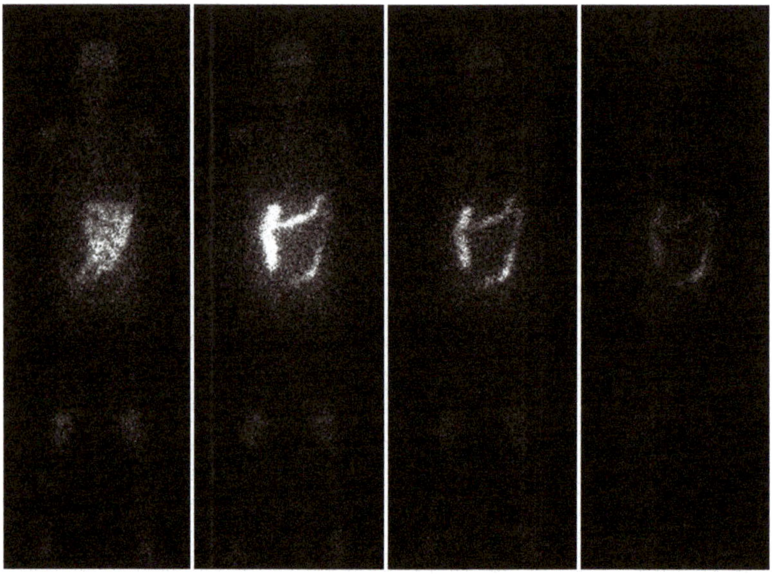

Figure 7.54 Sample noisy projections from an anterior view.

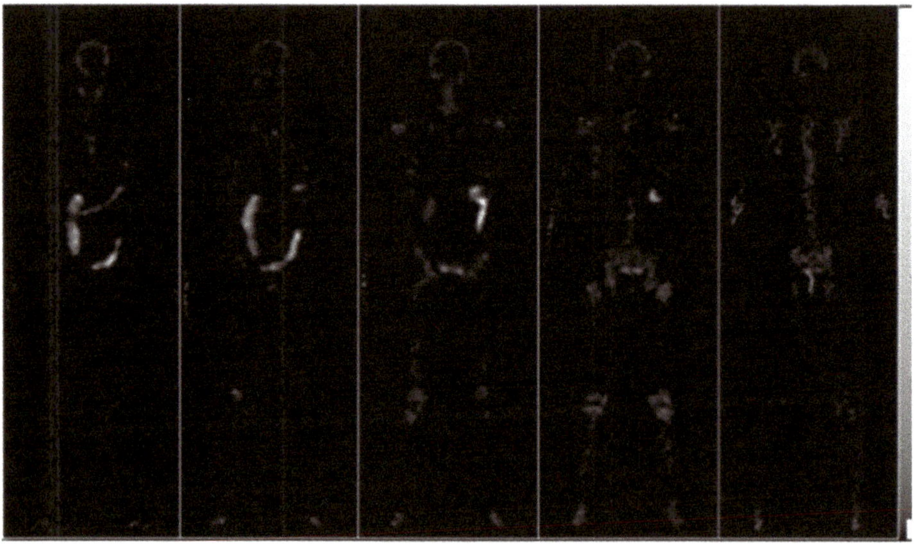

Figure 7.55 Five sample coronal slices through the reconstructed ^{223}Ra activity distribution at 144 h. The data are after 5 iterations and are shown on a logarithmic gray scale to emphasize background low-activity features.

Images were reconstructed using OS-EM and the MER-based projection model using 20 iterations (12 subsets per iteration). The pGTM partial volume compensation method was applied. The true organ VOIs were used, except for the small intestine where the VOI was modified to be a single 3D surface spanning the entire organ rather than following the detailed shape of the intestines, which could not be defined in practice. Five sample coronal slices from the 144 h images are shown in Figure 7.55. The means and standard deviations of organ activity estimates were computed and are shown in Table 7.12. The images are quite noisy and likely not suitable for visual interpretation. However, the errors in the activity concentrations are all better than 13% for all timepoints and organs.

7.10 Summary and Conclusions

This chapter has described the instrumentation of SPECT imaging systems and how these give rise to degradations in reconstructed SPECT images in the context of targeted radionuclide therapy. It has discussed considerations in selecting acquisition parameters needed to obtain reliable quantitative images. The SPECT images are degraded by factors including attenuation, scatter, the spatially varying collimator-detector response, and quantum noise. Limited resolution in the images gives rise to partial volume effects. All these effects can affect accuracy and precision of organ activity estimates obtained from SPECT and used in dosimetry calculations. Methods compensating for these effects by modeling them in statistical iterative reconstruction methods were described. Examples of applying these methods to phantom and realistic simulated data were presented to demonstrate the achievable accuracies and precisions. With careful attention to detail, errors of less than 5% for large organs are possible. With current technology, the reliability of activity estimates in objects less than the system spatial resolution is limited.

Acknowledgments

Many of the results and methods shown in this chapter were generated by and in collaboration with students, postdocs, colleages, and collaborators. This includes Dan Kadrmas, Wen-tung Wang, Xiyun Song, Yong Du, Bin He, Na Song, Lishui Cheng, Xing Rong, Jianting Yue, Michael Ghaly, Junyu Chen, Nadège Anizan, Thibault Mauxion, Esther Vicente, and Abhinav K. Jha. I also gratefully acknowledge years of mentoring and guidance by and collaboration with Benjamin M. W. Tsui, Grant T. Gullberg, Ronald J. Jaszczak, Michael Ljungberg, Freek Beekman, George Sgouros, and Harrison H. Barrett.

The work has been supported by numerous grants, including ones form the Whitaker Foundation and NHLBI, NCI, and NIBIB. Currently active grant support includes R44-CA213782, R01-CA240779, R01-EB031023, and U01-EB031798. The opinions and results expressed here do not necessarily express those of the funding agencies mentioned.

Disclosures

The reconstruction methods used to generate many of the results here are licensed by Johns Hopkins University to GE Healthcare and Rapid, LLC. The author receives a part of the royalties. These conflicts are being managed by Johns Hopkins under its Financial Conflict of Interest Policy. The author is a founder and part-time CSO of Rapid, LLC.

References

IAEA (2014). *Quantitative Nuclear Medicine Imaging: Concepts, Requirements and Methods.* International Atomic Energy Agency.

Bailey, D. L. and IAEA (2014). *Nuclear Medicine Physics: A Handbook for Teachers and Students.* International Atomic Energy Agency.

Cheng, L., R. F. Hobbs, P. W. Segars, G. Sgouros, and E. C. Frey (2013). "Improved dose-volume histogram estimates for radiopharmaceutical therapy by optimizing quantitative SPECT reconstruction parameters." *Phys. Med. Biol.* 58(11):3631–47.

Cherry, S. R., J. A. Sorenson, and M. E. Phelps (2012). *Physics in Nuclear Medicine E-Book.* Elsevier Health Sciences.

Dewaraja, Y. K., E. C. Frey, G. Sgouros, A. B. Brill, P. Roberson, P. B. Zanzonico and M. Ljungberg (2012). "MIRD pamphlet No. 23: quantitative SPECT for patient-specific 3-dimensional dosimetry in internal radionuclide therapy." *J. Nucl. Med.* 53(8):1310–25.

He, B., Y. Du, W. P. Segars, R. L. Wahl, G. Sgouros, H. Jacene, and E. C. Frey (2009). "Evaluation of quantitative imaging methods for organ activity and residence time estimation using a population of phantoms having realistic variations in anatomy and uptake." *Med. Phys.* 36(2):612–19.

He, B., Y. Du, X. Y. Song, W. P. Segars, and E. C. Frey (2005). "A Monte Carlo and physical phantom evaluation of quantitative In-111SPECT." *Phys. Med. Biol.* 50(17):4169–85.

He, B. and E. C. Frey (2006). "Comparison of conventional, model-based quantitative planar, and quantitative SPECT image processing methods for organ activity estimation using In-111 agents." *Phys. Med. Biol.* 51(16):3967–81.

He, B., R. L. Wahl, Y. Du, G. Sgouros, H. Jacene, I. Flinn, and E. C. Frey (2008). "Comparison of residence time estimation methods for radioimmunotherapy dosimetry and treatment planning—Monte Carlo simulation studies." *IEEE Trans. Med. Imaging* 27(4):521–30.

He, B., R. L. Wahl, G. Sgouros, Y. Du, H. Jacene, W. R. Kasecamp, I. Flinn, R. J. Hammes, J. Bianco, B. Kahl, and E. C. Frey (2009). "Comparison of organ residence time estimation methods for radioimmunotherapy dosimetry and treatment planning—patient studies. *Med. Phys.* 36(5):1595–601.

Song, N., Y. Du, B. He, and E. C. Frey (2011). "Development and evaluation of a model-based downscatter compensation method for quantitative I-131 SPECT." *Med. Phys.* 38(6):3193–204.

Chapter 8

PET/CT Quantitation and Uncertainties

Denis E. Bergeron, Ph.D.[1] and Brian E. Zimmerman, Ph.D.[1]

[1]Radiation Physics Division
National Institute of Standards and Technology
Gaithersburg, MD

8.1 Introduction

Positron emission tomography (PET) leverages the conservation of momentum in electron-positron annihilation to produce medical images with exceptional resolution. The imaging radiopharmaceuticals used in PET are based on radionuclides like ^{18}F that decay by positron emission. Depending on their initial energy, positrons travel up to a few millimeters in the body before colliding with an electron and annihilating. The 1.22 MeV of annihilation energy is usually shared by two annihilation photons, which travel in opposite directions, conserving momentum. In PET imaging, a patient is placed in a ring of detectors sensitive to the 511 keV annihilation photons (Figure 8.1); when coincident events are registered in detector blocks on opposite sides of the ring, a "line of response" can be drawn, upon which the annihilation event must have occurred. A PET image is a reconstruction of many millions of such lines.

With the development of biomolecular tracers like ^{18}F-fluorodeoxyglucose (^{18}F[FDG]), PET has become an important quantitative tool in the diagnosis and treatment of disease. By monitoring key metabolic indicators, informed assessments of status and progression of disease can be made, leading to more effective individualized treatment. The key to unlocking the potential of quantitative PET lies in two important factors:

1. the ability to faithfully represent the amount of radioactivity in a defined volume in an image in a way that is directly related to the administered activity, and
2. the ability to achieve reproducible performance characteristics for quantitation.

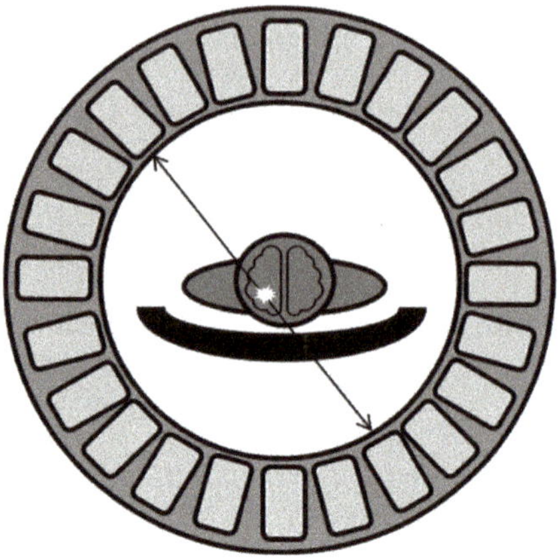

Figure 8.1 When a positron annihilates with an electron, two 511 keV photons are created, traveling in opposite directions. A PET scanner is a ring of detectors sensitive to these photons. When coincident events are registered in two detectors, a line of response upon which the annihilation event occurred is drawn. The PET image is reconstructed from many millions of these lines.

Simply said, the quantitative power of PET is realized only when there is a direct, defined, and reproducible link between the amount of radioactivity that is injected into the patient and the amount of radioactivity represented by the image. Since patient imaging data may be collected over relatively long time periods using different scanners—and sometimes even in different clinics—the only way to ensure that the above conditions are valid is through calibration of the scanners and radionuclide calibrators ("dose calibrators") to a common standard for radioactivity.

Calibration to a common standard is the foundation of good measurement, but it is only one component of the process of producing meaningful results in quantitative imaging. Without some expression of uncertainty, measurement results, including those from imaging studies, cannot be properly interpreted or compared. As an example of the role that uncertainty plays in a clinical context, consider the criteria set forth in the guidance document "Fluorodeoxyglucose (FDG) PET-CT as an Imaging Biomarker Measuring Response to Cancer Therapy" (Kinahan et al. 2020), developed by the Quantitative Imaging Biomarker Alliance of the Radiological Society of North America. This guidance claims that, under the constraints established in the protocol, FDG PET-CT can be used to parameterize tumor glycolysis activity (as maximum *standardized uptake value*, SUV_{max}) with a coefficient of variance of 10% to 12% for repeated measurements in the same patient. Further, the claim is that a change in the SUV_{max} value of ≥+39% or ≤–28% indicates a true change in tumor activity with 95% confidence. It should be noted that the uncertainty on a single SUV_{max} measurement was not considered.

The methodology used to derive the claim was limited by the quality of the available data, which used different metrics and different protocols to obtain the measured values. Given this, a value of 10% to 12% for an estimate of the reproducibility of SUV_{max} can be considered a reasonable conservative value. However, it is also possible that a robust assessment of the uncertainty could reveal that the magnitude is smaller, which could, in turn, support confidence in a smaller observed change. In any case, a rigorous uncertainty assessment brings with it the benefit of providing insight into what measurement variables can be optimized to make the uncertainty smaller.

This chapter deals broadly with the metrological concepts surrounding measurement as applied to PET imaging.

8.2 PET Imaging as Measurement

From the perspective of measurement science, quantitative PET imaging is a counting process. Events registered in pairs of detectors as outlined above and illustrated in Figure 8.1 are subjected to certain constraints (direction, energy, timing) and then counted. The resulting list of valid events is the raw imaging data, to which a set of corrections—including for individual detector efficiency, decay during acquisition, and decay to a reference time—is applied. The data are further processed with a reconstruction algorithm that includes corrections for photon attenuation and scattering based on the attenuation map derived from x-ray *computed tomography* (CT). The result is a series of two-dimensional images or "slices" that are stacked to give a three-dimensional image. After the corrections, the number of decay events attributed to specific volume elements (i.e., voxels) in the image can be counted. Dividing by the total acquisition time gives the activity[1] in that voxel.

The language of metrology is codified the International Vocabulary of Metrology (VIM; JCGM/WG2 2008) and the Guide to Uncertainty in Measurement (GUM; JCGM/WG 1 2008). From a formal perspective, the goal of any type of measurement is to provide an estimate of the value of a *measurand*, which in a quantitative imaging study is the activity (in Bq) in each voxel or the activity concentration (in Bq/mL) in a collection of voxels defining a *volume of interest* (VOI). To do this, we define a *measurement model*, which is a mathematical relationship between the quantities (observed and unobserved) in a measurement. For quantitative PET imaging, the typical measurement model can be simply expressed as:

$$A_0 = \frac{N_T G(T) D(T) B f_S f_A f_i \cdots}{T_{\text{acq}}} K, \tag{8.1}$$

where A_0 = the activity of the radionuclide in a given voxel at the beginning of acquisition;

N_T = the number of counts observed in the voxel during time T;

$G(T)$ = the decay correction to the beginning of image acquisition;

$D(T)$ = the correction for decay during acquisition;

B = the correction for positron branching and is specific to each radionuclide;

f_S, f_A, and $f_i \ldots$ = correction factors for scattering, attenuation, and other factors, respectively; and

K = the conversion factor that essentially represents the detection efficiency.

1. The SI unit for activity of a radionuclide is the becquerel (Bq), defined as decays per second.

In a PET scanner, each detector module has slightly different operating characteristics, so the *K* value is generally determined for each voxel in the sensitive volume as part of the system calibration.

The scattering corrections (f_S) are generally determined during acceptance testing and sometimes after major repairs using a specially designed phantom that is prepared with a known amount of ^{18}F (IAEA 2009). Attenuation corrections (f_A) are applied using x-ray CT images in conjunction with a simplified model based on assumed tissue type (and therefore density) and relatively simple linear extrapolations to account for differences in photon energies between the x-rays from the CT and the annihilation radiation (see Kinahan et al. 1998).

Of the variables that appear in Equation (8.1), the *K* factor is the main quantity that is regularly measured as part of the system calibration. The other factors are usually established by the manufacturer or are determined during acceptance testing and rarely change. *K* is the quantity that is most likely to change over time and is the focus of traceable scanner calibration. The calibration process generally involves the measurement of a uniform cylindrical phantom containing a known quantity of a solution of ^{18}F. Ideally, the activity of the ^{18}F solution been assayed using a radionuclide calibrator that has been calibrated with traceability to a national standard. In order to ensure that the traceability chain is maintained when preparing the phantom, it is necessary that the measured activity is *quantitatively* transferred to the phantom. That is, the amount of activity transferred from the container in which the measurement was performed needs to be known and an uncertainty on the amount transferred needs to be estimable. This can be achieved through several methods:

1. ***Complete transfer of the material from the measurement container to the phantom***. An example of this would be the case where the ^{18}F is first measured in a syringe (using the appropriate calibrator setting for the specific syringe type and filling volume). Then the contents of the syringe are transferred to the phantom, followed by several rinses of the syringe (by aspirating water into the syringe) that are also transferred to the phantom. It is assumed that after several rinses, nearly all of the ^{18}F has been transferred to the phantom. This should then be verified by a final measurement of the empty syringe in the radionuclide calibrator that indicates no residual activity. In cases where a source with traceable activity is received from the radiopharmacy, complete transfer is usually the preferred method for phantom preparation.

2. ***Transfer of the ^{18}F by mass into the phantom***. This method can provide the most accurate and traceable determination of the amount of ^{18}F in the phantom, but requires that the value of the massic activity (in Bq g^{-1}) of ^{18}F being transferred is known and traceable to a national standard. This is, in fact, the quantity that is typically certified by NIST when calibrating radioactive solutions, but it is rarely provided by radiopharmacies. This method also requires that the amount of solution transferred into the phantom be measured by mass on a balance or scale that has a traceable calibration and suitable precision. This mass determination can be made by weighing the phantom before

and after transfer, but since the mass of a full phantom may be much larger than the mass of ^{18}F solution added, it may be difficult to achieve sufficient precision. Alternatively, the mass determination can be made by weighing the transfer container (usually a syringe) before and after transfer. Regardless of which approach is used, the mass difference is equal to the mass of ^{18}F solution added, and the amount of added activity is the massic activity multiplied by the transferred mass.

3. ***Transfer of the ^{18}F by volume***. This method is considered the most difficult in terms of maintaining traceability because:

    - The measurement of the volume needs to be done in a traceable way. Syringe volume markings are not intended for precise measurements and should never be used as part of a quantitative phantom preparation. Typical laboratory micropipettes are more appropriate, provided they are traceably calibrated for volume.
    - Primary standards of radioactivity are rarely certified by national metrology institutes on the basis of volume because the value would be dependent on, among other factors, temperature.
    - In the event that a solution of ^{18}F is available only with certified massic activity, attempting to calculate the amount of activity transferred to the phantom using volumetric means would require that the value of the solution density is traceable, which is highly unlikely.

Once the phantom has been prepared with a traceable quantity of activity, a calibration protocol that is provided by the scanner manufacturer is used to determine the K factor. In general, PET scanners operate in a "black box" mode, with the details of operation (many of which are implemented with proprietary methods) hidden from the user. This is true for most aspects of the calibration protocols, but there are notable exceptions. The filling of the calibration phantom and provision of its activity (*vide supra*) are left to the user. The user must also define the *region of interest* (ROI). To avoid partial volume effects, the ROI is drawn far from the physical edges of the phantom—this is in stark contrast to the usual approach to ROI definition (i.e., segmentation) in clinical practice.

The calibration protocol typically spans several half-lives of ^{18}F, thereby allowing the activity to decay from a relatively high level to essentially background, and the dependence of K on other factors, such as system deadtime, can be determined. Corrections for deadtime can either be determined separately as a function of count rate or can be folded into K, thereby making it dependent on the count rate. Generally, such corrections for decay, deadtime, scattering, and more are hidden from the user, making it difficult to estimate associated uncertainties. Also hidden from the user is the application of the term B from Equation (8.1), which allows the K determined with ^{18}F to be applied to PET images acquired with other positron-emitting radionuclides. When the calibration is performed correctly, the uncertainty on K is expected to be dominated by the uncertainty on the activity in the phantom, which can be estimated by the user. The next section addresses the estimation of uncertainties in quantitative PET imaging.

8.3 Uncertainty and Bias

8.3.1 Definitions

Two important concepts necessary for judging the quality of a measurement result are its *uncertainty* and its *bias*. In simplified terms, the uncertainty refers to the amount of dispersion around a central measurement value (typically the mean). This is generally expressed as a standard deviation of a set of values or, in the case where a set of observed variables is used to derive an estimate, of the measurand, such as in Equation (1), the quadratic combination of the uncertainties on the individual variables. In the case where correlations between the uncertainty components are considered unimportant, an example of such a calculation using Equation (8.1) as the measurement model would be:

$$\frac{u_{A_0}}{A_0} = \sqrt{\left(\frac{{u_{N_T}}^2}{{N_T}^2}\right) + \left(\frac{{u_G}^2}{G^2}\right) + \left(\frac{{u_D}^2}{D^2}\right) + \left(\frac{{u_B}^2}{B^2}\right) + \left(\frac{{u_{f_S}}^2}{{f_S}^2}\right) + \left(\frac{{u_{f_A}}^2}{{f_A}^2}\right) + \left(\frac{{u_K}^2}{K^2}\right)\left(\frac{{u_{T_{acq}}}^2}{{T_{acq}}^2}\right) + \left(\frac{{u_{f_i}}^2}{f_i^2}\right) + \ldots} \tag{8.2}$$

Here, the variables are the same as those defined in Equation (8.1), and the u_x are the uncertainties on those variables. In practical applications, it is often impossible to treat all of the variables as mutually independent, so it is necessary to account for correlations. The prescriptions of the GUM (JCGM/WG1 2008) regarding uncertainty correlations are demonstrated nicely for image quantitation by Gear et al. (2018).

The presence of measurement *bias* is an indication of directional offset in the measurement result relative to the true value of the measurand. In the case of a radioactivity measurement, this is typically defined by the difference between that measurement result and the certified value of the standard used; bias is the result of uncorrected systematic effects. An example of this would be systematic low activity concentrations in a small volume due to uncorrected partial volume effects or systematic over- or underestimation of volumes during segmentation.

In cases where the desired measurand is the activity concentration (C_A) in a VOI, a term for the volume of the VOI (V) must be added to Equation (8.1) and the activities of the voxels within the VOI summed:

$$C_A = \frac{\sum A_{0,i}}{V}. \tag{8.3}$$

The uncertainty on the defined volume can be a significant contributor to the combined uncertainty on C_A, as has been demonstrated in studies of inter-operator variability in segmentation volumes (Krak et al. 2005; Fahey et al. 2010). Different approaches to estimating this uncertainty, both analytical and based on interobserver studies, have been discussed, with the appropriate method depending strongly on the approach to image segmentation and the available knowledge regarding that approach (Gear et al. 2018).

8.3.2 Assessing Uncertainty and Bias

It is possible to evaluate many of the common uncertainty components encountered in quantitative PET imaging through measurement, simulation, or informed estimation based on previous experience, expert knowledge, published values for similar measurement systems, etc. (Boellaard et al. 2008; Gear et al. 2018; Kinahan et al. 2020). However, the proprietary nature of most imaging systems and lack of user access to all of the operating characteristics of the scanner makes the evaluation of many components difficult. Of particular note is the uncertainty introduced by different implementations of the reconstruction process. A full accounting of all uncertainty components and sources of bias is not necessary for the typical clinical application. However, in any application where two or more measurements are to be compared, some knowledge about the realistic magnitude of the uncertainties is essential.

Several studies have been carried out to characterize uncertainty components in PET imaging (Boellaard et al. 2008; Boellaard 2009; Kinahan et al. 2020). Most of these focused on assessing interlaboratory variance, reproducibility (both short- and long-term), and repeatability. In several cases, some needed corrections—such as accounting for radioactive decay between the injection and the reference time for the imaging study (usually the start of acquisition, but may vary between manufacturers)—were identified and bias was reduced (Boellaard 2009; Lockart et al. 2011; AAPM 2012).

In clinical practice, the most significant source of bias is often from *partial volume effects* (PVE). As a result of image blurring ("spill out" of counts from the true object boundaries) and the mismatch of the PET voxel grid with the true object boundaries, counts from small objects are lost (Soret et al. 2007; Erlandsson et al. 2012). Numerous approaches to correcting this bias have been described, but most commonly, phantom studies are used to establish size-dependent (e.g, ROI diameter) fractional recovery coefficients that can be applied to subsequent images as a correction to the count rate (Erlandsson et al. 2012; Mille 2013; Dewaraja et al. 2013). In terms of Equation (8.1), the inverse of the fractional recovery coefficient would be included as an "f_i" term. The uncertainty associated with the PVE correction is typically determined in the phantom study, where the recovery coefficients are established and can be propagated according to Equation (8.2).

Table 8.1 presents some of the major uncertainty components encountered in quantitative PET imaging. Uncertainties related to lack of harmonization in patient preparation can be very significant (Boellaard et al. 2008; Boellaard 2009) but are not discussed here; nor are uncertainty components related to the "black box" components of a scanner system (attenuation correction, scattering correction, reconstruction algorithm and associated variables, and individual detector efficiency effects) which are extremely difficult to assess because of lack of access to the necessary information or control over the acquisition and analysis process. Only those factors that are primarily instrumental in nature and that are directly related to the measurement process are included. In most cases, the magnitude of the relative uncertainties

Table 8.1 A summary of some of the major instrument- and counting-related uncertainty components encountered in quantitative PET imaging

Uncertainty Component	Description	Magnitude (%)	Comments
Decay to a common reference time	Uncertainty in decay corrections due to uncertainty in time of day, counting live time, and radionuclide half-life	0.05% to 0.15%	
Poisson counting error	Uncertainty on activity measurement due to counting statistics or "Poisson counting error"	0.01% to 0.1%	Counting time and count rate dependent
Positron branching ratio	Uncertainty in activity due to uncertainty in positron branching ratio	0.2% (for ^{18}F)	
K factor/ calibration	Uncertainty in calibration factor due to uncertainty of activity in calibration phantom	0.4% to 3%	Directly related to precision of measured activity in the phantom
K factor/ calibration reproducibility	Uncertainty in scaling factor after recalibration; calibration and verification both performed with traceable sources	3% to 4%	Can be compensated for using calibrated long-lived phantom (see Zimmerman et al. 2018)
Measurement repeatability	Variability in recovered activity for successive scans over short time frame	0.5% to 2%	Partially dependent on count rate
Longitudinal homogeneity	Variability in recovered activity for same ROI propagated through axial slices of the phantom	0.5% to 1.5%	Partially dependent on count rate
ROI uniformity	Variability in recovered activity across all pixel values in a large (5 cm) circular ROI within a single slice in a 20 cm diameter phantom	0.1% to 3%	Highly dependent on count rate and total number of counts

given in Table 8.1 were determined by measuring absolute activity concentrations in a uniform phantom under controlled conditions at NIST (Zimmerman et al. 2013; Zimmerman et al. 2018). Under similar conditions, the relative magnitudes should be translatable to measurements of SUV, but these should be considered "best case" values. For a given clinical measurement, values for some of these uncertainty components may differ from those in Table 8.1 by orders of magnitude (Boellaard 2009), which underscores the importance of understanding the calibration and measurement process in order to minimize bias and uncertainty in individual quantitative PET imaging studies. With varied clinical applications and rapidly changing technologies, the future of quantitative PET imaging is one of promise—and challenge.

Acknowledgments

We are grateful to our NIST colleagues, Drs. Brittany Broder and Ryan Fitzgerald, for fruitful discussions and key comments and suggestions that improved this chapter.

References

American Association of Physicists in Medicine (AAPM) (2012). "The Selection, Use, Calibration, and Quality Assurance of Radionuclide Calibrators Used in Nuclear Medicine." College Park: AAPM, Report of AAPM Task Group 181.

Boellaard, R., W. J. G. Oyen, C. J. Hoekstra, O. S. Hoekstra, E. P. Visser, A. T. Willemsen, B. Arends, F. J. Verzijlbergen, J. Zijlstra, A. M. Paans, E. F. I. Comans, and J. Pruim (2008). "The Netherlands protocol for standardization and quantification of FDG whole body PET studies in multi-centre trials." *Eur. J. Nucl. Med. Mol. Imaging* 35:2320–33.

Boellaard, R. (2009). "Standards for PET Image Acquisition and Quantitative Data Analysis." *J. Nucl. Med.* 50:11S–20S.

Dewaraja, Y. K., M. Ljungberg, A. J. Green, P. B. Zanzonico, and E. C. Frey (2013). "SNMMI MIRD Committee MIRD pamphlet no. 24: guidelines for quantitative I-131 SPECT in dosimetry applications." *J. Nucl. Med.* 54:2182–88.

Erlandsson, K., I. Buvat, P. H. Pretorius, B. A. Thomas, and B. F. Hutton (2012). "A review of partial volume correction techniques for emission tomography and their applications in neurology, cardiology, and oncology." *Phys. Med. Biol.* 57:R119–59.

Fahey, F. H., P. E. Kinahan, R. K. Doot, M. Kocak, H. Thurston, and T. Y. Poussaint (2010). "Variability in PET quantitation within a multicenter consortium." *Med. Phys.* 37:3660–66. https://doi.org/10.1118/1.3455705.

Gear, J. I., M. G. Cox, J. Gustafsson, K. S. Gleisner, I. Murray, G. Glatting, M. Konijnenberg, and G. D. Flux (2018). "EANM practical guidance on uncertainty analysis for molecular radiotherapy absorbed dose calculations." *Eur. J. Nucl. Med. Mol. Imaging* 45:2456–74.

IAEA. *IAEA Human Health Series No. 1: Quality Assurance for PET and PET/CT Systems.* Vienna: International Atomic Energy Agency, 2009.

Joint Committee for Guides in Metrology Working Group 1 (JCGM/WG 1) (2008). *Evaluation of measurement data—Guide to the expression of uncertainty in measurement.* Sevres: BIPM, JCGM 100-2008.

Joint Committee for Guides in Metrology Working Group 2 (JCGM/WG 2) (2012). *International vocabulary of metrology—Basic and general concepts and associated terms (VIM)*, 3rd ed. Sevres: BIPM, JCGM 200-2012.

Kinahan, P. E., D. W. Townsend, T. Beyer, and D. Sashin (1998). "Attenuation correction for a combined 3D PET/CT scanner." *Med. Phys.* 25:2046–53.

Kinahan, P. E., E. S. Perlman, J. J. Sunderland, R. Subramaniam, S. D. Wollenweber, T. G. Turkington, M. A. Lodge, R. Boellaard, N. A. Obuchowski, and R. L. Wahl (2020). "The QIBA Profile for FDG PET/CT as an Imaging Biomarker Measuring Response to Cancer Therapy." *Radiology* 294:647–57.

Krak, N. C., R. Boellaard, O. S. Hoekstra, J. W. R. Twisk, C. J. Hoekstra, and A. A. Lammertsma (2005). "Effects of ROI Definition and Reconstruction Method on Quantitative Outcome and Applicability in a Response Monitoring Trial." *Eur. J. Nucl. Med. Mol. Imaging* 32:294–301.

Lockhart, C. M., L. R. MacDonald, A. M. Alessio, W. A. McDougald, R. K. Doot, and P. E. Kinahan (2011). "Quantifying and reducing effect of calibration error on variability of PET/CT standardized uptake value measurements." *J. Nucl. Med.* 52:218–24.

Mille, M. M. *A Study of shape-dependent partial volume corrections in PET imaging using ellipsoidal phantoms fabricated via rapid prototyping*. Ph.D. Thesis. Rensselaer Polytechnic Institute, UMI 3613294, Ann Arbor, MI: ProQuest, 2013.

Soret, M., S. L. Bacharach, and I. Buvat (2007). "Partial-Volume Effect in PET Tumor Imaging." *J. Nucl. Med.* 48:932–45.

Zimmerman, B. E., L. Pibida, L. E. King, D. E. Bergeron, J. T. Cessna, and M. M. Mille (2013). "Development of a calibration methodology for large-volume, solid ^{68}Ge phantoms for traceable measurements in positron emission tomography." *Appl. Radiat. Isot.* 87:5–9.

Zimmerman, B. E., D. E. Bergeron, and J. P. Edgerton (2018). "Assessing the absolute quantitative accuracy of Positron Emission Tomography for Cu-64 using traceable calibrated phantoms." *Appl. Radiat. Isot.* 134:68–73. https://doi.org/10.1016/j.apradiso.2017.09.016.

Chapter 9

Calibration for Quantitative SPECT/CT

Eric C. Frey, Ph.D., FSNMMI

Professor, Division of Radiological Physics
Russell H. Morgan Department of Radiology and Radiological Science
Baltimore, MD
and Founder and CEO
Rapid, LLC

9.1 The Need for Calibration

On many SPECT systems, images are output without units. There are a number of reasons for this, including the fact that SPECT images have primarily been used for visual interpretation, and because early SPECT reconstruction methods were not quantitative. For use as inputs to dosimetry, it is necessary to have organ or voxel activities in units of activity or activity concentration. Thus, there is a need to convert voxel values in reconstructed images to activities. Therefore, the most fundamental reason that calibration is needed is to do this conversion.

A second important reason is that SPECT reconstruction and compensation methods, even when designed for quantitative applications, may incompletely compensate for physical image-degrading effects or characteristics of the scanner. For example, images reconstructed with attenuation, patient scatter, and resolution recovery often do not model the sensitivity of the scanner. That is, the reconstructed image needs to be divided by the product of the sensitivity of the scanner (counts per time per activity) and the acquisition duration to produce an image in units of activity per voxel. Dividing this by the voxel volume will then give an activity concentration image.

Calibration is also needed when the reconstruction does not model a physical image-degrading effect. For example, resolution recovery sometimes does not model septal penetration and scatter. If the geometric sensitivity is used to convert to activity, then there will be a bias in activity estimates that can be improved by a calibration procedure.

Calibration is sometimes needed to measure a physical characteristic. For example, acquisitions with high count rates can suffer losses due to the finite resolving time of the gamma camera. In these cases, a calibration experiment is needed to measure the resolving time. Measuring the intrinsic energy or spatial resolution, the projection bin size, the projected position of the center of rotation, and alignment of the CT and SPECT scanners are other examples of calibrations that are needed to achieve good quantification.

Finally, calibration is needed when compensation for image-degrading factors is imperfect due to approximations in the compensation method or imperfect knowledge of scanner characteristics. For example, the model of the *collimator-detector response function* (CDRF) may be based on a manufacturer's specified collimator parameters (hole diameter, length, and septal thickness) or camera characteristics (intrinsic spatial and energy resolution). These will, in general, vary on different instances of a given camera model.

This chapter discusses calibration procedures needed for quantification that are not typically part of manufacturer quality control processes. For example, camera uniformity, center of rotation, and SPECT/CT registration are not discussed. Measurement of detector intrinsic spatial and energy resolution are also not discussed as they have relatively minor effects on compensation methods, and published manufacturer specifications are generally adequate. The next section discusses several kinds of calibrations and how they can be performed. There is the potential that calibrations can change over time and from scanner to scanner. Some data on this is presented. Finally, a key element in calibrations is filling phantoms. Some general recommendations are given for filling phantoms in order to obtain reproducible and accurate calibration results.

Table 9.1 Total sensitivity values from various systems and collimators

System	Collimator	Radionuclide	Crystal Thickness	Sensitivity (cps/MBq)	Sensitivity (cpm/μCi)
GE Discovery NM 670	LEUHR	Tc-99m	3/8"	37.4	83
GE Discovery NM 670	LEHR	Tc-99m	3/8"	72.1	160
GE Discovery NM 670	ELEGP	Tc-99m	3/8"	144.1	320
GE Discovery NM 670	MEGP	Ga-67	3/8"	64.9	144
GE Discovery NM 670	HEGP	I-131	3/8"	43.7	97
GE Discovery NM 670	LEHR	Tc-99m	5/8"	74.3	165
GE Discovery NM 670	ELEGP	Tc-99m	5/8"	148.6	330
GE Discovery NM 670	MEGP	Ga-67	5/8"	67.6	150
GE Discovery NM 670	HEGP	I-131	5/8"	74.3	165
Siemens Symbia	LEHS	Tc-99m	3/8"	459.5	1020
Siemens Symbia	LEAP	Tc-99m	3/8"	148.6	330
Siemens Symbia	LEHR	Tc-99m	3/8"	91.0	202
Siemens Symbia	LEUHR	Tc-99m	3/8"	45.0	100
Siemens Symbia	MEGP	Ga-67	3/8"	123.9	275
Siemens Symbia	HEGP	I-131	3/8"	60.8	135

Sensitivity values are from manufacturer specification sheets. Values specified in cpm/μCi were used and converted to cps/MBq.

9.2 Types of Phantom Measurements for Quantitative Imaging

9.2.1 Sensitivity Measurement

9.2.1.1 Definition of Sensitivity

The sensitivity of a SPECT system is a measure of the efficiency of the scanner. It is typically specified in units of counts per time per activity, e.g., cps/MBq in SI units or cpm/μCi in traditional units. Sensitivity depends on the radionuclide, collimator, crystal thickness, and energy resolution. Values of system sensitivity are given for a range of collimators for Siemens and GE systems in Table 9.1. These sensitivities are from manufacturer specification sheets and were thus measured based on NEMA standards. This includes using a disk-shaped solution containing the radionuclide, the recommended energy windows for the indicated radionuclide, and summing all the counts inside the field of view.

9.2.1.2 Geometric and Total Sensitivity

Two kinds of sensitivity are relevant for converting quantitative SPECT voxel values to activity: the *total sensitivity* and the *geometric sensitivity.*

- ***The total sensitivity*** is the sum of all counts, including septal penetration and scatter photons.

- ***The geometric sensitivity*** includes only counts resulting from photons that passed through collimator holes.

If the reconstructed image does not compensate for the full CDRF (including septal penetration and scatter), then the total sensitivity should be used to convert to activity; if the reconstructed images are compensated for septal penetration and scatter, then the geometric sensitivity should be used. One difficulty is that the geometric sensitivity cannot be measured directly as photons that penetrate the septa cannot be perfectly distinguished from geometrically collimated photons.

9.2.1.3 Source Geometries for Sensitivity Measurements

The NEMA standard specifies measuring the total sensitivity using a 15-cm-diameter Petri dish with 2–3 mm of activity-containing solution. This minimizes the effect of self-attenuation in the phantom and solution. However, managing the Petri dish, especially with long-lived radionuclides or ^{131}I, is a potential safety issue for regular clinical practice. An alternative is to use a small syringe containing the solution. One downside of this is that the potential of lateral position dependence for collimators with thick septa. Using a volume of activity such that the length is greater than several hole diameters plus septal thickness values can reduce the importance of this effect. Since the activity of the syringe can be measured directly, this method reduces the need to transfer activity and correct for residual.

As indicated, the geometric sensitivity cannot be measured directly when septal penetration is present. One alternative is to use radionuclides with lower-energy photons (e.g., ^{241}Am). This would require corrections for the effect of photon energy as well as the difference in photon yield. An alternative used by the author is to use Monte Carlo simulations to estimate the fraction of septal penetration that is included in a VOI of a given size and correct the measured result based on this.

Using the total sensitivity for reconstructions that do not include compensation for septal penetration and scatter also has some limitations. First, the ratio of the total to geometric photons varies as a function of distance. Also, the septal penetration and scatter photons contribute to the reconstructed image in a complicated and spatially varying way. Thus, for radionuclides where high-energy photons and septal penetration and scatter are a consideration, the use of a quantitative calibration phantom, as described below, is a better choice.

9.2.1.4 Sensitivity Measurements for Quality Control

Sensitivity measurements, especially using a syringe, are relatively quick and easy to make. As a result, they are useful to monitor the quantitative stability of the system. For example, imaging the same syringe of activity at each timepoint for a given patient study can be used to check for—and, if needed, correct for—variations in sensitivity over time and since the last calibration.

9.2.1.5 Sensitivity Measurement Phantom Preparation Acquisition Parameters

Care should be used in measuring the activity. The dial settings for activity meters (dose calibrators) for therapeutic radionuclides are sometimes not accurate. Ideally, they should be calibrated using sources traceable to national standards laboratories

Table 9.2 Percent of counts lost for different count rates for a paralyzable counting system

Maximum Count Rate (kcps)	Resolving Time (µs)	Fraction of Counts Lost (%) at True Count Rates of		
		20 kcps	10 kcps	5 kcps
200	1.84	3.90	1.90	0.90
300	1.23	2.50	1.20	0.62
400	0.92	1.90	0.90	0.46

(see Chapter 4). The size of the vial or syringe used should be consistent with the calibration, and consistent positioning of sources inside the instrument is important.

The same energy windows used for the SPECT acquisition should be used for the sensitivity measurement. The source should be positioned at a consistent distance from the face of the collimator. The NEMA measurement specifies a distance of 10 cm. There should be no additional attenuating medium between the source and the camera whose sensitivity is measured. Thus, when a Petri dish source is used, the sensitivity should be measured for only one camera at a time. For a syringe source, the syringe can be suspended between the cameras (e.g., by taping it to a meter stick) so that the support does not attenuate photons and so that it is at the desired distance from both cameras. In this case, the sensitivity can be measured for both cameras at once. In order to avoid errors due to attenuation, the source should not be placed on the camera bed, and data should not be acquired for the camera under the bed.

The activity and acquisition duration used for the measurement affects the precision and accuracy of the sensitivity measurement: the product of the activity and duration affects the total count rate, which determines the precision, while the activity affects the count rate, which affects accuracy due to count rate loss effects. The NEMA specification suggests count rates less than 20 kcps and at least 1 million total counts. One million total counts results in a *coefficient of variance* (COV) of the measurement due to counting statistics of $\sqrt{10^6} = 0.1\%$, which is a reasonable target. At 20 kcps, this means an acquisition duration of 50s. However, this count rate can result in appreciable fractional count rate losses. Table 9.2 shows the fraction loss of counts for paralyzable counting systems with different maximum count rates at measured count rates of 20, 10, and 5 kcps. Note that at 20 kcps 2% to 4% of counts are lost depending on the count rate performance of the system. Using a count rate of 5 kcps reduces the number of lost counts to under 1% in all cases. A count rate of 5 kcps requires an acquisition duration of 3.3 minutes. Note that since counts outside the photopeak energy windows contribute to count rate losses, 5 kcps is the total count rate over the entire energy spectrum.

The activity needed to achieve this count rate depends on the photon abundance for the radionuclide and the sensitivity of the collimator-detector combination. For example, consider the case of ^{177}Lu. Lutetium-177 has two photopeaks that are used for imaging with energies of 113 and 208 keV having abundances of 6.4 and 11 photons per 100 decays, respectively. If a medium-energy collimator is used on a GE Discovery 670 with a 9.525-cm-thick crystal, the sensitivity is 65 cps/MBq for ^{67}Ga. Assuming that 93, 185, and 300 keV photopeaks were used for the ^{67}Ga measurement, the ^{67}Ga photon abundance is about 77 photons per 100 decays. The activity needed to achieve a total count rate in both energy windows of 5 kcps is thus [(5000/65) × (77/17)] = 348 MBq. A lower activity could be used with a longer acquisition duration or lower total counts (and corresponding worse precision of the sensitivity).

Another consideration is background counts. The background count rate should be checked, and if it is not lower by a factor of 1000 than the count rate with the source, a background acquisition should be acquired and applied.

A final consideration is the *region of interest* (ROI) used to determine the counts. Even for the total sensitivity, it may be preferable to use an ROI smaller than the entire field-of-view, especially if a syringe source is used. This is because, for collimators well matched to the radionuclide under consideration, the septal penetration and scatter component will not extend far from the edge of the projected position of the source. The part of the ROI outside this will thus only contribute background counts. Therefore, using a smaller ROI can reduce the need for background correction. However, using a manually drawn ROI to measure sensitivity can result in increased variability due to subjectivity in defining the ROI in the sensitivity measurement.

9.2.2 SPECT Calibration Phantom

For therapeutic radionuclides, the use of an in-air sensitivity measurement is often suboptimal when converting SPECT voxel values to activity. This is for several reasons, including the difficulty of measuring the geometric sensitivity, spatial variance of the total sensitivity, incomplete knowledge of scanner parameters, and approximations in or limitations of compensation methods. Thus, the use of a calibration phantom is recommended. Note that the calibration factor obtained from this phantom is not intended to be used for partial volume compensation; considerations for measuring recovery coefficients are given below.

9.2.2.1 Phantom Selection

It is desirable that the calibration factor be as accurate as possible. This means that the bias in activity estimates obtained by using it for a wide range of patients and organs should be small. This might argue for using a phantom that mimics the activity distribution of interest. However, ease of preparing the phantom and analyzing the data are also important considerations. Taking these factors into account, use of a cylindrical phantom with either circular, elliptical, or body-shaped cross sections, a cold background, and a single compartment containing radioactivity is suggested. It makes sense to have the size of the phantom be similar in size to the region of the body of interest. Water-filled phantoms with spherical inserts are readily available and appropriate for use in this application. However, solid phantoms with other shaped inserts are also acceptable as they can ease phantom preparation and thus reduce the chance of errors when filling the phantom.

Spherical inserts for use with cylindrical phantoms are available from Data Spectrum Corporation (www.spect.com). Larger compartments are preferable, especially for use with general-purpose collimators. Spheres with diameters of a minimum 3 cm, and preferably 6 cm, should be used to reduce partial volume effects.

9.2.2.2 Phantom Preparation

To simplify quantification of activity in the sphere, it is recommended that the region surrounding the sphere have no activity in it and be filled with water. The activity in the sphere should be measured carefully (see below). See the section below on tips for filling calibration phantoms.

The principle for specifying the activity in the phantom is like that for the source in air: the count rate should be kept low enough to avoid count-rate losses. Since it is the total counts in the entire energy spectrum that determines the counts, scatter counts should be considered. An upper bound on the impact on the count rate is thus the count rate if the activity was imaged in-air, as was the case for the sensitivity measurement.

9.2.2.3 Acquisition Parameters

The same energy window, collimator, number of projections, etc. should be used as for the patient acquisitions. A body-contouring orbit is desirable to minimize partial volume effects. The acquisition duration can be different and should be selected to provide sufficiently low quantum noise. The total of the reconstructed voxel values (counts) is not, in general, Poisson distributed; this is especially true when compensations for scatter and attenuation are included. A rough approximation is to consider the total counts in projections to be Poisson distributed and that the COV for the sum of the voxel values to be one over the square root of the total counts.

9.2.2.4 Data Analysis

There are two approaches to calculating the calibration factor:

1. Use the ratio of the true activity to the measured total counts in the sphere.
2. Use the ratio of true activity concentration to the measured activity concentration.

The former requires careful definition of the sphere VOI, which the latter avoids. However, using activity concentration requires careful placement of the VOI due to the presence of noise spikes and ringing due to partial volume effects. The author considers it preferable to draw a VOI that is slightly larger than the sphere to minimize partial volume effects. This can be done objectively using the physical size of the sphere to define the physical shape of the object and dilating the resulting VOI by the number of voxels equal to the approximate FWHM resolution of the system at the center of the sphere. This approach relies on having no activity in the region surrounding the sphere.

9.2.3 Recovery Coefficient

A commonly used approach for partial volume compensation is to use *recovery coefficients* (RCs). For a full discussion of partial volume compensation methods, see Chapter 7 in this volume. In this approach, the ratio of the estimated divided by the true activity (i.e., the recovery coefficient) is measured for a set of spheres of known diameters. The purpose of this section is to describe experiments that can be used for this purpose.

9.2.3.1 Phantom Selection

Typically, a phantom containing six spheres of various sizes is used. The NEMA IEC PET Body Phantom (available from Data Spectrum or Capintec) has 6 spheres and a phantom that is approximately the cross section of a human torso. It can be used for RC measurements. The sphere diameters in cm are nominally 1.0, 1.3, 1.7, 2.2, 2.8, and 3.7. The smallest two spheres are likely too small for reliable activity estimation when using medium- and high-energy collimators. A large central cylinder is available (total volume ~0.26 l). For partial volume compensation, this cylinder would typically be removed or could be filled with a known activity and used for calibration. Another alternative is the Jaszczak cylindrical phantom. Either the circular cross section or the elliptical cross section can be used, though the circular cross section gives a more equal environment for the spheres. The phantom can be ordered without the rod inserts if these are not needed. A set of hollow spheres is sold that has internal diameters of 0.99, 1.24, 1.54, 1.98, 2.48, and 3.13 cm. A sphere with a diameter of 6 cm is also available and may be more relevant for medium- and high-energy collimators.

9.2.3.2 Phantom Preparation

There are two approaches to filling the phantom: using known activities or known activity concentrations. The former approach requires accurately measuring activity in each of the spheres; the latter approach requires a single activity measurement and a set of mass measurements (see the tips section below for details). It is generally easier and more accurate to measure masses, so the author generally prefers that approach. The spheres should generally be positioned so the distances between the sphere edges are similar for the different spheres, rather than simply in order of increasing size. Spinal tap or 4” blunt-tipped dispensing needles (available online) are useful for filling the NEMA IEC PET phantom as they are long enough to span the entire stem. Shorter needles can be used for the Data Spectrum sphere set. Care should be taken not to put activity in the stems of the spheres and, as much as possible, to avoid air bubbles. The volume can be easily measured by weighing the spheres before and after filling them. Generally, no background activity is used. However, the RC depends on the object-to-background activity ratio. Thus, if this is known for the objects of interest, using an appropriate background activity would be reasonable.

9.2.3.3 Acquisition Parameters

The same considerations for acquisition parameters apply as for the calibration phantom. The selection of body contouring or a circular orbit affects the image resolution and thus the RC, so specifying an orbit like the average expected for patients would be a reasonable choice. It is suggested that a diagnostic-quality CT be acquired in order to better visualize the sphere locations for data analysis.

9.2.3.4 Data Analysis

The VOI should be around the true physical volume of the sphere and should thus be defined based on a fused CT image. It is expected that some of the activity will be outside the VOI, and this is, in fact, what the RC measures. Some image analysis software allows specifying spherical VOIs with a given size. This can be done based on the expected diameter of the sphere (ideally estimated from the mass of the solution in the sphere). Positioning of the VOI is critical, especially for small spheres, and the center may not lie at the center of a voxel or on a voxel boundary. Positioning the VOI by small increments until it matches the position in a fused CT image is one effective strategy, assuming good registration of the CT and SPECT image. The total counts in each of the sphere VOIs is then divided by the corresponding true activity in the VOI.

9.3 Variability of Calibration Factors

9.3.1 Variability of Calibration over Time

One important consideration is how frequently to obtain the calibration factor. We have published two studies evaluating the stability of geometric sensitivity measurements over time. One study used data from ^{131}I Bexxar patient scans over a four-year period (Anizan, Wang et al. 2014). These involved wholebody scans at three different times. The same source was used for the three measurements for the same patient, but different sources were used for each patient. The sources were prepared in-house and measured before each acquisition. It was found that the COV of the calibration factor was 2.3% over this period. If the mean decay-corrected activity was used for each of the measurements from the three sources, the COV was reduced to 1.8%. An analysis of variance was performed and indicated that the largest source of variability was preparation of the different calibration sources, representing 77% of the overall variability. Based on the results of this study, great care should be taken in preparing and measuring the calibration source. Averaging multiple measurements of the source and carefully positioning the source are important.

A second study involved the use of ^{133}Ba sealed sources prepared by the United States National Institute of Standards and Technology (NIST) (Anizan, Wang et al. 2015). The use of a sealed source eliminates the variability associated with source preparation. We studied various factors, including the source positioning and ROI definition. This study demonstrated that the use of consistent ROI size is important and suggested an objective method for defining the VOI size based on size of the source d, the collimator hole diameter h and length L, the crystal thickness t, and the source-to-collimator distance S. The size of the VOI, D, in this method is given by:

$$D = d + 2\left(\frac{h(s + L + \frac{t}{2})}{l}\right). \tag{9.1}$$

The source-to-camera distance should be reproducible to within 0.5 cm. Using a cylindrical source with a diameter of 2.9 cm reduced the effect of lateral position on the calibration factor measurement.

The fact that the systems were stable over a longer period (four years) suggests doing infrequent, careful calibration. However, experience and the value of the data indicate that the calibration should be monitored more frequently, perhaps on any day that a patient scan is acquired. The monitoring can be done using a sensitivity measurement, even if calibration is done using a SPECT acquisition of a phantom. The results also emphasize the importance of reproducibly measuring the calibration factor. These data suggest that manufacturers develop automated procedures for this data analysis.

9.3.2 Variability of Calibration across Centers

There have been, to my knowledge, no publications of the variation of calibration factors over multiple centers for SPECT. This is an important problem for multi-center trials. Sensitivities for a group of 13 GE and 12 Siemens SPECT/CT systems were compiled from a set of multiple multi-center clinical trials. The measurements were performed using a syringe in-air using 20%-wide energy windows centered at 113 and 208 keV. Activities were measured at each site using their dose calibration procedure and syringes filled, measured, and imaged at the site by local personnel without external supervision. The sensitivities were calculated from the image data using activity measurement data from the sites by a centralized laboratory. Table 9.3 gives the coefficients of variations of the total sensitivities over all sites for both energy windows and scanner manufacturers. Several different models of scanners are included for each manufacturer. For GE there were 3 Infinia and 10 Discovery 670 scanners; for Siemens there were 11 Symbia scanners and 1 Intevo scanner. There was no obvious model-dependent variation in sensitivity for the same manufacturer. All the cameras had 9.525-cm-thick crystals, and cameras from each manufacturer used MEGP collimators with the same hole diameter and lengths. The collimator designs were different for the two manufacturers. The coefficients of variation for all the scanners for the same manufacturer and energy window are shown in Table 9.3 and are in the range 8% to 10%. This represents all sources of variation, and it is unclear to what extent it represents errors or variations in individual scanners. This is clearly an area where more attention is needed.

Table 9.3 Coefficients of variation in total sensitivities

	G.E.	Siemens
N	13	12
113 keV	9.4%	8.2%
208 keV	10.4%	8.0%

9.4 Tips for Filling Calibration Phantoms

9.4.1 Measuring Activity

Activities in nuclear medicine are measured using *dose calibrators*, more properly known as *activity meters*. These are ionization chambers that surround the source and have essentially no ability to measure energy. As a result, they are unable to distinguish particles based on energy. They are thus dependent on the positioning of the source in the chamber and the size and volume of the source being measured. It is thus highly desirable to measure using the same-sized container with the same volume of solution in the same position as was used to calibrate the instrument.

Activity meters typically have setting dials with values that are provided by manufacturers for different radionuclides. The dial settings should not, in general, be trusted, especially for therapeutic radionuclides. Ideally, they should be calibrated using sources traceable to standards agencies such as the NIST. Regular quality control should be performed to measure constancy, etc. Background should be checked prior to measuring activities and accounted for if large enough. This is especially important for small-activity sources. One alternative is to use a source from the radioactivity provider to calibrate the activity meter. However, these measurements are not always accurate.

It is very important to account for all sources of activity when filling phantoms. For example, always measure the residual activity in a syringe after using it to fill a phantom compartment. Be sure to include anything, including the needle or syringe cover, that may have activity on it. Record all these and subtract them from the activity placed in the phantom, correcting for decay if necessary.

9.4.2 Measuring Activity Concentration

It is generally easier and more accurate to measure masses than volumes. Electronic scales with various capacities and precisions are available. Checking the calibration (linearity and offset) using standard masses is advisable. Unless heated water ($T > 30$ °C) is used, measuring water temperature and accounting for density variation is not necessary. Measuring phantom compartments before and after filling them with water can be used to measure the mass of water inside a compartment. If this method is used, the compartments should be carefully prepared so that they can dry several days before the experiment.

It is sometimes desirable or convenient to fill a set of compartments (e.g., spheres in a recovery coefficient phantom) with the same activity concentration. One way to do this is to create a solution with a slightly larger volume of water than is needed to fill all the compartments. Again, measuring the volume via mass is often easier and more accurate. Accounting for the mass of the radioactive solution is important, so it should be added before the final mass measurement.

9.4.3 Filling Phantoms

Using water that is approximately at room temperature is desirable to avoid damage to phantoms due to thermal expansion or contraction. Letting water sit for a day prior to use is one way to achieve this, and this has the added effect of reducing absorbed gasses that can produce bubbles in the phantom. One way to do this is to fill the phantom background compartment most of the way full and fill other containers with enough water to fill the rest of the background and any other compartments used. Distilled water is often available in wet labs and should be used if available. This is especially important when using radioiodine due to the potential of oxidation by chlorine in tap water (Maguire 1980).

9.4.4 Buffer for Radioiodine

When performing experiments with iodine, it is important—for both accuracy and safety reasons—to prevent the formation of iodine gas. This can be done using an appropriate iodine solution. The idea is to keep the pH relatively high and to have a molar excess of cold iodine atoms greater by a factor of ~10^4 than the concentration of radioactive iodine. This prevents volatilization and adsorption on glass surfaces. The following is appropriate for a 200-MBq/g radioiodine solution, with distilled water called for as the diluent.

KI	0.5 mg/g solution
LiOH	0.2 mg/g solution
Na_2SO_3	0.3 mg/g solution

9.4.5 Chelator for Radiometals

Radiometals such as ^{90}Y and ^{111}In can adsorb onto the walls of plastic phantoms, resulting in a nonuniform distribution of activity in the compartments. This can be reduced by adding the activity to a solution containing the chelator ethylenediaminetetraacetate EDTA (Park, Mahmood et al. 2008). That study found that including ~700 EDTA molecules for each molecule of the radiometal eliminated adsorption. (Note that ringing associated with CDR compensation at the edges of objects with higher activities than the background can manifest as hot rims on spheres, depending on the sphere size, even for uniform activity distributions).

Acknowledgments

The results, data, and opinions presented here were generated while working with postdocs, colleagues, and collaborators. These include Dan Kadrmas, Wen-tung Wang, Xiyun Song, Yong Du, Bin He, Na Song, Lishui Cheng, Xing Rong, Jianting Yue, Michael Ghaly, Junyu Chen, Nadège Anizan, Thibault Mauxion, Esther Vicente, and Abhinav K. Jha. I also gratefully acknowledge years of mentoring and guidance by and collaboration with Benjamin M. W. Tsui, Grant T. Gullberg, Ronald J. Jaszczak, Michael Ljungberg, Freek Beekman, George Sgouros, and Harrison H. Barrett.

The work has been supported by numerous grants, including ones from the Whitaker Foundation and NHLBI, NCI, and NIBIB. Currently active grant support includes R44-CA213782, R01-CA240779, R01-EB031023, and U01-EB031798. The opinions and results expressed here do not necessarily express those of those funding agencies.

Disclosures

The reconstruction methods used to generate many of the results here are licensed by Johns Hopkins University to GE Healthcare and Rapid, LLC. The author receives a port of the royalties. These conflicts are being managed by Johns Hopkins under its Financial Conflict of Interest Policy.

References

Anizan, N., H. Wang, X. Zhou, R. Hobbs, R. Wahl, and E. Frey (2014). "Factors affecting the stability and repeatability of gamma camera calibration for quantitative imaging applications based on a retrospective review of clinical data." *EJNMMI Research* 4(1):67.

Anizan, N., H. Wang, X. C. Zhou, R. L. Wahl, and E. C. Frey (2015). "Factors affecting the repeatability of gamma camera calibration for quantitative imaging applications using a sealed source." *Phys. Med. Biol.* 60(3):1325–37.

Maguire, W. (1980). "A precaution for minimizing radiation exposure from iodine vaporization." *J. Nucl. Med. Technol.* 8(2):90–93.

Park, M.-A., A. Mahmood, R. E. Zimmerman, N. Limpa-Amara, G. M. Makrigiorgos and S. C. Moore (2008). "Adsorption of metallic radionuclides on plastic phantom walls." *Med. Phys.* 35(4):1606–10.

Chapter 10

Quality Assurance for SPECT/CT: Making the Transition from Diagnostic to Therapeutic Applications

Stephen A. Graves, Ph.D., DABR

Assistant Professor of Radiology
Division of Nuclear Medicine
University of Iowa
Iowa City, IA

10.1 Introduction

Single photon emission computed tomography (SPECT) has been developed and advanced over a period of approximately 60 years as of 2023. Notable improvements over this period include:

- the introduction of simultaneous 2D imaging techniques (as opposed to prior rectilinear-based projection generation),
- the development of energy-specific parallel-hole collimators,
- substantial increases to the imaging field of view,

- the development of digital signal processing (resulting in improved count rate performance, uniformity, and intrinsic resolution),
- improved intrinsic efficiency and sensitivity,
- the introduction of feature optimization and protocol management,
- physical coupling to anatomic imaging modalities (transmission scanning or x-ray computed tomography), and
- significantly improved image reconstruction techniques to correct for degrading factors.

The design of a SPECT/CT system has not changed much over the last 20 years, with a typical system having a 3/8" to 5/8" NaI(Tl) crystal, 2" to 3" PMTs, a Pb parallel hole collimator, and a 40 cm to 60 cm useful FoV. Innovative designs have been introduced for cardiac imaging and, more recently, for general-purpose imaging. Alternative detector materials (CsI, CZT) are increasing in utilization.

Despite the notable progress in SPECT/CT instrumentation and associated software over time, it is important to acknowledge that system design, system implementation, and literature guidance has largely been focused on meeting the needs of the diagnostic imaging community. To a limited extent, quantitative imaging of ^{131}I has played a role in cancer therapy, however adoption of SPECT dosimetry-based therapy administrations has been limited in scope. Utilization of quantitative SPECT imaging for radiation treatment planning and evaluation constitutes a meaningful departure from human observer-based metrics of system performance. Therefore, a different (but complimentary) quality assurance program is needed to ensure adequate end-to-end accuracy and safety. The purpose of this chapter is to (1) describe the status quo for SPECT/CT quality assurance and quality control and (2) suggest additional quality measures in support of dosimetry-guided radiopharmaceutical therapy.

10.2 SPECT/CT Major Components

SPECT/CT scanners typically have two primary detector heads mounted to a gantry that is coupled to a conventional CT scanner design—one x-ray tube and one detector panel within a closed gantry.[1] The primary detectors are most commonly equipped with either 5/8" or 3/8" thick NaI(Tl) scintillation-based detectors with PMT-based signal transduction. As an alternative to NaI(Tl), CsI scintillators have also been utilized to provide slightly better theoretical intrinsic resolution for the same intrinsic efficiency due to their higher density (4.51 g/cm^3 vs. 3.67 g/cm^3) and the higher Z_{eff} (54 vs. 51). CsI scintillators are enabling thinner crystal designs, albeit with a trade-off of a somewhat worse count rate performance or energy resolution, depending on pulse integration time. Avalanche photodiodes have been utilized in lieu of photomultiplier tubes for signal transduction, however adoption has been limited. Solid-state

1. Notable exceptions to this design at the time of writing include the Spectrum Dynamics Veriton-CT and the GE Healthcare Starguide. The design of these systems radically differ from conventional SPECT/CT systems and, therefore, they will not be discussed in detail in this chapter.

direct conversion using *cadmium zinc telluride* (CZT) has been implemented in lieu of a scintillation-based detectors for a limited number of system designs. The advantages of these systems include significant improvements to intrinsic spatial resolution, significant improvement in count rate performance, and moderate improvement in scatter rejection via energy resolution improvement.

Between the intrinsic detection technology and the patient resides some sort of photon collimation to constrain the incident photon flux to a particular projection geometry. Common choices include:

1. a 2D array of cylindrical or hexagonal channels contained within lead or tungsten septa (so-called "parallel hole" collimators), and
2. a single "pin-hole" in a shielding material, fixed at a particular distance from the detector.

The latter pinhole collimators are typically used for imaging applications where magnification is beneficial, such as thyroid imaging. For most radiopharmaceutical therapy-related applications, a parallel-hole collimator geometry would be considered standard. These collimators have a distance-dependent spatial resolution (closer to the face of the collimator results in better extrinsic resolution), and a relatively invariant sensitivity as a function of distance from the face of the collimator. Sensitivity invariance is a very convenient property in the context of quantitation, however distance-dependent spatial resolution degradation must be included as a feature in the iterative image reconstruction.

In addition to the detector and collimation technology, scanners have a typical patient bed that can transfer the patient between SPECT and CT position. Beds can also move at a continuous rate for whole-body single photon planar imaging. The SPECT detector's heads typically have "touch guards" at several locations to prevent accidental collision. Most detectors have optical proximity sensors to enable the scanner to automatically position the detectors as close as possible to the patient surface, thereby minimizing the distance-dependent spatial resolution degradation of parallel hole collimators. Positioning lasers are available on most scanners (similar to other PET and CT scanner designs), but the use of external positioning lasers (such as would be present in a CT simulation environment) would be atypical.

10.3 SPECT/CT Quality Assurance for Diagnostic Imaging

10.3.1 National Electrical Manufacturers Association (NEMA) and the International Electrotechnical Commission (IEC)

Standardized methods for gamma camera performance evaluation are described in NEMA and IEC technical documents, which are routinely updated (Waszyngton, N. 2012; IEC 2005; IEC 2015). The primary purpose of these technical documents is to provide particular and highly prescriptive measurement techniques, whereby the results for different scanners and manufacturers can be intercompared without bias.

This goal is pro-consumer and enables objective evaluation of system performance under various conditions.

Although effective for establishing a common standard, the measurements described by NEMA and IEC documents are often difficult to achieve in clinical practice by a clinical medical physicist, so alternative testing methodologies have been proposed by the AAPM.

10.3.2 AAPM Task Group 177

Practical alternatives to the testing prescribed by NEMA and IEC standards (Waszyngton, N. 2012; IEC 2005; IEC 2015) are provided in the report of AAPM Task Group 177 (AAPM 2019). Performance characteristics for which testing methodologies are recommended include:

- intrinsic and extrinsic flood-field uniformity,
- intrinsic off-peak uniformity,
- intrinsic count rate performance,
- extrinsic planar sensitivity,
- intrinsic and extrinsic spatial resolution (including WB scanning spatial resolution),
- spatial linearity,
- energy resolution,
- center of rotation calibration,
- multiple-energy window spatial registration,
- SPECT image quality evaluation (resolution, contrast, uniformity),
- SPECT spatial resolution,
- CT alignment and dosimetry,
- safety interlocks, and
- gamma detector shielding performance.

A comprehensive description of these tests is outside the scope of this chapter. However, in general, a majority of tests are performed using a limited array of equipment. Minimally, a physicist might utilize point, syringe, and line sources of isotopes used clinically (^{99m}Tc, ^{131}I, ^{111}In, etc.); copper plates for attenuation; a Co-57 sheet source or fillable liquid sheet source; a line pair ("bar") phantom; a fillable SPECT image quality phantom; and a scanner-specific source holder for multiple head registration, center of rotation calibration, and SPECT–CT coincidence calibration.

10.3.3 Scanner Manufacturers

Certain aspects of scanner quality control are subject to manufacturer-recommended techniques or automated processes. One example is daily uniformity evaluation, which on a Siemens scanner may be performed by positioning a point source near the uncollimated detectors rather than at a significant distance. This introduces inverse-square-related nonuniformity within the measured flood image, which is corrected for by vendor-provided software prior to uniformity evaluation.

Similarly, modern scanners have some automated daily and weekly QC tests for employee-specific sources and source strengths in geometries that deviate from NEMA and TG-177 recommendations. Additionally, calibrations required for routine camera operation are subject to manufacturer-provided protocols for acquisition, describing the source placement, source strengths, and acquisition time/counts. In general, users should familiarize themselves with manufacturer recommendations in addition to the testing recommended by other groups.

10.4 SPECT/CT Performance Characteristics for Therapy Applications[1]

Although the tests provided by TG-177 (or those by NEMA or IEC) provide a clear indication of gamma camera performance relative to manufacturer specifications, the testing does not necessarily provide the information that would be relevant for a dosimetry-guided RPT treatment program. Notable omissions include:

- evaluation of tomographic uniformity,
- monitoring for stability of scanner sensitivity calibration,
- reviewing image acquisition and reconstruction protocols for constancy and appropriateness,
- calibration for scanner dead-time correction,
- tomographic spatial resolution for partial volume correction,
- CT-to-density calibration, and
- end-to-end testing to evaluate for dosimetric accuracy.

10.4.1 Tomographic Uniformity

Reconstructed SPECT uniformity is relevant for RPT dosimetry, as nonuniformities can translate into dosimetric bias. Local nonuniformities may apply to certain structures *in vivo*, with intermittent appearance depending on patient positioning for each imaging session. Factors impacting SPECT uniformity, as well as quantitative measures of uniformity, were described in detail by Madsen (1997). Factors listed include:

1. statistical fluctuations resulting from the inherent Poisson noise in the projections and the reconstruction algorithm,
2. detector and collimator nonuniformities,
3. errors and statistical fluctuations in uniformity correction data,
4. inaccurate attenuation coefficient or pixel size applied to the attenuation coefficient, and

1. Portions of this section were adapted from a forthcoming task group report (TG-381) that is expected to be published in *Medical Physics* in 2024.

5. inaccurate boundary selection applied to the attenuation correction.

With the advent of CT-derived attenuation correction, uncertainty regarding inaccurate attenuation coefficients and boundary selection have abated somewhat, however CT-NM registration and CT noise remain factors that are relevant to reconstructed SPECT uniformity. Although not routinely evaluated in diagnostic gamma camera QC, sites may consider evaluating tomographic integral uniformity (Madsen 1997) as a measure of potential bias in dosimetric calculations.

10.4.2 Dead Time Calibration

Despite count rate performance being a system characteristic that is easily evaluated (decaying source method, copper attenuator method, peak count rate method), correction for dead time effects in the context of quantitative SPECT imaging requires substantially more effort. Although one would want to derive a scan-specific correction factor from the primary imaging photopeak—or possibly the photopeak combined with scatter windows—the system dead time may only weakly correlate with these measured data within a given scan. As discussed by Uribe and colleagues (2018), the system dead time is best described by the total system event rate, which is most appropriately evaluated by an end user by using a wide energy window to capture all registered count events. In practice, this requires:

1. characterizing the system dead time as a function of wide-window count rate and
2. adding additional upper and lower energy windows to the isotope-specific imaging protocol to allow for broad spectrum summing to determine the observed count rate during patient imaging.

More recent SPECT/CT models offered by Siemens (and possibly other manufacturers) have the ability to perform automated dead time correction (Siemens "True Calc") based upon known count losses. These electronically determined count losses, remedied by virtual count-injection, are likely more accurate than a window-based approximation and, therefore, they should be preferred over manual correction whenever possible.

System dead time as a function of event rate tends to be minimal for CZT-based SPECT systems. This is a function of the segmented detector design, with each detector element having a much lower overall event rate compared with a large, single scintillation crystal. As a result, dead time effects are negligible under typical imaging conditions with CZT-based imaging detectors.

10.4.3 Tomographic Spatial Resolution and Partial Volume Correction

Tomographic spatial resolution is largely a product of the collimator, the average detector distance to a location in the reconstructed image, and the reconstruction technique. If activity recovery coefficients are being utilized for dosimetry, it is recommended that the recovery coefficients be measured at least annually, or upon replacing a collimator, to evaluate for changes in resolution properties of the imaging system. Recovery coefficient evaluation is also advisable following major software updates to systems being used to perform image reconstruction.

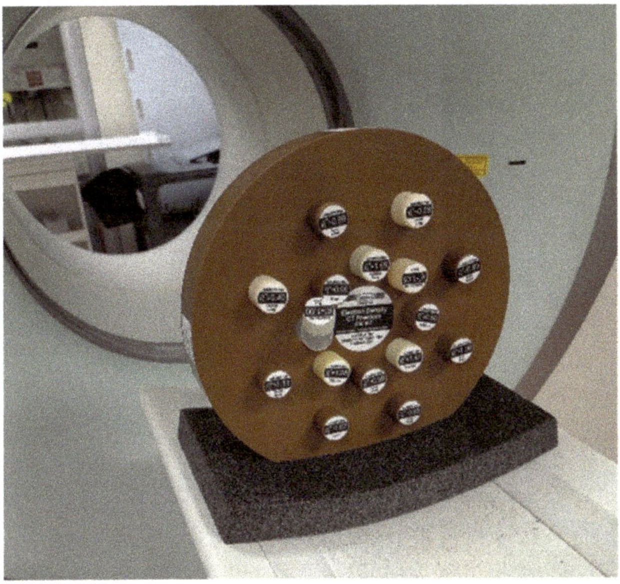

Figure 10.1 Electron density phantom utilized for conversion of CT-derived Hounsfield units into physical or electron density for dose calculation.

10.4.4 CT-to-Density Calibration

In addition to routine CT quality control (Samei et al. 2019), the CT-number-to-physical-density calibration should be verified annually for congruence with the calibration that is utilized in the clinical dosimetry software. Ideally, this verification should be performed using an electron density phantom, such as the one pictured in Figure 10.1. The SPECT/CT protocol for each therapeutic radioisotope should be verified to utilize the correct tube potential (kV) in comparison to baseline scan protocol commissioning. Spatial coincidence and spatial linearity of the SPECT and CT images should be evaluated at least annually using point sources containing a mixture of ^{99m}Tc and iodine contrast material (i.e., 2 to 3 drops of concentrated ^{99m}Tc solution and ~5 drops of iodine contrast material per sealed point source). Spatial coincidence can be quantitatively evaluated by measuring the distance of the SPECT point source centroid to that on CT. Spatial linearity can be evaluated by measuring the distance between point sources in the image in comparison to physical measurements made when aligning the sources for scanning. Often, the SPECT/CT multiple head registration and center of rotation calibration jig will provide a convenient way to position point sources for NM/CT coincidence testing.

10.4.5 End-to-End Testing

End-to-end testing is recommended prior to initiating a dosimetry-guided therapy program, as well as annually thereafter. Starting with a known quantity (as measured in a geometry that is traceable to the manufacturer or NIST) of the therapeutic isotope or surrogate isotope, a dilution can be prepared with known activity concentration of, e.g., ^{177}Lu. Approximately 20 μCi/L in 50 mM citric acid solution is adequate for phantom filling and will have enough acidity to prevent hydrolyzation or precipitation of unchelated ^{177}Lu trivalent cations. This solution can be placed in a phantom containing test objects, such as spheres or anthropomorphic organ volumes. A known quantity of the activity can be placed in the phantom background, or testing can be performed without background activity—the latter is less realistic but simplifies the determination of "known absorbed dose" based on the test geometries and known activity concentration. The phantom should subsequently be scanned by SPECT/CT using the clinical protocol at multiple timepoints over 4 to 10 days. Within the resultant images, and using previously characterized system sensitivity, spatial resolution (i.e., recovery coefficients), and dead time, the physicist should confirm accuracy of:

1. absolute activity measured within the test objects,
2. absolute activity concentration measured within the test objects,
3. mass of the test objects on CT imaging,
4. measured effective half-life relative to the physical half-life of ^{177}Lu (thereby evaluating system linearity), and
5. absorbed dose values for each test object within 5% of ground truth based on Monte Carlo or derived dose values for the test objects from look-up tables. Minimally, these tests should be performed for > 4 object sizes (i.e., 17 mm, 22 mm, 28 mm, 37 mm, but ideally more and larger-sized objects). If dosimetric error in excess of 5% is noted, other test parameters (absolute activity, concentration, mass, effective half-life) may inform of a systematic error in the quantification process. Absent an error in the time-integrated activity coefficient determination, systematic error in absorbed dose calculation (incorrect kernel, not including all decay modes in MC calculation) may result in dosimetric discrepancy from ground truth.

10.5 CT Performance Characteristics for RPT Dosimetry

10.5.1 Routine Testing

Routine CT testing typically evaluates the following system characteristics:

- spatial resolution,
- spatial linearity,
- noise,
- uniformity,
- contrast resolution, and
- CT dose.

These performance criteria are discussed elsewhere at length (Samei et al. 2019), however for the purposes of radiopharmaceutical dosimetry, one should verify that routine (in particular, daily) testing evaluates most of these characteristics. Often these routine tests are performed using a vendor-provided phantom and proprietary software work flows for scanning.

10.5.2 CT Quantitative Performance (HU Accuracy and HU-to-Density Mapping)

As discussed in Section 4.4, the HU-to-density calibration is required for accurate tissue mass determination, and it should be explicitly included in the denominator of all radiopharmaceutical dosimetry calculations (absorbed dose = energy / mass). Therefore, inaccuracies in the reported HU values, and subsequently tissue mass measurements, propagate directly into a dosimetric error. This sensitivity is even more significant than for *external beam radiotherapy* (EBRT) planning, because for EBRT planning, the relevant measurement is effective pathlength and associated photon attenuation, rather than direct voxel or tissue mass. For example, a 5% systematic error in reported HU values would induce a ~5% error in RPT dosimetry, in comparison to only a ~1.5% dose calculation error in EBRT (assuming ~10 cm depth and a 6-MV photon beam). For RPT dosimetry, in addition to the over- or under-estimation of tissue mass resulting from erroneous HU values, one must also consider the effect of over- or under-estimating photon attenuation during image reconstruction, which is perhaps an even larger effect (~13% per cm of effective pathlength error for 200-keV photons).

An accepted tolerance for CT HU accuracy is ±5 ([–5,5 HU]) for water, which is quite achievable when performing manufacturer-recommended daily calibration and QC. Similarly, the HU-to-density curve is reproducible, predictable, and largely x-ray energy-invariant, below a physical density of ~1.05 g/cm^3 in the human body. At higher densities (effectively, higher average atomic number) photoelectric effect plays a comparatively larger role than in water-like tissues. For this reason, the ratio of HU/ρ will increase approximately linearly beyond a density of 1 g/cm^3, giving rise to an approximately bi-linear relationship between HU and physical density of human tissues. This is illustrated in Figure 10.2. For this reason, drift in tube potential can result in changes in the measured HU values for a given tissue. Although this may be inconsequential in a diagnostic setting, the accuracy and constancy of HU values is highly relevant to RPT dosimetry. For this reason, it is advisable to evaluate the tube kV annually or, preferably, directly measure the HU density curve annually for each tube potential in use for hybrid RPT imaging.

In addition to considerations relating to the HU density calibration, one must also consider the impact of "missing tissue" when scanning larger patients. As shown in Figure 10.3, when patient body habitus extends beyond the fan-beam defined CT geometry, incomplete projection data is acquired. As a result, a typical CT reconstruction will inadvertently attribute the "missing attenuation" to the periphery of the reconstruction volume, resulting in artificially inflated CT numbers at the periphery of the reconstruction volume. Some scanners have the ability to reconstruct an "extended field of view" CT image despite the under-sampled projection data outside

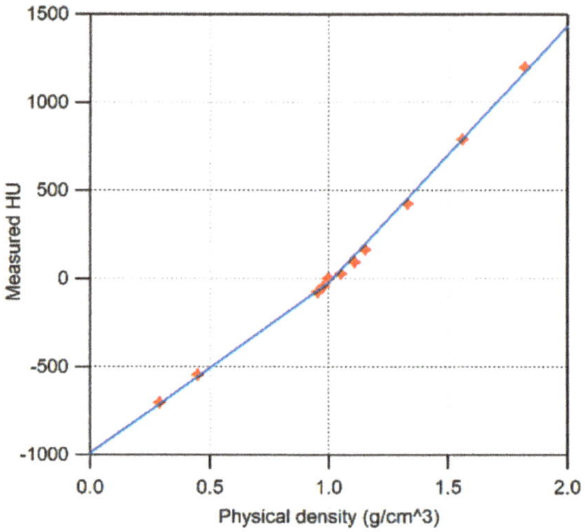

Figure 10.2 HU density data for a Siemens Intevo 2 SPECT/CT at 130 kV.

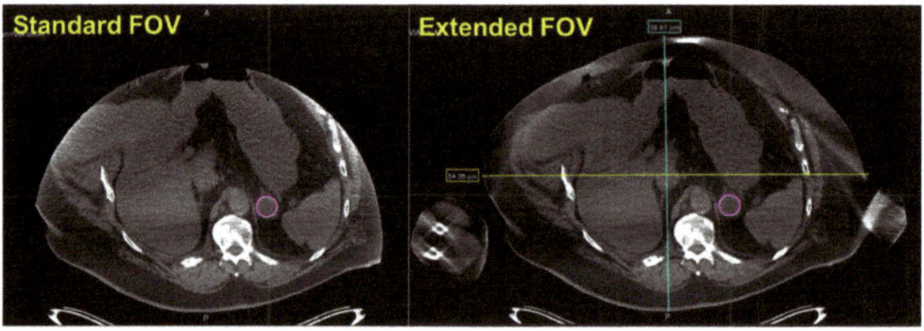

Figure 10.3 Standard (left) and extended (right) field of view for a same patient acquisition.

the central fan-beam region, allowing for improved HU accuracy within the normal field of view and, therefore, improved accuracy of attenuation correction and dosimetry. In cases where patients extend beyond the normal imaging FOV, it is advisable to utilize the extended field of view reconstruction, when possible, for quantitative applications, including image reconstruction and tissue mass determination.

10.5.3 Spatial Coincidence of CT-SPECT and Motion Artifacts

Spatial alignment of the reconstructed SPECT volume and CT volume can impact the dosimetry calculations modestly in the case of organ-level calculations with corrections for disparate NM and CT volumes. However, the impact can be much greater for some voxelwise approaches, without correction for local or global misalignment. The

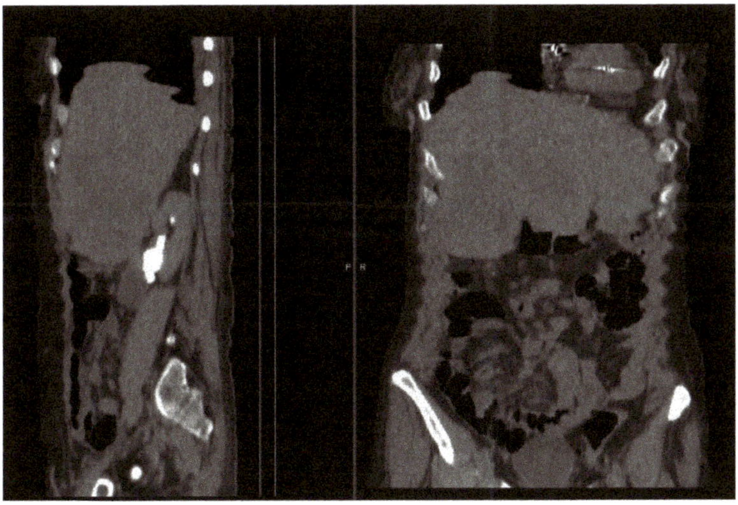

Figure 10.4 Motion artifacts due to extended CT acquisition.

physicist can ensure that the CT-NM coincidence calibration is maintained and is accurate, but this does not preclude patient motion between SPECT and CT imaging or movement during imaging.

Conventional hybrid single photon imaging systems have had comparatively low-quality CT hardware, characterized by slow gantry rotation speeds, limited tube current, and a small number of detector rows (and, therefore, slices per rotation). Until recently, a typical SPECT/CT system might have two CT slices and require more than one minute to cover a ~40 cm scan length. This scan speed inevitably results in respiratory motion artifacts in the reconstructed anatomy, as shown in Figure 10.4. These artifacts have minimal impact on the attenuation correction accuracy (except potentially near the diaphragm), but registration of the SPECT and CT volumes can be somewhat problematic, as well as have possibly distorted organ volumes (and, therefore, erroneous mass measures). For this reason, in cases where dosimetric accuracy is critical, multi-timepoint imaging may provide some statistical averaging toward the correct organ mass and, therefore, the correct overall absorbed dose determination.

10.6 Image Reconstruction and DICOM Formatting

SPECT image reconstruction is addressed elsewhere in detail, but in general, reconstruction parameters for dosimetry-related applications will utilize less smoothing and more iterative updates, resulting in a higher-resolution reconstruction with substantially increased noise compared to images typically used for diagnostic interpretation. As a consequence, the number of counts per reconstructed voxel can be very high, in some cases exceeding the bit-depth of a typical DICOM integer format (i.e., exceed-

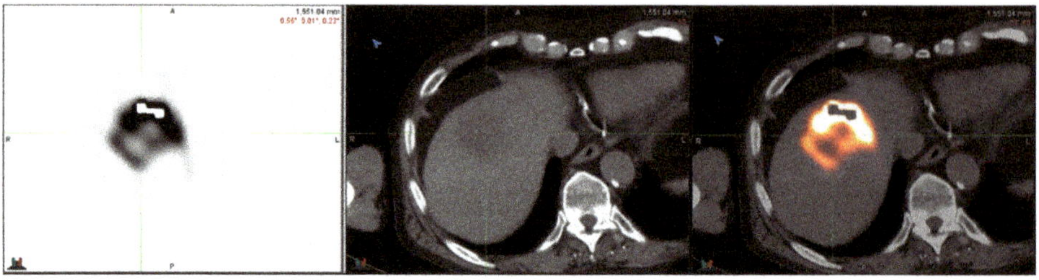

Figure 10.5 ^{99m}Tc-MAA hepatic mapping in preparation for Y-90 radioembolization. Image was reconstructed using Siemens Flash3D, 12 iterations, 8 subsets, and 4 mm Gaussian smoothing. In this case, the wrap artifact was resolved either by scaling the projections by a factor of 0.1 or by reconstructing the data with a third-party software product (MIM SPECTRA Quant®).

ing 65,536 counts in a single voxel), resulting in "wrap-around" artifact. An example of this is shown in Figure 10.5. There are multiple solutions to this problem, one of which is to scale the projection data by a factor of 0.1 to reduce the count density. However, one must scale the system sensitivity by an equivalent factor when converting images to activity per voxel, or activity per mL. Alternatively, one can utilize a platform that supports storing reconstructed data as floating point information (similar to conventional PET DICOM files).

10.7 Failure Modes

Potential failure modes in the quantitative imaging and dosimetry workflow are numerous. One must consider the entire chain, starting with activity receipt and dose calibrator assay, all the way to dosimetry report generation. Aspects of the process relating to SPECT/CT imaging, which can be particularly problematic and difficult to detect, include the following:

- ***Inaccurate dose calibrator assay***. This results from incorrect dial setting determination, deviating from calibration geometry, elevated or over-compensated background reading, and response drift over time due to loss of pressure.
- ***Inaccurate scanner sensitivity calibration***. This results from erroneous dose calibrator measurements during sensitivity determination or from measurement of a sensitivity factor that does not closely align with how the sensitivity factor is being utilized (i.e., point source in air vs. volume source in a scatter- and attenuation-inducing body).
- ***Low-quality reconstructed images***. This results from poor scanner uniformity, improper MHR/COR calibration, excessive count rate, or incorrect reconstruction technique.

- ***Change in imaging protocol***. This results from changes to the energy window, collimator, or reconstruction technique following commissioning and can produce significant quantitative errors.
- ***Data transfer integrity***. This results from loss of DICOM tags or improperly configured dosimetry/analysis software. This can be particularly problematic when transferring the images to a PACS system, followed by retrieval to the third-party analysis software.

These considerations are not comprehensive, but rather, are intended to give the reader a few things to consider when developing a robust clinical workflow.

10.8 Conclusion

Conventional utilization of SPECT imaging and hybrid SPECT/CT imaging has not required the same level of quantitative accuracy and rigor as that demanded when implementing a dosimetry-guided RPT program. Physicists should carefully consider the functionality of all system components and workflow steps individually, as well as holistically through end-to-end testing. Forthcoming innovations in scanner design and features may soon simplify the quantitative SPECT/CT imaging and dosimetry process.

References

AAPM. The Report of AAPM Task Group 177. "Acceptance Testing and Annual Physics Survey Recommendations for Gamma Camera, SPECT, and SPECT/CT Systems." Alexandria, VA: American Association of Physicists in Medicine, 2019.

IEC. IEC 60789. *Medical Electrical Equipment Characteristics and Test Conditions of Radionuclide Imaging Devices Anger Type Gamma Cameras*. Geneva: International Electrotechnical Commission, 2005.

IEC. International Electrotechnical Commission publication IEC 61675–2. *Characteristics and Test Conditions of Radionuclide Imaging Devices. Part 2: Single Photon Emission Computed Tomography*. Geneva: International Electrotechnical Commission, 2015.

Madsen, M. T. (1997). "A method for quantifying SPECT uniformity." *Med. Phys.* 24(11):1696–1700.

Samei, E., D. Bakalyar, K. L. Boedeker, S. Brady, J. Fan, S. Leng, K. J. Myers, L. M. Popescu, J. C. R. Giraldo, F. Ranallo, J. Solomon, J. Vaishnav, and J. Wang (2019). "Performance evaluation of computed tomography systems: summary of AAPM task group 233." *Med. Phys.* 46(11):e735–56.

Uribe, C. F., P. L. Esquinas, M. Gonzalez, W. Zhao, J. Tanguay, and A. Celler (2018). "Dead-time effects in quantification of ^{177}Lu activity for radionuclide therapy." *EJNMMI Phys.* 5(1):1–16.

Waszyngton, N. NEMA standards publication NU 1-2012: *Performance Measurements of Gamma Cameras*. National Electrical Manufacturers Association, 2012.

Chapter 11

Surrogate Imaging Dosimetry

Adam Leon Kesner, Ph.D.

Associate Attending, Nuclear Imaging Physics
Department of Medical Physics
Memorial Sloan Kettering Cancer Center
New York, NY

11.1 Introduction to Time Integration of Activity for Calculating Dosimetry

11.1.1 Introduction to Radiopharmaceutical Dosimetry

Radioactive decay is a well characterized phenomenon which naturally occurs in all atoms. Unstable atoms can spontaneously rearrange their proton/neutron configuration to get to a more stable state. In this process, known as *radioactive decay*, α/β particles, γ rays, or neutrons may be emitted.

The field of radiation dosimetry covers the study of the interaction of high-energy photons and particles with human tissue. These particles carry energy, and that energy can be transferred to the medium it is passing via interactions with the tissue. These interactions can alter cells and cellular organelles, causing damage or cell death.

There are many mechanisms for generating radiation. In the field of nuclear medicine, particle emissions come from radioactive medical isotopes that are introduced to patients for purposes of diagnostic or therapeutic care. Where chemistry permits, photon-emitting isotopes can be conjugated with different pharmaceuticals and imaged using detector systems, e.g., gamma cameras, to provide information on the distribution of those pharmaceuticals in the body. Alpha- and beta-emitting isotopes can be conjugated with molecules that have an affinity to localize at or near a biological target to deliver radiation to that target. Radioisotopes are selected specifically for their radioactive decay properties: types of radioactive emissions, energies of the emissions, and physical half-lives. Furthermore, different isotopes exhibit different atomic chemical properties, making them more or less suited to be integrated into a biologically relevant molecule.

Because nuclear medicine utilizes radiopharmaceuticals—pharmaceuticals that incorporate radioactive atoms—dosimetry associated with the radiopharmaceutical is specific to the decay properties of the incorporated isotope. Isotopes produce radioactive emissions when they undergo decay. The rate of particle decay is known as *activity*, defined as the emission of ionizing radiation or particles caused by the spontaneous disintegration of atomic nuclei. This decay, or decrease in the number of unstable particles, is strictly proportionate to the number of unstable atoms, with a proportionality known as the *decay constant* that is specific to each isotope. As a mathematical consequence of this proportionality, the number of unstable atoms in a radioactive material decreases exponentially with time.

11.1.2 The Concept of Time-Integrated Activity

Absorbed dose, defined as energy per mass, is directly proportional to the number of radioactive emissions in a given system. Thus, quantifying the amount of source radiation in a dosimetric scenario is critical, i.e., more radiation will deliver more absorbed dose to the tissue it interacts with. To characterize the amount of radiation introduced to a system by an isotope integrated within a radiopharmaceutical, a metric known as *time-integrated activity* has been introduced. The time-integrated activity represents the total emissions of an isotope radiation source over time, and it provides a fundamental input for dosimetry calculations. Mathematically it can be calculated by integrating the area under the time-activity curve, and is discussed more in Chapter 15.

11.1.3 How Time-Integrated Activity is Used in Dosimetry Calculations

Defining the source of radioactive emissions from a radiopharmaceutical in terms of strength and location is essential for calculating the effects of its radiation with surrounding tissue. The time-integrated activity is a quantity that is used to characterize multidimensional measurement data into a single number representing source strength. This measurement data is most commonly derived using nuclear medicine imaging techniques (PET or SPECT), but could also come from *ex vivo* (e.g., blood) measurements or more general single-probe measurements. Time-integrated activity can be modeled and presented at small scales, for example representing image voxels, or larger scales, for example representing organs of interest. By calculating the time-integrated activities for an administered radiopharmaceutical, we are able to characterize the complex time-space biodistribution of the tracer using a smaller set of geometrically defined source strengths.

In surrogate imaging, the topic of this chapter, the concept of time activity integration is relevant because time activity integration can be calculated/derived from using a surrogate (imaging) isotope, and then projected to characterize the activity and resultant dosimetry that would be present in a different (therapy) isotope using corrections for the relative half-lives.

11.2 Models for Calculating Time-Integrated Activity Coefficients

Time-integrated activity coefficients (TIACs) represent time-integrated activity normalized by the administered activity. They represent a quantitation of the total amount of radiation that comes from a region per unit of administered activity. Accordingly, dosimetry calculated from those TIACs are also referred to as *coefficients*, namely *dose coefficients*, which are the absorbed doses per unit of administered activity. Thus, source radiation distribution information gathered and contained in TIACs can be used to calculate dose coefficients, which can then be used during treatment planning by aiding in the determination of an ideal administration activity.

11.2.1 Non-Imaging Reference TIACs

For many radiopharmaceuticals that are used in clinical nuclear medicine, reference TIACs have been compiled, at the organ level, expressing typical biodistributions of different radiopharmaceutical agents. These TIACs are often derived from imaging studies that measure and report the radiopharmaceutical distribution in a cohort of patients. Several publications have performed the work of collecting this data and collating it in one place. For example, ICRP 128 reports TIAC data for 60 radiopharmaceuticals and in a variety of subject ages (Mattsson et al. 2015). It should be noted that the use of reference TIACS is only advised for diagnostic radiopharmaceuticals; therapeutic agents require individual measurement because small differences in radiopharmaceutical uptake between patients may result in large consequential differences with respect to patient dosimetry.

11.2.2 Imaging-Based Strategies for Deriving TIACs

The gold standard for acquiring detailed, patient-specific *in vivo* measurements of radiopharmaceutical biodistributions is to measure them directly from serial images acquired with emission imaging modalities, namely PET, SPECT, or gamma camera scintigraphy. A typical organ-level dosimetry protocol is:

1. Patients are administered a small test dose of a radiopharmaceutical, or a surrogate with similar physiological uptake.
2. Patients are imaged at multiple timepoints to capture "snapshots" of the biodistribution of a radiopharmaceutical over time.

3. Images are segmented to delineate the different organs, such that the organ activity measured from the images can be characterized at each imaging timepoint.
4. TIACs can be calculated for each organ region by plotting the organ uptake measurements vs. time and integrating to get the area under the curve using models or trapezoidal integration.
5. From a set of TIACs, a full accounting of source radiation is achieved. A full radiopharmaceutical biodistribution accounting for the temporal-spatial distribution of the isotope is articulated in the set of TIAC numbers. This data can then be used to derive dosimetry using dosimetry software based on mathematical and computational phantom models. These software packages include Olinda/EXM and MIRDcalc.

An alternative to organ-level dosimetry protocols are voxel-level protocols that are also commonly employed in the field. With voxel-level dosimetry, every voxel in an image is essentially considered a separate region for which activity and dosimetry is considered. This entails initial careful alignment of serial images so that all voxels represent a consistent location in the patient's body (see Chapter19 on image registration). Activity is quantified per voxel at each imaging timepoint, and voxel-level dose maps are then created using Monte Carlo or point kernel dosimetry calculation methods.

This simplest and most straightforward protocol for implementing dosimetry-guided radiopharmaceutical therapy is to use a small amount of the radiopharmaceutical for initial dosimetry measurements, and then adjust the administration activity of the planned therapy treatment using the initially derived dosimetry. This could work if the treatment isotope emits both imaging and therapy emissions. An alternative approach is to use a surrogate imaging isotope, one with better quantitative imaging capacities, to characterize the anticipated biodistribution of the therapeutic isotope. This is done by adjusting the uptake measurements from the imaging surrogate to account for differences in the half-lives between the intended surrogate and therapeutic isotopes. This approach can be employed if/when the therapy isotope does not have ideal imaging characteristics, and if there is a chemically accommodating surrogate agent, as discussed more below. A drawback of this approach, however, is that the strategy requires an assumption that the surrogate isotope-pharmaceutical combination will have similar biodistribution properties to the intended therapeutic agent. Furthermore, in cases where the biological half-life of the radiopharmaceutical is long and the physical half-life of the imaging surrogate is short, the predicted washout kinetics may be incorrect.

Imaging for Biodistribution Characterization

The spatio-temporal distribution of a radiopharmaceutical is characterized using serial emission imaging, ideally 3D imaging using PET or SPECT. The serial imaging provides a map of a particular radiopharmaceutical within a particular patient over time. This information is then used for analysis and dosimetry.

PET is preferable for quantification of the biodistribution of an RPT, from a technical standpoint, because it is a more quantitative modality. PET images are attenuation- and scatter-corrected, and they inherently provide quantitative data. However, gamma cameras are commonly used for dosimetry as many/most therapeutically relevant isotopes are not positron emitters that are needed for PET scanners. Furthermore, the chemistry of RPT production often makes photon-emitting isotopes favorable for conjugation with chemically relevant compounds.

11.2.3 Time-Integrated Activity Curve Modeling

To characterize a distribution of a radiopharmaceutical within a patient after administration of the agent, serial images can be acquired using emission imaging. Activity estimates can be determined from these images by segmenting the signal on the images and using the segmented data for analysis. In the case of preclinical research, imaging and image analysis can be replaced with direct measurement of dissected (*ex vivo*) organs to acquire biodistribution measurements. Irrespective of the measurement method, the biodistribution data provide time-specific characterizations of the tracer's expression in the body. A complete accounting of the radiotracer that summarizes the full involvement of the radiopharmaceutical in the patient—ranging from time of administration out to full isotope decay—can be generated using curve modeling of the measurements in a region over time. This process can be repeated across regions to generate an accounting of all the activity resulting from a radiopharmaceutical administration.

Typically, TIACs used for dosimetry calculations are derived by first creating time-activity graphs where normalized uptake measurements are plotted against time on an XY chart for a given source region, as shown in Figure 11.1. From this representation of the data, the TIACs can be understood as the integration of the activity measurements on the time-activity chart, i.e., the area under the curve. The ideal method of integration often depends on your data or the biology of uptake. In the simplest form, no biological models are required, and integration can be achieved with trapezoidal integration between the first and last measurement data points. In order to account for all of a tracer's emission, i.e., from time 0 to infinity, assumptions must be made to model the activity at times not captured in the images, namely the pre-measurement timepoints (nose) and the post-measurement timepoints (tail). However, since kinetics of many tracers can be modeled using known biological uptake/clearance parameters, time-activity data is often integrated using mathematical models that conform to the theoretical shape of a curve. For example, we may expect that uptake and washout of a radiopharmaceutical in a tumor to exhibit exponential rate-behaviors. In this case, we may choose to model our curve (and its integration) using a bi-exponential equation with the aid of curve-fitting software. Once an equation for a curve-matching uptake measurement is modeled with a known equation, the integration of the curve to get the TIAC can be acquired using mathematical integration.

The principles of prospective treatment planning are discussed further in Chapter 12, while curve fitting is discussed in greater depth in Chapter 15.

TIACs can be used for dosimetry of an isotope being imaged/studied. However, it is not difficult to extrapolate biodistributions defined by a set of uptake measurements

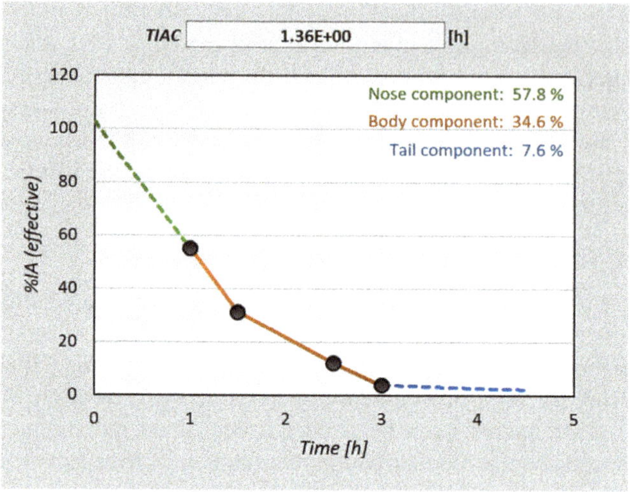

Figure 11.1 An example time-activity curve. The figure shows a case of ^{18}F measurements where measurement data is shown as percent injected activity vs. time. Example is a screenshot from MIRDfit curve-fitting software, available for free download at www.mirdsoft.org.

from a surrogate isotope using corrections for the different M using Equation (11.1). Essentially, extrapolation requires decay-correcting activity measurements from the surrogate imaging isotope to essentially derive percent uptake values, and then decay projecting those measurements to the therapy isotope to achieve the anticipated activity of the therapy isotope at that timepoint. This process is summarized in Equation (11.1).

$$A_t(r,t) = A_i(r)e^{t\ln 2/T_i}e^{-(t\ln 2/T_t)} \tag{11.1}$$

where A_t and A_i are the activities of the therapeutic and imaging radiopharmaceuticals in region r at time t, respectively. T_i and T_t are the physical half-lives of the imaging and therapeutic isotopes, respectively.

11.2.4 Monte Carlo Dose Rate Modeling

While TIACs are used principally for organ-level dosimetry, they can also be used for voxel-level dosimetry by assigning a TIAC to each voxel and deriving a dose distribution based upon the source TIACs in every voxel. This can be achieved using dose-point kernel convolution or Monte Carlo dose modeling.

With voxel-level dosimetry, there may be situations where it is advantageous to calculate dose rate at the times of measurement—when modeling biologically effective dose, for example. In these cases, surrogate isotopes can be accommodated by assigning percent injected activity to voxels (correcting for decay) and modifying the dose deposition parameters (e.g., dose-point kernel) to match the appropriate physical dose rate for the surrogate isotope. In this circumstance, dose rate in organs can be assessed for biological significance, and the cumulative dose can be modeled in the voxels by integrating the calculated dose rates.

11.2.5 Single-Timepoint Strategies

Multi-time-point dosimetry places additional burden on patients and requires added resources, time, and mathematical complexity. Accordingly, there is significant interest in the field to supply simplified dosimetry protocols. To achieve this, *single-time-point* dosimetry strategies are being investigated.

The principle behind a single-time-point dosimetry strategy is that models for a tracer's kinetics can be initially derived from a large patient cohort using multi-time-point imaging. This model can then be utilized and extrapolated to new patients receiving only a single-time-point uptake measurement. This means that patients are only required to participate in one additional imaging session to support the dosimetry protocol, and their dosimetry will be calculated based on personalized calibration of population-derived tracer behavior.

An example of this comes from Hänscheid and colleagues (2018), who derived a single-time-point dosimetry strategy for ^{177}Lu-DOTATE therapy treatments.

Multi-time-point imaging is more accurate, but dosimetry with one timepoint is more practical, with greater uncertainties, but still likely of benefit compared to fixed-activity implementation.

11.3 Surrogate Imaging Models for Theranostic Dosimetry

11.3.1 Principles

The concept of radiopharmaceutical theranostics entails a combination of diagnostic imaging and therapy using tracers that have the same affinity for the same molecular target. The *theranostic principle* is to "see what we treat and treat what we see" (Turner 2018). Imaging with a diagnostic radiopharmaceutical can provide assessment of disease and state of the disease for an individual patient. Some therapeutic isotopes emit both imageable radiation as well as therapeutic radiation, and they can be used for both imaging and therapy. One example is ^{131}I, which has a predominant gamma emission at 364 keV (82% abundance) that can be used for photon imaging with a gamma camera, as well as a β emission which can deliver a therapeutically relevant radiation dose.

However, not all therapeutic isotopes are imageable, and replacing therapeutic isotopes with surrogate imaging isotopes often provides better images/information. When paired with an analogous therapeutic surrogate molecule, an isotope molecule labeled as diagnostic can provide imaging that can elucidate the anticipated distribution of the therapeutic agent, and it will appropriately provide target-specific coverage. By transforming the diagnostic imaging information into quantitative dosimetric information, patient-specific treatment planning can be employed to determine the optimum amount of radioactivity of the therapeutic agent to use for treatment, i.e., surrogate imaging dosimetry. This patient-specific activity determination is guided by anticipated dose calculations to diseased and healthy tissue and the associated dose thresholds for treatment efficacy. The ultimate goal of dosimetry-guided radiotheranostics is to optimize the absorbed dose of therapeutic-labeled compounds relative to disease and healthy tissue.

11.3.2 Physical Considerations

According to a recent IAEA report, there are just over a dozen therapeutic isotopes emitting α and β particles that are now being applied for therapy routinely in clinical practice (2015). Different elements and isotopes have different chemical properties that make them more or less suitable to be integrated with molecular vectors. Once a pharmaceutical and isotope are selected to make a therapeutic radiopharmaceutical, consideration for surrogate imaging/dosimetry becomes prudent if the agent is to be used in a theranostic paradigm. Surrogate imaging isotopes must decay with useful imaging emissions, have practical/relevant half-lives, and are selected to have similar chemical properties to the therapeutic isotope to the extent possible. Furthermore, the surrogate imaging compound, made with a surrogate isotope, will need to be tested to ensure it has similar pharmacologic properties to the therapeutic compound.

11.3.3 Biological Considerations

Surrogate imaging and dosimetry for therapy works on the presumption that the paired imaging and therapeutic molecules have similar pharmacology and are thus treated the same by a patient's body when administered. Changing the radionuclide used to label a pharmaceutical can alter many physical properties of the compound. Ultimately, the two radiopharmaceuticals must have similar biodistributions, and similar kinetics. Classes of theranostic pair isotopes include "same-element isotope pairs," "different-element pairs," and "different-pharmaceutical pairs" (Miller et al. 2022). Technically speaking, positron-emitting radionuclides are ideal for providing the best diagnostic information possible, as PET is a more quantitatively accurate modality than scintigraphy or SPECT. Chemically speaking, if the radioactive element of a diagnostic and therapeutic surrogate pair are identical, e.g., a same-element isotope pairs, the two will be pharmacologically similar, and imaging data can be used directly to extract the absorbed doses of the therapeutic version. This is commonly denoted by the nomenclature of "matched pairs," where the diagnostic and therapeutic radionuclides belong to one and the same chemical element. An alternative surrogate imaging workflow may use chemically similar, but different isotopes.

An example of matched pairs include $^{86}Y/^{90}Y$. Yttrium-90 is a therapeutic isotope that has been integrated with many targeting compounds in the literature. In the investigation by Herzog and colleagues (1993), discussed in greater detail below, patients being treated with ^{90}Y-citrate were first imaged using PET with an analog ^{86}Y-citrate. Patient treatment was then tailored to the dosimetry calculations derived from PET imaging. With respect to uptake, biodistribution, and metabolism, ^{90}Y and ^{86}Y or compounds labeled with ^{90}Y and ^{86}Y, respectively, do not differ. What does differ, however, are the decay properties of the two isotopes, and these differences must be accounted for when calculating dosimetry (Equation (11.1)).

11.3.4 Anatomy of a Surrogate Dosimetry Workflow

A dosimetry-driven surrogate imaging/therapy workflow contains the following principal components (see Figure 11.2):

1. The patient is first imaged using an imaging surrogate isotope-labeled compound.
2. Data is analyzed to characterize a summary biodistribution of the imaging tracer (via the time-integrated activity calculations described above).
3. Biodistribution of the therapeutic surrogate isotope can be projected from the imaging isotope using Equation (11.1) using the assumption that the two compounds distribute the same in the patient.
4. Dosimetry coefficients (Gy/mCi or Gy/MBq) are projected for organs at risk and tumor sites.

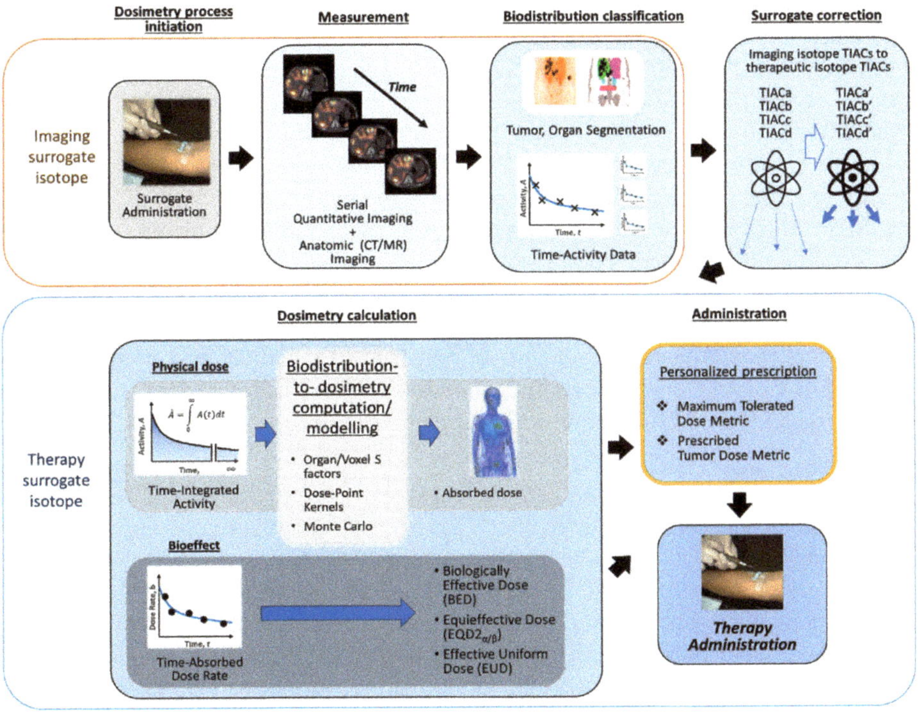

Figure 11.2 Anatomy of a surrogate imaging/therapy radiopharmaceutical therapy workflow. Figure is from J. O'Donoghue, P. Zanzonico, J. Humm, and A. Kesner (2022). "Dosimetry in Radiopharmaceutical Therapy." *J. Nucl. Med.* 63(10):1467–74. © SNMMI.

5. A dosimetry-derived administration activity is chosen to optimally accommodate absorbed dose to tumor and healthy tissue.
6. The patient is treated.

11.4 Clinical/Lab Examples of Surrogate Dosimetry for Theranostics

11.4.1 ^{23}I or ^{124}I /^{131}I PET/Therapeutic Radionuclide Pair in Metastatic Thyroid Cancer

Radioactive iodine (RAI) therapy with the isotope ^{131}I is used for treating hyperthyroidism, therapeutic ablation, adjuvant treatment, and locoregional or metastasis of well-differentiated thyroid cancer. Iodine is an ideal tracer for thyroid treatment because when iodine is introduced into the body, it is hyper-selectively absorbed in the thyroid gland cells. The treatment works so well that RAI therapy has become a classic use case for radiopharmaceutical therapy. Iodine-131 is a β emitter and is capable of delivering therapeutic radiation to the site of uptake, and as well having some imageable photon emissions. Other isotopes of iodine, namely ^{123}I and ^{124}I, can be used as surrogates in place of their therapeutic counterpart for imaging with SPECT/gamma camera or PET, respectively.

Because these isotopes are so chemically similar, the distributions of them in the body is also very similar, enabling an ideal imaging/therapeutic surrogate pair. The American Thyroid Association provides several indications for use of diagnostic whole-body imaging of iodine distribution to assess thyroid update, and it recommends performing these scans using one of the imaging surrogates "because it delivers lower radiation doses to the body and provides better quality images" (Haugen et al. 2016). However, one notable drawback of using the imaging surrogates is that the half-lives of the ^{123}I (13.2 hours) is shorter than the half-life for the therapeutic ^{131}I (8 days), and it is difficult to model the projected washout phase of the therapy. Thus, this imaging isotope is useful for detection of thyroid tissue and metastatic thyroid cancer as a purely qualitative diagnostic, but it has limitations as a pre-therapeutic. The half-life of ^{124}I is 4.18 days, and while shorter than that of ^{131}I, it is on a similar scale and better suited for theranostic surrogate imaging.

11.4.2 ^{86}Y/^{90}Y

Yttrium-90 is a β-emitting isotope which has been shown to have demonstrated utility in therapeutic applications in benign bone disease and cancer. Earlier attempts were made to predict dosimetry of ^{90}Y-labeled *monoclonal antibodies* (MAb) using the SPECT-compatible imaging surrogate ^{111}In, as both isotopes are trivalent metals and have similar molecular properties (Zimmer et al. 1992). However, it was found in this case that the radionuclide pair was not well suited for quantification of biodistribution and patient individualized dosimetry calculations.

In the 1990s, the idea was proposed to use a pair of radionuclides of the same element, ^{86}Y and ^{90}Y, which allowed the combination of PET and internal radiotherapy.

As PET is a quantitative imaging modality, this combination could be ideal. Because Yttrium-86 decays with a positron yield of 33%, it is not as efficient for imaging as ^{18}F, for example, which has a positron yield of 97%. Furthermore, ^{86}Y also decays by gamma-emitting photons in or above the typical PET detector energy windows (approx. 450 to 640 keV), which degrades the quantitative capacity of the PET image. Nevertheless, ^{86}Y has enough signal to make it a suitable ^{90}Y surrogate for PET imaging, and it is still preferable to a gamma camera analog.

The first human study that utilized ^{86}Y as an analog for determining the distribution of a ^{90}Y compound was in a breast cancer patient in 1993 (Herzog et al. 1993). The aim of this study was to use an ^{86}Y-citrate imaging-based biodistribution study to personalize the planned treatment of ^{90}Y-citrate. After characterizing the distribution of the imaging surrogate, the clinicians were able to account for the differences in the decay properties of the isotopes and project anticipated absorbed doses from the ^{90}Y-citrate injection.

Other examples of the ^{86}Y/^{90}Y surrogate theranostic combinations include the research from Rösch et. al. (1999), who investigated a potentially promising agent for somatostatin receptor-expressing tumors, ^{90}Y-DOTA-DPhe^{1}-Tyr^{3}-octreotide, testing the protocol in primates. The authors demonstrated relatively low absorbed doses to normal organs and were able to use the projected dosimetry data to support clinical phase 1 trials. A couple years later, Förster and colleagues (2001) reported "preliminary data on biodistribution and dosimetry for therapy planning of somatostatin receptor positive tumors" using ^{86}Y-DOTA-DPhe^{1}-Tyr^{3}-octreotide (^{86}Y-DOTATOC), and they showed that biodistributions for the tracer did not differ significantly between humans and nonhuman primates. Further surrogate imaging with ^{86}Y/^{90}Y labeled DOTATOC showed superiority to other (^{111}In) imaging analogs. Ultimately, ^{90}Y-DOTATOC was approved for clinical treatment of neuroendocrine tumors, and the DOTATOC and similar DOTATATE agents are used widely today, with their approvals and protocol optimizations being based on biodistribution data and dosimetry provided by the imaging surrogate pair.

Despite the utility, imaging with ^{86}Y-labeled peptides and ^{86}Y-labeled antibodies remains most widely reported for preclinical investigations. Comprehensive information on ^{90}Y as a therapeutic agent can be found in the IAEA guidance document (2015).

11.4.3 Tc-^{99}m MAA and ^{90}Y Microspheres

In recent decades, ^{90}Y microsphere brachytherapy has become an established treatment for liver cancer. Cancers of the liver, including metastasized secondary cancers in the liver, occur at significant rates among cancers and have limited treatment options. Yttrium-90 radioembolization is a strategy which takes advantage of the differences in vasculature properties between the liver and tumors. In the procedure, microspheres infused with ^{90}Y are injected into a patient. The microspheres are small enough to become lodged in the arterioles within the growing rim of the tumor, but they are too large to pass through the capillaries and into the venous system. Once lodged (aka embolization), they deliver radiation therapy to the tumor via the β emissions from the ^{90}Y isotope. Technically speaking, ^{90}Y microspheres are a form of brachytherapy. However, because the spheres are very small, 20 to 60 microns in diameter, they often behave and are treated as a liquid radiopharmaceutical.

Yttrium-90 is a β emitter. It can be imaged directly, either by SPECT or PET, although neither is ideal. For gamma imaging, secondary gamma rays created via bremsstrahlung interaction of the principal electron emissions can be imaged with a gamma camera. Because the strategy involves imaging bremsstrahlung radiation rather than spectral peaks of photon emissions, energy discrimination cannot be usefully employed, and the images will have low resolution and poorer quantitative capacity. Alternatively, there is a small fraction (3.2×10^{-5}) of ^{90}Y decays that result in the emission of positrons, and PET can be used to acquire images of these emissions. This can and is used for clinical quantification of ^{90}Y radiation, but it is also not ideal as long PET imaging times are required to acquire images, and the images have much less statistical quality than most diagnostic PET scans.

The design and function of ^{90}Y microspheres are not so different than the common technetium-based imaging analogue ^{99m}Tc-micro-aggregated albumin (MAA). Tecnetium-99m-MAA is also widely available and relatively inexpensive. It has been found that ^{99m}Tc-MAA can be a useful imaging surrogate for ^{90}Y microspheres. They are routinely used within the vendor-prescribed protocols as an initial indicator of the anticipated shunting fraction of spheres in a patient's lungs—a major safety concern of the treatment. Measurements of ^{99m}Tc-MAA lung shunting after liver administrations help provide assurance that injected spheres will not migrate to the lungs and cause unintended lung damage. In addition, the measured fraction of spheres that migrate to the lungs can be integrated into treatment planning dosimetry calculations. More recently ^{99m}Tc-MAA is being used as a direct surrogate for ^{90}Y microsphere dosimetry within the liver and tumor.

11.4.4 ^{132}La/^{225}Ac

Actinium-225 is a promising radionuclide to be utilized for targeted radionuclide therapy. It is an α emitter, with a half-life of 9.29 days, long enough that it can be produced and globally distributed, and short enough that it will not persist in a patient beyond a number of months. Its physical and chemical properties make it widely suitable for integration with radiopharmaceuticals that encompass small molecules, peptides, antibodies, and antibody fragments.

A drawback of ^{225}Ac, as with most α emitters, is that it is not well suited as an imaging isotope, making the noninvasive characterization of ^{225}Ac pharmaceuticals' biodistribution difficult. Several surrogate isotopes have been considered, for example ^{68}Ga, but it has been shown that ^{68}Ga is chemically dissimilar and, therefore, has limitations in addition to having a short half-life (68 min).

Aluicio-Sarduy and colleagues (2021) have proposed and tested a lanthanum surrogate for ^{225}Ac, ^{132}La. The ^{132}La isotope is a positron-emitting isotope with a half-life of 4.59 days. The authors were able to compare the relative biodistributions of a DOTA-based, tumor-targeting alkylphosphocholine (NM600), labeled with ^{225}Ac and ^{132}La. The comparisons were made using *ex vivo* quantitation of the biodistribution of the compound in mice with implanted tumors. The authors found that measurements of ^{132}La-NM600 organs and tumors were highly correlated with measurements of

^{225}Ac-NM600 across 14 organ regions plus tumor. They concluded that there is promise and potential for establishing ^{132}La as an imaging surrogate for ^{225}Ac, exemplifying the future opportunities we may have for development in the field.

11.5 Clinical Examples of Imaging Only Surrogates

11.5.1 ^{177}Lu-DOTATATE and ^{111}In-DTPA-Octreotide or ^{68}Ga-DOTATOC

Peptide receptor radionuclide therapy (PRRT) utilizes radiolabeled peptides to target with high affinity and specificity receptors that are over-expressed in tumors. Somatostatin receptor antagonists DOTATOC or DOTATE labeled with theragnostic isotopes ^{90}Y or ^{177}Lu have been successfully employed globally for many years (Zaknun et al. 2013). The ^{177}Lu-DOTATATE product Lutathera® gained EMA and FDA approvals in 2017 and 2018, respectively, for the treatment of somatostatin receptor positive gastroenteropancreatic neuroendocrine tumors (Hennrich and Kopka 2019).

As a precursor for PRRT, patients are screened for receptor overexpression, requisite for treatment efficacy. This is achieved using scintigraphy or PET imaging of surrogate labeled isotopes, either ^{111}In-DTPA-Octreotide (OctreoScan®) or ^{68}Ga-DOTATOC, (SomaKit TOC®). These surrogates are utilized to capture patient suitability for treatment, but not for dosimetry.

11.5.2 ^{177}Lu-PSMA-617 and ^{68}Ga- or ^{18}F-labeled PSMA

In terms of potential patient numbers, RPT for prostate cancer is likely to dominate the field for the foreseeable future. Recently reported results of randomized clinical trials (Hofman et al. 2021; Sartor et al. 2021) show that ^{177}Lu-PSMA-617 can produce significant improvements in outcome, and the agent is now marketed as under the brand name Pluvicto®. Like the above-described ^{177}Lu-DOTATATE, prospective patients are first screened using an imaging analog: ^{68}Ga- or ^{18}F-labeled PSMA. The use of this imaging surrogate provides the necessary indications for a patient to be deemed suitable for a therapeutic treatment. It has been suggested in the literature that these surrogate imaging scans may be useful for determining dosimetry (Violet et al. 2019).

References

Aluicio-Sarduy, E., T. E. Barnhart, J. Weichert, R. Hernandez, and J. W. Engle (2021). "Cyclotron-Produced ^{132}La as a PET Imaging Surrogate for Therapeutic ^{225}Ac." *J. Nucl. Med.* 62:1012–15.

Förster, G. J., M. J. Engelbach, J. J. Brockmann, H. J. Reber, H. G. Buchholz, H. R. Mäcke, F. R. Rösch, H. R. Herzog, and P. R. Bartenstein (2001). "Preliminary data on biodistribution and dosimetry for therapy planning of somatostatin receptor positive tumours: comparison of (86)Y-DOTATOC and (111)In-DTPA-octreotide." *Eur. J. Nucl. Med.* 28:1743–50.

Hänscheid, H., C. Lapa, A. K. Buck, M. Lassmann, and R. A. Werner (2018). "Dose Mapping After Endoradiotherapy with (177)Lu-DOTATATE/DOTATOC by a Single Measurement After 4 Days." *J. Nucl. Med.* 59:75–81.

Haugen, B. R., E. K. Alexander, K. C. Bible, G. M. Doherty, S. J. Mandel, Y. E. Nikiforov, F. Pacini, G. W. Randolph, A. M. Sawka, M. Schlumberger, K. G. Schuff, S. I. Sherman, J. A. Sosa, D. L. Steward, R. M. Tuttle, and L. Wartofsky (2016). "2015 American Thyroid Association Management Guidelines for Adult Patients with Thyroid Nodules and Differentiated Thyroid Cancer: The American Thyroid Association Guidelines Task Force on Thyroid Nodules and Differentiated Thyroid Cancer." *Thyroid* 26:1–133.

Hennrich, U. and K. Kopka (2019). "Lutathera(®): The First FDA- and EMA-Approved Radiopharmaceutical for Peptide Receptor Radionuclide Therapy." *Pharmaceuticals (Basel, Switzerland)* 12:114.

Herzog, H., F. Rösch, G. Stöcklin, C. Lueders, S. M. Qaim, and L. E. Feinendegen (1993). "Measurement of pharmacokinetics of yttrium-86 radiopharmaceuticals with PET and radiation dose calculation of analogous yttrium-90 radiotherapeutics." *J. Nucl. Med.* 34:2222–26.

Hofman, M. S., L. Emmett, S. Sandhu, A. Iravani, A. M. Joshua, J. C. Goh, D. A. Pattison, T. H. Tan, I. D. Kirkwood, S. Ng, R. J. Francis, C. Gedye, N. K. Rutherford, A. Weickhardt, A. M. Scott, S. T. Lee, E. M. Kwan, A. A. Azad, S. Ramdave, A. D. Redfern, W. Macdonald, A. Guminski, E. Hsiao, W. Chua, P. Lin, A. Y. Zhang, M. M. McJannett, M. R. Stockler, J. A. Violet, S. G. Williams, A. J. Martin, I. D. Davis, TheraP Trial Investigators, and the Australian and New Zealand Urogenital Prostate Cancer Trials Group (2021). "[^{177}Lu] Lu-PSMA-617 versus cabazitaxel in patients with metastatic castration-resistant prostate cancer (TheraP): a randomised, open-label, phase 2 trial." *Lancet* 397:797–804.

IAEA. *Yttrium-90 and Rhenium-188 Radiopharmaceuticals for Radionuclide Therapy.* Vienna: IAEA, 2015.

Mattsson, S., L. Johansson, S. Leide-Svegborn, J. Liniecki, D. Noßke, and K. Riklund (2015). "ICRP Publication 128: Radiation dose to patients from radiopharmaceuticals: a compendium of current information related to frequently used substances." *Ann. ICRP* 44:7–321.

Miller, C., J. Rousseau, C. F. Ramogida, A. Celler, A. Rahmim, and C. F. Uribe (2022). "Implications of physics, chemistry and biology for dosimetry calculations using theranostic pairs." *Theranostics* 12:232–59.

Rösch, F., H. Herzog, B. Stolz, J. Brockmann, M. Köhle, H. Mühlensiepen, P. Marbach, and H. W. Müller-Gärtner (1999). "Uptake kinetics of the somatostatin receptor ligand [86Y] DOTA-DPhe1-Tyr3-octreotide ([86Y] SMT487) using positron emission tomography in non-human primates and calculation of radiation doses of the 90Y-labelled analogue." *Eur. J. Nucl. Med.* 26:358–66.

Sartor, O., J. de Bono, K. N. Chi, K. Fizazi, K. Herrmann, K. Rahbar, S. T. Tagawa, L. T. Nordquist, N. Vaishampayan, G. El-Haddad, C. H. Park, T. M. Beer, A. Armour, W. J. Perez-Contreras, M. DeSilvio, E. Kpamegan, G. Gericke, R. A. Messmann, M. J. Morris, B. J. Krause, and Vision Investigators (2021). "Lutetium-177-PSMA-617 for Metastatic Castration-Resistant Prostate Cancer." *N. Engl. J. Med.* 385:1091–103.

Turner, J. H. (2018). "Recent advances in theranostics and challenges for the future." *Br. J. Radiol.* 91:20170893.

Violet, J., P. Jackson, J. Ferdinandus, S. Sandhu, T. Akhurst, A. Iravani, G. Kong, A. R. Kumar, S. P. Thang, P. Eu, M. Scalzo, D. Murphy, S. Williams, R. J. Hicks, and M. S. Hofman (2019). "Dosimetry of (177)Lu-PSMA-617 in Metastatic Castration-Resistant Prostate Cancer: Correlations Between Pretherapeutic Imaging and Whole-Body Tumor Dosimetry with Treatment Outcomes." *J. Nucl. Med.* 60:517–23.

Zaknun, J. J., L. Bodei, J. Mueller-Brand, M. E. Pavel, R. P. Baum, D. Hörsch, M. S. O'Dorisio, T. M. O'Dorisiol, J. R. Howe, M. Cremonesi, and D. J. Kwekkeboom (2013). "The joint IAEA, EANM, and SNMMI practical guidance on peptide receptor radionuclide therapy (PRRNT) in neuroendocrine tumours." *Eur. J. Nucl. Med. Mol. Imaging* 40:800–16.

Zimmer, A. M., T. M. Kuzel, E. Samuelson, S. M. Spies, S. Rosen, and R. T. Maguire (1992). "Comparative pharmacokinetics of In-111 and Y-90 B72. 3 in patients following single dose intravenous administration." *Antibody Immunoconj. Radiopharm.* 5:285–94.

Chapter 12

Principles of Prospective Treatment Planning

Robert F. Hobbs, Ph.D., DABR

Associate Professor
Department of Radiation Oncology
Johns Hopkins Medical Institute
Baltimore, MD

12.1 The Current State of RPT

Radiopharmaceutical therapy (RPT) has grown rapidly over the last decade for treating numerous cancer types. Current standard practice for RPTs is to treat with fixed activity or mass-based treatment, often over multiple fractions as presented in Chapter 1, with few exceptions. The rationale for this approach comes from chemotherapy, where fixed or ad hoc methods for determining amounts of drugs to administer are used and no personalized quantification of potential toxicity or efficacy is currently available. The amount of drug administered is likely to be tolerated by most patients, and the fractionation enables physicians to monitor potential developing toxicity in patients and modify or curtail the treatment as needed. This approach has made its way into radiopharmaceutical therapy, perhaps because the method of administration is essentially the same. The amount of activity determined per cycle is typically established in phase II clinical trials, where the amount of activity per cycle is escalated until toxicities appear in patients. The fractionation is likely established preliminarily by consensus, possibly based on preclinical considerations or physicians' expectations of toxicity versus efficacy. As such, the patients with the most susceptibility to

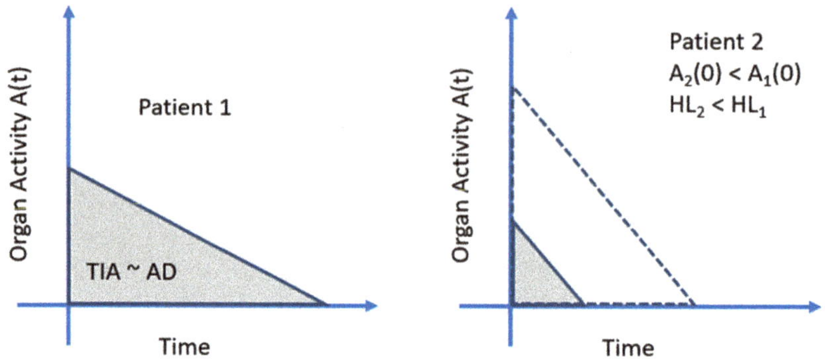

Figure 12.1 Activity escalation paradigm. Fixed activities are set for all patients based on the kinetics of Patient 1. AD to the dose-limiting organ (DLO) is roughly equal to the grey area under the curve, the time-integrated activity, TIA. Patient 2 has a lower uptake, $A_2(0) < A_1(0)$, and a faster clearance rate, $HL_2 < HL_1$, and is severely underdosed. The dashed line shows a TIA approximately equal to that of Patient 1, which would allow for more than twice the administered activity of Patient 1.

the toxic side effects of these drugs set the limit for how much is administered to all, as illustrated in Figure 12.1.

In RPT, as opposed to chemotherapy, damage to both normal organs and target lesions may be quantified by the physical quantity absorbed dose. This quantity depends on the pharmacokinetic distribution of the drug throughout the body as a function of time. One could assume that damage to organs at risk in chemotherapy also depend on the pharmacokinetics of the administered drug, but the differences with RPT are threefold:

1. The exact physical mechanism of damage is known in RPT (the absorbed dose).
2. Since the drug is radioactive, it can be imaged using SPECT/CT or PET/CT as a function of time and the individual's pharmacokinetics obtained and quantified.
3. The damage (absorbed dose) to patient organs and lesions can be obtained using computer Monte Carlo simulation of radioactive decay within the patient.

In Figure 12.1, the area under the time-activity curve represents the number of decays of radioactive material in the organ at risk and (to first order) the absorbed dose to that organ. The administered activity for all patients is determined by patient 1, who has a greater uptake in the organ at risk and a greater retention of activity than patient 2. As with *external beam radiation therapy* (EBRT), treatment efficacy or normal tissue toxicity is caused by radiation and may be quantified by absorbed dose thresholds. Patient 2 could potentially benefit from an increased amount of activity while staying below the threshold of toxicity for the organ at risk.

12.2 Personalized Dosimetry-Based Prospective Treatment Planning

Exploiting these particularities intrinsic to radiopharmaceutical therapy to tailor therapy to each patient would operate as follows:

1. Injection of a small amount of therapeutic activity, known as a tracer or pre-therapeutic amount, typically on the order of 2–5 mCi, or of an appropriate surrogate (see Chapter 11).
2. Imaging of the activity as a function of time at judiciously chosen time points (see Chapter 15).
3. Conversion of activity as a function of time to absorbed dose per unit activity, known as the *dose coefficient*, *d*, in organs at risk and lesions.
4. Scale the administered activity, A_0, to the Maximum Tolerated absorbed Dose per normal organ at risk, *MTD*:

$$A_0(organ) = \frac{MTD(organ)}{d(organ)} \tag{12.1}$$

5. Perform procedure (4) for all organs at risk and select the lowest value. This is the *maximum tolerated activity* (MTA) for the specific patient.

$$MTA = \min\{A_0(organ)\} \tag{12.2}$$

6. Administering the MTA will allow for the most activity and absorbed dose to the lesions. This may be reduced if the activity needed to reach a *target therapeutic mean absorbed dose* (TD) in the lesions is less than the MTA:

$$A_0(tumor) = \frac{TD(tumor)}{d(tumor)} \tag{12.3}$$

7. If applicable, calculate the *minimum effective activity* (MEA), which is the largest of the activities needed to treat the tumors, i.e., it is the activity needed to deliver the target absorbed dose or higher to all tumors.

$$MEA = \max\{A_0(tumor)\} \tag{12.4}$$

8. Select the appropriate administered therapeutic activity. If MEA < MTA, then there is a therapeutic window for the patient where any activity administered between the two values, MEA < A_0 < MTA, will deliver safe and effective treatment. However, since RPT often treats systemic disease, there may be invisible disseminated disease, and treating to MTA may still be the better option. On the other end, the MTA may create clinical logistic complications (inpatient vs. outpatient, higher risk of long term stochastic risk) such that a lower value may indeed prove to be the better choice.

In order for a change of activity administration paradigm to a personalized approach to be beneficial—and to put into practice the exploitation of the particularities of radiopharmaceutical therapy vis-a-vis chemotherapy—several conditions are required:

1. ***There needs to be substantial variability in individual patient pharmacokinetics to warrant the time, money, and effort to image and perform dosimetry on individuals***. The amount of activity administered to each patient would vary considerably, similarly to the schematic example shown in Figure 12.1. If not, then dosimetry collected from a small population could serve as the basis for determining the dosimetry for all patients, in a manner similar to the way in which effective dose is calculated for stochastic risk assessment from diagnostic radiopharmaceuticals from a small cohort of individuals. To that point, multiple studies have shown that *absorbed dose* (AD) to normal organs frequently varies between patients from a same modality by factors of 2, 3 and up to an order of magnitude. As an example—and one that we will go into in further detail later in the chapter—is the variability of activity needed to deliver 75 cGy of absorbed dose to the whole body from ^{131}I-tositumomab (Bexxar) (Wahl 2014), where a somewhat Gaussian distribution of 50–150 mCi to patients with a tail of up to 150 mCi is needed for the 75 cGy of whole body absorbed dose. This type of variability is consistent across modalities, showing that the vast majority of patients in RPT are seriously underdosed and would benefit from personalized treatment planning.
2. ***The individual patient's pharmacokinetics needs to be consistent.*** That is, the pharmacokinetics of the tracer, or pre-therapeutic, activity must adequately predict the therapeutic pharmacokinetics. Fewer studies have been done to validate this point, which is generally less contentious, especially for normal organs. Tumor dosimetry predictability is considered to be less stable. However, Dewaraja et al. (2014) studied predicted tumor absorbed dose from pre-therapeutic imaging versus therapeutic tumor absorbed in 130 patients treated with ^{131}I-tositumomab and found that the pre-therapeutic absorbed dose reasonably predicted the therapeutic absorbed dose (248 cGy on average vs. 270 cGy, with 80% of the values within 25%). This relative uncertainty in the predictability of the tumor absorbed dose may be another reason to treat closer to MTA. There is some question as to whether repeated (fractionated) therapies change the character of the pharmacokinetics in particular for tumors where expression modification may occur. Similar questions have been voiced concerning the administration of the pre-therapeutic modifying the uptake of the therapeutic. If pharmacokinetics were unreliable or variable, then dosimetry could only be performed retrospectively, which would be desired for absorbed dose verification and reporting. Verification would be similar to that performed in brachytherapy, where seed implants are planned dosimetrically, but post-implant imaging and dosimetry are still performed to compare to and validate the plan, as well as to catch any

unforeseen and potentially detrimental deviations. This post-therapy validation is not done in RPT routinely, but should be done in conjunction with pre-therapeutic dosimetry-based treatment planning for the same reasons as in brachytherapy. Additionally, if pre-therapeutic imaging or an adequate surrogate is not available, post-dosimetric analysis of the first fraction of a treatment may be used as the pre-therapeutic for subsequent treatment.

12.3 Approaches

In the procedure listed in the above section, one item deserves a little more attention, namely item (3): *Conversion of activity as a function of time to absorbed dose per unit activity, known as the dose coefficient, d, in organs at risk and lesions*. The step of moving from activity as a function of time to dose coefficient is the core of the dosimetric process. Two general methods are available, the *MIRD S value method* and the *personalized Monte Carlo-based method*, also known as "voxelized" dosimetry, and these will be discussed in more detail in Chapters 13 and 18, but we give here the main steps and highlight some of the commonalities, as well as the main differences (Graves et al. 2021).

1. ***Register the CT scans and the associated emission (SPECT or PET) images.*** Note that this is not strictly necessary for S value-based dosimetry, as contours could be drawn over the regions of interest at the different time-points. This is more complicated and adds a level of uncertainty. In the past, however, before reliable deformable registration, this approach would actually have been more precise for small regions of interest, such as tumors.
2. ***Segment organs, tumors, or other regions of interest.***
3. ***Monte Carlo method only: run the decay simulations using the patient-specific images, collect the energy in the regions of interest, and divide by the mass of the region of interest.*** This can be done at the region level and at the voxel level. The result is a dose rate per region or voxel.
4. ***Model and integrate the activity (S value method) or dose rate (Monte Carlo method) as a function of time.*** Doing this obtains the time-integrated activity (S value method) or absorbed dose (Monte Carlo method), respectively.
5. ***S value method only: convert time-integrated activity in source regions to absorbed dose in target regions using.*** This is done using established matrix conversion values.

The approaches do not differ significantly, especially when one consider that the S values used in step (5) were obtained using Monte Carlo simulations of decays, albeit in anthropomorphic phantoms rather than on individual patients. The steps are very similar, but not necessarily in the same order. Voxelized dosimetry offers a higher level of personalization, with the ability to better integrate tumor dosimetry with the normal organ dosimetry. S values do not exist between tumors and normal tissues, as the phantoms are of healthy subjects, such that tumor dosimetry typically uses a spherical model for self-dose, potentially supplemented by a whole body to whole body photon absorbed dose. However, the S value method, though typically applied to whole organs, is, in fact, more flexible than the voxelized approach and has been used to perform dosimetry at scales that are below imaging resolution.

12.4 Roadblocks

Historically, the concept of quantitative personalized determination of the administered activity for radiopharmaceuticals and dosimetry in general has been mostly ignored by the nuclear medicine community. As interest in RPT has grown recently, this view is changing, and dosimetry is now seen as having a definite potential. Push-back has also been strong from industry, as companies have been afraid of the consequences of using dosimetry, such as more regulation for the different pharmaceuticals, delays in bringing products to market, and a loss of customers for whom the dosimetric prognosis could be poor. Here, too, companies are realizing the potential benefits for the patients and are cautiously eyeing the developments in RPT dosimetry or even embracing the change.

12.4.1 Lack of Dosimetric Evidence

One of the major objections to RPT dosimetry has traditionally been the lack of evidence to support the use of dosimetry, in particular with calls for multi-center two-arm clinical trials to compare outcomes from standard fixed activity-based therapy to personalized dosimetry-based therapy. However, because of a reluctance to fund such trials, RPT dosimetry has found itself in a catch-22 of not being able to prove it is beneficial and not having its use desired until said benefit was proven. Nevertheless, review of the literature reveals a rather large body of more focused dosimetric-response relationships for both normal organs and tumors. In particular, the reader is directed to Strigari et al. (2013) and Graves and Hobbs (2020), who performed such reviews and found extensive evidence of correlations between absorbed dose and clinical endpoints in a majority of studies examined: 48 of 79 in the Strigari et al. review and 32 of 39 in the Graves review.

Some of the more notable or famous examples include the study by Barone et al. (2005) who published a seminal paper showing correlation between AD (and BED) and renal toxicity as measured by creatinine clearance loss/year in *^{90}Y-peptide receptor radiation therapy* (PRRT). Bodei et al. (2008) later established toxicity thresholds in ^{90}Y-PRRT at 40 Gy of BED for healthy kidneys and 28 Gy of BED for diseased or damaged kidneys. In thyroid cancer, Maxon et al. (1983) published a study that showed a threshold for complete remission of metastatic tumors (80 Gy) and another threshold for thyroid ablation (300 Gy). In ^{131}I-tositumomab therapy of lymphoma, higher whole-body dose correlated with a statistically significant survival improvement (Wahl et al. 1998; Kaminski 2005). Dewaraja et al. (2014) also showed a tumor absorbed dose-response relationship for ^{131}I-tositumomab, with a survival of 13.6 months for patients whose lesions received greater than 200 cGy, versus 1.9 months for those patients whose lesions received less than 200 cGy.

More recently, ^{90}Y-microsphere radioembolization has seen a flurry of dosimetric publications, where Garin et al. (2013; 2015) have led the way by showing significantly improved response rates among patients with *hepatocellular carcinoma* (HCC) with a threshold determined to be 205 Gy. A response rate of 92.1% was found for lesions with a tumor absorbed dose greater than 205 Gy (105 of 114), whereas tumors that received less than 205 Gy had a response rate of only 5.5% (1 of 18). The same group published results of a two-arm trial comparing endpoints from traditional therapy versus personalized normal organ and tumor threshold-based dosimetry-guided treatment planning and showed a difference of 26.6 months of overall survival for the personalized dosimetry group versus 10.7 months for the standard treatment planning group (Garin et al. 2020).

In more recent PRRT studies, Garske-Roman et al. (2018) performed a prospective observational trial among 200 patients with advanced metastatic NETs, whereby treatment cycles were added until the cumulative renal dose reached 23 Gy, or treatment was halted due to toxicity. They found that median overall survival was 54 months in patients who reached a renal dose of 23 Gy as opposed to 25 months in patients who did not reach this target. Jahn et al. (2020) reported an excellent relationship between tumor absorbed dose and treatment response for pancreatic NETs.

While most of these studies only show relationships for secondary endpoints rather than overall survival, the accumulation of evidence is beginning to sway even the most recalcitrant of nonbelievers such that the idea that dosimetry in RPT could be beneficial is a believable proposition.

12.4.2 Inconvenience to Patients

The question then becomes: if dosimetry is a beneficial factor, how can it be implemented realistically, practically? One of the major objections is that multiple timepoint imaging is a major time inconvenience to patients, compared to a simple injection. The counter to this argument is rather straightforward. If one's reference is to typical diagnostic nuclear medicine procedures, then the time commitment may seem onerous. However, if one compares to typical EBRT therapeutic regimens, then a week of imaging is a small price to pay. Moreover, the benefits of the additional time commitment need to be well explained to the patient. If information about absorbed dose-response data akin to that presented in the previous section is also presented to the patient, that patient is much more likely to be receptive to more extended pre-therapeutic hospital visits.

12.4.3 Lack of Resources (SPECT Cameras, Physicists)

A returning question surrounding the ability of departments to consistently employ dosimetry-based treatment planning is the lack of resources needed to implement this paradigm. Typically, smaller nuclear medicine departments employ one, if any, medical physicist who would be needed to perform, or at least oversee, the dosimetry. A similar problem exits with the SPECT/CT machines needed. If the machines exist, it is unlikely that there is enough free time available to plan for multiple patients with multiple timepoints of imaging on a regular basis. This dearth of existing resources has been the impetus for studies examining the viability of single timepoint imaging, which would reduce the burden and cost needed to run dosimetry-based treatment planning for a heavy schedule of patients. The answer may lie in part outside the nuclear medicine departments, in the sister discipline of radiation oncology, where medical physicists are more plentiful and versed in therapeutic dosimetry. Even if a direct collaboration or utilization of EBRT resources is not utilized, the structure of workflow in those departments could be appropriated in nuclear medicine. In particular, where dosimetrists perform the dosimetry and physicists perform QA I [in?] radiation oncology, a similar workflow where nuclear medicine technologists perform the dosimetry and physicists control for QA could be imagined. Moreover, should dosimetry-based treatment planning prove to be beneficial to the patients, it is most likely that this in and of itself would be sufficient to drive both nuclear medicine and radiation oncology departments intent on pursuing RPT programs with dosimetry-based treatment to invest in more SPECT/CT machines, as well as physicists.

12.4.4 Lack of Expertise/Consistency across Sites

A newer, more serious objection is that there is currently a very heterogeneous mix of RPT dosimetry expertise in clinics around the country. This is exemplified by the recently published SNMMI dosimetry challenge (Uribe et al. 2021), where volunteers participated in performing dosimetric tasks using a common ^{177}Lu-DOTATATE patient data set. The results were less than overwhelming, with a spread of approximately 200% in time-integrated activity calculations and 60% in absorbed dose values. This study was done in several parts, and the exercises in which the participants were allowed the most leeway in how they performed the dosimetry resulted in the greatest variability. Similarly, the IAEA engaged in a Dosimetry Coordinated Research Project with participants from nine different countries performing dosimetry on a common dataset using the same software and a common protocol with the idea of examining the variability of results. After three years of efforts and multiple training sessions—with iterations of group meetings where results were presented and discussed and apparent outliers examined for errors—it became apparent that the results were still a mixture of expected user choice variations and actual errors. These errors could be due to transcription from software output or methodological errors, but it was also clear that the participants' lack of experience played a significant role in the ability to identify, diagnose, and correct common errors.

Most societies engaged in RPT—such as the AAPM, SNMMI, ASTRO, EANM, NCI, etc.—are devoting substantial time and resources to developing training programs, guideline documents, and providing opportunities such as this summer school to educate medical physicists, physicians, and technologists in RPT dosimetry. The MIRD Committee at SNMMI has published the MIRD 2020 Primer (Bartlett et al. 2022), the IAEA has published "Dosimetry for Radiopharmaceutical Therapy," and the ICRU has published Report 96 on Radiopharmaceutical Therapy and Dosimetry (Sgouros et al. 2021) (to be discussed in Chapter 22). Various MIRD, SNMMI, and EANM pamphlets and reports exist on more specific topics within RPT, and the AAPM has also begun producing task groups and medical physics practice guidelines. What is still lacking—and what will be a major focus of this Summer School—will be the QA/QC, checklists, failure modes and error analyses, standards, and methods to validate each institution's dosimetry. This is an active area of work, and it is imperative to have these items in place before dosimetry-based treatment planning can be available to any and all sites.

12.5 Examples

Clinical cases where dosimetry-based treatment planning has been used prospectively in RPT are relatively rare. We will look at three examples, in chronological order.

12.5.1 ^{131}I-tositumomab for Lymphoma

Iodine-131-tositumomab, also known as Bexxar, was radioimmunotherapy for the treatment of patients with CD20-positive relapsed or refractory, low grade, follicular, or transformed non-Hodgkin's lymphoma having progressed during or after rituximab therapy. It was given in two parts, the first of which was cold tositumomab coupled with a pre-therapeutic (5 mCi) ^{131}I-tositumomab dosage. The therapeutic administration was optimized using patient-specific dosimetry of the pre-therapeutic, albeit of the simplest kind imaginable in order to be implemented as widely as possible. The organ at risk was the bone marrow, and studies had shown that whole-body absorbed dose could be a valid surrogate for bone marrow dosimetry for this RPT in that it correlated with hematotoxicity, with the threshold set at 75 cGy, but with possible lowering of that value as a function of patient blood counts. The patient was imaged using whole-body anterior and posterior planar scans at three different timepoints: (1) on Day 0; (2) on Day 2, 3, or 4; and (3) on Day 6 or 7. This flexibility also allowed for a maximum of patient and scheduling convenience. The first set of images was taken before the patient as allowed to void, so that the antero-posterior mean (AP) counts at Day 0, c(0), gave the baseline for relative dosimetry. That is, no activity quantification was necessary; the activity at each succeeding timepoint was simply the administered activity modulated by the ratio of AP counts at the timepoint to the AP counts at Day 0:

$$A(t) = A(0) \times \frac{c(t)}{c(0)}. \tag{12.5}$$

No calculations by the user were necessary, as a protected Excel spreadsheet distributed by the company with input value boxes and formulas for the dosimetry was used in order to reduce the risk of user error. The activities at the three timepoints were fit with a mono-exponential (shown as a straight line on a log-linear plot), and the decay constant λ (or equivalently, the effective half-life) was obtained. The absorbed dose per unit activity, the dose coefficient, d, to the patient was then:

$$d(wb) = \frac{1}{\lambda} \times S(wb \leftarrow wb). \tag{12.6}$$

The administered activity was then scaled to the MTD of 75 cGy.

$$A_0 = \frac{75 \text{ cGy}}{d(wb)}. \tag{12.7}$$

The advantage of such a system was its simplicity and fool-proof design that bypassed the need for technical dosimetric expertise.

Another important point was the use of a surrogate for bone marrow toxicity. In general, bone marrow absorbed dose is much more difficult to determine than absorbed dose for other organs due to its disseminated nature. Several different techniques have evolved to account for this difficulty. The use of blood as a surrogate was common for a while, sometimes modulated by an ad hoc factor to account for the different concentration of RPT in the bone marrow as compared to the blood (0.4 is used for antibody-based radiopharmaceuticals). A common model is to use a two-compartment model for calculating absorbed dose to the blood. Blood samples can be taken at different timepoints to obtain the blood activity concentration as a function of time and the self-dose of blood to blood while using a standard reference total blood volume for the patient or a total blood volume based on patient mass. An additional component of the whole-body photon component may be added:

$$D(blood) = \left[\tilde{A}(blood) \times S(blood \leftarrow blood)\right] + \left[\tilde{A}(wb) \times S_{\varphi}(wb \leftarrow wb)\right], \quad (12.8)$$

where S_{φ} is the photon-only whole-body to whole-body S value.

With the more recent development of voxelized dosimetry, contouring of specific regions known to be marrow-rich is often used, in particular the trabecular bone of the L2-L4 lumbar vertebrae. The assumption is then that the absorbed dose to that region is representative of the bone marrow absorbed dose as a whole. More sophisticated methodologies exist that account for the marrow cavity versus bone structure within the contoured voxels.

12.5.2 [^{131}I]NaI for Thyroid

In 2008 a patient was treated at Johns Hopkins for metastatic papillary thyroid carcinoma after thyroidectomy using a fully personalized three-dimensional dosimetric prospective in-house treatment planning software, 3D-RD (Prideaux et al. 2007), the forerunner of the now-commercially available 3D-RD-S. The details were published in Hobbs et al. (2009), and we present a brief synopsis here.

The patient presented with extensive pulmonary metastases as well as two iodine-avid lesions in the brain. The concern over pulmonary toxicity resulted in the course of action chosen. A dosage of 2.5 mCi of ^{124}I was administered as a pre-therapeutic, and the patient was imaged at five timepoints, once on each of the first five days with PET/CT. The images were registered across time using CT-to-CT registration, and contours were drawn on the lungs. Monte Carlo simulations were run on each of the five PET/CT image sets by accounting for the differences in physical half-life between the pre-therapeutic ^{124}I and the therapeutic ^{131}I (for more on this process, see Chapter 11). The energy deposited in the lungs was tabulated and divided by the lung mass to obtain the dose rates in the lungs at the different timepoints. A function of the form

$$D(t) = C(1 - e^{-Kt})e^{-\lambda t} \quad (12.9)$$

was fit to the dose rate data, divided by the administered pre-therapeutic activity and the administered therapeutic activity scaled to the MTD of 27 Gy, a value obtained from the literature (Press et al. 1993). Pursuant to the calculation, the patient was administered the recommended activity of 5.1 GBq one week after the pre-therapeutic administration. The patient went into remission, with TSH numbers dropping dramatically and the brain lesions resorbing. The patient continues to live a healthy life with no re-treatments needed to date.

Although anecdotal, the successful application of the prospective dosimetry-based treatment planning can give confidence in the approach. A major learning point was the parallel S value dosimetry performed as a second check. Only two regions of interest were considered, the lungs and the rest of body. The absorbed dose to the lungs was then

$$D(lungs) = \left[\tilde{A}(lungs) \times S(lungs \leftarrow lungs)\right] + \left[\tilde{A}(rb) \times S(lungs \leftarrow rb)\right]. \quad (12.10)$$

Two complications arose. First, how does one determine the S value for the rest of body or the remainder of body (*rb*). This is actually given by the formula for the general case of a target region of interest, r_T, and several source regions of interest, r_S, as:

$$S\left(r_T \leftarrow rb\right) = S\left(r_T \leftarrow wb\right) \times \frac{m(wb)}{m(rb)} - \Sigma_{r_S} S\left(r_T \leftarrow r_S\right) \times \frac{m(r_S)}{m(rb)}. \quad (12.11)$$

In this instance, the formula becomes:

$$S(lungs \leftarrow rb) = \left[S(lungs \leftarrow wb) \times \frac{m(wb)}{m(rb)}\right] - \left[S(lungs \leftarrow lungs) \times \frac{m(lungs)}{m(rb)}\right]. \quad (12.12)$$

In general, this formulation is not needed, as S value software will calculate remainder of body S values automatically from the existing S values for the different regions as a function of the chosen source and target regions.

The second complication is: what phantom should be used for the S values? The patient was an 11-year-old female, so both the 10-year-old and 15-year-old phantoms were used, and the results taken as the interpolation between the two weighted by the mass of the patient and the masses of the two phantoms. However, the initial calculation did not agree with the Monte Carlo result. It wasn't until the phantom results were interpolated using the lung mass of the patient and the lung masses of the phantoms that agreement was found. This was because the tumor mass in the lung resulted in a lung mass that was disproportionately heavy compared to the patient body habitus. This highlighted the need for secondary absorbed dose calculation to corroborate results, as well as the importance of modulating the S value dosimetry with mass corrections.

12.5.3 ^{90}Y-Microspheres for HCC

Yttrium-90-microspheres are sometimes categorized as RPT, but they are fundamentally different in that they are nanoparticles that embolize in the vasculature, rather than molecules that circulate throughout the body as a function of time. As a result, the dosimetry is much simplified (see Chapter 14), making it an ideal modality for patient-specific treatment planning. Indeed the standard treatment paradigm for ^{90}Y-TheraSpheres is already close to personalized treatment, as the administered activity is determined based on a calculation that delivers a determined absorbed dose (typically 120 Gy) to the treated region in the liver using the target region's mass from patient images. However, this paradigm does not distinguish between lesion and normal tissue. The crucial step that was added in the two-arm trial by Garin et al. (2020) was to increase the specificity by delineating between normal liver and tumor and prescribing to normal tissue tolerance with an upper activity limit based on a threshold for tumor absorbed dose of 205 Gy. This approach, popularly known as the *partition model*, showed an increase in overall survival to 26.6 months compared to the 10.7 months using the standard single-region approach.

This result shows that a higher level of personalization can give a dramatic difference in outcome and encourages similar approaches in other RPT modalities, as well as moving RPT to more frontline treatment.

References

Barone, R., et al. (2005). "Patient-specific dosimetry in predicting renal toxicity with (90)Y-DOTATOC: relevance of kidney volume and dose rate in finding a dose-effect relationship." *J. Nucl. Med.* 46(Suppl 1):99S–106S.

Bartlett, R. M., et al. *MIRD Primer 2022: A Complete Guide to Radiopharmaceutical Dosimetry.* Reston, VA: Society of Nuclear Medicine and Molecular Imaging, 2022.

Bodei, L., et al. (2008). "Long-term evaluation of renal toxicity after peptide receptor radionuclide therapy with 90Y-DOTATOC and 177Lu-DOTATATE: the role of associated risk factors." *Eur. J. Nucl. Med. Mol. Imaging* 35(10):1847–56.

Dewaraja, Y. K. et al. (2014). "Tumor-Absorbed Dose Predicts Progression-Free Survival Following (131)I-Tositumomab Radioimmunotherapy." *J. Nucl. Med.* 55(7):1047–53.

Garin, E., et al. (2013). "Boosted selective internal radiation therapy with 90Y-loaded glass microspheres (B-SIRT) for hepatocellular carcinoma patients: a new personalized promising concept." *Eur. J. Nucl. Med. Mol. Imaging* 40(7):1057–68.

Garin, E., et al. (2015). "Personalized dosimetry with intensification using 90Y-loaded glass microsphere radioembolization induces prolonged overall survival in hepatocellular carcinoma patients with portal vein thrombosis." *J. Nucl. Med.* 56(3):339–46.

Garin, E., et al. (2020). "Personalised versus standard dosimetry approach of selective internal radiation therapy in patients with locally advanced hepatocellular carcinoma (DOSISPHERE-01): a randomised, multicentre, open-label phase 2 trial." *Lancet Gastroenterol. Hepatol.* 6(1):17–29.

Garske-Roman, U., et al. (2018). "Prospective observational study of (177)Lu-DOTA-octreotate therapy in 200 patients with advanced metastasized neuroendocrine tumours (NETs): feasibility and impact of a dosimetry-guided study protocol on outcome and toxicity." *Eur. J. Nucl. Med. Mol. Imaging* 45(6):970–88.

Graves, S. A. and R. F. Hobbs (2021). "Dosimetry for Optimized, Personalized Radiopharmaceutical Therapy." *Semin. Radiat. Oncol.* 31(1):37–44.

Hobbs, R. F., et al. (2009). "124I PET-based 3D-RD dosimetry for a pediatric thyroid cancer patient: real-time treatment planning and methodologic comparison." *J. Nucl. Med.* 50(11):1844–47.

Jahn, U. et al. (2020). "177Lu-DOTATATE Peptide Receptor Radionuclide Therapy: Dose Response in Small Intestinal Neuroendocrine Tumors." *Neuroendocrinology* 110(7–8):662–70.
Kaminski, M. S. et al. (2005). "131I-tositumomab therapy as initial treatment for follicular lymphoma." *N. Engl. J. Med.* 352(5):441–49.
Maxon, H. R. et al. (1983). "Relation between effective radiation dose and outcome of radioiodine therapy for thyroid cancer." *N. Engl. J. Med.* 309(16):937–41.
Press, O. W. et al. (1993). "Radiolabeled-antibody therapy of B-cell lymphoma with autologous bone marrow support." *N. Engl. J. Med.* 329(17):1219–24.
Prideaux, A. R. et al. (2007). "Three-dimensional radiobiologic dosimetry: application of radiobiologic modeling to patient-specific 3-dimensional imaging-based internal dosimetry." *J. Nucl. Med.* 48(6):1008–16.
Sgouros, G. et al. (2021). "ICRU Report 96: Dosimetry-guided radiopharmaceutical therapy." *J. ICRU 21:*1–212.
Strigari, L. et al. (2014). "The evidence base for the use of internal dosimetry in the clinical practice of molecular radiotherapy." *Eur. J. Nucl. Med. Mol. Imaging* 41(10):1976–88.
Uribe, C. et al. (2021). "An International Study of Factors Affecting Variability of Dosimetry Calculations, Part 1: Design and Early Results of the SNMMI Dosimetry Challenge." *J. Nucl. Med.* 62(Suppl 3):36S–47S.
Wahl, R. L., S. Kroll, and K. R. Zasadny (1998). "Patient-specific whole-body dosimetry: principles and a simplified method for clinical implementation." *J. Nucl. Med.* 39(8 Suppl):14S–20S.
Wahl, R. L. (2003). "The clinical importance of dosimetry in radioimmunotherapy with tositumomab and iodine I 131 tositumomab." *Semin. Oncol.* 30(2 Suppl 4):31–38.

Chapter 13

The MIRD Schema and Computational Human Phantoms

Wesley E. Bolch, Ph.D., PE, FAAPM

Director, Advanced Laboratory for Radiation Dosimetry Studies
J. Crayton Pruitt Family Department of Biomedical Engineering
University of Florida
Gainesville, FL

13.1 Introduction

The dosimetry schema developed by the Medical Internal Radiation Dose (MIRD) Committee of the Society of Nuclear Medicine and Molecular Imaging (SNMMI) has served as a primary method for organ and tissue absorbed dose calculation in the field of nuclear medicine from the early 1960s to the present day. The MIRD schema was first introduced in 1968 (Loevinger and Berman 1968b), with a major revision released by the MIRD committee in 1976 (Loevinger and Berman 1976). Further details with example absorbed dose calculations were later assembled by the committee within its monograph entitled "The MIRD Primer," which was released originally in 1988 (Loevinger et al. 1988) and revised in 1991 (Loevinger et al. 1991). In 2009, a substantial revision to the MIRD schema nomenclature was put forward within "MIRD Pamphlet No. 21" (Bolch et al. 2009).

The most recent version of the Primer—*MIRD Primer 2022: A Complete Guide to Radiopharmaceutical Dosimetry*—is presently available through the SNMMI website (MIRD Committee et al. 2021). It includes expanded sections on radiobiology and quantitative imaging, a regulatory guide for radiopharmaceutical development and clinical trials, and a greatly expanded list of worked dosimetry case studies.

For many decades, internal dose computations in the field of radiological protection were performed under a schema developed independently of the MIRD committee by the International Commission on Radiological Protection (ICRP), even though the foundational aspects of these calculations (to include the concept of the absorbed fraction as well as source and target regions) were based upon those originally put forward by Robert Loevinger and Mones Berman in MIRD Pamphlet No. 1 and its precursor documents (Loevinger and Berman 1968a; Loevinger et al. 1965). A harmonization of the MIRD schema and the ICRP internal dosimetry schema was first established with the release of ICRP Publication 130 in 2015 (ICRP 2015b). Further adoption of the MIRD schema formation—as per "MIRD Pamphlet No. 21"—was made by the International Commission on Radiation Quantities and Units (ICRU) with the publication of ICRU Report No. 96, *Dosimetry-Guided Radiopharmaceutical Therapy*, in 2021 (ICRU 2021).

In this chapter, we review the fundamentals of the MIRD schema for assessing absorbed dose to organ and tissue regions from internalized radiopharmaceuticals. Computational human phantoms—computerized models of human anatomy—are an integral component to radiation transport simulations of internalized radioactivity, and thus we review their various forms—format types and morphometric categories—with a further summary of the three fundamental format types: stylized anatomic models, voxel-type anatomic models, and mesh-type anatomic models.

13.2 The MIRD Schema for Internal Dosimetry

13.2.1 Source and Target Regions

A key concept in the MIRD schema is the definition of source regions and target regions. *Source regions* are tissue compartments where the radiopharmaceutical is localized in a time-dependent manner via an array of different physiologic or biochemical processes. Source regions include blood, as many radiopharmaceuticals are administered by intravenous injection. In many cases, there will be a general background of radiopharmaceutical activity in the body outside readily identifiable source organs (apparent in 2D or 3D nuclear medicine images), and this "rest-of-body" activity is also a source region of activity. For a given organ, such as the liver, observable activity (e.g., through 3D SPECT imaging) will necessarily include radiopharmaceutical activity taken up by the organ parenchyma, but will also include activity present in liver blood vasculature during the period of image acquisition. Separating these components of the liver source will involve estimates of overall blood activity and its fractional volume with that organ.

Target regions, in contrast, are those tissues and organs for which a dosimetric estimate is sought. For the purpose of radiological protection in diagnostic nuclear medicine, the organs that are defined in the ICRP's protection quantity—the effective dose—are typically designated as target organs. For radiopharmaceutical therapy, two groups of target regions are of interest:

1. ***Normal organs and tissues***, for which the absorbed dose estimate is needed in the prediction of possible organ toxicity, and
2. ***Diseased tissue or cancerous tumors***, for which cell kill is the therapeutic intent.

In both cases, quantitative nuclear medicine imaging techniques are required to assemble activity versus time profiles.

It is a common misperception that the MIRD schema is applicable only at the level of the whole organ. As shown below, source regions (designated as r_S) and target regions (designated as r_T) are by design generic and completely user-defined. Yes, they can be assigned to whole organs in a computational human phantom of the radiopharmaceutical therapy patient, but they can also be defined as organ sub-regions, tissue cellular layers, multi-cellular clusters, single cells, and even sub-cellular compartments (Bolch et al. 2009; MIRD Committee et al. 2021). For nonuniform activity distributions within source tissues, the MIRD schema has found widespread application at the voxel level in which each source region corresponds to each imaging voxel in a quantitative 3D SPECT or PET image (Bolch et al. 1999; Dieudonne et al. 2008; Dieudonne et al. 2010; Lanconelli et al. 2012).

13.2.2 Mean Absorbed Dose Rate

The mean absorbed dose is defined as the mean energy imparted to target tissue (or region) per unit tissue mass. The rate at which the absorbed dose is delivered to target tissue within a patient from a radiopharmaceutical distributed uniformly within source tissue at time t following radiopharmaceutical administration is given as:

$$\dot{D}(r_T,t) = \sum_{r_S} A(r_S,t)\, S(r_T \leftarrow r_S,t) \tag{13.1}$$

where $A(r_S,t)$ is the time-dependent activity of the radiopharmaceutical in source tissue and $S(r_T \leftarrow r_S,t)$ is a quantity called the *radionuclide S value* (or *S coefficient*). We first note that this expression involves a summation over all possible source regions that might contribute absorbed dose to the target region r_T. For short-range radiations emitted by the radionuclide, the relevant source tissue may only be the target tissue itself (i.e., $r_T = r_S$). For source tissues that emit photons (gamma-rays or x-rays), the relevant source tissues may be adjacent to or even at a distance from the target tissue (i.e., $r_T \neq r_S$) and may also include the target tissue itself (i.e., $r_T = r_S$). The time-dependent activity represents the rate of nuclear decays, while the S value represents the absorbed dose rate to the target tissue per unit activity present in source tissue at time t.

The S value is characteristic of the radionuclide and the anatomic model chosen to represent the internal body anatomy of the patient. In many cases, "reference" anatomic models are used to compute the S value. These may be anatomic models of either the average-sized male or female patients at fixed ages—newborn, 1-year-old, 5-year-old, 10-year-old, 15-year-old, and adult. This choice of a reference patient is perfectly adequate for patient organ dosimetry under the MIRD schema for diagnostic nuclear medicine. In radionuclide therapy, however, it may be desirable to compute S values that are truly unique to the individual organ anatomy of the patient based upon CT or MR images. In general, however, the source and target regions r_S and r_T, respectively, are those defined within the anatomic model, and may represent the full range of configurations, including

- whole organs,
- sub-organ tissue regions,
- voxels from SPECT or PET images,
- tumors and cell clusters,
- individual cells, or
- cell components.

13.2.3 Mean Absorbed Dose: Time-Independent Formulation

In Equation (13.1), the absorbed dose rate to the target tissue is shown to be given as the sum of the product of two terms—the activity of the radiopharmaceutical activity in the source tissue $A(r_S,t)$ and the radionuclide S value for the source/target organ pair of interest $S(r_T \leftarrow r_S,t)$. While the first term is clearly a function of time since administration, the second term—the S value—may typically be considered not to vary with time during the period when the absorbed dose is delivered. Thus, the corresponding expression for the mean absorbed dose to target tissue may be written as:

$$D(r_T,\tau) = \int_0^{\tau} \dot{D}(r_T,t)dt = \sum_{r_S} \tilde{A}(r_S,\tau) S(r_T \leftarrow r_S) \qquad (13.2)$$

with

$$\tilde{A}(r_S,\tau) = \int_0^{\tau} A(r_S,t)dt \qquad (13.3)$$

where $\tilde{A}(r_S,\tau)$ is the time-integrated activity (TIA) in source tissue r_S over the dose-integration period. The time-integrated activity represents the total number of decays with the source organ.

The magnitude of the organ absorbed dose $D(r_T,\tau)$ is directly proportional to the activity of the radiopharmaceutical administered to the patient, A_0. Thus, doubling or tripling A_0 will then double or triple the organ absorbed doses resulting from that nuclear medicine study. It is then convenient to normalize organ absorbed dose by the value of A_0, thus reporting the organ absorbed dose per activity administered (mGy per MBq). If the ratio of $A(r_S,t)$ to the administrated activity A_0 is denoted as $a(r_S,t)$, then we can define the absorbed dose coefficient $d(r_T,\tau)$ in the target tissue as:

$$d(r_T,\tau)=\sum_{r_S}\tilde{a}(r_S,\tau)S(r_T\leftarrow r_S) \tag{13.4}$$

with

$$\tilde{a}(r_S,\tau)=\int_0^\tau \frac{A(r_S,t)}{A_0}dt=\int_0^\tau a(r_S,t)\,dt \tag{13.5}$$

where $a(r_S,t)$ is the fraction of administered activity in the source tissue r_S at time t after administration, and the quantity $\tilde{a}(r_S,\tau)$ is the time-integrated activity coefficient (TIAC). The time-dependent activity in source tissues of the patient may be obtained directly via quantitative imaging, including 2D planar imaging, 3D SPECT, 3D PET, or by tissue sampling (e.g., blood or urine collection). In the case of preclinical animal studies, this may also be determined by the counting of individual organs or tissues of the experimental animals. Alternatively, in the absence of direct measurement, the time-dependent activity in the source tissue may be obtained by numeric solution of a set of first-order coupled differential equations defined by compartment models for all organs and sub-organ tissues of interest.

The second term in equations (13.2) and (13.4) is the S value, a quantity in the MIRD schema that is specific to the radionuclide and to the computational anatomic model defining the spatial relationship and tissue compositions of r_S and r_T, as well as their intervening tissues. In equation form, the S value is defined as:

$$S(r_T\leftarrow r_S)=\frac{1}{M_T}\sum_i E_i\,Y_i\,\phi(r_T\leftarrow r_S,E_i)=\frac{1}{M_T}\sum_i \Delta_i\,\phi(r_T\leftarrow r_S,E_i) \tag{13.6}$$

where E_i = the mean (or individual) energy per particle of the ith radiation emitted;
Y_i = the number of the ith radiation emitted per nuclear decay;
Δ_i = the product (mean energy of the ith radiation emitted per nuclear decay);
$\phi(r_T\leftarrow r_S,E_i)$ = the absorbed fraction (AF) (fraction of E_i emitted within r_S that is absorbed in the target tissue r_T); and
M_T = the mass of the target tissue r_T in the model.

The specific absorbed fraction (SAF) $\phi(r_T\leftarrow r_S,E_i)$ is further defined as the ratio of the absorbed fraction and the target mass:

$$\Phi(r_T\leftarrow r_S,E_i)=\frac{\Phi(r_T\leftarrow r_S,E_i)}{M(r_T)} \tag{13.7}$$

such that

$$S(r_T\leftarrow r_S)=\sum_i \Delta_i\,\Phi(r_T\leftarrow r_S,E_i) \tag{13.8}$$

13.2.4 Mean Absorbed Dose: Time-Dependent Formulation

In the vast majority of the applications of the MIRD schema for radiopharmaceutical dosimetry, the absorbed dose integration period τ is taken to be infinity, as radionuclides of general use in nuclear medicine have relatively short physical half-lives. However, there may be instances in radiobiological modeling where a non-infinite time-integration period would be adopted, one dictated by time scales of DNA repair and cell repopulation or, alternatively, limited to a time point beyond which the radiobiological effect (cell kill) of the dose rate is low compared to cell repopulation. Furthermore, there may be situations—such as in radiopharmaceutical therapy—where the target tissue mass M_T or the source tissue mass M_S changes during the absorbed dose integration period. In such cases, the radionuclide S value itself might be a function of time during the period of dosimetric assessment. We can thus update the MIRD schema equations above to include a time dependence of both the target mass and radionuclide S value:

$$D(r_T,\tau) = \int_0^{\tau} \dot{D}(r_T,t)\,dt = \sum_{r_S} \int_0^{\tau} A(r_S,t)\,S(r_T \leftarrow r_S,t)\,dt \tag{13.9}$$

$$d(r_T,\tau) = \frac{D(r_T,\tau)}{A_0} = \sum_{r_S} \int_0^{\tau} a(r_S,t)\,S(r_T \leftarrow r_S,t)\,dt \tag{13.10}$$

with

$$\begin{aligned} S(r_T \leftarrow r_S,t) &= \frac{1}{M_T(t)} \sum_i E_i\, Y_i\, \phi(r_T \leftarrow r_S,E_i,t) \\ &= \frac{1}{M_T(t)} \sum_i \Delta_i\, \phi(r_T \leftarrow r_S,E_i,t) \end{aligned} \tag{13.11}$$

13.3 Computational Human Phantoms

13.3.1 Phantom Format Types

There are two general attributes of *computational human phantoms*, or CHPs: (1) *format types* and (2) *morphometric categories* (Bolch et al. 2010). Format types describe the computational or mathematical techniques applied to describe both the virtual patient's outer body contour as well as the patient's internal organ structure. To date, CHPs have been produced along three format types:

- *stylized* (or mathematical),
- *voxel* (or tomographic), and
- *mesh* (polygon-mesh, tetrahedral-mesh, or nonuniform B-spline surfaces or NURBS).

The earliest first-generation CHPs were introduced in the early 1960s and termed *stylized phantoms* where organ/body surfaces are described via 3D geometric expressions (spheres, ellipsoids, toroids, truncated cones, etc.). Beginning in the late 1980s, and coupled with the clinical maturation of CT and MR imaging, computational human phantoms appeared in their second generation of format type in which all body regions and internal organs are represented as contiguous collections of cuboidal volume elements (voxels), each with a unique organ/tissue identifier for elemental composition and mass density. In many respects, CT-based patient models used in external beam radiotherapy are voxel phantoms of the cancer patient to be treated.

Finally, in the early 2000s, mesh-based phantoms emerged as the third generation of CHPs where body contours and organ boundaries are described either by an array of control points (defining NURBS surfaces) or by arrays of triangles (defining a polygon-mesh surface). Another variation of the mesh-type phantom is one in which the organ and body region volumes—initially defined by their polygon-mesh surfaces—are filled in 3D by tetrahedrons defining a tetrahedral-mesh phantom (Han et al. 2020).

There was a period of time between the early 2000s and the early 2010s when mesh-type phantoms were available, but radiation transport codes could not accommodate the CHP in this phantom format, so these mesh-type phantoms had to be converted into a corresponding voxel-type phantom for radiation transport simulations. This process, however, allowed the user full control over the final voxel size, which is not the case for a CT/MR-based voxel phantom where the voxel size is predetermined by the original patient image resolution. Fortunately, nearly all Monte Carlo-based radiation transport codes can now efficiently handle computational human phantoms directly in their mesh-type formats, thus making obsolete the need for the voxelization process (Furuta et al. 2017).

13.3.2 Phantom Morphometric Categories

For each of the three phantom format types, one can design a computational human phantom along one of four different morphometric categories. These are

1. reference phantoms,
2. patient-dependent phantoms,
3. patient-sculpted phantoms, and
4. patient-specific phantoms.

Reference phantoms were defined originally for radiation protection purposes by the International Commission on Radiological Protection (ICRP). ICRP reference phantoms include males and females of only a few select ages: 0-year (newborn), 1-year, 5-year, 10-year, 15-year, and adult (nominally 35 years of age). Furthermore, body sizes of reference phantoms are defined by standing heights, total body masses, and internal organ masses fixed at roughly their 50th percentile values across select populations (mostly North Americans and Western Europeans) as given in ICRP Publication 23 (ICRP 1975) and the later ICRP Publication 89 (ICRP 2002).

With the development of an increasing number of voxel phantoms, and more recently with the scalability of mesh phantoms, these limitations on 50th percentile phantom heights/weights could be lifted, thus permitting the development of *patient-dependent phantoms* of different combinations of non-50th percentile heights and weights (Johnson et al. 2009). By definition, patient-dependent phantoms are members of phantom libraries, from which precomputed databases of organ dose coefficients can be generated across various medical imaging modalities (CT, fluoroscopy, nuclear medicine, and radiography). Use of these phantoms or dose coefficient libraries in patient dosimetry are made through a morphometric matching of the individual patient to select members of the phantom library (Stepusin et al. 2017). The simplest and, in many cases, most accurate matching process is by closest pairing of patient-to-phantom by standing height and total body weight. Precomputed dose coefficient libraries are limited to those of reference and patient-dependent phantoms.

The third morphometric category is the *patient-sculpted phantom*. Here, a member of a phantom library is first selected to represent the medical patient, but a further step is then imposed—3D sculpting of the outer body contour to closely, or even exactly, match that of the real patient. In so doing, internal organs of the patient-dependent phantom are either uniformly enlarged or reduced as different body circumferences of the real patient are matched. In this process, the final phantom appears on the exterior to fully represent the patient, though no guarantee is made that internal organ structures are equivalently matched. Residual differences in organ size, shape, and internal position remain, with resulting baseline uncertainties in organ doses from medical imaging or therapy procedures (~10% in the case of CT imaging (Stepusin et al. 2017)).

Of course, a full match of patient-to-phantom appears in the *patient-specific phantom*, one in which both the outer body shape and the internal organ anatomy of the phantom exactly match those of the real patient. The patient-specific phantom—which can only be developed from 3D imaging of the patient (CT or MR) followed by detailed organ and body surface contouring—is thus the true *virtual twin* to the actual medical patient. In modern-day radiopharmaceutical treatment planning—in which SPECT/CT and PET/CT images are used to both define radiopharmaceutical time-activity curves and patient organ anatomy—organ dosimetry is indeed based upon patient-specific phantoms. Limitations exist, however:

1. The CT or MR image of the patient might only be covering a portion of the total body (thus complicating assessment of absorbed dose to organs out of the imaging field as well as to distributed organs, such as the active bone marrow)

2. Automated image segmentation of the full complement of internal organs might not be realized.

These limitations, however, might eventually be overcome. One way would be to use phantoms constructed as a fusion of preexisting computational phantom anatomy and inserted partial-body, patient-specific CT/MR-based anatomy (Kuzmin et al. 2018). Another approach would be to use AI-driven optimized algorithms for rapid and accurate anatomy image segmentation (Xie and Zaidi 2019). Nevertheless, short of whole-body imaging and fully automated organ segmentation, radiopharmaceutical therapy treatment planning will continue make use of either patient-dependent phantom libraries or patient-sculpting of select members of these virtual patient libraries.

13.4 Stylized Anatomic Models

13.4.1 The MIRD Phantom

The *MIRD Phantom* was developed by Walter Snyder's research group at the Oak Ridge National Laboratory (ORNL) (Snyder et al. 1969) and, even though Snyder chose to call the phantom a "standard man approximation," it was based on information being compiled for ICRP Publication 23 (ICRP 1975) on Reference Man. In reality, there was a parallel effort at the time to provide improved estimates of absorbed dose and dose equivalent for the International Commission on Radiological Protection (ICRP). These estimates were later published as the ICRP Publication 30 series of dose coefficients (ICRP 1979).

The heterogeneous phantom contained three principal idealizations:

1. ***It had simple geometrical shapes that approximated the organ and body shapes and dimensions***. Twenty-two internal organs were included in the design, but other unimportant structures (the nose, hands, feet, etc.) were ignored. It was assumed that each organ or tissue was homogeneous in terms of composition and density. However, different compositions and densities were used for the skeletal region, the lungs, and the remainder of the body (soft tissue).
2. ***The monoenergetic radiation source was assumed to be uniformly distributed in one of the organs (i.e., the source organ)***.
3. ***The Monte Carlo method was applied in the calculations to obtain a set of photon histories upon which estimates of the energy deposited in the organs and tissues of the phantom could be based***. Only photon transport was considered, and energy transferred to electrons by photon interactions was assumed to be deposited locally (i.e., the kerma approximation).

Fifteen source organs and 25 target organs or regions were included in the set of calculations. Twelve photon energies were considered, ranging from 0.01 to 4.0 MeV. A limited number of histories were followed in each calculation (25,000 to 50,000 histories) and, for low-energy photons, the absorbed fractions of energy for many organs were considered unreliable and were not included in the tabulations.

In 1978, Snyder et al. (1978) published the results of an extensive set of calculations of specific absorbed fractions of energy using an improved heterogeneous phantom. As with previous calculations, these results were for monoenergetic photon sources uniformly distributed in organs of the heterogeneous phantom. However, significant improvements had been made in the phantom. These improvements included:

1. The head section was represented by a right elliptical cylinder topped by half an ellipsoid.
2. The leg regions consisted of the frustums of two circular cones.
3. The male genitalia were moved to a position outside the trunk and on the front of the revised model of the legs.
4. Detailed descriptions of the scapulae and the clavicles were included.

5. The stomach and upper and lower large intestine were modeled as organs with walls and contents.
6. A source and target region called the "skin" was added to the phantom.

Twenty source organs and 20 target organs or regions were included in the set of calculations. Twelve photon energies were considered, ranging from 0.01 to 4.0 MeV. The number of histories followed was increased to 60,000 in hopes of increasing the reliability of the results. However, as before, calculations for low-energy photons were very unreliable, and other methods were used to provide an estimate of the specific absorbed fractions in these cases.

Over the years, a number of improvements to the heterogeneous phantom have been made. However, the fundamental use of the phantom for internal dose calculations has been essentially unchanged since its inception.

13.4.2 The ORNL Phantom Series

The Oak Ridge National Laboratory (ORNL) in 1980 developed a new series of stylized models of various ages (Cristy 1980). The series included an adult male, a newborn, and individuals of ages 1, 5, 10, and 15 years developed from anthropological data (legs, trunk, and head) and from age-specific organ masses published in ICRP Publication 23 (ICRP 1975). Figure 13.1 shows frontal views of the entire series, while Figure 13.2 shows representative transverse cross-sectional views of the adult and newborn models. Although some of the organ shapes and centroids were still obtained using the similitude rule from the Snyder-Fisher adult model, these phantoms represented a great improvement for pediatric dosimetry over the similitude pediatric phantoms. These phantoms also presented new regions and improvements, such as a new stylized breast tissue region for all ages, the inclusion of the new model of the heart developed by Coffey (Coffey et al. 1981; Coffey and Watson 1981), and a new model of the thyroid. While the ORNL pediatric model series was initially published in 1981 (Cristy 1980), these models were not readily utilized until 1987 with the publication of ORNL/TM-8381 (Cristy and Eckerman 1987). In this report, the only major change in the phantom series was that the 15-year-old model was assumed also to represent the average adult female. For this purpose, the breast, uterus and ovaries were modified according to published reference average values (ICRP 1975). The phantoms were used with the ETRAN Monte Carlo photon transport code (Seltzer 1988; 1991) to calculate specific absorbed fractions of energy in all five pediatric phantoms, as well as in the adult male, for 12 photon energies (0.01 to 4.0 MeV). Electron transport was not considered in these simulations, and electron energy was assumed to be locally deposited.

13.4.3 Updates to the ORNL Stylized Phantoms

Other modifications and additions to the ORNL and MIRD stylized models include a peritoneal cavity (Watson et al. 1989), a revised model of the nasal cavity and major airway (Deloar et al. 1997), and a new model of the long bones (Clariand et al. 2000). The peritoneal cavity model was developed in 1989 by Watson et al. (1989). They modified the MIRD phantom to include a region representing the peritoneal cavity to

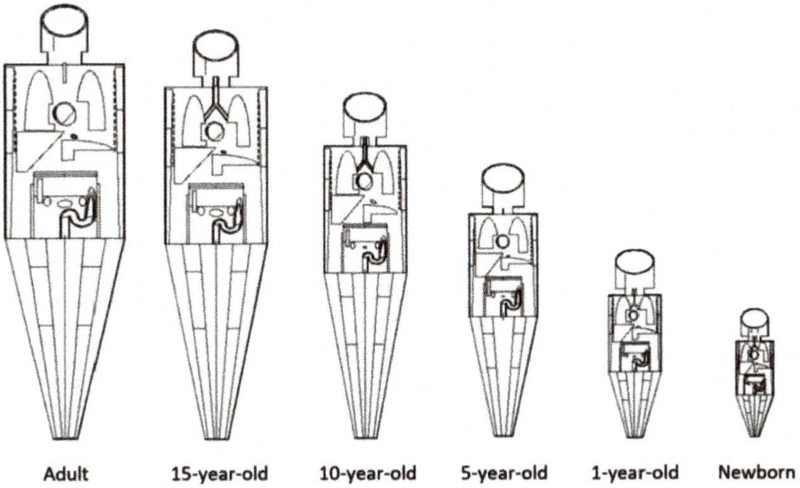

Figure 13.1 Front views of the series of ORNL stylized hermaphrodite computational phantoms.

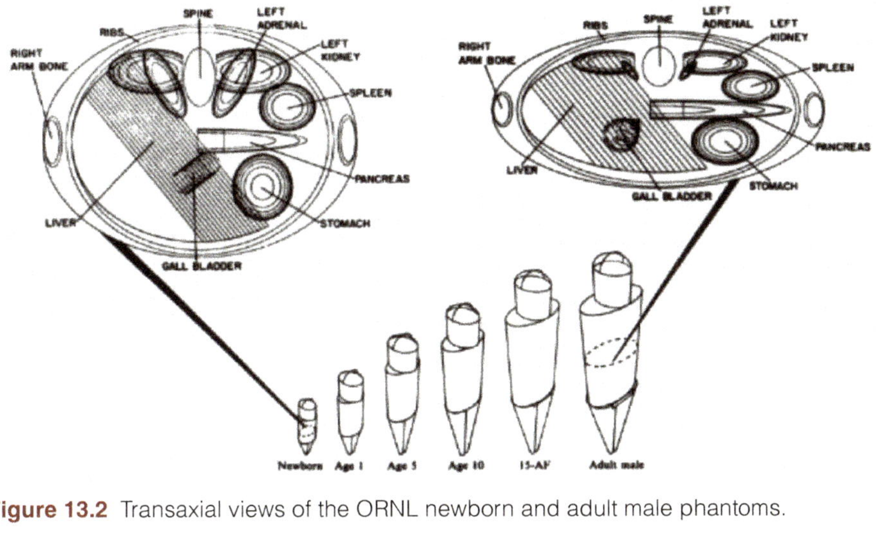

Figure 13.2 Transaxial views of the ORNL newborn and adult male phantoms.

support the dosimetry associated with several therapeutic and diagnostic techniques involving injection of radioactive material into the peritoneal cavity. Similarly, Deloar et al. (1997) developed a model of the nasal cavity and major airway to support the dosimetry associated with ^{15}O-labeled gases PET imaging agents. Clairand et al. (2000) modified the model of the long bones of the ORNL phantom series to properly

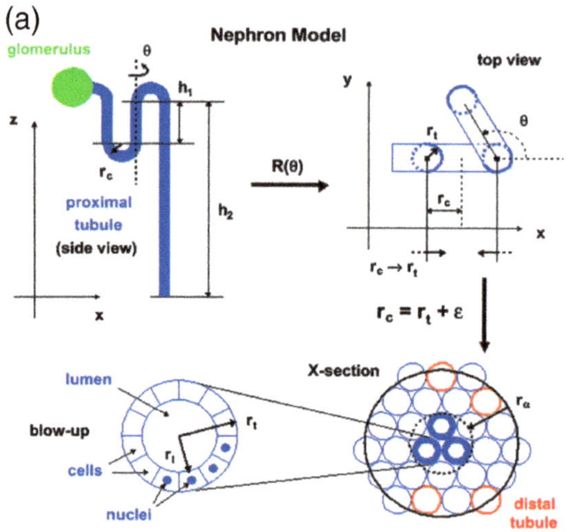

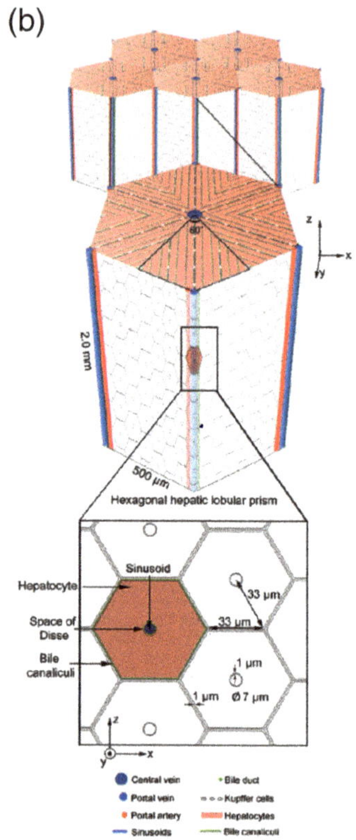

Figure 13.3 (a) Stylized small-scale model of the kidney nephron from Hobbs et al. (2012) and (b) stylized small-scale model of the liver from the study of Stenvall et al. (2014).

differentiate the cortical bone, trabecular bone, and medullary cavity. This effort was undertaken to support bone dosimetry for photon sources and to allow more accurate marrow dose calculations for children. Because the active marrow in children is not only found in the trabecular bone but also in the medullary cavities, the stylized models of the long bones of the legs and arms (truncated circular cones) of the ORNL phantom series did not allow for accurate bone dosimetry calculations.

Han et al. (2006) published extensive revisions to the adult and pediatric series of ORNL stylized phantoms, making the following revisions:

- incorporating recent developments in stylized models of the head, brain, kidneys, rectosigmoid colon, and extra-pulmonary airways;
- incorporating new models of the salivary glands and the mucosa layer of the urinary bladder, alimentary tract organs, and respiratory airways;
- adopting reference values of elemental tissue compositions and mass densities from ICRP Publication 89 and ICRU Report 46;
- providing for explicit treatment of left and right organs within organ pairs;

- providing for a systematic tabulation of electron absorbed fractions as a function of energy and subject age for all internal organs; and
- providing for methods of deriving patient-specific values of the specific absorbed fraction for both electrons and photons through interpolation/extrapolation of their phantom-derived values.

At the time of its publication, it was argued that while voxel computational phantoms (see Section 13.5.4) provided improved anatomic realism given the CT or MR image sets used in their construction, there did not yet exist a comprehensive series of reference pediatric voxel phantoms, nor the ability to simulate very fine anatomic structures as can be modeled via mathematical approximation. The authors noted that these revised stylized pediatric phantoms would fill this data need in medical dosimetry.

13.4.4 Use of Stylized Models in Radiopharmaceutical Therapy

With the publication of ORNL TM-8381 by Cristy and Eckerman (1987), *photon-specific absorbed fractions* (SAFs) were made available to both the nuclear medicine and radiological protection communities worldwide, and thus for several decades internal dosimetry computations via the MIRD schema were dominated by use of Monte Carlo radiation transport results in the ORNL series of stylized computational phantoms. In many respects, this continues to be the case for diagnostic nuclear medicine dose calculations (ICRP 2015a) and for those in radiological protection (EPA 1999; 2019).

In radiopharmaceutical therapy, however, whole-body stylized models of the phantoms are no longer deemed appropriate nor adequate for the demands of organ dosimetry accuracy. As noted below, direct CT (or possibly MR) imaging of the patient form the basis for voxelized patient models for dosimetric analysis. Still, stylized models continue to play a key and important role in therapy nuclear medicine, particularly for small-scale or micro-scale tissue regions thought to be important for assessing the absorbed doses to tissue regions and organ cell populations needed for the prediction of organ toxicity. Stylized models of tumors are also used extensively, especially for computational of the tumor self-dose.

In 2012, Hobbs et al. (2012) introduced an extension of the MIRD schema to the microscale of tissue histology, being primarily motivated by the need to support tissue dosimetry for predictions of toxicity in normal organs for alpha-emitting radiopharmaceuticals. The methodology—termed *macro-to-macro* (or M2μ) *dosimetry*—initially applies the MIRD schema at a microscale level of the organ of interest:

$$D(r_{T-micro}) = \sum_{r_{S-micro}} \tilde{A}(r_{S-micro})\ S(r_{T-micro} \leftarrow r_{S-micro}) \tag{13.12}$$

The subscript modifier *micro* denotes that the calculation is now to be performed at the microscopic or cellular level. In this instance, the target region $r_{T\text{-}micro}$ might refer to the walls of the proximal convoluted tubules, while the source region $r_{S\text{-}micro}$ might refer to radiopharmaceutical activity within the blood vessels and interstitial fluid spaces of the renal cortex. With an appropriate cellular-level tissue model, decay sites of α-particle emission may be sampled within $r_{S\text{-}micro}$, and radiation transport simulations may be used to compute the values of the absorbed fraction $\phi(r_{T\text{-}micro} \leftarrow r_{S\text{-}micro})$ To support the M2μ dosimetry approach, Hobbs and colleagues developed a stylized model of the kidney nephron, as shown in Figure 13.3a. Microscale S values were computed for source regions comprising the glomeruli and the proximal convoluted tubules (content, wall surface, or wall volumes), with target regions being the glomeruli and proximal convoluted tubule walls.

This work was later followed by the modeling study of Stenvall et al. (2014) who presented stylized microscale models of the liver, as shown in Figure 13.3b. Following radiation transport simulations, microscale S values were reported for several therapeutic radionuclides for a variety of source regions (Kupffer cells, portal artery, portal vein, central vein, and blood sinusoids) and a variety of target regions (Kupffer cells, hepatocytes, Space of Disse, and blood regions within the portal artery, portal vein, central vein, and sinusoids). Once microscale S values are established, the application of Equation (13.12) further required the time-integrated activity within these same microscale source regions—a requirement that is not feasible in a given radiopharmaceutical therapy patient. Resultantly, the M2μ dosimetry approach formulates a link between observable radiopharmaceutical time-integrated activity at the macroscale $\tilde{A}(r_{S-macro})$ (e.g., the whole kidney or the whole liver) and the desired TIA at the microscale $\tilde{A}(r_{S-micro})$:

$$\tilde{A}(r_{S-micro}) = \tilde{A}(r_{S-macro}) f_{M2\mu} \tag{13.13}$$

where $f_{M2\mu}$ is defined as the time-integrated activity apportionment factor which defines what fraction of observable macroscale activity is localized to given source regions at the microscale. Values of the TIA apportionment factor may be obtained through micro-imaging techniques in a preclinical animal model, as described and presented in Hobbs et al. (2012).

Stylized models of tumors are widely used to provide rapid assessments of self-dose, providing that quantitative SPECT or PET imaging is able to provide time-integrated activities with reasonable accuracy and precision. In the study by Olguin et al. (2000), a series of spherical tumors of varying size was used to compute both (1) monoenergetic *specific absorbed fractions* (SAFs) for monoenergetic photons, electrons, and alpha particles, and (2) some 22 different therapeutic radionuclides, including some 14 additional alpha-emitter progeny. A unique feature of this study was the consideration of tissue composition, whereby each tumor was modeled along 5 different elemental mixtures representing 100% soft tissue (ST), 100% mineral bone (MB), and three mixtures of 25/75, 50/50, and 75/25 percentages of ST and MB, respectively. Table 13.1 presents S values for all radionuclides at three of these compositions for tumors of 5 mm in radius.

Table 13.1 The S values for 22 radionuclides in tumors of 0.5 cm radius with varying material composition, with 50% ST corresponding to a tumor of 50% soft tissue (ST) and 50% mineral bone (MB)

Radionuclide	Mineral Bone (Gy Bq^{-1} s^{-1})	50% ST (Gy Bq^{-1} s^{-1})	Soft Tissue (Gy Bq^{-1} s^{-1})	% Difference (MB–ST)/ST (%)
Alpha Emitters				
At-211	4.00E-10	5.20E-10	7.44E-10	–46%
Bi-212	4.18E-10	5.40E-10	7.60E-10	–45%
Bi-213	8.29E-11	1.05E-10	1.45E-10	–43%
Ra-223	9.32E-10	1.21E-09	1.73E-09	–46%
Ac-225	9.42E-10	1.23E-09	1.75E-09	–46%
Th-227	9.66E-10	1.26E-09	1.80E-09	–46%
Beta Emitters				
Sr-89	8.05E-11	1.01E-10	1.34E-10	–40%
Y-90	1.15E-10	1.40E-10	1.75E-10	–34%
I-124	3.05E-11	3.63E-11	4.50E-11	–32%
I-131	3.11E-11	3.97E-11	5.54E-11	–44%
Sm-153	4.30E-11	5.46E-11	7.60E-11	–43%
Ho-166	9.44E-11	1.17E-10	1.54E-10	–39%
Lu-177	2.34E-11	3.02E-11	4.27E-11	–45%
Re-186	5.01E-11	6.38E-11	8.81E-11	–43%
Re-188	1.02E-10	1.26E-10	1.62E-10	–37%
Auger Electron Emitters				
Pd-103	8.97E-12	1.10E-11	1.39E-11	–35%
In-111	9.08E-12	1.05E-11	1.23E-11	–26%
Sn-117m	2.78E-11	3.51E-11	4.83E-11	–42%
I-123	7.26E-12	8.13E-12	9.50E-12	–24%
I-125	6.86E-12	7.04E-12	6.96E-12	–1%
Pt-193m	2.24E-11	2.91E-11	4.13E-11	–46%
Pt-195m	3.13E-11	4.01E-11	5.63E-11	–44%

13.5 Voxel-Type Anatomic Models

As discussed above, stylized mathematical models of human anatomy provide a relatively efficient geometry for use with Monte Carlo radiation transport codes. Nevertheless, these models are only approximations to the true anatomical features of individuals for which absorbed dose estimates are required. An alternative class of anatomic models is based upon 3D imaging techniques, such as *magnetic resonance* (MR) imaging or *computed tomography* (CT). These voxel (or tomographic) models of human anatomy represent large arrays of image voxels that are individually assigned both a tissue type (e.g., soft tissue, bone, air, etc.) and an organ identity (heart wall, femur shaft, tracheal airway, etc.). Thus, image segmentation is needed to process the original image into a format acceptable for radiation transport using codes.

13.5.1 Methods of Construction

Two general anatomic sources exist for the construction of tomographic models of human anatomy: (1) cadavers and (2) live subjects, typically medical patients. Each data source has its distinct advantages and disadvantages. Cadaver imaging generally offers a substantially improved opportunity for full anatomic coverage, including the extremities. With CT image acquisition, higher-resolution scans can be performed, as radiation dose considerations and patient motion are not of concern. For these same reasons, multiple scans on the same cadaver can be performed using different technique factors (kV, mA, filtration, etc.). Problems associated with cadaver imaging include tissue shrinkage, body fluid pooling, air introduction to the gastrointestinal tract, collapse of the lungs, and general organ settling. Perhaps the greatest disadvantage, however, is that cadaver imaging will most likely not involve the use of CT tissue contrast agents needed for soft-tissue image segmentation.

Computed tomography is perhaps the imaging modality of choice for construction of full-body tomographic computational models. Skeletal tissues are more readily defined under CT imaging, whereas image distortions of skeletal tissues are problematic under MR imaging. With live patient imaging, CT contrast agents provide acceptable definitions of soft tissue borders needed for image segmentation, and thus MR does not offer a distinct advantage over CT for this data source. In cadaver imaging, the lack of contrast agents hinders border definitions for soft-tissue organs. To compensate, MR imaging of the cadaver might be considered. However, the frozen state of the cadaver tissues can alter both spin-lattice and spin-spin relaxation times (T1 and T2, respectively), complicating and distorting MR images of the subject. Thus, CT imaging is recommended for both live and deceased subjects upon whom tomographic models are to be constructed. With modern-day multi-detector, multi-slice CT imaging, scan times are minimized, offering improved opportunities for live-subject imaging as potential data sources. CT-MR image fusion is another possibility for input data to image segmentation. However, patient motion artifacts within MR images, and their generally lower image resolutions, currently restrict opportunities for CT-MR co-registration to the head region where patient motion is more easily minimized, as is the case for radiosurgery treatment planning of brain tumors.

Following image acquisition, the next step in model construction is image segmentation. In cases where the CT image provides strong tissue contrast, automated pixel-growing methods of image segmentation can be applied to rapidly delineate organ boundaries. In these methods, a central pixel is tagged, and the algorithm groups all neighboring pixels within a defined gray-level interval (e.g., skeletal regions, air lumen of nasal passages, and certain soft-tissue organs under CT contrast). For those organs with poor tissue contrast, organ boundaries must be segmented manually. Final review of the tomographic model by medical personnel trained in radiographic anatomy is highly recommended. Standardized software packages, such as IDL (Research Systems, Boulder, CO) and 3D Slicer (https://www.slicer.org/), can offer all necessary tools for construction of voxel-based computational models.

13.5.2 Review of Representative Models

Several research groups have been involved in the development of voxel computational models over the past three decades. Most efforts have been directed toward adult models, but a few important studies have resulted in pediatric tomographic models for radiation dosimetry studies.

In 1994, Zubal and colleagues published their work in manually segmenting 129 CT transverse slices of a living male adult (Zubal et al. 1994). Their goal was to create a tomographic computational model for use in simulating nuclear medicine imaging data (Zubal and Harell 1992). A total of 36 tissue regions were identified within the 3D voxel array. The original CT images were obtained in two imaging sessions: one of 78 slices from the neck to mid-thigh with 10-mm slice thicknesses (pixel size of 1 mm), and one of 51 slices from the head to the neck region with 5-mm slices (pixel size of 0.5 mm). Image processing techniques were applied to achieve a final isotropic cubic voxel resolution of 4 mm on a side. Similar techniques also have been applied in the creation of a high-resolution, MRI-based voxel head model (Zubal et al. 1996). Stabin et al. (1999) subsequently coupled the Zubal total-body model to the MCNP radiation transport code. Yoriyaz and Stabin later reported estimates of cross-organ photon dose based on MCNP calculations within the Zubal model (Yoriyaz et al. 2000; Stabin and Yoriyaz 2002). Tagesson (1996) used the segmented Zubal head model to estimate absorbed fractions and radionuclide S values for a variety of radiopharmaceuticals used in neuroimaging.

In 1997, Jones reported research performed at the National Radiation Protection Board (NRPB) on a voxel-based computational model of the adult (NORMAN) (Jones 1997). This model was used to estimate organ doses from external photon sources over a range of energies and irradiation geometries. When comparing the results to those obtained using a MIRD-type stylized model, differences in organ absorbed doses were found to range from a few percent to over 100% at photon energies between 10 and 100 keV; differences were less than 20% at photon energies exceeding 100 keV. A subsequent study found that estimates of organ absorbed doses from internal photon sources differed widely as assessed using two different models, and differences in cross-organ absorbed doses generally increased with decreasing photon energy (Jones 1998).

The National Library of Medicine's Visible Human Project also provides an excellent data source from which one may construct tomographic computational models for use in diagnostic x-ray simulations for adult patients. This data set consists of MRI, CT, and anatomical images. The CT data consist of axial CT scans of the entire body taken at 1-mm intervals at a resolution of 512 × 512 pixels. Dr. George Xu and his students reported on their efforts to merge this data set with both the MCNP and EGS4 radiation transport codes (Xu et al. 2000). Computational results were subsequently reported for internal electron sources (Chao and Xu 2001), external photon sources (Chao et al. 2001a), neutrons (Bozkurt et al. 2000; 2001), and electrons (Chao et al. 2001b).

Some of the very first pediatric voxel computational phantoms were developed at the GSF Laboratory in Munich, Germany, in the late 1980s. These models were constructed for specific use in pediatric dosimetry studies (Zankl et al. 1988; Zankl et al. 1994). Two of these are the BABY and CHILD models developed at by Zankl and colleagues. BABY is constructed from a whole-body CT exam of an 8-week-old female 24 hours after death. CHILD was constructed from a CT exam of a live 7-year-old female who was undergoing subsequent whole-body irradiation prior to bone marrow transplantation. Image segmentation was performed that assigned voxels to 54 regions in the baby and 64 regions in the child. The models have been used to determine pediatric organ doses in a variety of medical exposure scenarios, including nuclear medicine (Petoussi-Henss and Zankl 1998), projection radiography (Petoussi-Henss et al. 1995; Veit and Zankl 1992; 1993; Zankl 1993), and computed tomography (Kalender et al. 1999; Zankl et al. 1995). Another early pediatric model, ADELAIDE, was based upon 44 consecutive CT scans of a 14-year-old female (Caon et al. 1999). This particular model is restricted to the torso only, and has been used with the EGS4 transport code to assess organ doses received during CT exams.

The first true newborn voxel phantom was developed at the University of Florida (UF) and published in 2002 (Nipper et al. 2002). This model was used to explore organ dosimetry in both fluoroscopy and CT imaging with organ absorbed doses compared to those given by the ORNL stylized newborn phantom (Staton and Bolch 2004; Staton et al. 2003; Staton et al. 2007; Pazik et al. 2007). Additional pediatric voxel phantoms were developed at UF to include the UF Series A and Series B pediatric voxel phantoms.

In 2005, the UF Series A voxel phantoms were published to include five phantoms of different ages: 9-month male, 4-year female, 8-year female, 11-year male, and a 14-year male) (Lee et al. 2005). They were constructed from CT imaging data of live patients using an IDL-based image segmentation tool. Lungs, bones, and adipose tissue were automatically segmented through window leveling of the CT numbers. Additional organs were segmented either semiautomatically or manually with the aid of both anatomic knowledge and available image-processing techniques. The phantoms were created from fused images taken from head and chest-abdomen-pelvis CT exams of the same individuals or of two different individuals of the same sex and similar age. As a result, they were partial body phantoms without arms and legs (which were outside the original scan range).

In 2006, the UF Series B pediatric voxel phantoms were published to include (1) inclusion of arms and legs via rescaling of the extremities of a health Korean adult model and (2) resizing of internal organs and body height/weight as guided by age/sex-dependent interpolation of ICRP Publication 89 reference values. The UF Series B phantoms were later used as the basis for the creation of the UF series of hybrid phantoms (see Section 13.6) and later the ICRP Publication 143 reference voxel phantoms (ICRP 2020a).

13.5.3 The ICRP Voxel-Based Reference Computational Phantoms (VCRPs)

Nuclear medicine dose coefficients for various radiopharmaceuticals have been computed and tabulated by the ICRP within its Task Group 36 based upon (1) development of reference models of the radiopharmaceutical biokinetics (mostly in the form of sums of exponential terms) and (2) radionuclide S-values derived from the source-target organ-specific absorbed fractions (SAFs) coming from radiation transport simulations in the ORNL stylized phantom series (Bolch et al. 2016). While not fully considered ICRP-endorsed computational phantoms of the ICRP reference persons, the ICRP, in effect, had adopted the ORNL phantoms for that purpose. Even the most recent compilation of nuclear medicine dose coefficients (organ absorbed dose per administered activity) in ICRP Publication 128 (ICRP 2015a) continue to use photon SAFs from the original 1987 transport results in ORNL TM-8381 (Cristy and Eckerman 1987) and to assume (for most source/target organs) that all the energy from all beta particles and electrons is locally deposited in the source organs.

In the late 1990s, the ICRP Main Commission made the decision to move forward with the development—within ICRP Committee 2—of its own reference computational phantom in the voxel-type format. This work was primarily conducted by Maria Zankl and colleagues at the GSF laboratory in Munich, Germany, as part of a task group of ICRP Committee 2. The source images of what would eventually become the ICRP Publication 110 *voxel-type reference computational phantoms* (VRCPs) of the ICRP Publication 89-defined adult male and adult female were the following (ICRP 2009).

- ***The adult male model (called Golem)*** was made from whole-body clinical CT images of a 38-year-old male of height 176 cm and body mass of slightly under 70 kg was selected. This individual had been treated by whole-body radiation therapy for leukemia, yet did not have obvious signs of anatomical abnormalities that would have precluded his use in model development. The CT image set was of the man lying supine with his arms parallel to his body torso. The image data set consisted of 220 CT image slices of an in-plane pixel matrix of 256 by 256. The original voxel size was 8 mm along the z-axis (cranial-to-caudal direction), with a resulting voxel volume of 34.6 mm^3.
- ***The adult female reference voxel phantom (called Laura)*** was constructed from a CT image set of a 43-year-old individual of standing height 167 cm and total body mass of 59 kg. This data set consisted of 174 CT image slices 5 mm in voxel height (head and trunk) and 43 CT image slices 20 mm in voxel height (legs), with both having an in-plane image matrix of 256 by 256 pixels. Voxel volumes were 17.6 mm^3. Segmentation of the image sets were performed by both in-house and commercial software.

Following a series of voxel-based adjustments to the Golem and Laura voxels, the official ICRP-adopted voxel-based reference computational phantoms were finalized to include 141 different internal tissue structures, to include a skeletal tissue compartmentalized to trabecular spongiosa, cortical bone, and medullary marrow cavities for the long bone shafts. Figure 13.4 shows frontal views of both the ICRP adult male and adult female reference voxel phantoms, along with internal views of the male model.

These adult reference voxel phantoms from ICRP have been used in a series of subsequent publications, including the reporting of dose coefficients for external radiation fields found in occupational settings (ICRP 2010), in space activities within

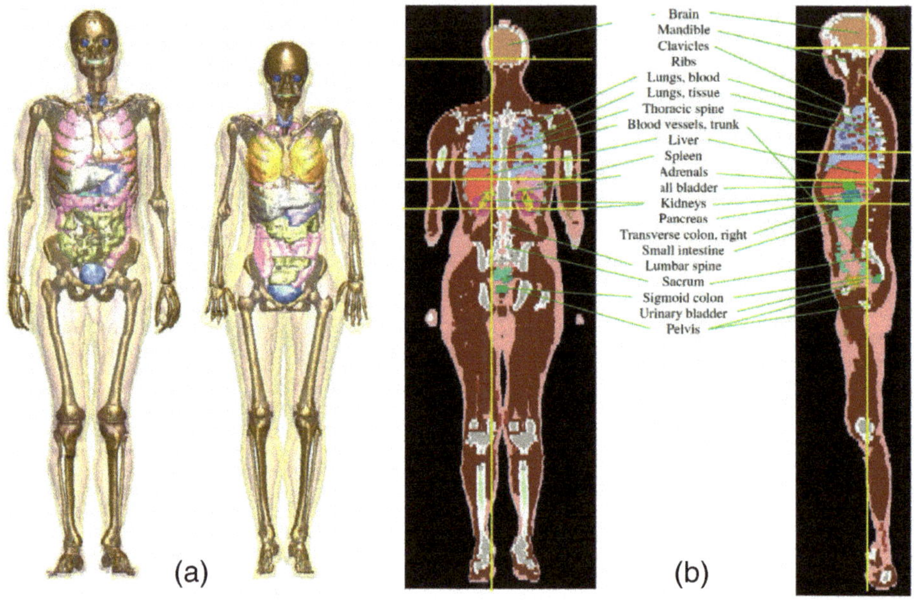

Figure 13.4 (a) Frontal views of the ICRP Publication 110 adult voxel reference computational phantoms (VRCPs) with the adult male on the left and the adult female on the right. (b) Selected coronal and sagittal views of the adult male phantom.

low-earth orbit (ICRP 2013), in commercial aviation (ICRP 2016a), and in regions of environmental radionuclide contamination (ICRP 2020b). For internal exposures, the ICRP 110 voxel phantoms were used to compute a comprehensive series of energy-dependent SAFs for internally emitted photons, electrons, alpha particles, and neutrons (the latter to accommodate internal intake of radionuclides that decay in part through spontaneous fission). This work is summarized in ICRP Publication 133 (ICRP 2016b) and was used to assemble a wide series of dose coefficients for internally inhaled or ingested radionuclides. These internal dose coefficients have appeared in a series of ICRP publications under the title OIR (Occupational Intakes of Radionuclides) (ICRP 2015b; 2016c 2017; 2019). ICRP Publication 133 included a major departure from the prior SAF datasets used by ICRP from ORNL TM-8381 in that, for the first time, an explicit accounting for electron transport—to include both collisional losses and bremsstrahlung x-ray production—was made, thus avoiding the simplistic assumption of full-energy absorption of beta particles and electrons in the source organ.

As of this writing, ICRP Task Group 36 is undertaking a major revision to the nuclear medicine dose coefficients in its Publication 128 to include both (1) updated SAFs from ICRP Publication 133 and (2) a revision of all biokinetic models of radiopharmaceuticals, from simple sums of exponential terms to full compartmental models. These models will include explicit consideration of circulating blood as a separate and independent source region for dose estimation.

13.5.4 Use of Voxel Phantoms in Radiopharmaceutical Therapy

In many forms of radiation cancer therapy—including external beam radiotherapy and brachytherapy—radiation dosimetry is an integral component of treatment planning, where the objective is to maximize the tumor absorbed dose while avoiding or minimizing normal tissue toxicities. In therapeutic nuclear medicine, tumor dosimetry may be problematic owing to:

1. lack of imaging data to define the tumor, especially for disseminated and diffuse disease,
2. the dynamic nature of radiopharmaceutical uptake, retention, and washout,
3. nonuniformities in the spatial distribution of the agent at the cellular level, and
4. time-dependent dose rates.

Consequently, there is a paucity of data on tumor dose-response relationships, upon which values of prescribed tumor absorbed dose may be assessed. Exceptions do occur, such as in the treatment of solid tumors and malignant and benign thyroid disease, but even here, there are no standardized clinically accepted protocols for radionuclide treatment planning based upon delivery of an absorbed dose prescription to the tumor.

Resultantly, current dosimetry practice in therapeutic nuclear medicine is to assess the absorbed dose to radiosensitive tissues, and based upon a general understanding of toxicity thresholds (taken primarily from previous experience in external beam radiotherapy), adjust administered activities to the cancer patient to maximize uptake and absorbed dose to the tumor while avoiding normal organ toxicities. The absorbed dose to the tumor is neither quantified nor prescribed. In contrast to diagnostic nuclear medicine dosimetry, more sophisticated methods of dosimetry are generally required in therapeutic nuclear medicine in order to provide predictive indices of tissue toxicities. In the context of therapy, patient variability in terms of both agent pharmacokinetics and body morphometry must be explicitly considered on an individual patient basis to assure optimized treatment.

Prior to the era of hybrid nuclear medicine imaging, organ dosimetry needed for prediction of normal organ toxicity was performed via the MIRD schema and precomputed radionuclide S values derived originally from stylized anatomic models, but later by individual voxel models created from image segmentation of CT or MR image sets. Obviously, the accuracy of these S values for the individual patient could be questioned owing to anatomical differences between organ size, shape, and location in the cancer patient and those within the voxel phantom, from which the S values were derived. With the introduction of hybrid imaging like SPECT/CT and PET/CT, the possibility for constructing patient-specific voxel phantoms using the CT portion of the hybrid image became possible. As described in the various chapters of this text, a number of different research and commercial software codes are based upon the CT-derived patient-specific voxel phantom resulting from image segmentation of the CT image. Tissue heterogeneity is accounted for via variations in voxel-specific CT numbers of the image. The issues still remain that (1) many of these CT images are partial body, thus complicating the assignment of tissue dose to out-of-scan and distributed organs, and (2) the boundaries of voxelized organs of interest must be segmented either manually, automatically, or through combination of both techniques. These two approaches to nuclear medicine organ dosimetry—the use of precomputed S values from whole-body stylized/voxel phantoms versus the direct computation of organ absorbed dose via Monte Carlo radiation transport or dose-point kernel convolution in a CT-devised patient-specific voxel phantom—are clearly justified in the context of diagnostic versus therapeutic uses of radiopharmaceuticals.

13.6 Mesh-Type Anatomic Models

13.6.1 Methods of Construction

As noted previously, mesh-type phantoms have been available since the early 2000s. They allow for the scalability and deformability of the modeled anatomy similar to that provided by stylized phantoms, yet they retain the anatomical realism of voxel phantoms. In the coupling of mesh phantoms to radiation transport codes, however, a final step of voxelization had to be performed as the particle tracking algorithms employed at that time did not recognize NURBS or polygon mesh surfaces. Mesh phantom voxelization thus entailed filling these surfaces with an array of voxels of user-defined dimensions.

A second advantage of mesh phantom voxelization was a resolution of potential surface overlaps and intersections introduced during phantom construction, rescaling, or deformation. The voxelization processes, by definition, eliminated these tissue incongruities. Within the past few years, however, significant advances have been made in particle tracking algorithms that now enable the direct use of meshed geometries during MC radiation transport simulation. These developments were initially introduced into the MCNP code in 2009 (Werner et al. 2018), into the GEANT4 code in 2013 (Collaboration 2017), and into the PHITS code in 2015 (Sato et al. 2018; Furuta et al. 2017).

There are, in general, three methods by which a mesh-type phantom can be created.

1. ***Segment the medical images of either live subjects or cadavers***, which is identical to that for the construction of a voxel-type computational phantom. Here, body contour and internal organ boundaries are delineated using segmentation tools within either a commercial program (such as 3D-Doctor, https://www.3d-doctor.com/) or open-source code (such as 3D-Slicer, https://www.slicer.org/). Next, these contour surfaces may be represented as either NURBS or polygon-mesh surfaces using either a commercial program (such as Rhinoceros 3D, https://www.rhino3d.com/) or open-source code (such as Blender, https://www.blender.org/).
2. ***Convert a previously constructed voxel-type phantom into a mesh-based format***. A significant example of this process is described in ICRP Publica-

tion 145, in which members of ICRP Task Group 103 developed mesh-based twins of the reference voxel phantoms from ICRP Publication 110 (see below in Section 13.6.6). This process was very laborious and time-consuming. Consequently, researchers have sought more robust and automated means of converting voxel phantoms to a polygon-mesh structure. One such approach was outlined by Brown et al. (2019) in which a boundary detection algorithm was implemented in conjunction with polygon detection to form high-quality mesh data suitable for radiation transport simulations.

3. ***Adopt an existing publicly or commercially available model, and then alter that model via combinations of uniform and nonuniform scaling***. An example of this approach is the construction of the RPI-AM (adult male) and RPI-AF (adult female) reference phantoms created at Rensselaer Polytechnic Institute (RPI) and reported by Zhang et al. (2009). These mesh-type phantoms were created from a set of 140 organ-specific surface mesh models provided by the Anatomium 3D P1 V2.0 model library (500 models were available for this purpose). Organ masses were carefully adjusted to agree within 0.5% error to reference organ mass given in ICRP Publication 89 (ICRP 2002).

Below we explore a small sample of the many mesh-type phantoms that have been developed from the early 2000s to the present day. These are given for illustration purposes only; the reader is encouraged to refer to more comprehensive reviews by Xu (2014) and Kainz et al. (2019).

13.6.2 Review of Representative Models

The very first mesh-type anatomic model applied to medical imaging was the NCAT phantom, developed and conceived by Paul Segars within his Ph.D. dissertation at the University of North Carolina in 2001 (Segars 2001). NCAT, which stands for *NURBS-based Cardiac-Torso* phantom, was an extension from the previous MCAT (stylized or mathematically based) phantom used to model nuclear medicine images from a known torso anatomy under the influence of both cardiac and respiratory motion. This phantom has been extended over the years at both Johns Hopkins University and now at Duke University to include anatomy beyond the torso and to include a library of patient-specific NURBS-based models called the XCAT (eXtended CArdiac-Torso) phantoms (Segars et al. 2013). The XCAT phantoms have been developed for a variety of multi-modality imaging research applications, including PET, SPECT, and CT imaging, as well as fluoroscopic and tomosynthesis imaging.

In 2010, the University of Florida series of mesh-type phantoms was presented by Lee et al. (2010). This phantom series was constructed as a combination of NURBS surfaces with more complex organ structures (such as the spine) left in polygon mesh format. In a collaboration with the National Cancer Institute (NCI), the UF series of reference phantoms was rescaled to present a population of anatomic models across a broad range of heights and weights. The resulting UF/NCI phantom library as reported by Geyer et al. (2014) consists of four sub-libraries: 100 adult males, 93 adult females, 85 pediatric males, and 73 pediatric females. The UF/NCI phantom library has been utilized in a variety of dosimetry studies, including CT imaging (Stepusin 2016), fluoroscopic imaging (Marshall et al. 2019), and diagnostic nuclear medicine (Sgouros et al. 2011).

In a collaboration between the IT'IS Foundation, the U.S. Food and Drug Administration, and other groups, a series of mesh-based anatomic models within the Virtual Family and Virtual Population were developed to include male and female adults and children, an obese male, an elderly male, a pregnant female, and a newborn child (Gosselin et al. 2014). These models were all developed from MR data of healthy volunteers. The latest version of the Virtual Population series is based on higher-resolution segmentation (0.5 mm voxels). Members of the series are able to be morphed and repositioned based upon realistic biomechanical models to simulate extremity articulation and tissue deformation, expansion, and shrinkage.

Under the direction of Dr. George Xu, researchers at RPI extended their work with image data sets of the Visible Human Project to convert their VIP-Man voxel model to a mesh-based format with the ability to model both respiratory and cardiac motion (Xu et al. 2000). Following the development of the RPI-AM and RPI-AF polygon-mesh reference phantoms (Zhang et al. 2009), Ding extending the anatomic modeling developed to produce a set of 10 phantoms (5 males and 5 females) covering body mass indices from 23.5 kg/m^2 (normal body weight) to 46.4 kg/m^2 (morbidly obese body weight).

13.6.3 The ICRP Mesh-Based Reference Computational Phantoms (MCRPs)

In May of 2016, the ICRP established its Task Group 103 with the intended goal of extending its adult reference computational phantoms from a voxel-based format to a mesh-based format. This work was completed in 2020, driven to a significant extent by Dr. Chan Kim and his students at Hanyang University in the Republic of Korea. The result was the release of ICRP Publication 145 (ICRP 2020c). These *mesh-type reference computational phantoms* (MRCPs) of the ICRP reference adult male and adult female were to be mesh-based twins of their original *voxel-based reference computational phantoms* (VRCPs) given in Publication 110 (ICRP 2009) and shown in Figure 13.5.

Three key developments were noted by the Task Group in the course of the voxel-to-mesh conversion process.

1. ***The newer technology would improve the anatomical integrity of the anatomic model***. Since the VCRPs in Publication 110 were created from direct segmentation of contiguous CT image slices of fairly course slice thickness (8 mm for the male and 5 mm for the female), stair-step artifacts were apparent in the original voxel-based format to include (a) gaps in the voxelized skin model with external exposure of skeletal muscle and breast tissue, and (b) gaps in the voxelized cortical bone model with internal exposure of the underlying regions of trabecular spongiosa.
2. ***The ICRP had previously relied on supplemental stylized models of very thin target tissue layers along the respiratory and alimentary tracts, which***

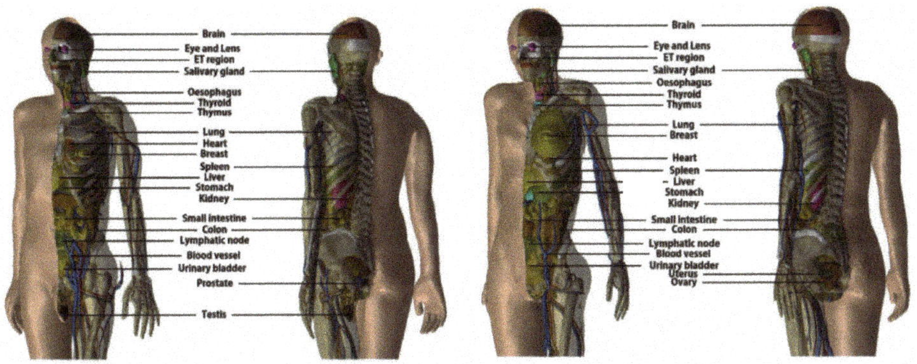

Figure 13.5 Mesh-type ICRP adult male (left) and adult female (right) computational reference phantoms (MCRP).

with the mesh-based technology could now be accommodated within the whole-body phantom.

3. ***It became apparent that all previous phantoms developed to represent the ICRP reference series had intentionally targeted internal organ volumes that, when coupled with reference tissue mass densities, would agree with reference organ masses given in ICRP Publication 89*** (ICRP 2002). Unfortunately for prior phantom modelers, it only recently became apparent that Publication 89 reference organ masses were to represent organ parenchyma only, and thus these reference masses did not account for organ blood content. Consequently, the organ volumes with the VCRPs in ICRP Publication 110 were too small, and the self-dose SAFs given in Publication 133 (ICRP 2016b) had to be mass adjusted to account for the added organ blood masses. Thus, in the conversion from voxel to mesh-based formats under Task Group 103, all internal organ volumes could now be properly adjusted to reflect their true *in vivo* volumes.

In 2020, Choi et al. at the Hanyang University Radiation Engineering Laboratory (HUREL) reported on the development of 108 adult male and 104 adult female mesh-based anatomic models covering most body sizes representative of Caucasian and Asian populations (Choi et al. 2020). These models were developed using an in-house program for automatic phantom adjustment using the ICRP Publication 145 MRCPs as anchor (or source) phantoms. Ten secondary anthropometric parameters with respect to standing height and body mass were derived from various databases and applied to the phantom library construction. Representative images of the HUREL library of mesh-type phantoms are shown in Figure 13.6.

With the successful conversion and expansion of the ICRP VRCPs from ICRP Publication 110 to a high-quality polygon-mesh format as given in ICRP Publication 145, Task Group 103 is presently extending this voxel-to-mesh conversion work to the ICRP pediatric phantom series. Work in the mesh-based skeletal models of ICRP

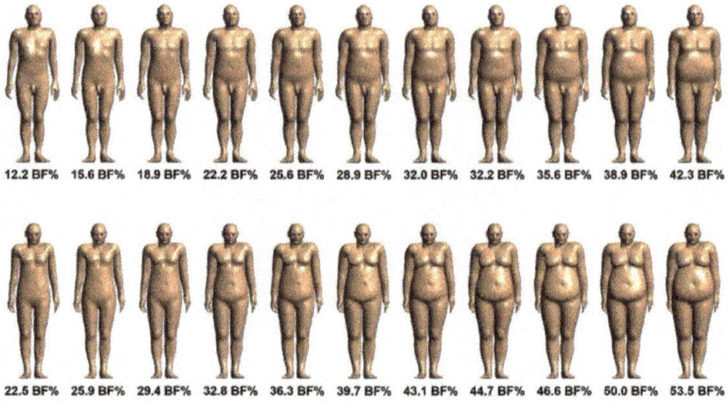

Figure 13.6 HUREL mesh-type phantom library for varying body-fat percentages (BF%s). Male adult phantoms (top row) and female adult phantoms (bottom row).

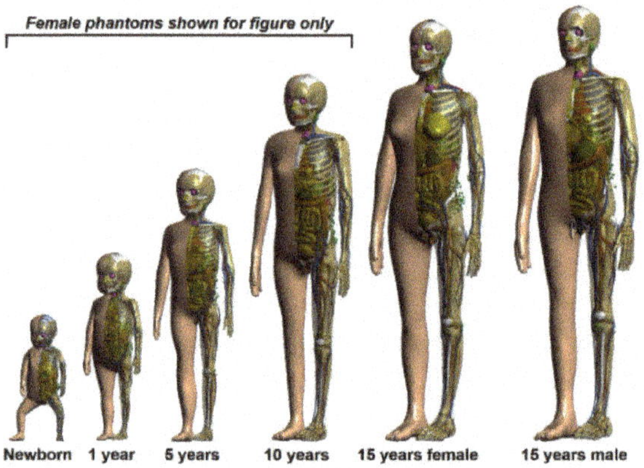

Figure 13.7 Provisional mesh-type ICRP pediatric reference phantoms prepared by ICRP Task Group 96.

pediatric phantoms has been reported by Choi et al. (2021a). The current status of the full series of mesh-type pediatric reference phantoms was presented at the 2020 meeting of the International Radiation Protection Association (IRPA) and is described in Choi et al. (2021b). Oblique views of the mesh-type ICRP pediatric phantoms in their provision state are shown in Figure 13.7. It is anticipated that they will be published by the ICRP in an upcoming report released in 2023.

13.6.4 Use of Mesh Phantoms in Therapy Nuclear Medicine

For assessment of absorbed dose to normal organs and tissues of concern for radiogenic toxicity, the morphometric category of phantom most appropriate to the task is that of the *patient-specific* phantom. For organ dosimetry in diagnostic medical imaging, including nuclear medicine, radiation doses are relatively small, are generally well below thresholds for organ toxicity, and thus the level of dose accuracy is not demandingly high. Consequently, use of *patient-dependent* phantom libraries (morphometrically matched to the patient) or *patient-sculpted* phantoms (with residual uncertainties in internal organ structure) are sufficient for image optimization and management of stochastic risk, which themselves have large uncertainties. In radiopharmaceutical therapy, however, the requirement for dose accuracy is high, as critical decisions on administered activity (and possibly dosing sequence) are dependent upon reliable models of dose response.

As noted previously, state-of-the-art dosimetry software for radiopharmaceutical therapy are widely based upon voxel-type patient-specific models originating from CT or MR images of the therapy patient. As in other forms of radiation therapy—to include external beam and brachytherapy—voxel-based CT-derived patient models allow one to score absorbed dose in individual voxels so as to assemble dose-volume histograms across both the treated tumor and the normal organs at risk. Providing voxel-based organ dosimetry in radiopharmaceutical therapy allows for the combination of DVHs across different therapy modalities.

Given the utility of voxel-based patient-specific anatomic models, what are the possible reasons to move toward mesh-based patient-specific anatomic models? Two factors come into this decision.

- The first is that for a CT-based patient-specific model, the resulting phantom is perhaps never of the entire body, thus limiting absorbed dose assessment to organs within the scan range and limiting the computation of absorbed dose to distributed organs such as the active bone marrow.
- The second is the need to delineate which array of CT voxels belong to different organs (organ segmentation). In a mesh-type phantom, all tissues and organs are predefined and available for tissue dosimetry, which is in contrast to the need for organ boundary segmentation for CT-based models.

One possible solution is the use of intensity-based deformable image registration between a preexisting whole-body mesh phantoms (reference or patient-dependent) and patient CT images. In a recent study by Carter et al. (2021), a linear attenuation coefficient phantom was generated by resampling the PSRK-Man Korean reference phantom onto a voxel grid and defining organ regions with corresponding *Hounsfield unit* (HU) reference values. The resulting HU phantom was co-registered with a patient CT scan using Plastimatch B-spline deformable registration (Figure 13.8). In parallel, major organs were manually contoured to generate a "ground truth" patient-specific phantom for comparisons. Monte Carlo-derived S values, which support nuclear medicine dosimetry, were calculated using both approaches and compared. The study found that deformable registration techniques can be used to create a personalized mesh-type phantom to better support nuclear medicine organ dosimetry. The method was also shown to be easier and faster than manual segmentation, factors that were helpful in creating a patient-specific voxel phantom. While the study focused on the diagnostic imaging agent ^{18}F-FDG, the application of this approach to radiopharmaceutical therapy is obvious and promising. Clearly, an additional step not

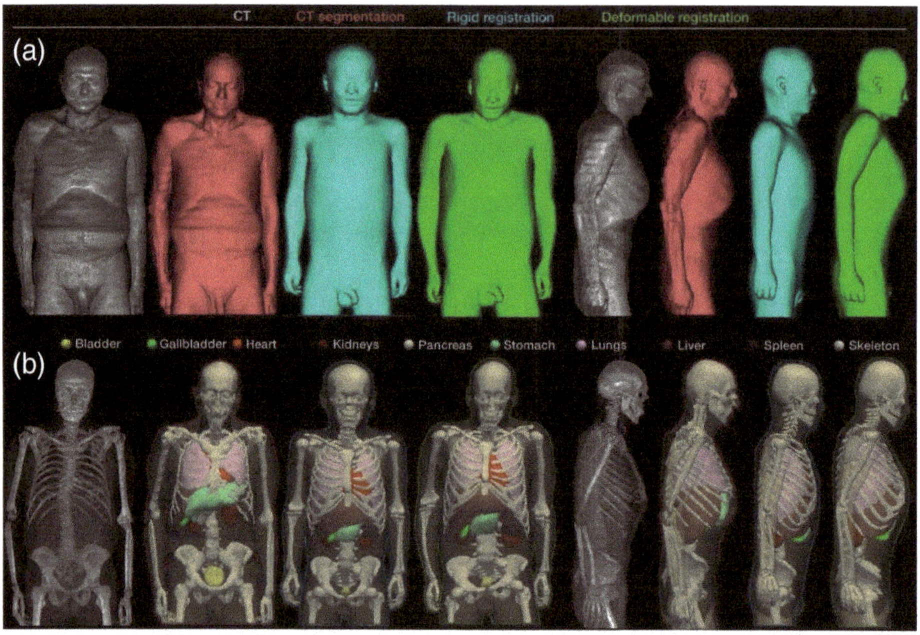

Figure 13.8 Phantoms derived from PET/CT segmentation, and morphing of a reference phantom via deformable image registration. (a) Volume-rendered CT and exterior surfaces of segmented phantom (red), rigidly co-registered reference phantom (blue), and deformably co-registered reference phantom (green). Note the improved match of exterior body contour evident by visual assessment of the deformed phantom relative to the original (rigidly co-registered) phantom. (b) Volume-rendered CT windowed for bone contrast; interior organs of each phantom.

discussed would be the insertion of modeled tumors, as the starting mesh-type phantom would have been tumor-free by design.

13.7 Summary

The MIRD schema of internal dosimetry was originally developed for assessment of radiation dose at the whole organ level to satisfy the needs of diagnostic nuclear medicine imaging. In its original implementation, values of absorbed fractions and radionuclide S values were computed using the first generation of computational human phantoms based upon geometric surface equations (i.e., stylized phantoms). With the significant advancements seen in the therapeutic application of radiopharmaceuticals for disease treatment and cancer therapy, the flexibility of the MIRD schema has ensured its continued utility. Source and target regions have evolved from the whole-organ level to now include organ subregions, imaging voxels, cellular clusters, and micrometastases. Applications of the MIRD schema through voxel-based S values now permit the determination of dose volume histograms (DVHs), which can be combined with those computed routinely in external beam radiotherapy treatment planning systems. Similarly, the movement from stylized to voxel to now mesh-based anatomic phantoms has moved MIRD schema dosimetry toward an ever-increasing path of improved patient specificity. As such, polygon mesh technology provides for nonuniform scaling of a preassembled whole-body phantom to one that can match the body contour of the individual patient (patient-sculpted phantoms). Additionally, when coupled to segmented CT images of patient internal anatomy, they can give us truly patient-specific models of our therapy patients.

References

Bolch, W., C. Lee, M. Wayson, and P. Johnson (2010)."Hybrid computational phantoms for medical dose reconstruction." *Radiat. Environ. Biophys.* 49:155–68.

Bolch, W. E., L. G. Bouchet, J. S. Robertson, B. W. Wessels, J. A. Siegel, R. W. Howell, A. K. Erdi, B. Aydogan, S. Costes, and E. E. Watson (1999). "MIRD Pamphlet No. 17: The dosimetry of nonuniform activity distributions —Radionuclide S Values at the voxel level." *J. Nucl. Med.* 40:11S–36S.

Bolch, W. E., K. F. Eckerman, G. Sgouros, and S. R. Thomas (2009). "MIRD Pamphlet No. 21: A generalized schema for radiopharmaceutical dosimetry: Standardization of nomenclature." *J. Nucl. Med.* 50:477–84.

Bolch, W. E., N. Petoussi-Henss, F. Paquet, and J. Harrison (2016). "ICRP dose coefficients: computational development and current status." *Ann. ICRP* 45:156–77.

Bozkurt, A., T. C. Chao, and X. G. Xu (2000). "Fluence-to-dose conversion coefficients from monoenergetic neutrons below 20 MeV based on the VIP-man anatomical model. *Phys. Med. Biol.* 45:3059–79.

Bozkurt, A., T. C. Chao, and X. G. Xu. (2001). "Fluence-to-dose conversion coefficients based on the VIP-Man anatomical model and MCNPX code for monoenergetic neutrons above 20 MeV." *Health Phys.* 81:184–202.

Brown, J. L., T. Furuta, and W. E. Bolch (2019). "A Robust Algorithm for Voxel-to-Polygon Mesh Phantoms." In *Brain and Human Body Modeling: Computational Human Modeling at EMBC 2018*, S. Markarov, M. Horner, and G. Noetscher, eds. Zurich, Switzerland: Springer Open, 2019.

Caon, M., G. Bibbo, and J. Pattison (1999). "An EGS4-ready tomographic computational model of a 14-year-old female torso for calculating organ doses from CT examinations." *Phys. Med. Biol.* 44:2213–25.

Carter, L. M., J. C. Ocampo-Ramos, W. E. Bolch, J. S. Lewis, and A. L. Kesner (2021). "Technical Note: Patient-morphed mesh-type phantoms to support personalized nuclear medicine dosimetry—a proof of concept study." *Med. Phys.* 48:2018–26.

Chao, T. C., A. Bozkurt, and X. G. Xu (2001a). "Conversion coefficients based on the VIP-Man anatomical model and EGS4." *Health Phys.* 81:163–83.

Chao, T. C., A. Bozkurt, and X. G. Xu (2001b). "Organ dose conversion coefficients for 0.1–10 MeV electrons calculated for the VIP-Man tomographic model." *Health Phys.* 81:203–14.

Chao, T. C. and X. G. Xu (2001). "Specific absorbed fractions from the image-based VIP-Man body model and EGS4-VLSI Monte Carlo code: internal electron emitters." *Phys. Med. Biol.* 46:901–27.

Choi, C., B. Shin, Y. S. Yeom, H. Han, S. Ha, S. Moon, G. Son, T. T. Nguyen, C. H. Kim, B. S. Chung, and W. E. Bolch (2021a). "Development of skeletal systems for ICRP pediatric mesh-type reference computational phantoms." *J. Radiol. Prot.* 41(2):139–61.

Choi C., B. Shin, Y. S. Yeom, T. T. Nguyen, H. Han, S. Ha, B. S. Chung, W. E. Bolch, and C. H. Kim (2021b). "Development of pediatric mesh-type reference computational phantom series of international commission on radiological protection." *J. Radiol. Prot.* 41(3):S160–70.

Choi, C, Y. S. Yeom, H. Lee, H. Han, B. Shin, T. T. Nguyen, and C. H. Kim (2020). "Body-size-dependent phantom library constructed from ICRP mesh-type reference computational phantoms." *Phys. Med. Biol.* 65:125014.

Clariand, I., L. G. Bouchet, and W. E. Bolch (2000). "A new macroscopic model of the long bones for skeletal dosimetry—Abstract." *J. Nucl. Med.* 41:1062.

Coffey, J. L., M. Cristy, and G. G. Warner (1981). "MIRD Pamphlet No. 13: Specific absorbed fractions for photon sources uniformly distributed in the heart chambers and heart wall of a heterogeneous phantom." *J. Nucl. Med.* 22:65–71.

Coffey, J. L. and E. E. Watson. "S-values for selected radionuclides and organs with the heart wall and heart contents as source organs." In *Third International Radiopharmaceutical Dosimetry Symposium*, E. E. Watson, A. T. Schlafke-Stelson, J. L. Coffey, and R. J. Cloutier, eds. Rockville, MD: US Department of Health & Human Services, 1981.

Collaboration (2017). Introduction of GEANT4 Release 10.4. Geneva, Switzerland: CERN.

Cristy, M. (1980). "Mathematical phantoms representing children of various ages for use in estimates of internal dose." Oak Ridge, TN: Oak Ridge National Laboratory.

Cristy, M. and K. F. Eckerman (1987). "Specific absorbed fractions of energy at various ages from internal photon sources." Oak Ridge, TN: Oak Ridge National Laboratory.

Deloar, H. M., H. Watabe, T. Nakamura, Y. Narita, A. Yamadera, T. Fujiwara, and M. Itoh (1997). "Internal dose estimation including the nasal cavity and major airway for continuous inhalation of C15O2, 15O2 and C15O using the thermoluminescent dosimeter method." *J. Nucl. Med.* 38:1603–13.

Dieudonne, A., I. Gardin, P. Zhang, W. E. Bolch, K. Assie, and I. Buvat (2008). "VoxelDose: A software for 3D Dosimetry in Targeted Radiotherapy using S-Values at the voxel level." *Eur. J. Nucl. Med. Mol. Imaging* 35:S135–36.

Dieudonne, A., R. F. Hobbs, W. E. Bolch, G. Sgouros, and I. Gardin (2010). "Fine-resolution voxel S values for constructing absorbed dose distributions at variable voxel size." *J. Nucl. Med.* 51:1600–07.

EPA. *Federal Guidance Report No. 13: Cancer risk coefficients for environmental exposures to radionuclides.* Washington, DC: US Environmental Protection Agency, 1999.

EPA. *Federal Guidance Report No. 15: External exposure to radionuclides in air, water, and soil.* Washington, DC: U.S. Environmental Protection Agency, 2019.

Furuta, T., T. Sato, M. C. Han, Y. S. Yeom, C. H. Kim, J. L. Brown, and W. E. Bolch (2017). "Implementation of tetrahedral-mesh geometry in Monte Carlo radiation transport code PHITS." *Phys. Med. Biol.* 62:4798–810.

Geyer, A. M., S. O'Reilly, C. Lee, D. J. Long, and W. E. Bolch (2014). "The UF/NCI family of hybrid computational phantoms representing the current US population of male and female children, adolescents, and adults—application to CT dosimetry." *Phys. Med. Biol.* 59:5225–42.

Gosselin, M. C., E. Neufeld, H. Moser, E. Huber, S. Farcito, L. Gerber, M. Jedensjo, I. Hilber, F. Di Gennaro, B. Lloyd, E. Cherubini, D. Szczerba, W. Kainz, and N. Kuster (2014). "Development of a new generation of high-resolution anatomical models for medical device evaluation: the Virtual Population 3.0." *Phys. Med. Biol.* 59:5287–303.

Han, E. Y., W. E. Bolch, and K. F. Eckerman (2006). "Revisions to the ORNL series of adult and pediatric computational phantoms for use with the MIRD schema." *Health Phys.* 90:337–56.

Han, H., Y. S. Yeom, C. Choi, S. Moon, B. Shin, S. Ha, and C. H. Kim (2020). "POLY2TET: a computer program for conversion of computational human phantoms from polygonal mesh to tetrahedral mesh." *J. Radiol. Prot.* 40:962–79.

Hobbs R F, Song H, Huso D L, Sundel M H and Sgouros G 2012 A nephron-based model of the kidneys for macro-to-micro alpha-particle dosimetry Phys. Med. Biol. 57 4403-4424

ICRP (1975). ICRP Publication 23: Report on the Task Group on Reference Man. In: Annals of the ICRP, (Pergamon Press, Oxford, UK: International Commission on Radiological Protection) pp 1-480

ICRP (1979). "ICRP Publication 30, Part 1: Limits for intakes of radionuclides by workers." Ann. ICRP 3:1–116.

ICRP (2002). "ICRP Publication 89: Basic anatomical and physiological data for use in radiological protection—reference values." *Ann. ICRP* 32:1–277.

ICRP (2009). "ICRP Publication 110: Adult reference computational phantoms." *Ann. ICRP* 39:1–165.

ICRP (2010). "ICRP Publication 116: Conversion coefficients for radiological protection quantities for external radiation exposures."*Ann. ICRP* 40:1–257.

ICRP (2013). "ICRP Publication 123: Assessment of Radiation Exposure of Astronauts in Space." *Ann. ICRP* 42:1–339.

ICRP (2015a). "ICRP Publication 128: Radiation Dose to Patients from Radiopharmaceuticals–A Compendium of Current Information Related to Frequently Used Substances." *Ann. ICRP* 44:1–321.

ICRP (2015b). "ICRP Publication 130: Occupational Intakes of Radionuclides, Part 1."*Ann. ICRP* 44:1–188.

ICRP (2016a). "ICRP Publication 132: Radiological protection from cosmic radiation in aviation." *Ann. ICRP* 45:1–48.

ICRP (2016b). "ICRP Publication 133: The ICRP computational framework for internal dose assessment for reference adults: specific absorbed fractions." *Ann. ICRP* 45:1-74.

ICRP (2016c). "ICRP Publication 134: Occupational Intakes of Radionuclides, Part 2." *Ann. ICRP* 45:1–352.

ICRP (2017). "ICRP Publication 137: Occupational intakes of radionuclides, Part 3." *Ann. ICRP* 46 1–486.

ICRP (2019). "ICRP Publication 141: Occupational intakes of radionuclides, Part 4." *Ann. ICRP* 48:1–497.

ICRP (2020a). "ICRP Publication 143: Paediatric reference computational phantoms." *Ann. ICRP* 49:1–299.

ICRP (2020b). "ICRP Publication 144: Age-dependent dose coefficients for external exposures to environmental sources." *Ann. ICRP* 49:1–147.

ICRP (2020c). "ICRP Publication 145: Adult mesh-type reference computational phantoms." *Ann. ICRP* 49:1–203.

ICRU (2021). "ICRU Report 96: Dosimetry-guided radiopharmaceutical therapy." *J. ICRU* 21:1–216.

Johnson, P., S. Whalen, M. Wayson, B. Juneja, C. Lee, and W. Bolch (2009). "Hybrid patient-dependent phantoms covering statistical distributions of body morphometry in the US adult and pediatric population." *Proc. IEEE* 97:2060–75.

Jones, D. G. (1997). "A realistic anthropomorphic phantom for calculating organ doses arising from external photon irradiation." *Radiat. Prot. Dosim.* 72:21–29.

Jones, D. G. (1998). "A realistic anthropomorphic phantom for calculating specific absorbed fractions of energy deposited from internal gamma emitters." *Radiat. Prot. Dosim.* 79:411–14.

Kainz, W., E. Neufeld, W. E. Bolch, C. G. Graff, C. H. Kim, N. Kuster, B. Lloyd, T. Morrison, P. Segars, Y. S. Yeom, M. Zankl, X. G. Xu, and B. M. W. Tsui (2019). "Advances in Computational Human Phantoms and Their Applications in Biomedical Engineering—A Topical Review." *IEEE Trans. Radiat. Plasma Med. Sci.* 3:1–23.

Kalender, W. A., B. Schmidt, M. Zankl, and M.Schmidt (1999). "A PC program for estimating organ dose and effective dose values in computed tomography." *Eur. Radiol.* 9:555–62.

Kuzmin, G. A., M. M. Mille, J. W. Jung, C. Lee, C. Pelletier, G. Akabani, and C. Lee (2018). "A novel method to extend a partial-body CT for the reconstruction of dose to organs beyond the scan range." *Radiat. Res.* 189:618–26.

Lanconelli, N., M. Pacilio, S. Lo Meo, F. Botta, A. Di Dia, A. Torres Aroche, M. A. Coca Pérez, and M. Cremonesi (2012). "A free database of radionuclide voxel S values for the dosimetry of nonuniform activity distributions." *Phys. Med. Biol.* 57:517–33.

Lee, C., D. Lodwick, J. Hurtado, D. Pafundi, J. L. Williams, and W. E. Bolch (2010). "The UF family of reference hybrid phantoms for computational radiation dosimetry." *Phys. Med. Biol.* 55:339–63.

Lee, C., J. Williams, C. Lee, and W. Bolch (2005). "The UF series of tomographic computational phantoms of pediatric patients." Med. Phys. 32:3537–48.

Loevinger, R. and M. Berman (1968a). "A formalism for calculation of absorbed dose from radionuclides." *Phys. Med. Biol.* 13:205–17.

Loevinger, R. and M. Berman. *MIRD Pamphlet No. 1: A schema for absorbed-dose calculations for biologically distributed radionuclides*. New York: Society of Nuclear Medicine, 1968b.

Loevinger, R. and M. Berman. *MIRD Pamphlet No. 1, Revised: A revised schema for calculating the absorbed dose from biologically distributed radionuclides.* New York: Society of Nuclear Medicine, 1976.

Loevinger, R., T. F. Budinger, and E. E. Watson. *MIRD Primer for Absorbed Dose Calculation.* New York: Society of Nuclear Medicine, 1988.

Loevinger, R., T. F. Budinger, and E. E. Watson. *MIRD Primer for Absorbed Dose Calculations.* New York: Society of Nuclear Medicine, 1991.

Loevinger, R., J. G. Hiolt, and G. J. Hine. "Internally administered radionuclides." In *Radiation Dosimetry*, G. J. Hine and G. L. Brownell, eds. New York: Academic Press, 1965.

Marshall, E., D. Rajderkar, J. Brown, E. Stepusin, D. Borrego, and W. Bolch (2019). "A scalable database of organ doses for common diagnostic fluoroscopy examinations of children: Procedures of current practice at the University of Florida." *Phys. Med. Biol.* 64:134023.

MIRD Committee, R. M. Bartlett, W. E. Bolch, A. B. Brill, Y. K. Dewaraja, F. H. Fahey, D. R. Fisher, R. F. Hobbs, R. W. Howell, R. F. Meredith, J. G. Rajendran, G. Sgouros, and P. Zanzonico. *MIRD Primer 2022: A Complete Guide to Radiopharmaceutical Dosimetry.* Reston, VA: The Society of Nuclear Medicine and Molecular Imaging (SNMMI), 2021.

Nipper, J. C., J. L. Williams, and E. E. Bolch (2002). "Creation of two tomographic voxel models of paediatric patients in the first year of life." *Phys. Med. Biol.* 47:3143–64.

Olguin, E., B. President, M. Ghaly, E. Frey, G. Sgouros, and W. E. Bolch (2020). "Specific absorbed fractions and radionuclide S-values for tumors of varying size and composition." *Phys. Med. Biol.* 65(23):235015.

Pazik, F. D., R. J. Staton, D. E. Hintenlang, M. M. Arreola, J. L. Williams, and W. E. Bolch (2007). "Organ and effective doses in newborns and infants undergoing voiding cystourethrograms (VCUG): A comparison of stylized and tomographic phantoms." *Med. Phys.* 34:294–306.

Petoussi-Henss, N., W. Panzer, M. Zankl, and G. Drexler (1995). "Dose-area product and body doses." *Radiat. Prot. Dosim.* 57:363–66.

Petoussi-Henss, N. and M. Zankl (1998). "Voxel anthropomorphic models as a tool for internal dosimetry." *Radiat. Prot. Dosim.* 79:415–18.

Sato, T., Y. Iwamoto, S. Hashimoto, H. Ogawa, T. Furuta, S. Abe, T. Kai, P. Tasi, N. Matsuda, H. Iwase, N. Shigyo, L. Sihver, and K. Nitta (2018). "Features of Particle and Heavy Ion Transport code system (PHITS) version 3.02." *J. Nucl. Sci. Tech.* 55 684–90.

Segars, W. P. (2001). "Development and application of the new dynamic NURBS-based cardiac-torso (NCAT) phantom." Biomedical Engineering, University of North Carolina, Ph.D. thesis.

Segars, W. P., J. Bond, J. Frush, S. Hon, C. Eckersley, C. H. Williams, J. Feng, D. J. Tward, J. T. Ratnanather, M. I. Miller, D. Frush, and E. Samei (2013). "Population of anatomically variable 4D XCAT adult phantoms for imaging research and optimization." *Med. Phys.* 40:043701.

Seltzer, S. M. "An overview of ETRAN Monte Carlo Methods." In *Monte Carlo Transport of Electrons and Photons*, T. M. Jenkins, W. R. Nelson, A. Rindi, A. E. Nahum, and D. W. O. Rogers, eds. New York: Plenum Press, 1988.

Seltzer, S. M. (1991) "Electron-photon Monte Carlo calculations: the ETRAN code." *Appl. Radiat. Isot.* 42:917–41.

Sgouros, G., E. C. Frey, W. E. Bolch, M. B. Wayson, A. F. Abadia, and S. T. Treves (2011). "An approach for balancing diagnostic image quality with cancer risk: application to pediatric diagnostic imaging of 99mTc-dimercaptosuccinic acid." *J. Nucl. Med.* 52:1923–29.

Snyder, W. S., M. R. Ford, G. G. Warner, and H. L. Fisher. *MIRD Pamphlet No. 5: Estimates of absorbed fractions for monoenergetic photon sources uniformly distributed in various organs of a heterogeneous phantom.* New York: Society of Nuclear Medicine, 1969.

Snyder, W. S., M. R. Ford, and G. G. Warner. *MIRD Pamphlet No. 5, Revised: Estimates of specific absorbed fractions for photon sources uniformly distributed in various organs of a heterogeneous phantom.* New York: Society of Nuclear Medicine, 1978.

Stabin, M. G. and H. Yoriyaz (2002). "Photon specific absorbed fractions calculated in the trunk of an adult male voxel-based phantom." *Health Phys.* 82:21–44.
Stabin, M. G., H. Yoriyaz, A. B. Brill, and M. Dawant (1999). "Monte Carlo calculation of dose conversion factors for a new generation of dosimetry phantoms." *J. Nucl. Med.* 40:310P–11P.
Staton, R. J. and W. E. Bolch (2004). "Implementation of a newborn tomography computational model for dose assessment in pediatric radiology [Abstract]." *Med. Phys.* 31:1757.
Staton, R. J., F. D. Pazik, J. C. Nipper, J. L. Williams, and W. E. Bolch (2003). "A comparison of newborn stylized and tomographic models for dose assessment in pediatric radiology." *Phys. Med. Biol.* 48:805–20.
Staton, R. J., J. L. Williams, M. M. Arreola, D. E. Hintenlang, and W. E. Bolch (2007). "Organ and effective doses in infants undergoing upper gastrointestinal (UGI) fluoroscopic examination." *Med. Phys.* 34:703–10.
Stenvall, A., E. Larsson, S. E. Strand, and B. A. Jonsson (2014). "A small-scale anatomical dosimetry model of the liver." *Phys. Med. Biol.* 59:3353–71.
Stepusin, E. J. (2016). "Pre-Computed Monte Carlo Dosimetry and Reporting for Computed Tomography." Dept. of Biomedical Engineering, University of Florida, Ph.D. dissertation.
Stepusin, E. J., D. J. Long, E. L. Marshall, and W. E. Bolch (2017). "Assessment of different patient-to-phantom matching criteria applied to a Monte Carlo-based CT organ dose library." *Med. Phys.* 44:5498–508.
Tagesson, M., I. G. Zubal, M. Ljungberg, and S. E. Strand. "S Values for Subregions in the Brain." In *Proceedings of the Sixth International Radiopharmaceutical Dosimetry Symposium,* Oak Ridge, TN, January 1999. A. T. S-Stelson, M. G. Stabin, and R. B. Sparks, eds. Sponsored by Oak Ridge Associated Universities, U.S. Food and Drug Administration, and U.S. Department of Energy.
Veit, R. and M. Zankl (1992). "Influence of patient size on organ doses in diagnostic radiology." *Radiat. Prot. Dosim.* 43:241–43.
Veit, R. and M. Zankl (1993). "Variation of organ doses in pediatric radiology due to patient diameter calculated with phantoms of varying voxel size." *Radiat. Prot. Dosim.* 49:353–56.
Watson, E. E., M. G. Stabin, J. L. Davis, and K. F. Eckerman (1989). "A model of the peritoneal cavity for use in internal dosimetry." *J. Nucl. Med.* 30:2002–11.
Werner, C. J., J. S. Bull, C. J. Solomon, F. B. Brown, G. W. McKinney, M. E. Rishing, R. A. Dixon, R. L. Martz, H. G. Hughes, L. J. Cox, A. J. Zukaitis, J. C. Armstrong, R. A. Forster, and L. Casswell. *MCNP Version 6.2 Release Notes.* Los Alamos, NM: Los Alamos National Laboratory, 2018.
Xie, T. and H. Zaidi (2019). "Estimation of the radiation dose in pregnancy: an automated patient-specific model using convolutional neural networks." *Eur. Radiol.* 29:6805–15.
Xu, X. G. (2014). "An exponential growth of computational phantom research in radiation protection, imaging, and radiotherapy: a review of the fifty-year history." *Phys. Med. Biol.* 59:R233–R302.
Xu, X. G., T. C. Chao, and A. Bozkurt (2000). "VIP-Man: an image-based whole-body adult male model constructed from color photographs of the Visible Human Project for multi-particle Monte Carlo calculations." *Health Phys.* 78:476–86.
Yoriyaz, H., A. dos Santos, M. G. Stabin, and R. Cabezas (2000). "Absorbed fractions in a voxel-based phantom calculated with the MCNP-4B code." *Med. Phys.* 27:1555–62.
Zankl, M. (1993). "Computational models employed for dose assessment in diagnostic radiology." *Radiat. Prot. Dosim.* 49:339–44.
Zankl, M., W. Panzer, H. Petoussi-Henns, and G. Drexler (1995). "Organ doses for children from computer tomographic examinations." *Radiat. Prot. Dosim.* 57:393–96.
Zankl, M., R. Veit, N. Petoussi, E. Mannweiler, A. Wittmann, and G. Drexler (1994). "Realistic computerized human phantoms." *Adv. Space Res.* 14:423–31.
Zankl, M., R. Veit, G. Williams, K. Schneider, H. Fendel, N. Petoussi, and G. Drexler (1988). "The construction of computer tomographic phantoms and their application in radiology and radiation protection." *Radiat. Env. Biophys.* 27:153–64.
Zhang, J., Y. H. Na, P. F. Caracappa, and X. G. Xu (2009). "RPI-AM and RPI-AF, a pair of mesh-based, size-adjustable adult male and female computational phantoms using ICRP-89 parameters and their calculations for organ doses from monoenergetic photon beams." *Phys. Med. Biol.* 54:5885–908.

Zubal, I. G. and C. H. Harell (1992). "Voxel based Monte Carlo calculations of nuclear medicine images and applied variance reduction techniques." *Image and Vision Computing* 10: 342–48.

Zubal, I. G., C. R. Harrell, E. O. Smith, Z. Rattner, G. Gindi, and P. B. Hoffer (1994). "Computerized three-dimensional segmented human anatomy." *Med. Phys.* 21:299–302.

Zubal, I. G., C. R. Harrell, E. O. Smith, A. L. Smith, and P. Krischlunas. "High Resolution, MRI-based, Segmented, Computerized Head Phantom." In *Proceedings of the Sixth International Radiopharmaceutical Dosimetry Symposium,* Oak Ridge, TN, January 1999. A. T. S-Stelson, M. G. Stabin, and R. B. Sparks, eds. Sponsored by Oak Ridge Associated Universities, U.S. Food and Drug Administration, and U.S. Department of Energy.

Chapter 14

Yttrium-90 Microspheres: Treatment Planning using the MIRD Formalism

Diane Alvarez[1], MSc, DABR
and Robert F. Hobbs[2], Ph.D., DABR

[1]Department of Radiation Oncology
Miami Cancer Institute at Baptist Hospital
Miami, FL

[2]Department of Radiology and Radiological Sciences
Johns Hopkins Medicine
Baltimore, MD

14.1 Introduction

14.1.1 Rationale of Radiomicrospheres Therapy

Trans-arterial radioembolization (TARE), *radiomicrosphere therapy* (RMT). and *selective internal radiation therapy* (SIRT) are some of the more common names that describe the same radiotherapy technique available to patients suffering from liver cancer. The technique, herein forth referred to as TARE, entails a physician inserting a micro-catheter within a patient's (typically femoral) artery which is then navigated via the vasculature of the liver to a region containing the lesion(s), often a liver lobe or segment. Once localized, millions of radiomicrospheres made of tissue-compatible resin or glass are then administered through the micro-catheter allowing infusion within the designated tissue. The radiomicrospheres become lodged within the patient's capillaries, where they will remain fixed indefinitely and irradiate the surrounding tissues. The intent of the treatment is usually local control of the malignancy, but it can also be used for palliation.

The treatment is based on the principle that liver tumors are primarily vascularized by the hepatic artery, whereas healthy liver tissues receive the majority of its blood supply from the portal vein. Therefore, following administration in the hepatic artery, radiomicrospheres will be carried preferentially toward tumors (Pasciak, Bradley, and McKinney 2017).

The most commonly used radioisotope for this application is yttrium-90 (^{90}Y). Preparation for ^{90}Y TARE requires a number of steps, one of which is pretreatment planning. This entails the use of different imaging modalities in computing suitable ^{90}Y activity to prescribe. There are competing activity calculation models that vary in complexity due to their respective imaging requirements and the complex interaction with the specific anatomy of the patient which, in turn, intertwines with the level of patient specificity. Given the current flux of imaging technologies across different facilities, it is difficult to determine the impact of these competing calculation models in pretreatment activity calculation since each model has certain imaging requirements. The importance of activity calculations directly relates to the absorbed dose deposited into the patient's tissues. Furthermore, the quality of the image studies themselves, as well as the quality of image registrations, directly impact the accuracy of the volumes and spatial location needed for activity calculation and, thus, impacts dosimetry. The intent of this chapter is to present activity calculation by way of the *Medical Internal Radiation Dose* (MIRD) formalism and the importance of its potential impact on patient outcome.

14.1.2 Brief Overview of ^{90}Y Microspheres

Yttrium-90 is a nearly pure (99.99%) beta emitter with a mean energy of 926.7 keV and maximum energy of 2.28 MeV. It has a short soft-tissue range (mean 2.5 mm and max 11 mm) and a 64.1-hour half-life. It decays to stable zirconium-90 (^{90}Zn). Yttrium-90 has a small branching to the excited state of ^{90}Zn at 1.76 MeV in which 32 out of one million decays result in internal electron-positron pair production (Dezarn et al. 2011). Using functional imaging, the scintillator crystals found in nuclear medicine imaging systems can detect both the bremsstrahlung photons as well as the photons from the low-branching ratio positrons. Planar and *single-photon computerized tomography* (SPECT) can detect the continuous bremsstrahlung radiation from the high-energy beta emissions, while *positron emission tomography* (PET) can detect the positron from the internal pair production (Wright et al. 2015).

Currently, there are two products that are FDA-approved in the United States for clinical use:

1. ^{90}Y SIR-Spheres® resin microspheres (Sirtex Medical Inc., Woburn, MA) and
2. Therasphere™ glass microspheres (Boston Scientific Corp., Marlborough, MA).

There are other products available, although they are considered investigational devices in the United States, such as:

- ^{166}Ho *poly-l-lactic acid* (PLLA) microspheres (CE-cleared for use in Europe) called QuiremSpheres^{TM} (Terumo EMEA, Belgium),
- ^{90}Y radiopaque glass microspheres called Eye90 (ABK Biomedical, Halifax, NS, Canada), and
- PLLA ^{188}Re, which are in preclinical investigation phase (Gulec and McGoron 2022).

14.1.3 Summary Overview of Procedure

There are several steps taken prior to a patient's ^{90}Y TARE treatment before activity can be calculated. Briefly, the patient diagnosed with primary or metastatic liver cancer consults an interventional radiologist (IR). After evaluating contraindications and liver function lab tests (Weber et al. 2022), if the IR considers the patient a potential candidate for liver TARE, the patient receives a diagnostic procedure called *mapping*. In this procedure, the patient will receive an angiogram, which maps the vasculature of the patient's liver. The reader is referred to Pasciak, Bradley, and McKinney (2017) for in-depth hepatic vascular anatomy. The IR determines the vessels responsible for feeding the lesions by performing a simulation procedure using contrast and fluoroscopic imaging to visualize and determine whether they can reach these vessels using a micro-catheter. The IR also assesses whether there are any aberrant vessels that may shunt into the surrounding organs, such as gastrointestinal or other surrounding organs (Padia et al. 2017). If such aberrant vessels are found, the IR attempts to embolize these vessels to close off the passage of microspheres to undesirable regions of the body. If the IR cannot successfully avoid extrahepatic shunting or if other medical criteria are not met—such as healthy kidney and liver function and acceptable lung shunting—the patient is considered a poor candidate for this treatment modality (Kim et al. 2019).

When available, in-suite *cone-beam computed tomography* (CBCT) is used by the IR to image the patient during the mapping procedure. CBCT is considered the gold standard and preferred method to define the target volume when available (Salem et al. 2019). Two CBCTs attained sequentially within seconds of each other are acquired of the target region while injecting contrast to capture the perfused volume within the liver during the arterial and venous phases (Ertreo et al. 2018; O'Connor et al. 2020). The arterial (or early) phase best captures the perfusion of the target region, such hepatic lobes or whichever segment or selective area one desires to treat. The venous (or late) phase best captures the location of the *gross tumor volumes* (GTVs) within the perfused volume. The exact information gleaned from the dual-phase CBCT may also depend on the disease type. These images may be utilized in pretreatment planning (O'Connor et al. 2020).

After the mapping procedure is completed, the patient is then injected with *technicium-99m labeled macro-aggregated albumin* (^{99m}Tc-MAA) and imaged using SPECT/CT. The ^{99m}Tc-MAA SPECT/CT study has three purposes.

1. The study's radiotracer is, in theory, a surrogate for the ^{90}Y microspheres, predicting their distribution within the liver. Thus it can be used for more detailed dosimetric treatment planning (Kappadath et al. 2021a).
2. The study can be used to assess any extrahepatic shunting that may not have been discernible in the mapping procedure (Georgiou et al. 2021).
3. The study enables the calculation of the *lung shunt fraction* (LSF), which has been historically calculated using planar imaging (Kim et al. 2019). LSF is the predicted fractional amount of radioactive exposure expected to the lungs. This is typically measured using planar scintigraphy and calculated using Equation (14.1):

$$LSF = \frac{\text{Counts}_{Lung}^{Posterior} - \text{Counts}_{Lung}^{Background}}{\left(\text{Counts}_{Lung}^{Posterior} - \text{Counts}_{Lung}^{Background}\right) + \left(\text{Counts}_{Liver}^{Anterior} - \text{Counts}_{Liver}^{Background}\right)} \tag{14.1}$$

This method of calculating LSF has yielded a long history of safe absorbed dose tolerance constraints to the lungs: 30 Gy mean absorbed dose limit in a given ^{90}Y TARE, with a cumulative maximum of 50 Gy for all treatments (Levillain et al. 2021; Kappadath et al. 2021b). Guidelines recommend that an LSF above 20% contraindicates a patient's candidacy for ^{90}Y TARE (Dezarn et al. 2011). It is possible to measure LSF with data from the SPECT/CT and, in fact, it has been shown that LSF is overestimated when calculated using planar image data (Kappadath et al. 2021a; Elsayed et al. 2021; Georgiou et al. 2021). This method of calculating LSF is considered more accurate, but it is yet to be widely used. Because of the potentially large difference in values—depending on whether the values are calculated using the data from planar vs. SPECT images—it is suspected that absorbed dose tolerances to the lungs may differ on SPECT, necessitating new modality-specific absorbed dose constraint guidelines that are yet to be established (Kappadath et al. 2021a).

14.2 Pretreatment Planning Dosimetry

After the patient receives the ^{99m}Tc MAA SPECT/CT study, a pretreatment plan is created that involves delineating liver volumes for the purpose of calculating administered activity. Depending on the *instructions for use* (IFU) from the manufacturer of the microspheres, differing activity models may be recommended. Furthermore, the required volumes necessary to calculate activity may differ depending on the activity model selected.

14.2.1 Body Surface Activity (BSA) Model

For patients prescribed ^{90}Y resin microspheres (SIR-Spheres®), the manufacturer recommends the *body surface activity* (BSA) model for activity calculation (Sirtex Medical Limited 2019). This is a modified model based on the empiric method, which instead of treating the whole liver at once, is structured to treat the hepatic lobes (or specific segments) at separate occasions to avoid liver toxicity. Additionally, the BSA model scales the activity to nominally take into account the patient's size in relation to the liver size:

$$A_{BSA} = \left(0.20247 \times H^{0.725}[\mathrm{m}] \times W^{0.425}[\mathrm{kg}] - 0.2\right) \times \left[\frac{V_{Tumor}}{V_{Tumor} + V_{Liver}}\right] \quad (14.2)$$

where A_{BSA} = calculated BSA activity,
H = height in meters,
W = weight in kilograms,
V_{Tumor} = volume of tumor within treatment area, and
V_{Liver} = volume of liver within treatment area.

The BSA model has several limitations. It does not apply when patient size is not relatively proportional to their liver size. Patients with larger livers require more activity but are relatively underdosed, and patients with smaller livers can potentially receive more activity than necessary (Kao et al. 2011; Lam et al. 2014). The empirically derived data for safe lung absorbed dose limit is based on having lungs with a mass of 1 kg (ICRP 23 1975), which is a gross approximation at best for most people. Furthermore, the manufacturer recommends activity reductions based on the LSF, with a recommendation of no treatment if the LSF is > 20% (Sirtex Medical Limited 2019). The BSA model does not account for the desired absorbed dose to be delivered to the target, thus it is a non-dosimetric model, nor does it consider the volume of normal liver in the irradiated tissue (Kao et al. 2011).

14.2.2 Mono-Compartment Model (MIRD Method)

The *mono-compartment model* (sometimes referred to as *MIRD method*) is a dosimetric model that is FDA-approved for the calculation of activity for patients prescribed ^{90}Y glass microspheres (Therasphere^{TM}) (Boston Scientific 2021). A more scientifically derived model, the mono-compartment model, is based on the MIRD formalism (Weber 1991) and is driven by the desired prescribed absorbed mean dose to the entire treated volume, $\bar{D}(Liver)$, with no distinction between disease and normal tissue, as shown in Equation (14.3) (Kim et al. 2019). It requires the perfused mass of liver (often a lobe or a segment) that is intended for treatment, *m(Liver)*, and the activity A_0 that will theoretically deliver the desired absorbed dose:

$$\bar{D}(Liver)[\mathrm{Gy}] = 50\left[\frac{\mathrm{J}}{\mathrm{GBq}}\right] \times \frac{A_0\,[\mathrm{GBq}]}{m(Liver)[\mathrm{kg}]} \quad (14.3)$$

Clinically, the LSF is subtracted from Equation (14.3):

$$\bar{D}(Liver)[\mathrm{Gy}] = 50\left[\frac{\mathrm{J}}{\mathrm{GBq}}\right] \times \frac{A_0\,[\mathrm{GBq}]}{m(Liver)[\mathrm{kg}]} \times (1 - LSF). \quad (14.4)$$

Equation (14.4) gives the absorbed dose to the treated volume for a given administered activity. By rearranging Equation (14.4), the administered activity needed to obtain the target absorbed dose may be obtained:

$$A_0\,[\text{GBq}] = \frac{\bar{D}(Liver)[\text{Gy}]\times m(Liver)[\text{kg}]}{50\left[\dfrac{\text{J}}{\text{GBq}}\right]\times(1-LSF)} \tag{14.5}$$

The absorbed dose to the lungs may then be obtained:

$$\bar{D}(Lungs)[\text{Gy}] = 50\left[\frac{\text{J}}{\text{GBq}}\right]\times\frac{A_0\,[\text{GBq}]}{m(Lungs)[\text{kg}]}\times LSF. \tag{14.6}$$

Minor variations of these formulas may be found in the literature where volume is used instead of mass for the liver, often with a density factor incorporated into the numeric conversion factor (i.e., the 50 value is slightly different). Sometimes the conversion factor is slightly different with more precise values, and historically the lung mass is taken to be the ICRP reference value.

The mono-compartment model calculates an administered activity for the entire treated volume, even though the activity distribution invariably has a high level of nonuniformity (Snyder et al. 1975). The coefficient of 50 in Equation (14.5) comes about from the assumption that 1 GBq of ^{90}Y activity per unit mass of tissue deposits approximately 50 Gy (Kim et al. 2019). The derivation of Equation (14.3) will be discussed in Section 14.3.1. Note that this model does *not* consider the amount of activity that may be delivered to normal liver tissue.

14.2.3 Multi-Compartment Model (Partition Method)

For historical reasons, the *multi-compartment model*, also known as the *partition method* or model, is often presented as an alternative model to the mono-compartment model with a different formulation. In this alternate model, a delineation is made between tumor tissue and normal liver.

However, the MIRD formalism and the associated equations derived are just as easily applied to any compartment in the multi-compartment model. The only two additional complications are (a) the segmentation of the tissue types and (b) the measurement of the relative amount of pre-therapeutic ^{99m}Tc activity in the different compartments (similar to the LSF for the lungs). This approach allows *authorized users* (AUs) to tailor the activity based on the patient's absorption of ^{90}Y microspheres within tumors (for efficacy) or normal tissue (for safety), but it requires 3D activity quantification from SPECT/CT rather than planar images.

SPECT collects the counts within the liver or the different compartments, and CT reveals the spatial locations of the tumor, normal liver, and lungs. This requires accurate image registration between SPECT and CT. If volumes are drawn using a different study other than the CT acquired during the SPECT scan, the contours must be transferred over to the SPECT/CT study, with the possibility of misregistration, which can lead to false counts within tumor and normal liver. A misregistration may be a result of cumulative factors, such as a small offset in the isocenter coincidence between CT and gamma detectors on a SPECT/CT scanner, or from an imperfect image registration between planning CT and diagnostic CT, MRI, or CBCT. Consider

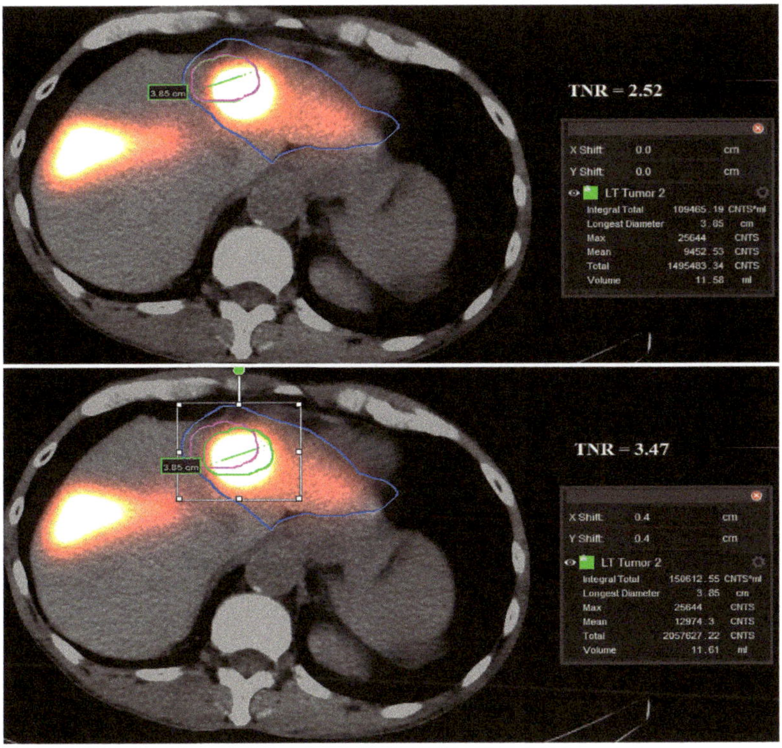

Figure 14.1 Demonstration of the effects of image registration on TNR. (a) Representative axial slice of a clinical tumor volume with a TNR of 2.52. (b) A 4-mm shift was introduced in the x- and y-directions causing the TNR to change to 3.47.

the image registrations demonstrated in Figure 14.1. The image in Figure 14.1(a) shows an axial slice of a patient's clinical tumor volume. A 4-mm shift in the x- and y-directions, representative of a misregistration, results in a difference in *tumor-to-normal-tissue ratio* (TNR), as shown in Figure 14.1(b).

The MIRD multi-compartment model can potentially be applied to either resin or glass ^{90}Y microsphere products, although the absorbed dose values for tumor efficacy and normal liver tolerance are likely different between the two types of microspheres because of factors such as differing specific activities, microsphere density, composition, and diameter (Pasciak, Bradley, and McKinney 2017). This is exemplified by the traditional prescribed absorbed dose ranges of the two different ^{90}Y products: 80 Gy to 150 Gy for Therasphere and 100 Gy to 120 Gy for the SIRspheres (Levillian et al. 2021). As it is a more complex model to use, with additional steps required, the multi-compartment model is less commonly used (Kao et al. 2011), although historical reasons and the lack of dosimetric expertise certainly play a role. Use of the multi-compartmental model and prescribing to a combination of normal organ tissue tolerance and tumor efficacy threshold was shown to have an improved overall survival benefit of 26.6 months vs. 10.7 months with the mono-compartment model in a two-armed randomized clinical trial (Garin et al. 2020).

For purposes of quantification, it is assumed that once deposited, the microspheres do not migrate. It is also assumed that they are distributed only within the liver (tumor and normal liver) and lungs. Using the volumes derived from segmentation of the normal liver and tumor(s) on the CT from the ^{99m}Tc-MAA SPECT/CT study and the bremsstrahlung counts within these structures derived from the SPECT of the same study, the activity corresponding to the desired mean absorbed dose to the tumor, or limited by the absorbed dose to the normal liver or lungs, is:

$$A_0\,[\text{GBq}] = \frac{\bar{D}(roi)[\text{Gy}] \times m(roi)[\text{kg}]}{50 \times f_{roi}} \tag{14.7}$$

where f_{roi} is the fractional uptake of ^{99m}Tc-MAA in the *region of interest* (roi) (i.e., lungs, tumor, normal liver) as measured by the SPECT/CT, which is assumed to be the same fractional uptake of the ^{90}Y-microspheres.

The formula can be reversed to make sure that safe mean absorbed doses to the *organs at risk* (OARs)—either the normal liver, the lungs, or both—are not exceeded:

$$\bar{D}(roi)[\text{Gy}] = \frac{50 \times A_0\,[\text{GBq}] \times f_{roi}}{m(roi)[\text{kg}]} \tag{14.8}$$

Equation (14.8) is also used for post-treatment dosimetric verification. The astute observer will notice that these two formulas are the exact same as for the mono-compartment model. It is simply a question of delineating any *roi* and quantifying the relative activity, f_{roi}, in that *roi*. Should $\bar{D}(normal\ liver)$ or $\bar{D}(lungs)$ exceed their safe tolerances, then A_0 must be reduced to meet the maximum safe dose of the limiting organ. This can be achieved by entering the OAR constraint dose and solving for A_0. The above equations will be derived in Section 14.3.1.

As with all MIRD applications, a uniform activity within the source region and a uniform absorbed dose within the target region is assumed. In reality, it has been reported that the microspheres preferentially get localized in the periphery of tumors (Kao et al. 2011). MIRD formalism also assumes that ^{99m}Tc-MAA particle distribution accurately simulates that of the ^{90}Y microspheres. This assumption is somewhat controversial, as many studies have been done leading to contradictory results (Pasciak, Bradley, and McKinney 2017), particularly those studies focused on the localization of the micro-catheter for activity administration. In general, it has become accepted that gross delineation between normal tissue and the tumor is well predicted, while specific voxelized distributions are less predictable.

Historically, the TNR, the ratio of uptake in the tumor and normal tissue regions, has been used as a parameter for predicting efficacy, while formulations for the dosimetry have been developed that specifically use the TNR as input quantities. The user is encouraged to use the ratio of tumor-to-normal-tissue absorbed *doses*.

The multi-compartment model calculates the estimated absorbed *mean* dose to the tumor, normal liver, and lung, which is limited information compared to voxel-based dosimetry. For ^{90}Y microspheres, voxel-based dosimetry may be obtained by extending the MIRD formalism to each voxel to create a three-dimensional representation of activity distribution that is, in turn, converted to an absorbed dose distribution (Capala et al. 2021). In the case of pretreatment planning, the ^{99m}Tc-MAA SPECT/CT is used to compute predictive absorbed *dose volume histograms* (DVHs) for tumor, normal liver, and lungs by using the same formulas for the compartments at the voxel level. Alternatively, Monte Carlo or dose kernel convolution methods may be used on the voxel activity data to simulate energy deposition, which then collect the deposited energy for DVHs in a more classical "voxelized" dosimetry approach (Pasciak, Bradley, and McKinney 2017).

Although promising, the practice of voxel-based dosimetry (be it predictive or post-treatment verification) is still considered optional in the United States. While its implementation is recommended to complement organ-level dosimetry (i.e., absorbed mean dose calculation using MIRD formulation), its superiority is still under debate in the radiopharmaceutical community (Chiesa, Bardiès, and Zaidi 2019), essentially due to the uncertainty in localization associated with voxel-level activity values. For microspheres, this spill-out effect is on the order of the spread energy deposition, such that it is questionable whether Monte Carlo or dose kernel convolutions truly improve dosimetric precision. The reader is referred to the works of Capala et al. (2021), Chiesa et al. (2021), and Thomas et al. (2020), while subsequent chapters in this book have further information on voxel-based dosimetry.

14.3 MIRD Formalism

As mentioned earlier, the mono- and multi-compartment models for activity calculation are derived from MIRD formalism. The following sections will describe how the properties of ^{90}Y simplify the general MIRD equations to those presented in Equations (14.3) through (14.8) (Gulec, Mesoloras, and Stabin 2006; Dezarn et al. 2011; ICRU 2021).

14.3.1 MIRD Paradigm Overview and Derivation for ^{90}Y Microspheres

The MIRD committee, which was formed in the 1960s, has worked on absorbed dose calculation methods for the nuclear medicine community for more than 50 years. The committee has produced a number of pamphlets, books, reports, and software tools, within which standard methods, models, assumptions, and mathematical models have been developed for various radioisotopes (Weber 1991; Howell et al. 1999).

The MIRD method, in its most general form, involves considering the mean absorbed dose, D, to a target region, r_T, as the sum of all the absorbed dose contributions from activity in all the potential source regions, r_S, where the activity, A, is assumed uniform in each individual source region. The formalism is:

$$D(r_T) = \sum_{r_s} \tilde{A}(r_s)\, S(r_T \leftarrow r_s) \tag{14.9}$$

$$\tilde{A}(r_s) = \int_0^{\infty} A(r_s,t)\,dt \tag{14.10}$$

Here $\tilde{A}$ is the time-integrated activity in the source region, that is, the number of radioactive decays in that region. The variable S stands for the *S values*, which are matrix multipliers that convert the time-integrated activity to absorbed dose, most often determined from Monte Carlo simulations of activity in idealized computer phantoms. The S value can be broken down into:

$$S(r_T \leftarrow r_S) = \frac{1}{m(r_T)} \sum_i E_i\, Y_i\, \phi(r_t \leftarrow r_S, E_i), \tag{14.11}$$

where E_i = the mean (or individual) energy per particle of the ith radiation emitted;
Y_i = the number of the ith radiation emitted per nuclear decay;
$\phi\ (r_T \leftarrow r_s, E_i)$ = the absorbed fraction, i.e., the fraction of E_i emitted within source tissue r_S that is absorbed in target tissue r_T, all specific to a given isotope; and
$m(r_T)$ = the mass of the target tissue r_T.

The MIRD formalism applied to ^{99m}Tc-MAA/^{90}Y-microsphere dosimetry benefits from three major simplifying assumptions:

1. ***Radioembolization implies that the activity does not re-localize as a function of time.*** Therefore, the activity as a function of time is a mono-exponential, characterized by the ^{90}Y physical decay constant, λ_{Y90}:

$$A(r_S,t) = A(r_S,0)e^{-\lambda_{Y90} \times t}, \tag{14.12}$$

   where $A(r_s,0)$ is the activity in the source region at the time of administration ($t = 0$). Consequently:

$$\tilde{A}(r_S) = \int_0^{\infty} A(r_S,0)e^{-\lambda_{Y90} \times t} = \frac{A(r_S,0)}{\lambda_{Y90}}. \tag{14.13}$$

2. ***^{90}Y is a "pure" beta-emitter, with a very low branching ratio of prompt gammas and some bremsstrahlung photons.*** To a reasonable approximation, all energy from the decay can be assumed to be deposited locally. In MIRD terms, the absorbed fraction ϕ is 1 for source and target regions that are the same (self-dose) and 0 for all others. This removes the need to distinguish between source and target sources as defined in the MIRD formalism (henceforth $roi = r_S = r_T$).

Substituting these assumptions into the MIRD formalism results in:

$$D(roi) = \frac{\bar{E} \times A(roi,0)}{\lambda_{Y90} \times m(roi)[\text{kg}]}, \tag{14.14}$$

where $\bar{E}$ is the average energy deposited per decay. Both $\bar{E}$ and λ_{Y90} are constants, and the numerical values may be substituted. Numerically, this gives a value of:

$$0.9267\left[\frac{\text{MeV}}{\text{dis}}\right] \times \frac{2.6684}{\ln(2)}[\text{day}] \times \frac{86400}{1}\left[\frac{\text{s}}{\text{day}}\right] \times \frac{10^9}{1}\left[\frac{\text{Bq}}{\text{GBq}}\right] \times \frac{1\,{}^{\text{dis}}\!/\!_{\text{s}}}{\text{Bq}} \times 1.6022 \times 10^{-13}\left[\frac{\text{J}}{\text{MeV}}\right]$$
$$= 49.38\left[\frac{\text{J}}{\text{GBq}}\right] \tag{14.15}$$

thus:

$$D(roi)[\text{Gy}] \cong \frac{50 \times A(roi,0)[\text{GBq}]}{m(roi)[\text{kg}]} \tag{14.16}$$

3. ***Relative (instead of absolute) dosimetry is generally used because of the lack of pharmacokinetics and the limited number of organs where the activity is expected to localize.*** For purposes of quantification, it is assumed that microspheres are deposited solely within the liver and the lungs. This reduces to determining the fraction of activity, f_{roi}, of the entire administered activity, A_0, in a region equal to the fraction of counts present in the region derived from either ^{99m}Tc-MAA planar or SPECT/CT images.

$$f(roi) = \frac{c_{roi}\,[\text{counts}]}{\sum_1^n c_{roi}\,[\text{counts}]} \tag{14.17}$$

Note that Equation (14.17) equates to the LSF equation when the denominator consists of c_{lungs} and c_{liver} and the numerator is c_{lungs}, i.e., f_{lungs} = LSF. The activity in the *roi*, A_{roi}, is thus related to the administered activity, A_0, by the fraction of counts present in the region derived from either ^{99m}Tc-MAA planar or SPECT images, f_{roi}.

$$A_{roi} = f_{roi} \times A_0 \tag{14.18}$$

These three simplifications combined mean that fractional activity measured in any defined *roi* at any time point on the pre-therapeutic ^{99m}Tc-MAA is expected to be equal to the fractional activity and fractional absorbed dose at every time point for the ^{90}Y-microsphere treatment for the same region. Substituting these into the MIRD formalism reported to two significant figures results in Equation (14.8).

Equation (14.8) is a formula that is valid for all approaches, including the mono-compartment, partition, and voxel-based local deposition methods and can be used to find the absorbed dose for all *roi*: liver, hepatic lobe, lung, individual tumor, collective tumors, normal tissue, voxel, etc.

14.3.2 Assumptions and Limitations of the MIRD Formalism

As with all formalisms, there are certain assumptions made and limitations to the model that will lead to inaccuracies if they don't hold true. First, the assumption that activity is spatially distributed homogeneously within r_S is particularly limiting if activity is deposited only within a localized region within r_S, or if the user wishes to calculate mean dose to r_T if it is a paired organ. For example, if r_S is liver and r_T is lungs, then the mean dose calculated would apply to both lungs, when in fact the right lung will have higher dose than the left lung given its proximity to the liver (Snyder et al. 1975).

An additional assumption is that the patient for whom the dose is being calculated is similar in dimensions and anatomy (i.e., has similar organ masses) to the specific phantom used in the computation of the S values. In fact, the phantom known as the "MIRD-5 phantom" (Figure 14.2) (Snyder et al. 1974) was used in the Monte Carlo computation of the tabulated S values published in MIRD Pamphlet #11 (Sny-

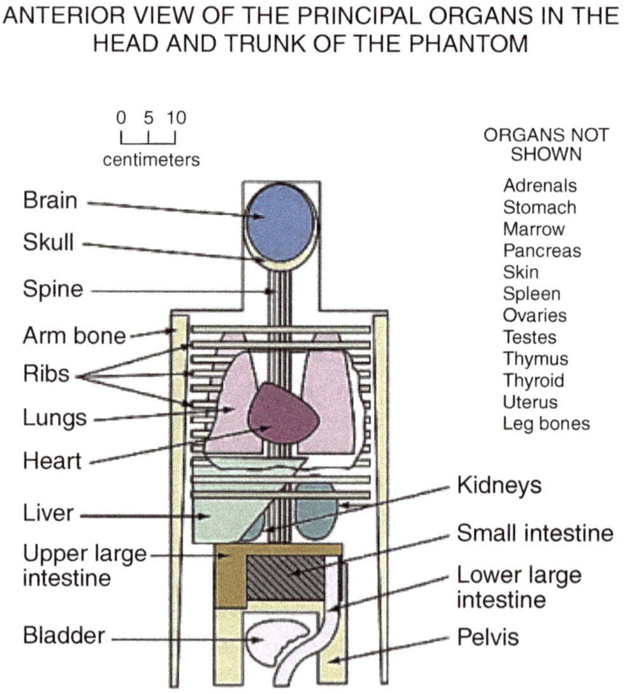

Figure 14.2 The MIRD-5 phantom. (Image adapted from: MIRD Pamphlet No. 5, Revised: Estimates of Absorbed Fractions for Photon Sources Uniformly Distributed in Various Organs of a Heterogeneous Phantom (1978) © SNMMI.)

der et al. 1975). It was based on the Reference Man, a simplified model of a 70-kg adult male (ICRP 1975). Recent advances in phantoms have given them more realistic shapes, as well as the ability to tailor the masses of the patient's organs in the calculations.

14.4 Conclusion

The advancement of ^{90}Y TARE from a mostly palliative treatment to a more frontline therapeutic treatment will depend heavily on our ability to calculate the appropriate amount of therapeutic activity that is deliverable to hepatic tumors (Pasciak, Bradley, and Mckinney 2017). Based on the prescribed activity, knowing the absorbed doses following ^{90}Y TARE in tumors, normal liver, and lungs are a critical after-care component of managing patient follow-up and predicting outcomes. Using the MIRD-derived multi-compartment model currently represents the most patient-specific methodology in the pretreatment planning process to calculate the activity that may lead to tumorcidal doses. The potential for TARE goes well beyond palliation and depends on reproducible activity and dose calculations that will allow us to optimize treatments in a consistent manner.

References

Bevelacqua, J. J. (2005). "Internal dosimetry primer." *Radiat. Prot. Manage.* 22(5):7–17.

Boston Scientific (2021). "TheraSphere Instructions For Use (IFU)." Internet. doi: https://www.bostonscientific.com/content/dam/elabeling/pi/OTT-00221_Rev1_TheraSphere_en_s.pdf. [Cited on 12/7/2022.].

Capala, J., S. A. Graves, A. Scott, G. Sgouros, S. St. James, P. Zanzonico, and B. E. Zimmerman (2021). "Dosimetry for radiopharmaceutical therapy: current practices and commercial resources." *J. Nucl. Med.* 62 (Supplement 3):3S–11S.

Chiesa, C., M. Bardiès, and H. Zaidi (2019). "Voxel-based dosimetry is superior to mean absorbed dose approach for establishing dose-effect relationship in targeted radionuclide therapy." *Med. Phys.* 46(12):5403–06. doi: 10.1002/mp.13851.

Chiesa, C, K Sjogreen-Gleisner, S. Walrand, L. Strigari, G. Flux, J. Gear, C. Stokke, P. M. Gabina, P. Bernhardt, and M. Konijnenberg (2021). "EANM dosimetry committee series on standard operational procedures: a unified methodology for 99mTc-MAA pre- and 90Y peri-therapy dosimetry in liver radioembolization with 90Y microspheres." *EJNMMI Physics* 8(1):77. doi: 10.1186/s40658-021-00394-3.

Dezarn, W. A., J. T. Cessna, L. A. DeWerd, W. Feng, V. L. Gates, J. Halama, A. S. Kennedy, S. Nag, M. Sarfaraz, V. Sehgal, R. Selwyn, M. G. Stabin, B. R. Thomadsen, L. E. Williams, and R. Salem (2011). "Recommendations of the American Association of Physicists in Medicine on dosimetry, imaging, and quality assurance procedures for 90Y microsphere brachytherapy in the treatment of hepatic malignancies." *Med. Phys.* 38(8):4824-45. doi: 10.1118/1.3608909.

Elsayed, M., B. Cheng, M. Xing, I. Sethi, D. Brandon, D. M. Schuster, Z. Bercu, J. Galt, B. Barron, and N. Kokabi (2021). "Comparison of Tc-99m MAA Planar Versus SPECT/CT Imaging for Lung Shunt Fraction Evaluation Prior to Y-90 Radioembolization: Are We Overestimating Lung Shunt Fraction?" *Cardiovasc. Intervent. Radiol.* 44(2):254–60. doi: 10.1007/s00270-020-02638-8.

Ertreo, M., H. Choi, D. Field, J. W. Lischalk, E. Cohen, G. E. Lynskey, T. Caridi, D. Buckley, K. Unger, and A. Y. Kim (2018). "Comparison of Cone-Beam Tomography and Cross-Sectional Imaging for Volumetric and Dosimetric Calculations in Resin Yttrium-90 Radioembolization." *Cardiovasc. Intervent. Radiol.* 41(12):1857–66. doi: 10.1007/s00270-018-2030-0.

Garin, E., T. Lambros, B. Guiu, J. Chalaye, J. Edeline, et al. (2021). "Personalised versus standard dosimetry approach of selective internal radiation therapy in patients with locally advanced hepatocellular carcinoma (DOSISPHERE-01): a randomised, multicentre, open-label phase 2 trial." *Lancet Gastroenterol. Hepatol.* 6(1):17–29.

Georgiou, M. F., R. A. Kuker, M. T. Studenski, P. P. Ahlman, M. Witte, and L. Portelance (2021). "Lung shunt fraction calculation using (99m)Tc-MAA SPECT/CT imaging for (90)Y microsphere selective internal radiation therapy of liver tumors." *EJNMMI Res.* 11(1):96. doi: 10.1186/s13550-021-00837-z.

Gulec, S. A, and A. J. McGoron (2022). "Radiomicrosphere Dosimetry: Principles and Current State of the Art." *Semin. Nucl. Med.* 52:215–28. doi: https://doi.org/10.1053/j.semnuclmed.2021.12.009.

Gulec, S. A., G. Mesoloras, and M. Stabin (2006). "Dosimetric Techniques in 90Y-Microsphere Therapy of Liver Cancer: The MIRD Equations for Dose Calculations." *J. Nucl. Med.* 47(7):1209.

Ho, S., W. Y. Lau, T. W. T. Leung, M. Chan, Y. K. Ngar, P. J. Johnson, and A. K. C. Li (1996). "Partition model for estimating radiation doses from yttrium-90 microspheres in treating hepatic tumours." *Eur. J. Nucl. Med.* 23(8):947–52. doi: 10.1007/BF01084369.

Howell, R. W., B. W. Wessels, R. Loevinger, E. E. Watson, W. E. Bolch, A. B. Brill, N. D. Charkes, D. R. Fisher, M. T. Hays, J. S. Robertson, J. A. Siegel, and S. R. Thomas (1999). "The MIRD perspective 1999. Medical Internal Radiation Dose Committee." *J. Nucl. Med.* 40(1):3s–10s.

ICRP 23. *Report of the Task Group on Reference Man.* ICRP Publication 23. London: Pergamon Press, 1975.

ICRU (2021). "Dosimetry-guided radiopharmaceutical therapy." Report 96. *J. ICRU* 21(1):1–212.

Kao, Y. H., E. H. Tan, C. E. Ng, and S. W. Goh (2011). "Clinical implications of the body surface area method versus partition model dosimetry for yttrium-90 radioembolization using resin microspheres: a technical review." *Ann. Nucl. Med.* 25(7):455–61. doi: 10.1007/s12149-011-0499-6.

Kappadath, S. C., B. P. Lopez, R. Salem, and M. G. Lam (2021a). "Lung shunt and lung dose calculation methods for radioembolization treatment planning." *Q. J. Nucl. Med. Mol. Imaging* 65(1):32–42. doi: 10.23736/s1824-4785.20.03287-2.

Kappadath, S. C., B. P. Lopez, R. Salem, and M. Lam (2021b). "Reassessment of the lung dose limits for radioembolization." *Nucl. Med. Commun.* 42(10):1064–75. doi: 10.1097/mnm.0000000000001439.

Kim, S. P., C. Cohalan, N. Kopek, and S. A. Enger (2019). "A guide to 90Y radioembolization and its dosimetry." *Physica Medica* 68:132–45. doi: https://doi.org/10.1016/j.ejmp.2019.09.236.

Lam, M. G. E. H., J. D. Louie, M. H. K. Abdelmaksoud, G. A. Fisher, C. D. Cho-Phan, and D. Y. Sze (2014). "Limitations of Body Surface Area-Based Activity Calculation for Radioembolization of Hepatic Metastases in Colorectal Cancer." *J. Vasc. Interv. Radiol.* 25(7):1085–93. doi: https://doi.org/10.1016/j.jvir.2013.11.018.

Levillain, H., O. Bagni, C. M. Deroose, A. Dieudonné, S. Gnesin, O. S. Grosser, S. C. Kappadath, A. Kennedy, N. Kokabi, D. M. Liu, D. C. Madoff, A. Mahvash, A. Martinez de la Cuesta, D. C. E. Ng, P. M. Paprottka, C. Pettinato, M. Rodríguez-Fraile, R. Salem, B. Sangro, L. Strigari, D. Y. Sze, B. J. de Wit van der Veen, and P. Flamen (2021). "International recommendations for personalised selective internal radiation therapy of primary and metastatic liver diseases with yttrium-90 resin microspheres." *Eur. J. Nucl. Med. Mol. Imaging* 48(5):1570–84. doi: 10.1007/s00259-020-05163-5.

O'Connor, P. J., S. D. Pasik, I. M. van der Bom, V. Bishay, A. Radaelli, and E. Kim (2020). "Feasibility of Yttrium-90 Radioembolization Dose Calculation Utilizing Intra-procedural Open Trajectory Cone Beam CT." *Cardiovasc. Intervent. Radiol.* 43(2):295–301. doi: 10.1007/s00270-019-02198-6.

Padia, S. A., R. J. Lewandowski, G. E. Johnson, D. Y. Sze, T. J. Ward, R. C. Gaba, M. O. Baerlocher, V. L. Gates, A. Riaz, D. B. Brown, N. H. Siddiqi, T. G. Walker, J. E. Silberzweig, J. W. Mitchell, B. Nikolic, and R. Salem (2017). "Radioembolization of Hepatic Malignancies: Background, Quality Improvement Guidelines, and Future Directions." *J. Vasc. Interv. Radiol.* 28(1):1–15. doi: https://doi.org/10.1016/j.jvir.2016.09.024.

Pasciak, A. S., Y. Bradley, and J. M. McKinney. *Handbook of Radioembolization: Physics, Biology, Nuclear Medicine, and Imaging.* Boca Raton, FL: CRC Press, 2017.

Salem, R., S. A. Padia, M. Lam, J. Bell, C. Chiesa, K. Fowers, B. Hamilton, J. Herman, S. C. Kappadath, T. Leung, L. Portelance, D. Sze, and E. Garin (2019). "Clinical and dosimetric considerations for Y90: recommendations from an international multidisciplinary working group." *Eur. J. Nucl. Med. Mol. Imaging* 46(8):1695–1704. doi: 10.1007/s00259-019-04340-5.

Sirtex Medical Limited (2019). "SIR-Spheres Instructions For Use (IFU). doi: https://www.sirtex.com/media/55rprxv2/ssl-us-14-sir-spheres-microspheres-ifu-us.pdf. Internet. [Cited on 12/7/2022.]

SNMMI [website] (2022). "Committee on Medical Internal Radiation Dose (MIRD)." Accessed 12/7/22. http://www.snmmi.org/AboutSNMMI/CommitteeContent.aspx?ItemNumber=12475&navItemNumber=763.

Snyder, W. S., M. R. Ford, G. G. Warner, and S. B. Watson. MIRD Pamphlet #11. *S Absorbed Dose per Unit Cumulated Activity for Selected Radionuclides and Organs.* New York: Society of Nuclear Medicine, 1975.

Snyder, W. S., M. R. Ford, and G. G. Warner. Revision of MIRD pamphlet No. 5. *Estimates of Absorbed Fractions for Monoenergetic Photon Sources Uniformly Distributed in Various Organs of a Heterogeneous Phantom.* United States: N. p., 1974. Web.

Thomas, M. A., A. Mahvash, M. Abdelsalam, A. O. Kaseb, and S. C. Kappadath (2020). "Planning dosimetry for (90) Y radioembolization with glass microspheres: Evaluating the fidelity of (99m) Tc-MAA and partition model predictions." *Med. Phys.* 47(10):5333–42. doi: 10.1002/mp.14452.

Weber, D. A. (1991). "The MIRD method of estimating absorbed dose." United States. https://www.osti.gov/servlets/purl/5915860.

Weber, M., M. Lam, C. Chiesa, M. Konijnenberg, M. Cremonesi, P. Flamen, S. Gnesin, L. Bodei, T. Kracmerova, M. Luster, E. Garin, and K. Herrmann (2022). "EANM procedure guideline for the treatment of liver cancer and liver metastases with intra-arterial radioactive compounds." *E. J. Nucl. Med. Mol. Imaging* 49(5):1682–99. doi: 10.1007/s00259-021-05600-z.

Wright, C. L., J. Zhang, M. F. Tweedle, M. V. Knopp, and N. C. Hall (2015). "Theranostic Imaging of Yttrium-90." *Biomed. Res. Int.* 2015:481279. doi: 10.1155/2015/481279.

Chapter 15

Curve Fitting

Joe O'Donoghue, Ph.D.

Department of Medical Physics
Memorial Sloan Kettering Cancer Center
New York, NY

15.1 Introduction

Radiopharmaceutical therapy (RPT) is characterized by the protracted irradiation of target and nontarget tissues by ionizing particles emitted internally by radionuclides. Activity concentrations in tissues and the corresponding absorbed dose-rates vary with time, reflecting the kinetic development of the bio-distribution. For any particular tissue volume, the relationship between activity (or dose-rate) and time is a smooth curve. However, in practice, there are limitations on how much data can be acquired. Each data point represents a patient measurement and may entail a significant commitment of time and resources, both from the patient and the healthcare facility.

One of the main applications of *curve-fitting* in RPT is to estimate the shapes of the smooth curves describing activity or dose-rate using only the limited amount of data available. This is usually achieved by fitting a mathematical function, typically composed of sums of exponentials, to the data. Such curves may be used to perform absorbed dose calculations (see Chapter 13 by Bolch and Chapter 18 by Graves) or radiobiological modeling (see Chapter 21 by Hobbs) and are relatively straightforward to convert from one radionuclide to another in cases where imaging surrogates are being used (see Chapter 11 by Kesner).

Typically, we wish to estimate the *areas under the curves* (AUC), as these are the key determinants of absorbed dose. AUC may be estimated by analytical integration of the fitted mathematical functions or by numerical integration of raw data. The area under the activity-time curve represents the total number of disintegrations and, while technically dimensionless, is often expressed in units of MBq h (or mCi h). The *time-*

integrated activity coefficient (TIAC) is the area under the activity-time curve divided by the administered activity, A_0, and has dimensions of time:

$$TIAC = \frac{AUC_{Activity-Time}}{A_0}. \tag{15.1}$$

Whole-organ estimates of TIAC (formerly known as *residence times*) are the basic input data for organ-level dosimetry as implemented in software applications such as OLINDA/EXM (Stabin et al. 2005), IDAC-Dose 2.1 (Andersson et al. 2017), or MIRDcalc (Kesner et al. 2018). Voxelized TIAC may be used in conjunction with voxel-level S values, dose point kernels, or Monte Carlo methods to estimate 3D voxelized absorbed dose distributions.

15.2 Biological Drivers of Kinetic Curves

The shapes of radiopharmaceutical kinetic curves are determined by a number of biological factors. The clearance rate of an agent from the circulation depends on its molecular weight, mass-dose, and the presence or absence of normal tissue or disease-related "sinks." However, in general, large molecules, such as antibodies, clear more slowly than small molecules, such as peptides. Uptake and clearance of radiopharmaceuticals in normal tissues and tumors depend on the serum concentration kinetics, the vascular permeability, the concentration and accessibility of molecular target, the rates of molecular association and dissociation, and how the agent is metabolized. Most of these factors are fixed by the choice of radiopharmaceutical or the nature of the clinical scenario. However, mass dose can be chosen, to some extent, and can have a significant impact on kinetics (Pandit-Taskar et al. 2008).

15.3 Activity-Time Curves

In most cases, patient measurements (e.g., ROI analysis of clinical images) are proportional to activity rather than the molar amount of material present and, therefore, depend on the physical half-time, T_p, of the radionuclide. If the biological kinetics are defined by some function, $f_b(t)$, the activity kinetics, $f_a(t)$, are given by

$$f_a(t) = f_b(t)\exp(-\lambda t), \tag{15.2}$$

where λ is the physical decay rate (= ln(2) / T_p). For example, if the biological clearance from a tissue is represented by a mono-exponential function with clearance rate k (= ln(2) / T_b),

$$f_b(t) = A_0 \exp(-kt). \tag{15.3}$$

Then the activity clearance is

$$f_a(t) = f_b(t)\exp(-\lambda t) = A_0 \exp(-kt)\exp(-\lambda t) = A_0 \exp\left[-(k+\lambda)t\right]. \tag{15.4}$$

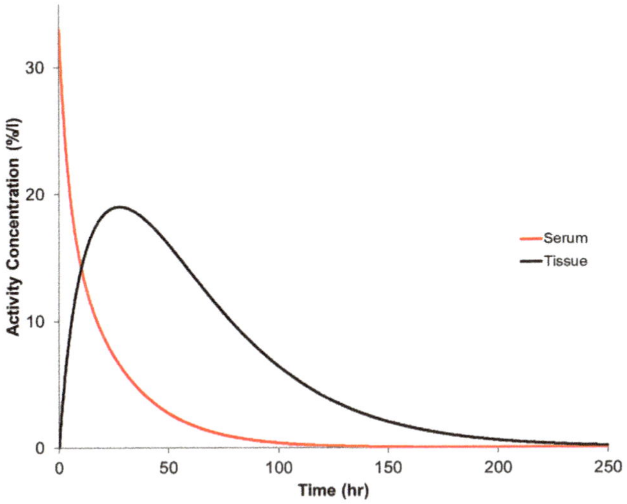

Figure 15.1 Example of activity-time curves for serum (monotonically decreasing) and a tissue with active uptake (rising then falling). The relatively slow kinetics illustrated here are typical for a large molecular vector such as an antibody.

Here, $k + \lambda$ is the effective clearance rate corresponding to an effective half-time, T_{eff}, given by:

$$T_{eff} = \frac{T_b T_p}{T_b + T_p}. \tag{15.5}$$

Activity-time data or fit parameters based on the combination of biological kinetics and physical decay are referred to as *effective* and are what is required for dosimetry calculations. However, in some cases, *biological* rather than effective kinetic curves are measured directly. This may occur when performing *ex vivo* measurements e.g., counting blood samples. Samples are collected at different post-injection times, and then all are counted at the same time. In such cases, each individual sample has experienced the same amount of physical decay, partly while inside the patient and partly while being stored in the lab. These types of data or functional fits must be converted to effective values (Equation (15.2)) for dosimetry calculations.

The use of *imaging surrogates* refers to where different radionuclides are used for image-based patient assessments and clinical therapy, and these surrogates are an important component of RPT. Often a therapeutic radionuclide is unsuitable or suboptimal for imaging. For example, only very low activities of the multi-alpha emitters ^{225}Ac or ^{227}Th may be given safely due to their extreme radiobiological potency. This makes imaging these agents very difficult. For dosimetry purposes, studies may be performed with the same molecular vectors labeled with ^{111}In for SPECT imaging or ^{89}Zr for PET imaging (Solomon et al. 2019; Nakata et al. 2022; Broer et al. 2022). Of course, detailed preclinical studies are required to verify how well the respective biodistributions correspond. Another example is the use of the γ-emitter ^{203}Pb (T_p = 51.9 h) as a surrogate for the α-emitter ^{212}Pb (T_p = 10.6 h) (Banerjee et al. 2020).

For dosimetry purposes we need to convert from the imaging activity kinetics, $f_{aI}(t)$, to the corresponding therapy values, $f_{aT}(t)$:

$$f_{aT}(t) = f_{aI}(t)\exp\left[-(\lambda_T - \lambda_I)t\right], \tag{15.6}$$

where λ_T is the physical decay rate for the therapeutic radionuclide (= ln(2) / T_{pT}) and λ_I is the physical decay rate for the imaging radionuclide (= ln(2) / T_{pI}).

The shapes of activity-time curves in tissues are generally simple, either monotonically decreasing (e.g., serum concentration) or rising-then-falling (e.g., tissues with active uptake). Figure 15.1 shows examples of these types of curves.

15.4 Curve Fitting

Typically, a mathematical function composed of sums of exponentials is fitted to the data and then integrated to estimate the AUC. Fitting models consist of a number of *adjustable parameters* (AP). For example, in a mono-exponential function of the form $f(t) = A_1\exp(-a_1 t)$ there are 2 AP. In general, at least one more data point than the number of AP is required to uniquely define the function. Some curve-fitting software applications will not allow model functions to be used if AP > #datapoints+1, but others will allow this. If a function with AP > #datapoints+1 is used, solutions will not be unique and may be highly dependent on initial parameter estimates. No matter what method is used, it is good practice to always examine the fit and decide if it looks reasonable (i.e., eyeball test). This is especially important in the context of the increasing tendency for dosimetry software applications to automatically generate dosimetry estimates without providing intermediate steps in the process.

All curve-fitting routines are iterative—an initial set of parameter values are selected, and an algorithm generates successively better values by minimizing an objective function, O, which is usually the sum of weighted mean square deviations between the actual data, y_i, and the model-fitted values, $f(t_i)$. The weights for the individual squared differences are the reciprocals of the variances. For radioactivity measurements described by Poisson statistics, the appropriate weights are $1/y_i$:

$$O = \sum \frac{\left[y_i - f(t_i)\right]^2}{y_i}. \tag{15.7}$$

The best-fit parameters are arrived at when further parameter adjustment no longer reduces the objective function.

Many different mathematical functions may be used for curve fitting (Kletting et al. 2013; Bartlett et al. 2022). These generally conform to either a monotonically decreasing sum of exponentials (e.g., for whole body or serum clearance) or a rising-then-falling shape (e.g., for tissues or disease sites with active uptake). An example of each of these, taken from the MIRD primer (Bartlett et al. 2022), are included below, with

$$f(t) = A_1 \exp(-a_1 t) + A_2 \exp(-a_2 t) \tag{15.8}$$

showing a monotonically decreasing sum of two exponentials for all parameters positive (4 AP) and

$$f(t) = \left[1 - A_1 \exp(-a_1 t)\right] A_2 \exp(-a_2 t) \tag{15.9}$$

showing a rising-then-falling combination of two exponentials for all parameters positive (4 AP). Note that Equation (15.9) could equally well be written in the form of Equation (15.8) if negative parameter values are allowed.

In general, the use of more AP (up to a maximum of #datapoints-1) will progressively minimize the objective function but may lead to overfitting. A good model should strike a balance between goodness-of-fit and parsimony of AP. To select the most appropriate model from a group of models, metrics such as the *Akaike information criterion* (AIC) or *Bayesian information criterion* (BIC) may be used—better models will have lower values of these metrics. AIC or BIC are calculated by several curve-fitting applications, e.g., SAAMII (Barrett et al. 1998), NUKFIT (Kletting et al. 2013), and the soon-to-be-released MIRDfit (Carter et al. 2022).

Mathematical functional fits may be integrated to determine the AUC:

$$AUC = \int_0^\infty f_a(t)\,dt. \tag{15.10}$$

Function parameters and AUC are typically provided by curve-fitting applications together with their associated uncertainties, estimated numerically. The uncertainty in AUC depends on the uncertainties associated with the data and the parameters of the fitting function (see Chapter 16 by Gear). Historically, it has usually been impractical to use the uncertainties in individual TIAC in the estimation of the overall uncertainty of absorbed dose estimates. However, more recent organ-level dosimetry software such as MIRDcalc (Kesner et al. 2018) incorporate treatment of error propagation and, in the future, the process is likely to be more straightforward.

As an alternative to fitting and integrating mathematical functions, AUC may be estimated numerically by *trapezoidal integration*. This approach requires the terminal AUC to be estimated separately, usually by extrapolation from the last measured value using the faster of apparent terminal clearance rate or physical decay.

15.5 An Example of Curve Fitting

In this exercise, some real clinical data are fit using the SAAMII application. The data represent the serum clearance of an ^{89}Zr-labeled antibody from a patient and are expressed in terms of percent of administered activity per liter, i.e., effective data uncorrected for physical decay. The aim is to derive a serum TIAC expressed in terms of h per liter.

Three exponential clearance functions are used:

1. Mono-exponential

$$f(t) = A_1 \exp(-a_1 t)$$

2. Bi-exponential

$$f(t) = A_1 \exp(-a_1 t) + A_2 \exp(-a_2 t)$$

3. Tri-exponential

$$f(t) = A_1 \exp(-a_1 t) + A_2 \exp(-a_2 t) + A_3 \exp(-a_3 t)$$

Table 15.1 Data used in the example

T/h	% IA/l
0.25	43.50
0.5	45.27
1.0	45.14
2.0	43.19
4.0	39.77
25.8	24.79
50.3	16.31
193.1	2.26
405.3	0.17

Table 15.1 shows the measured data that was used with SAAM II with a variance structure stated to be Poisson.

The functional fits are shown in Figure 15.2. On first examination, the bi- and tri-exponential functions both appear to fit to the data better than the mono-exponential function, where there are noticeable deviations between the data and the curve. The best-fit parameter estimates are shown in Table 15.2.

Examining the fit parameters reveals several points:

1. The tri-exponential curve results in the lowest value of the objective function (Obj). SAAM II uses a more sophisticated objective function than the sum of

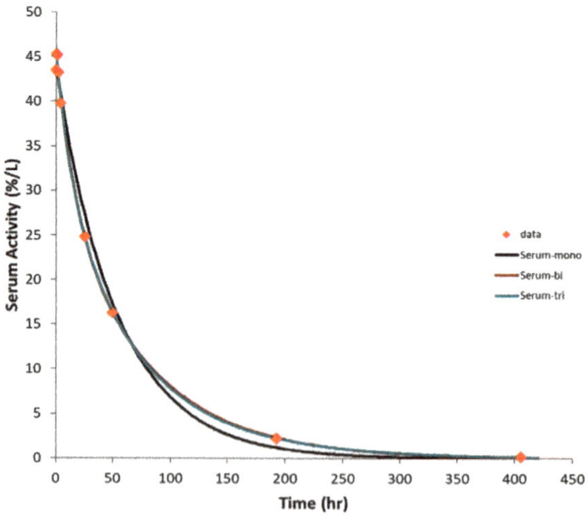

Figure 15.2 Functional fits to the data.

Table 15.2 Model parameter fits in the example

Mono-exponential Fit			Bi-exponential Fit			Tri-exponential Fit		
Parameter	**Value**	**SD**	**Parameter**	**Value**	**SD**	**Parameter**	**Value**	**SD**
A1	44.0	1.2	A1	15.0	3.9	A1	9.4	31.9
			A2	30.3	4.1	A2	16.0	31.5
						A3	19.9	56.6
a1	0.019	0.001	a1	0.059	0.024	a1	0.084	0.196
			a2	0.013	0.0009	a2	0.022	0.069
						a3	0.012	0.007
AUC	2364	167	AUC	2530	66	AUC	2516	120
Obj	1.554		Obj	–0.733		Obj	–0.817	
AIC	2.029		AIC	1.108		AIC	1.288	
BIC	2.062		BIC	1.163		BIC	1.365	

weighted mean square deviations; this can have negative values (Barrett 1998).

2. Both mono- and bi-exponential fit parameters are defined relatively precisely—the *standard deviations* (SD) are much less than the estimated parameter values. The tri-exponential parameters are not precisely defined—the standard deviations are greater than the estimated parameter values.
3. Both AIC and BIC metrics are smallest for the bi-exponential fit.

On this basis, the bi-exponential function

$$f(t) = 15.0\exp(-0.059t) + 30.3\exp(-0.013t) \tag{15.11}$$

is chosen as the best model for this dataset. It gives the most tightly defined estimate of the AUC, i.e., 2530 (SD 66) % h per liter. However, it should be noted that all three models give reasonably similar AUC values in this particular case. The best estimate for the serum TIAC is thus 25.3 (SD 0.66) h per liter.

Serum (or whole blood) TIAC is often used in the estimation of absorbed dose estimates for tissues such as red marrow or heart wall.

15.6 Considerations for Clinical Protocol Design

A data collection scheme for clinical absorbed dose estimation has to strike a balance between having enough information to define the main features of the biodistribution kinetics and the various costs entailed, as each data point may require a significant commitment of time and resources. This is especially the case if RPT is delivered on an outpatient basis, as each measurement requires the patient to travel to the treatment

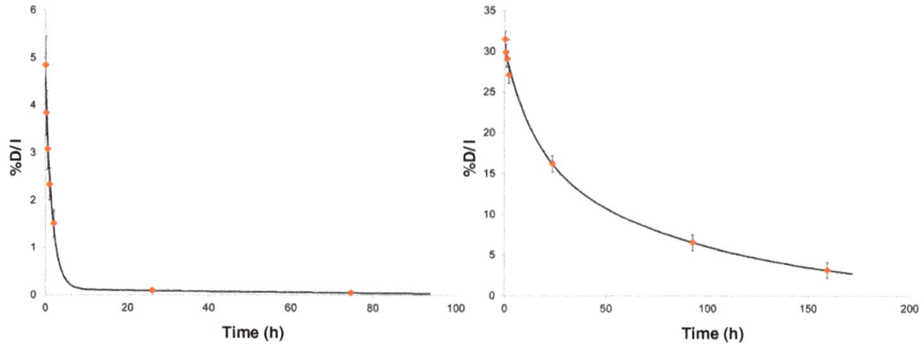

Figure 15.3 Serum clearance for a small molecule (left) and an antibody (right).

center. Where patients are treated as inpatients, it is easier to collect more data, especially in the first few days after RPT administration when patients are present in the center. MIRD Pamphlet 16 (Siegel et al.1999) discusses many of the issues related to data acquisition for human absorbed dose estimates, including temporal sampling.

Uptake in normal tissues and tumors is driven by the concentration of radiopharmaceutical in the serum. This differs between small and large molecules, as illustrated in Figure 15. 3, which compares serum clearance for a small molecule with that for an intact antibody. In this example, the small molecule has almost completely cleared from the circulation in less than 10 h, whereas antibody concentration remains significant for more than a week. We would, therefore, expect small-molecule uptake to peak early, and the main kinetic questions would relate to retention. For antibody-like molecules, both uptake and retention are important kinetic questions.

For data collection, a useful rule of thumb is that each exponential component of a multi-exponential kinetic curve requires 2 to 3 data points. However, for RPT dosimetry, the importance of these various components to the overall AUC must also be considered, and the accurate definition of a rapid uptake phase may not be strictly necessary, especially if there is prolonged retention.

Radionuclides used for RPT typically have relatively long physical half-lives, for example the half-life of ^{177}Lu is 6.7 days. For serial imaging of small-molecule RPT, such as ^{177}Lu-DOTATATE or ^{177}Lu-PSMA, data are typically acquired up to around 7 days post-administration (Huizing et al. 2018; Herrmann et al. 2020). This is generally adequate to define the shapes of retention-dominated curves. However, for molecules that clear slowly from the circulation, this may be insufficient. An active area of current clinical research involves modifying the kinetics of small molecules by incorporating an albumin-binding motif into their molecular structure (Wang et al. 2018; Kelly et al. 2019). In addition to potentially increasing tumor uptake, this results in much longer serum retention. For these sorts of molecules, as well as antibody-based RPT, imaging at significantly longer than 1 week may be necessary.

In the case of surrogate imaging, mismatches in half-life mean that it may not be possible to examine the full biodistribution kinetics of the therapeutic agent. The use of ^{111}In (T_p = 2.8 days) or ^{89}Zr (T_p = 3.3 days) as surrogates of ^{225}Ac (T_p = 10.0 days) or ^{227}Th (T_p = 18.7 days) falls into this category. In practical terms, imaging with typical administered activities (185 MBq for ^{111}In, 74 MBq for ^{89}Zr) is only possible for up to approximately one week. For small-molecule kinetics dominated by retention, this may be adequate. However, for slow-clearing agents, surrogate imaging may lead to underestimates of uptake in normal tissues and tumors. In such circumstances, compartmental modeling may provide better estimates of absorbed doses.

15.7 Conclusions

Calculating areas under the activity or dose rate versus time curve is a fundamental component of RPT dosimetry. It is usually achieved by fitting a mathematical function, composed of sums of exponentials, to the data and then integrating. Curve fitting routines are iterative, where an algorithm generates successively better parameter estimates by minimizing an objective function until additional adjustments lead to no further reduction. To select the most appropriate models from several possible choices, quantities such as the *Akaike information criterion* (AIC) or *Bayesian information criterion* (BIC) may be used.

Clinical protocol design is a compromise between data collection and cost/practicality. It is important to tailor data collection to the characteristics of the targeting molecule and the clinical situation.

References

Andersson, M., L. Johansson, K. Eckerman, and S. Mattsson (2017). "IDAC-Dose 2.1, an internal dosimetry program for diagnostic nuclear medicine based on the ICRP adult reference voxel phantoms." *EJNMMI Res.* 7:88.

Banerjee, S. R., I. Minn, V. Kumar, A. Josefsson, A. Lisok, M. Brummet, J. Chen, A. P. Kiess, K. Baidoo, C. Brayton, R. C. Mease, M. Brechbiel, G. Sgouros, R. F. Hobbs, and M. G. Pomper (2020). "Preclinical Evaluation of $^{203/212}$Pb-Labeled Low-Molecular-Weight Compounds for Targeted Radiopharmaceutical Therapy of Prostate." *J. Nucl. Med.* 61:80–88.

Barrett, P. H., B. M. Bell, C. Cobelli, H. Golde, A. Schumitzky, P. Vicini, and D. M. Foster (1998). "SAAM II: Simulation, Analysis, and Modeling Software for tracer and pharmacokinetic studies." *Metabolism* 47:484–92.

Bartlett, R. M., W. E. Bolch, A. B. Brill, Y. K. Dewaraja, F. H. Fahey, D. R. Fisher, R. F. Hobbs, R. W. Howell, R. F. Meredith, J. G. Rajendran, G. Sgouros, and P. Zanzonico. *MIRD Primer 2022.* Reston, VA: Society of Nuclear Medicine & Molecular Imaging, 2022.

Broer, L. N., D. G. Knapen, F. V. Suurs, I. Moen, D. Giesen, S. J. H. Waaijer, B. Indrevoll, C. Ellingsen, A. Kristian, A. S. Cuthbertson, D.-J. A. de Groot, P. E. Cole, E. G. E. de Vries, U. B. Hagemann, and M. N. Lub-de Hooge (2022). "^{89}Zr-3,2-HOPO-Mesothelin Antibody PET Imaging Reflects Tumor Uptake of Mesothelin-Targeted ^{227}Th-Conjugate Therapy in Mice." *J. Nucl. Med.* 63:1715–21.

Carter, L., J. C. Ramos, W. Bolch, P. Zanzonico, and A. Kesner (2022). "MIRDfit: A tool for fitting of biodistribution time-activity data for internal dosimetry." *J. Nucl. Med.* 63:2349.

Herrmann, K., K. Rahbar, M. Eiber, B. J. Krause, M. Lassmann, W. Jentzen, L. Blumenstein, P. Klein, J.-R. Basque, and J. Kurth (2020). "Dosimetry of ^{177}Lu-PSMA-617 for the treatment of metastatic castration-resistant prostate cancer: results from the VISION trial substudy." *J. Clin. Oncol.* 40(6):97.

Huizing, D. M. V., B. J. de Wit-van der Veen, M. Verheij, and M. P. M. Stokkel (2018). "Dosimetry methods and clinical applications in peptide receptor radionuclide therapy for neuroendocrine tumours: a literature review." *EJNMMI Res.* 8(1):89.

Kelly, J. M., A. Amor-Coarasa, S. Ponnala, A. Nikolopoulou, C. Williams Jr., S. G. DiMagno, and J. W. Babich (2019). "Albumin-Binding PSMA Ligands: Implications for Expanding the Therapeutic Window." *J. Nucl. Med.* 60:656–63.

Kesner, A, E. Olguin, P. Zanzonico, and W. Bolch (2018). "MIRDCalc V 1.0-A community spreadsheet tool for organ-level radiopharmaceutical absorbed dose calculations." *J. Nucl. Med.* 59(supplement):473.

Kletting, P., S. Schimmel, H. A. Kestler, H. Hänscheid, M. Luster, M. Fernández, J. H. Bröer, D. Nosske, M. Lassmann, and G. Glatting (2013). "Molecular radiotherapy: the NUKFIT software for calculating the time-integrated activity coefficient." *Med Phys.* 40:102504.

Nakata, N., N. Kobashi, Y. Okumura, M. Sato, M. Matono, K. Otsuki, A. Tanaka, and A. Hayashi (2022). "Radiation dosimetry and efficacy of an ^{89}Zr/^{225}Ac-labeled humanized anti-MUC5AC antibody." *Nucl. Med. Biol.* 108–109:33–43.

Pandit-Taskar, N, J. A. O'Donoghue, M. J. Morris, E. A. Wills, L. H. Schwartz, M. Gonen, H. I. Scher, S. M. Larson, and C. R. Divgi (2008). "Antibody mass escalation study in patients with castration-resistant prostate cancer using 111In-J591: lesion detectability and dosimetric projections for 90Y radioimmunotherapy." *J. Nucl. Med.* 49:1066–74.

Siegel, J. A., S. R. Thomas, J. B. Stubbs, M. G. Stabin, M. T. Hays, K. F. Koral, J. S. Robertson, R. W. Howell, B. W. Wessels, D. R. Fisher, D. A. Weber, and A. B. Brill (1999). "MIRD pamphlet no. 16: Techniques for quantitative radiopharmaceutical biodistribution data acquisition and analysis for use in human radiation dose estimates." *J. Nucl. Med.* 40:37S–61S.

Solomon, V. R., E. Alizadeh, W. Bernhard W, S. V. Hartimath, W. Hill, R. Chekol, K. M. Barreto, C. R. Geyer, and H. Fonge (2019). "^{111}In- and ^{225}Ac-Labeled Cixutumumab for Imaging and α-Particle Radiotherapy of IGF-1R Positive Triple-Negative Breast Cancer." *Mol. Pharm.* 16:4807–16.

Stabin, M. G., R. B. Sparks, and E. Crowe (2005). "OLINDA/EXM: the second-generation personal computer software for internal dose assessment in nuclear medicine." *J. Nucl. Med.* 46:1023–27.

Wang, H., Y. Cheng, J. Zhang, J. Zang, H. Li, Q. Liu, J. Wang, O. Jacobson, F. Li, Z. Zhu, and X. Chen (2018). "Response to Single Low-dose ^{177}Lu-DOTA-EB-TATE Treatment in Patients with Advanced Neuroendocrine Neoplasm: A Prospective Pilot Study." *Theranostics* 8:3308–16.

Chapter 16

Uncertainties and Error Propagation

Jonathan I. Gear, Ph.D.

Joint Department of Physics
Institute of Cancer Research and
Royal Marsden NHS Foundation Trust
United Kingdom

16.1 Introduction

16.1.1 Uncertainty and Error

For any measurement made for either scientific or medical purposes, it is important to understand the reliability of the calculated result. This is particularly true if that measurement is to be used to inform a patient treatment, as is the aim with absorbed dose calculations. With many nuclear medicine applications, measurement accuracy is tested using phantom acquisitions, whereby a "dummy" acquisition is performed on an imaging test object and the measured result compared to the "ground-truth" derived when the phantom is first prepared. The difference between the ground-truth and the measurement result is the error. While of interest, this phantom error measurement cannot necessarily infer the reliability of a similar measurement made on a patient and, therefore, a more comprehensive analysis of measurement uncertainty is required.

It is important at this early stage to give some definitions of common terms to make the distinction between uncertainty and error, as the two words are often used incorrectly to infer the same thing.

- ***Error*** is the absolute difference between a measurement and the true value of the measurand (the particular quantity being measured). By definition, if you know the error, you know the true value.
- ***Uncertainty*** means "doubt" and, therefore, "uncertainty of measurement" means doubt about the validity of a measurement result. A more formal definition is that of a parameter, associated with the measurement result, that characterizes an interval within which the true value lies with some probability. The uncertainty is sometimes given alongside a coverage factor which reflects the chosen level of probability with which the interval covers the true value. For a given estimate of the measurand quantity Y and a coverage factor of 1, the half-width of this interval is the standard uncertainty, $u(y)$, describing a coverage probability of 68%.

16.1.2 The Guide to Expression of Uncertainty in Measurement (General Principles and Pitfalls)

For a generic measurement model, $Y = f(X_1, X_2, ..., X_n)$, the standard uncertainty in the estimate y depends on the standard uncertainties in each of the estimated input quantities, $x_1, x_2, ..., x_n$. For absorbed dose, examples of input quantities would be time-integrated activity and S value. These in turn may be estimated using a measurement model with further input quantities, such as activity, absorbed fraction, or mass, etc. This hierarchical system of input and output quantities, therefore, requires a system of propagation of uncertainty estimates for each input quantity to the given output quantity. In addition, instances can occur where different quantities depend on the same input and correlations need to be considered when propagating these uncertainties further along the chain.

Propagation of uncertainty can be achieved by applying the *law of propagation of uncertainty* (LPU) in measurement, described extensively in the *guide for expression of uncertainty and measurements* (GUM) (BIPM et al. 2008a). This approach relies on simplifying the measurement model by describing the change of an output quantity, given a small change to the input quantity linearly as,

$$\Delta Y = \frac{\partial f}{\partial X} \Delta X \tag{16.1}$$

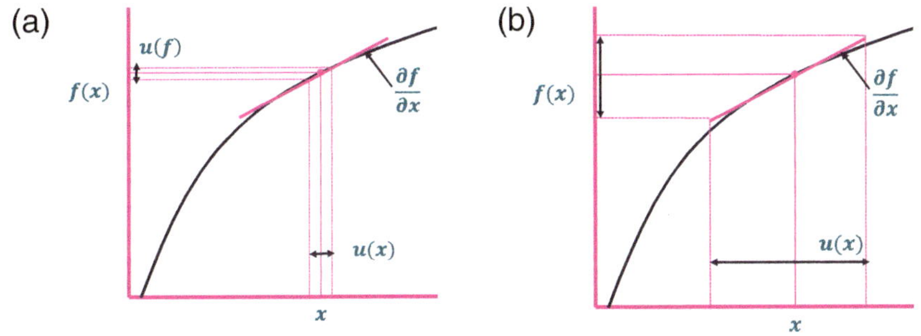

Figure 16.1 An illustrative example demonstrating propagation of uncertainty. The uncertainty in *x*, represented by the dashed line, is propagated to an uncertainty in *f*(*x*) based on the gradient of the function at *x* (a). However, if the uncertainty in *x* is very large, this simplification can underestimate the uncertainty (b).

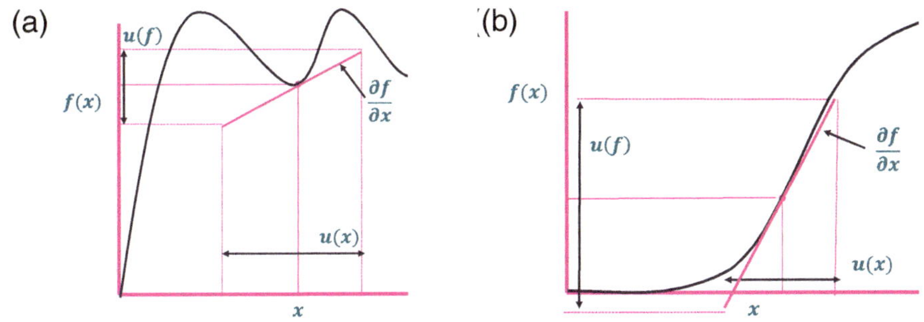

Figure 16.2 Illustrative examples showing s :uations where the linearization of the LPU fails.

Similarly, in the LPU, the uncertainty in an output quantity, $u(y)$, given an uncertainty in the input quantity $u(x)$ can is expressed using

$$u(y) = \frac{\partial f}{\partial x} u(x) \tag{16.2}$$

This concept is shown graphically in Figure 16.1. A function y = f(x) is used to determine a value of y given a measurement of x, illustrated in Figure 16.1a by reading off the graph at the value of x. The standard uncertainty in x is illustrated by the dotted lines. The gradient $\partial f / \partial x$ at the measured value of x is used to propagate an uncertainty u(y). It is important to be mindful at this point and recognize that the use of the gradient is an approximation and, as shown in Figure 16.1b, if the uncertainty associated with x is too large, the estimated uncertainty will no longer represent the true variation in y observed for values of x within the uncertainty bound.

This concept is further exaggerated when the gradient of the function is changing significantly within the range of uncertainty measured. As shown in Figure 16.2, this can lead to estimates of uncertainty significantly different from the likely values of y and, in some cases, impossible negative values.

It is, therefore, not unreasonable to ask: why use the gradient at all? Why not propagate uncertainty by simply evaluating value of y+(u)y and y–u(y) based on given values of x+(u)x and x–u(x)? This approach is indeed valid and is easily adopted when there is only a single input quantity. However, when multiple input quantities are involved, such an approach quickly becomes complex, as all the different combinations of uncertainty must be combined and accounted for. When applying such an approach, it is then common to assign a probability distribution to each of the input quantities and use a Monte Carlo method to evaluate repeated estimates of Y for different input quantities randomly sampled from the probability distribution (BIPM et al. 2008b). This method then requires additional resources that a physicist may not wish to dedicate to an uncertainty estimate. In such a case, the law of propagation of uncertainty—expanded to multiple input quantities—is more straightforward.

16.1.3 Fundamental Equations and Derivations

For a function involving multiple input quantities, the matrix form of the LPU can be used (BIPM et al. 2011):

$$u^2(y) = \boldsymbol{g}_x^{\top} \boldsymbol{V}_x \boldsymbol{g}_x, \tag{16.3}$$

where $u^2(y)$ represents the variance (squared standard uncertainty) associated with the estimate y, and

$$\boldsymbol{g}_x = \begin{bmatrix} \frac{\partial f}{\partial x_1} \\ \vdots \\ \frac{\partial f}{\partial x_n} \end{bmatrix} \tag{16.4}$$

is the gradient matrix in which the ith element denotes the partial derivative of f with respect to the quantity X_i and $\boldsymbol{V}_x$ is the input covariance matrix associated with the estimated input quantities.

$$\boldsymbol{V}_x = \begin{bmatrix} u^2(x_1) & \dots & u(x_1, x_n) \\ \vdots & \ddots & \vdots \\ u(x_n, x_1) & \dots & u^2(x_n) \end{bmatrix} \tag{16.5}$$

For a two-variable multiplicative function, after performing the matrix multiplication this expression becomes

$$u^2(f) = \frac{\partial f^2}{\partial x_1} u^2(x_1) + \frac{\partial f^2}{\partial x_2} u^2(x_2) + 2\frac{\partial f}{\partial x_1}\frac{\partial f}{\partial x_2} u(x_1, x_2) \tag{16.6}$$

which after solving the partial derivative,

$$\frac{\partial f}{\partial x_1} = x_2 = \frac{f(x)}{x_1} \text{ and } \frac{\partial f}{\partial x_2} = x_1 = \frac{f(x)}{x_2}, \tag{16.7}$$

can be written in the form:

$$\left[\frac{u(y)}{y}\right]^2 = \left[\frac{u(x_1)}{x_1}\right]^2 + \left[\frac{u(x_2)}{x_2}\right]^2 + 2\frac{u(x_1,x_2)}{x_1 x_2}. \tag{16.8}$$

If the two quantities X_1 and X_2 are mutually independent, the covariance term of Equation (16.8), $u(x_1, x_2)$, is zero, giving the familiar expression,

$$\left[\frac{u(y)}{y}\right]^2 = \left[\frac{u(x_1)}{x_1}\right]^2 + \left[\frac{u(x_2)}{x_2}\right]^2. \tag{16.9}$$

Similarly, for a two-parameter additive function with two independent input quantities, the expressions should be familiar to most:

$$u^2(y) = u^2(x_1) + u^2(x_2). \tag{16.10}$$

However, if the two input quantities are not independent, and each depend on some other input quantity, say z, the covariance of the input estimates x_1 and x_2 should not be ignored and can be further described using:

$$u(x_1,x_2) = \frac{\partial x_1}{\partial z}\frac{\partial x_2}{\partial z}u^2(z). \tag{16.11}$$

The importance of this term will be apparent in later sections.

16.2 Mean Absorbed Dose

16.2.1 The MIRD Equation

Currently the expressions described bear no relation to the topic of interest, namely dosimetry. However, when we consider the MIRD equation, written in its simplest form, with a single self-irradiating source, it can be seen that this is a simple two-parameter multiplicative function of two input parameters, time-integrated activity, $\tilde{A}$, and the S value,

$$D = \tilde{A}S. \tag{16.12}$$

Provided time-integrated activity and the S value are independent, estimation of uncertainty for absorbed dose is a straightforward implementation of Equation (16.9), propagating the uncertainty of the two parameters provides

$$\left[\frac{u(D)}{D}\right]^2 = \left[\frac{u(\tilde{A})}{\tilde{A}}\right]^2 + \left[\frac{u(S)}{S}\right]^2 \tag{16.13}$$

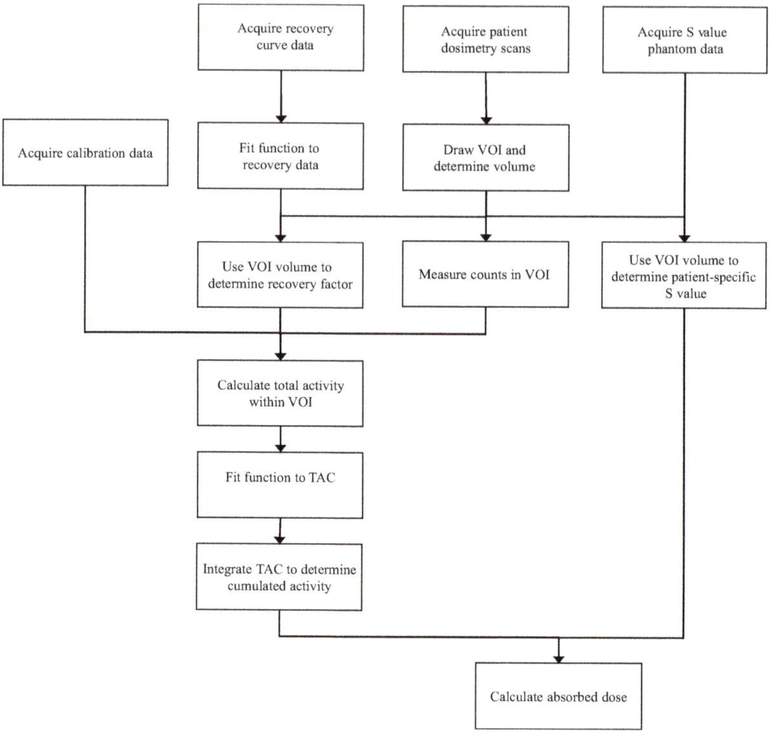

Figure 16.3 Flow chart showing the different steps in a typical absorbed dose calculation.

The challenge, therefore, is to determine the uncertainty in these two parameters and verify if they are truly independent of each other. The validity of this assumption will depend largely on the dosimetry methods used.

16.2.2 Dosimetry Methodology

To understand how uncertainties will propagate, it is useful to describe the different dosimetry steps using a flow chart to ascertain how a result or decision in one step will affect that in another. Figure 16.3 describes the approach to what is commonly referred to as *organ-level dosimetry*. The mean dose to a well-defined lesion or organ is determined assuming a uniform uptake of activity. In this example, multiple SPECT/CT images are acquired, and the volume of the organ is delineated on the appropriate image. The delineated volume is used to determine the number of reconstructed counts within that volume, which is converted to activity using a predetermined calibration or sensitivity factor. Secondary corrections for "spill out" of counts and "dead time" are also applied at this point.

Once the activity in the *volume of interest* (VOI) has been determined for each time point, a *time-activity curve* (TAC) is generated and a function fitted to the data that describes the observed uptake and retention. This function is finally integrated to obtain time-integrated activity.

To determined mean absorbed dose, an S value is also required. These are generally based on pre-evaluated estimates of reference models, which as a secondary correction can be scaled according to the mass of the volume identified on the SPECT/CT images. Methods to derive uncertainty at each of these steps and the propagation of uncertainty down the chain is described in each of the following sections.

16.3 Source and Target Delineation

16.3.1 Options and Limitations

Source and target delineation is arguably one of the most crucial aspects of dosimetry as it is used to determine mass, which, as will be seen, is required in latter steps of the dose calculation. The operator is faced with multiple options when choosing how to delineate a volume, including both which technique and which image to use for outlining. In most cases, outlining using an anatomical image dataset, such as CT or MR, is favored, although fusion with the functional SPECT or PET will help determine that the anatomical volume contains activity. In some instances, the contrast of a CT can be very poor (for example, liver lesions), and an operator may be required to outline on SPECT. In this instance, image resolution is often poor and the definition of edges difficult. The uncertainty attributed to such an outline will, therefore, be significantly higher. Figure 16.4 shows six images of kidneys from different image modalities and a phantom image. The edge definitions on all of these images are very different, so the defined uncertainty should be, too.

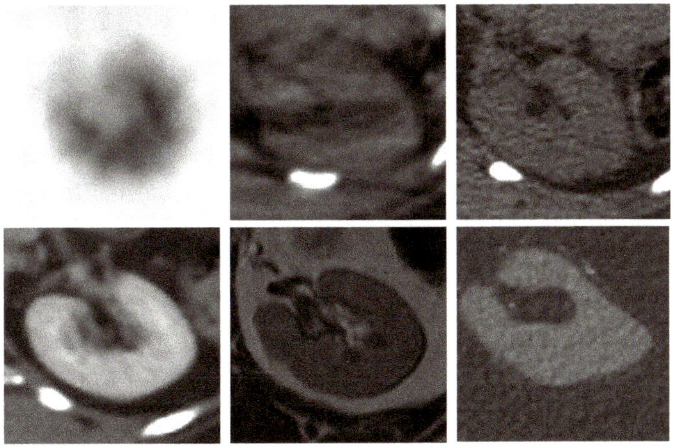

Figure 16.4 Example images of a kidney from different imaging modalities, from top left to bottom right, SPECT, cone-beam CT, low-dose CT, contrast-enhanced CT, MRI, and a phantom image.

The choice of outline will also affect the accuracy of the measured volume, and consideration should be given when ascertaining uncertainty in the different situations. Automatic outlining using AI or more simple threshold techniques may be quick and reproducible, but systematic errors can easily be introduced, particularly in the case of SPECT thresholding. Manual outlining by an operator can be time-consuming and prone to random errors between operators.

16.3.2 Analytical Uncertainty Equations

The EANM guidance for uncertainty estimation (Gear et al. 2018) provided some analytical expressions for evaluating an estimate of uncertainly in a drawn outline. These were based on the assumption that an operator is outlining a spherical volume of diameter d. Sources of uncertainty were introduced based on the voxel size

$$u_{vox}^2(d) = u^2(P_i - P_j) = \frac{a^2}{6}, \tag{16.14}$$

and resolution of the image,

$$u_{res}^2(d) = 2\sigma^2 = \left(\frac{FWHM}{2\sqrt{\ln 2}}\right)^2. \tag{16.15}$$

These uncertainties were combined and appropriately scaled to reflect an uncertainty in the outlined volume

$$\left[\frac{u(v)}{v}\right]^2 = \left[3\frac{u(d)}{d}\right]^2 = \left[3\frac{u_{vox}(d)}{d}\right]^2 + \left[3\frac{u_{res}(d)}{d}\right]^2 \tag{16.16}$$

16.3.3 Inter-operator Studies

In practice, the expression of uncertainty given in Equation (16.16) will likely underestimate the true uncertainty because other contributing factors, such as contrast, are not accounted for. It is, therefore, recommended, where possible, to base an uncertainty estimate on an *inter-operator study* that evaluates the variation in outline volume when multiple operators delineate the same volume. If multiple operators are each independently outlining the organ for which dosimetry is to be carried out, the standard uncertainty of the mean of the volumes can be derived from the standard deviation,

$$u^2(v_{hist_i}) = \frac{1}{n(n-1)}\sum_{i=1}^{n}(v_i - \overline{v})^2. \tag{16.17}$$

Alternatively, rather than have multiple operators outlining for each dosimetry, use *M* historical data sets to perform an inter-operator study. The coefficient of variation of the multiple outlines can then inform the likely uncertainty of future outlines for a similar organ and data set.

The standard uncertainty $u(v)$ associated with the newly drawn volume is then expressed as:

$$\left(\frac{u(v)}{v}\right)^2 = \frac{1}{M^2}\sum_{m=1}^{M}\left(\frac{s(v_{\text{hist}_i})}{v_{\text{hist}_i}}\right)^2, \tag{16.18}$$

where $v_{\text{hist}_i})$ is the average of operator volumes of the historical dataset i and $s(v_{\text{hist}_i})$ is the standard deviation of the historical dataset,

$$s^2(v_{\text{hist}_i}) = \frac{1}{N-1}\sum_{m=1}^{N}\left(v_m - v_{\text{hist}_i}\right)^2. \tag{16.19}$$

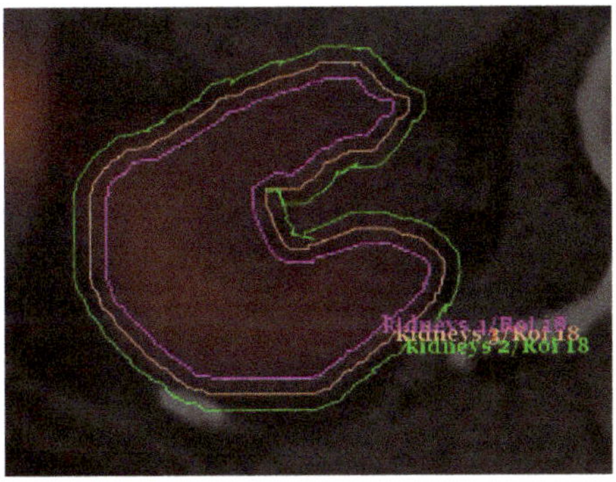

Figure 16.5 Example kidney outlines expanded and contracted by ±5% in volume.

For reasons previously discussed, care should be taken when choosing the historical datasets to match the current study as differences could lead to an inaccurate estimate of the final standard deviation. Using this approach, Marin et al. (2015) reported VOI kidney uncertainties of 5% based on variations in segmented organ mass at different cycles of a PRRT therapy. Figure 16.5 shows a similar kidney with three outlines, representing the initial operator outline and that expanded and contacted to given ±5% of the initial volume.

16.4 VOI Counts

For a non-quantified SPECT image, variation in the VOI used to delineate the source region will result in disparity in the total counts attributed to that source. Similarly, for a quantified SPECT, a VOI that is overestimated in size will potentially overestimate the activity within that VOI. EANM guidelines proposed an analytical equation that can be used for uncertainty propagation based on a Gaussian smoothed sphere model of radius, r:

$$u(C) = \frac{C_{\text{total}}}{2v}\left\{erf\left(\frac{2r}{\sigma\sqrt{2}}\right) - \frac{2\sigma}{r\sqrt{2\pi}}\left[1 - e^{-\left(\frac{2r^2}{\sigma^2}\right)}\right]\right\}u(v) \tag{16.20}$$

where σ is the measured standard deviation describing the width of the 3D Gaussian function. In hindsight, this complicated expression is still a simplified model of the problem. A more pragmatic and, arguably, accurate approach is to model the uncertainty using additional VOIs, both expanded and contracted (such as that shown in Figure 16.5) about the initial VOI such that,

$$u(C) = \frac{\partial C}{\partial v}u(v) \approx \frac{C_{v+\Delta v} - C_{v-\Delta v}}{v_{v+\Delta v} - v_{v-\Delta v}}u(v) \tag{16.21}$$

where $C_{v+\Delta v}$ are the counts measured in the expanded volume, $v_{v+\Delta v}$, and $C_{v-\Delta v}$ are the counts recorded in the contracted volume $v_{v-\Delta v}$.

16.5 Scanner Calibration

16.5.1 Sensitivity

Having determined the total counts within the segmented VOI, this value must then be converted to activity using a scanner calibration factor. Uncertainties associated with calibrating a SPECT system have been discussed in a number of publications (Anizan et al. 2015; D'Arienzo and Cox 2017; Marin et al. 2017) and can be derived from the uncertainty in the phantom activity concentration and the uncertainty in the counts measured during the analysis. Most radionuclide dose calibrators use an internal or user-defined calibration factor that will also have an associated uncertainty, often quoted by the manufacture or the metrology institute to which the primary standard is situated. Dose calibrator stability measurements can also be gathered and further accounted for, including any geometric corrections made when taking measurements.

However, for most radionuclides, the accuracy of activity measurements are sufficient that uncertainty in the preparation of a calibration phantom are insignificant. In this regard, it is more straightforward to determine the standard uncertainty associated with the system sensitivity factor Q from a series of repeat measurements. Uncertainty can then be obtained for the mean, $\bar{Q}$, of those measurements using the standard deviation of that series,

$$u(\bar{Q}) = \frac{s(Q)}{\sqrt{N}}. \tag{16.22}$$

16.5.2 Recovery Coefficients

Due to the limited spatial resolution of SPECT and PET systems, events known to have originated from a source can appear on the image outside the delineated anatomical volume. There are a number of correction methods to improve image resolution, but it is still common to apply a secondary correction based on recovery coefficients. A nice approach that allows propagation of uncertainty is to acquire data for a set of spherical phantom inserts and calculate activity concentration in each insert using a VOI matching the physical dimensions of that insert. The ratio of the measured concentration over the actual concentration is the recovery factors $R(v)$ at that volume. A plot of recovery factors against volume can then be used to generate a recovery curve, as shown in Figure 16.6. When fitting to such data, the fitting parameters are chosen to obtain a result that best fits the measured data. This fit is rarely perfect and, as such, the chosen fitting parameters are only a best estimate and carry an associated uncertainty.

An example of how these uncertainty values are determined is given in the upcoming section of time activity curve fitting. For the purposes here, it is sufficient to describe the process graphically, as demonstrated in Figure 16.6. In this instance, the uncertainty estimates on the fit parameters can be used to generate two additional lines representing the 68% confidence interval of the fitted function. In addition, as uncertainty exists in the estimate of source volume, this can also be incorporated into the uncertainty estimate of the recovery coefficient (shown as a dashed red line in Figure 16.6).

For a two-parameter fit function $f(p_1,p_2)$, the matrix form of the LPU, given in expression 3, can be used, with the additional inclusion of the volume uncertainty such that,

$$u^2(R) = \boldsymbol{g}_{p,v}^{\mathrm{T}} \boldsymbol{V}_{p,v} \boldsymbol{g}_{p,v} \quad (16.23)$$

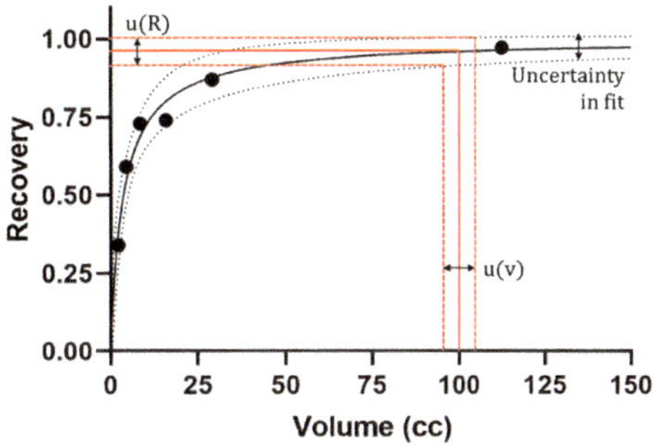

Figure 16.6 A typical recovery curve with representative lines indicating how uncertainty propagates when reading off the curve to determine a recovery coefficient for a given volume.

where

$$\boldsymbol{g}_{p,v}^{\mathrm{T}} = \left[\frac{\partial f}{\partial p_1}, \frac{\partial f}{\partial p_2}, \frac{\partial f}{\partial v} \right] \quad (16.24)$$

and

$$\boldsymbol{V}_{p,v} = \begin{pmatrix} u^2(p_1) & u(p_1,p_2) & 0 \\ u(p_2,p_1) & u^2(p_2) & 0 \\ 0 & 0 & u^2(v) \end{pmatrix}. \quad (16.25)$$

Note the zeros in the covariance matrix exist as the estimate of volume is independent from the determination in the fitting parameters; however, the two fitting parameters are not independent and, therefore, some covariance exists. The importance of this covariance will be seen later. The solution of the partial derivatives within the gradient matrix will obviously depend on the function chosen by the operator to fit to the data.

16.6 Time Activity Fitting

16.6.1 Activity

The expression relevant to assess the uncertainty associated with the activity A_i, determined from the measured count rate C_i in a target VOI at time t_i, scaled by the sensitivity factor Q and recovery factor R is,

$$A_i = \frac{C_i}{QR'} \quad (16.26)$$

This is a multiplicative measurement model and therefore the associated uncertainty can be shown to be:

$$\left[\frac{u(A_i)}{A_i}\right]^2 = \left[\frac{u(Q)}{Q}\right]^2 + \left[\frac{u(R)}{R}\right]^2 + \left[\frac{u(C_i)}{C_i}\right]^2 - 2\frac{u(R,C_i)}{RC_i}. \quad (16.27)$$

Note that of the three input parameters, the counts C_i and R were both derived based on the initial outline and volume estimate. These two parameters are, therefore, correlated, the covariance of which is given by

$$u(C_i, R) = \frac{\partial C_i}{\partial v}\frac{\partial R}{\partial v}u^2(v) = \frac{\partial R}{\partial v}u(v)u(C_i). \quad (16.28)$$

As above, the solution of the partial derivative $\partial R/\partial v$ will depend on the function chosen by the operator to fit to the recovery data.

In addition, when the same volume is used for all measurement time points, these fractional uncertainties are consistent for each activity measurement and can be considered as a systematic component of uncertainty.

16.6.2 TAC Uncertainty and Covariance

A similar method to that described above can be used to determine estimates of uncertainty when fitting to a *time activity curve* (TAC). Conventionally, a TAC is fitted by iteratively determining parameters of a function that minimize the square difference between the measured activity values A_i and the corresponding value of the fit function $f(t_i)$, i.e., minimizing:

$$x^2 = \sum\left[A_i - f(t_i)\right]^2. \quad (16.29)$$

A covariance matrix of the fit parameters can then be estimated using

$$\boldsymbol{V_p} = \frac{\boldsymbol{x^2}}{n-q}\left[\boldsymbol{J_p^{\mathrm{T}} J_p}\right]^{-1}. \quad (16.30)$$

where q is the number of parameters in the function $f(t_i)$ and n is the number of acquired data points. $\boldsymbol{J_p}$ is a matrix of firstorder partial derivatives of the TAC model with respect to the function parameters evaluated for each A_i. For the case of a single exponential function, the fit parameters are A_0 and λ giving

$$\boldsymbol{J_p} = \begin{bmatrix} \frac{\partial A_1}{\partial A_0} & \frac{\partial A_1}{\partial \lambda} \\ \vdots & \vdots \\ \frac{\partial A_n}{\partial A_0} & \frac{\partial A_n}{\partial \lambda} \end{bmatrix} = \begin{bmatrix} e^{-\lambda t_1} & -A_0 t_1 e^{-\lambda t_1} \\ \vdots & \vdots \\ e^{-\lambda t_n} & -A_0 t_n e^{-\lambda t_n} \end{bmatrix} \quad (16.31)$$

and

$$\boldsymbol{V_p} = \begin{bmatrix} u^2(A_0) & u(A_0,\lambda) \\ u(A_0,\lambda) & u^2(\lambda) \end{bmatrix}. \quad (16.32)$$

Many curve-fitting programs have facilities to estimate these matrix elements and, if not, they can always be determined outside of the program once the fitting parameters have been determined. As will be shown, the covariance term $u(A_0,\lambda)$ in this example is particularly important when propagating uncertainty and should not be ignored.

16.6.3 Time-Integrated Activity

Time-integrated activity, $\tilde{A}$, is defined as the integral of the TAC. For the single exponential example, when integrated from $t = 0$ to ∞, TIA can simply be expressed as the ratio of the TAC parameters, that is,

$$\tilde{A} = \int_0^\infty A(t)\mathrm{d}t = \int_0^\infty A_0 e^{-\lambda t}\mathrm{d}t = \frac{A_0}{\lambda} = A_0\lambda^{-1}. \tag{16.33}$$

Application of Equation (16.3) then allows the uncertainty in $\tilde{A}$ to be expressed as

$$\left[\frac{u_r(\tilde{A})}{\tilde{A}}\right]^2 = \left[\frac{u(A_0)}{A_0}\right]^2 + \left[\frac{u(\lambda)}{\lambda}\right]^2 - 2\frac{u(A_0,\lambda)}{A_0,\lambda}. \tag{16.34}$$

Note that because λ is expressed to the negative power in Equation (16.33), the covariance term in Equation (16.34) is negative. The implications of this can be seen in Figure 16.7, where the uncertainty bounds of a fit have been plotted alongside the fit function. In Figure 16.7a the covariance between the parameters has been ignored when estimating the uncertainty bound, resulting in an overly wide and unrealistic uncertainty. In Figure 16.7b the uncertainty bounds appear more plausible given the distribution of data about the best fit function.

The total uncertainty in TIA can then be determined by adding the contribution from the fitting and the systematic uncertainty in activity quantification, such that

$$\left[\frac{u_r(\tilde{A})}{\tilde{A}}\right]^2 = \left[\frac{u(A_0)}{A_0}\right]^2 + \left[\frac{u(\lambda)}{\lambda}\right]^2 - 2\frac{u(A_0,\lambda)}{A_0,\lambda} + \left[\frac{u(Q)}{Q}\right]^2 + \left[\frac{u(R)}{R}\right]^2 + \left[\frac{u(C_i)}{C_i}\right]^2 - 2\frac{u(R,C)}{RC}. \tag{16.35}$$

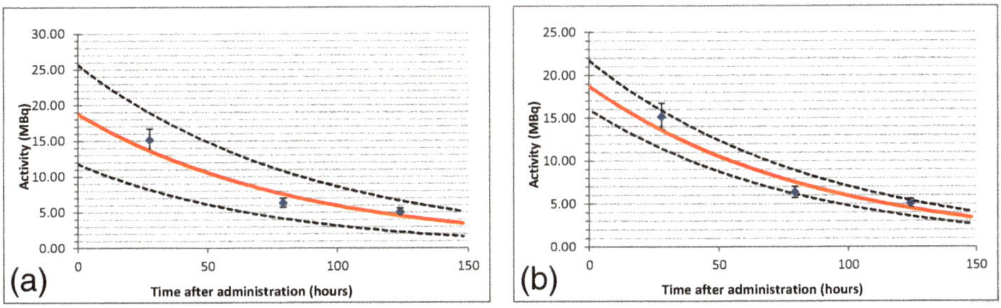

Figure 16.7 . Time activity curve showing line of best fit (red) and uncertainty bounds when covariance is (a) not taken into account and (b) is taken into account.

Table 16.1 Values of emission probabilities and uncertainty of annihilation photons in ^{90}Y decay

Reference	I_p ($\times 10^{-6}$)	Uncertainty ($\times 10^{-6}$)	% Uncertainty
Greenberg	36	9	25
Langoff	34	4	11.76
Selwyn	31.86	0.47	1.48
Stanicek	34.5	0.8	3.32
Adopted	**32.6**	**0.7**	**2.15**

16.7 S Values

16.7.1 Radiation Quantities

Uncertainty in S values can be estimated by examining the uncertainty of each parameter of the S value equation, namely,

$$S = \frac{\sum E_i n_i \phi_i}{m}. \tag{16.36}$$

Nuclear transition data for most common radionuclides are now very well known, and uncertainty in E_i and n_i and can usually be considered negligible. As an aside, there are always exceptions, and a good example of this is that of ^{90}Y. This isotope was originally considered a pure beta emitter. However, since the early 1960s, a minor branch to the first excited state followed by internal pair production was known to exist. The branching ratio of the preceding annihilation photons is extremely small, although application of ^{90}Y 0 PET imaging was demonstrated. For a long while the exact probability of the emissions was not known to a high degree of accuracy. Table 16.1 is taken from the Laboratoire National Henri Becquerel ICRM Nuclear Decay Data Working Group report (Bé, M.-M et al. 2006). For PET imaging the branching ratio is vital to relate the measured events to the count rate, such that even today the best estimate of an activity estimate using Y-90 PET cannot be calculated with an uncertainty less than 2.15%.

For calculating an S value, this uncertainty is still relatively negligible, as the proportion of energy emitted from the election/positron pair is significantly less than the primary beta emission and, as such, it is still fair to say that this aspect of uncertainty can largely be ignored. However, as more exotic radionuclides emerge, this aspect of uncertainty may need consideration in the future.

16.7.2 Absorbed Fraction

With the emission energies and probabilities well characterized, the next source of uncertainty is that associated with the absorbed and specific absorbed fractions. For the situation being discussed here—where the source and target are bulky solid tumors and organs—physics knowledge of radiation transport and advances in com-

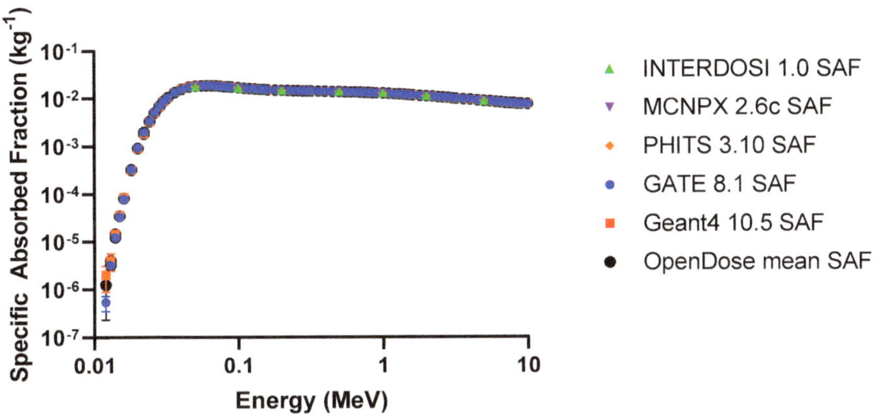

Figure 16.8 Plot of specific absorbed fractions for different photon energies calculated using different Monte Carlo codes. Data taken from https://www.opendose.org/. The source is the left kidney, and the target is the liver.

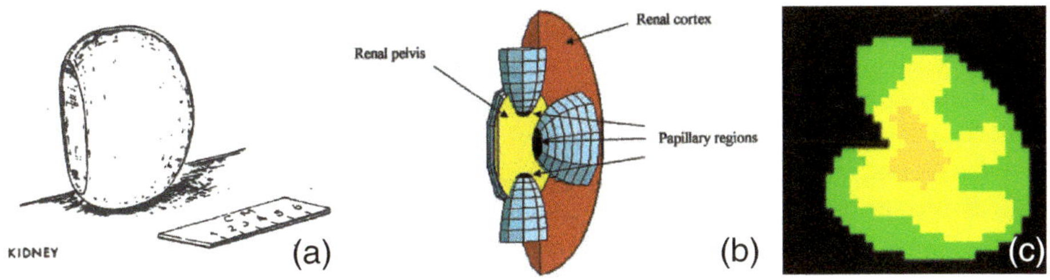

Figure 16.9 Illustrations of the different kidney phantoms: (a) MIRD 5, (b) MIRD 19, and (c) ICRP 110.

puting power allow Monte Carlo simulations to be conducted that calculate absorbed fractions within reference phantoms with a statistical uncertainty well below 1% (Chauvin et al. 2020). This uncertainty inevitably increases for situations where the source and target are separated and as the energy of the particles decrease. A good visual tool providing data of specific absorbed fractions of the ICRP adult male and female phantom is provided on the open dose website (https://www.opendose.org/). These specific absorbed fractions are calculated using different Monte Carlo programs and quoted with uncertainty values (see Figure 16.8).

16.7.3 Geometry Effects

The most significant source of error when using S values comes from the fact that the models used to derive the absorbed fractions do not match the specific patient geometry. The magnitude of this effect cannot easily be determined on an individual basis, but comparisons of different phantoms can give some indication of its significance.

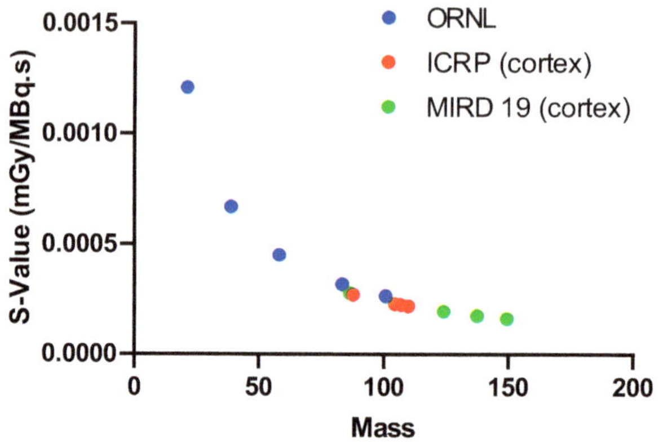

Figure 16.10 Plot of S value against mass for different reference phantom geometries.

In the first instance we can consider the kidney geometries of three "reference" phantoms (shown in Figure 16.9) from the ORNL phantom used in MIRD pamphlet 5 (Snyder et al. 1969), the compartmental model described in MIRD pamphlet 19 (Bouchet et al. 2003) and the ICRP110 kidney phantom (ICRP 2009). The original ORNL phantom only considers the kidneys as a single organ, whereas the latter two have different subsections to the organ (cortex, pelvis, medullary pyramids, etc.). The ICRP phantom also has different geometries for the left and right kidney, and all organs have age-dependent options. However, despite these differences, plots of S value for the different models are still largely similar (Figure 16.10).

16.7.4 Target Mass

From Figure 16.10 it can be observed that the S value varies as a function of volume more so than any variation in the different geometries. As a first approximation it is, therefore, common to use the reference S value and simply scale according to the mass ratio of the model and patient. As the geometry and mass of the reference is well known, the only other parameter where a significant uncertainty still exists is the mass of the patient's organ, so the uncertainty in S value can simply be expressed as

$$\frac{u(S)}{S} = \frac{u(m)}{m} = \frac{u(v)}{v}. \tag{16.37}$$

It was discussed at the beginning of the chapter that provided S value and time-integrated activity were determined independently, then the covariance term of Equation (16.8) could be ignored. It has hopefully become clear that using the methods described, both A and S rely on an accurate measurement of target volume (when source and target are the same). The covariance between these two parameters is then defied using

$$u(\tilde{A},S) = \frac{\partial \tilde{A}}{\partial v}\frac{\partial S}{\partial v}u^2(v). \tag{16.38}$$

As time integrated is derived from other input quantities that equally rely on volume, this equation can be further expanded and expressed in terms of activity, giving

$$u(\tilde{A},S)=\frac{\partial\tilde{A}}{\partial v}\frac{\partial S}{\partial v}u^2(v)\frac{\tilde{A}}{A_i}. \qquad (16.39)$$

And this equation can similarly be expanded again by considering the input quantities, Q and C, which were also derived from the VOI outline, giving

$$u(\tilde{A},S)=\frac{\tilde{A}}{A_i}\left(\frac{1}{QR}\frac{\partial C_i}{\partial v}-\frac{C_i}{\partial R^2}\frac{\partial R}{\partial v}\right)\times\frac{\partial S}{\partial v}u^2(v). \qquad (16.40)$$

A full derivation of this equation is given in the EANM guidance document (Gear et al. 2018).

16.8 Conclusions

This chapter has given a short overview of some of the key concepts to measurement uncertainty and how these concepts can be used in a dosimetry calculation. Application of these and similar approaches is still fairly new, and while of interest, there are still very few publications that apply uncertainty to every aspect of the dosimetry chain (Finocchiaro et al. 2020). The methods and equations used will vary depending on the method of dosimetry used. For example, the need to account for covariance between S value and time-integrated activity will not be required if an alternative approach to activity quantification is used. A recent publication using oversized VOIs that did not match the anatomical volume is a good example of this (Carnegie-Peake et al. 2022). Irrespective of the dosimetry method applied, the reader is encouraged wherever possible to make some attempt at reporting uncertainty, even if it is just for one aspect of the dose calculation. As a minimum, an estimation of volume uncertainty and that associated with the TAC fit should be achievable for all dosimetrists.

References

Anizan, N., et al. (2015). "Factors affecting the repeatability of gamma camera calibration for quantitative imaging applications using a sealed source." *Phys. Med. Biol.* 60(3):1325–37.

Bé, M.-M, et al. *Monographie BIPM-5: Table of Radionuclides (Vol. 3–A = 3 to 244).* Pavillon de Breteuil, F-92310 SÉVRES, France: Bureau International Des Poids Et Mesures, 2006.

BIPM, et al. (2008a). "Evaluation of measurement data—Guide to the expression of uncertainty in measurement." Joint Committee for Guides in Metrology (JCMG).

BIPM, et al. (2008b). "Evaluation of measurement data—Supplement 1 to the 'Guide to the expression of uncertainty in measurement'—Propagation of distributions using a Monte Carlo method." Joint Committee for Guides in Metrology (JCMG).

BIPM, et al. (2011). "Evaluation of measurement data—Supplement 2 to the 'Guide to the expression of uncertainty in measurement'—Extension to any number of output quantities." Joint Committee for Guides in Metrology (JCMG).

Bouchet, L. G., et al. (2003). "MIRD Pamphlet No 19: absorbed fractions and radionuclide S values for six age-dependent multiregion models of the kidney." *J. Nucl. Med.* 44(7):1113–47.

Carnegie-Peake, L., et al. (2022). "Quantification and dosimetry of small volumes including associated uncertainty estimation." *EJNMMI Phys.* 9(1)86.

Chauvin, M., et al. (2020) "OpenDose: Open-Access Resource for Nuclear Medicine Dosimetry." *J. Nucl. Med.* 61(10):1514–19.

D'Arienzo, M. and M. Cox (2017). "Uncertainty Analysis in the Calibration of an Emission Tomography System for Quantitative Imaging." *Comput. Math. Methods Med.* 2017:9830386.

Finocchiaro, D., et al. (2020). "Uncertainty analysis of tumour absorbed dose calculations in molecular radiotherapy." *EJNMMI Phys.* 7(1):63.

Gear, J. I., et al. (2018). "EANM practical guidance on uncertainty analysis for molecular radiotherapy absorbed dose calculations." *Eur. J. Nucl. Med. Mol. Imaging* 45(13):2456–74.

ICRP (2009). "Adult Reference Computational Phantoms." ICRP Publication 110. *Ann. ICRP* 39(2):1–164.

Marin, G., et al. (2015). "Towards a quality assurance for dosimetry in peptide receptor radionuclide therapy (PRRT) with 177Lu-DOTATATE." *Eur. J. Nucl. Med. Mol. Imaging* 42 Suppl 1:S159.

Marin, G., et al. (2017). "Accuracy and precision assessment for activity quantification in individualized dosimetry of Lu-177-DOTATATE therapy." *EJNMMI Phys.* 4:7. doi:10.1186/s40658-017-0174-7.

Snyder, W. W., et al. (1969). "MIRD Pamphlet 5: Estimates of Absorbed Fractions for Monoenergetic Photon Sources Uniformly Distributed in Various Organs of a Heterogeneous Phantom." *J. Nucl. Med.* 10(Suppliment 3):7–52.

Chapter 17

Consistency and Standardization

Rachel Barbee[1], Ph.D., DABR and David Barbee[2], Ph.D., DABR

[1]Associate Professor
Department of Imaging Physics
MD Anderson Cancer Center
Houston, TX
[2]Clinical Associate Professor
NYU Langone Health
New York, NY

17.1 Introduction

Radiopharmaceutical therapies (RPTs) continue to grow, both in the number of patients administered dosages and in the potential disease sites targeted using an ever-expanding range of novel compounds. Concurrent with this growth is an increase in the number of practitioners interested in RPT, as well as an increased desire to better understand the radiation dosimetry of the many available compounds in this arena.

Radiopharmaceutical therapy dosimetry calculates the absorbed dose to normal tissues and tumors from the energy released during decay of the radionuclide as the pharmaceutical distributes over time. The absorbed dose is calculated by observing the activity or dose rate of a specific target volume continuously over time or at a single or multiple time points, depending on the radiopharmaceutical and its dosimetry model. Integrated activities or dose rate observations over time are inserted into dosimetry models that can account for patient anatomy, patient biokinetics (radiopharmaceutical distribution), and radionuclide properties. However, measuring and understanding biokinetics is a complex process requiring measurement techniques specific to each compound and radionuclide, as well as detailed analysis and understanding to interpret results in order to assess treatment efficacy or inform administered activity.

Quantitative *nuclear medicine* (NM) imaging is the foundation for patient-specific dosimetry, wherein the absorbed dose in tissues, energy absorbed per unit mass of material, can be calculated volumetrically from PET or SPECT images. Patient-specific dosimetry can be performed using a pre-therapy diagnostic dosage (radiotracer) of the same or similar agents that target the same disease sites (theranostic agents) or, for multiple-cycle radiopharmaceutical therapies, dosages can be tailored for later cycles based on dosimetry of earlier cycles. Personalized dosimetry has the potential to optimize treatment planning and therapeutic response by ensuring the maximum possible absorbed dose is delivered to tumors, while at the same time ensuring absorbed doses to normal tissues remain below thresholds that could cause toxicity or adverse events (Sgouros et al. 1990; Strand et al. 1993; Eiber et al. 2018).

Radiopharmaceuticals differ in their biokinetic and radionuclide properties, making a single universal *standard operating procedure* (SOP) or workflow for all patients or radiopharmaceuticals challenging (Strand et al. 1993). Variability for the same radiotracer may exist in data acquisition, reconstruction technique, segmentation, and dose calculation techniques. This ultimately results in dosimetric estimates varying widely among practices as internal absorbed dose (BED, EQDX) estimates are derived from biodistribution measurements. A brief survey across the landscape of radiopharmaceutical therapy depicts a field of nonuniformity; even well-established radiopharmaceuticals in clinical use for decades can vary in how they are prescribed (Ovčariček et al. 2022). For example, prescribed ^{131}I activity to treat distant metastases of thyroid cancer can be either empirically based (Cooper et al. 2006), estimated using a combination of whole body imaging and blood sampling (Avram et al. 2022; Lassmann et al. 2008), or estimated using only whole body image data alone (Thomas et al. 1993).

As radiopharmaceutical therapies grow and efforts to integrate dosimetric calculations into the clinical routine move forward, there must likewise be an effort toward standardization. Standardization reduces errors and simultaneously harmonizes process output (Rozich et al. 2004; Zeitman et al. 2012). With the continued growth of the radiopharmaceutical therapy field, standardization will be a critical component in the safe expansion of this treatment modality.

The Role of Standardization

Standardization can be described as defining a workflow or process, such that it can be comfortably performed the same way regardless of equipment or operator, with the goal of having consistent results from the process. In the context of radiopharmaceutical therapy dosimetry, this means that we aim to prescribe, administer, measure, and assess each patient's radiopharmaceutical therapy in a consistent manner while following best practices at each step. Dosimetry is a multistep procedure, and the results from each step need to be accurately known in order to obtain an accurate estimate of the adsorbed dose and derived quantities.

To facilitate the growth of RPTs—particularly toward personalized dosimetric estimates for the purpose of RPT treatment planning—the efficacy of the personalized absorbed dose estimate must be established (Sgouros 2005). However, the value of patient-specific dosimetry for RPT is currently controversial (Flux et al. 2018). The need for dosimetry is mired in a catch-22: patient-specific dosimetry is not performed owing to limited data supporting its efficacy in improving patient outcomes, yet dose-response data are limited because dosimetry is not being performed (O'Donoghue et al. 2022). Thus, to grow RPT, dose (EQDX)-response relationships need to be consistently evaluated.

An analogous situation for improved dosimetry in radiation medicine leading to improved patient outcomes already exists in the field of *external beam radiation therapy* (EBRT). The field has seen significant change from the early days of dose-limiting skin erythema to the absorbed dose response tables of Emami, QUANTEC, and HyTEC. These advances in dose response understanding were made possible through the combination of 3D imaging and computational power capable of calculating increasingly accurate volumetric dose distributions over the past several decades, as well as the efforts in dose reporting and data aggregation.

However, the field has repeatedly seen the need for improved standardization, such as:

- dose calculation accuracy in lung SBRT (Xiao et al. 2009; Ding et al. 2007; Zhuang et al. 2013),
- machine output calibration procedures (Taylor et al. 2003; McEwen et al. 2014), and
- implementation of *intensity-modulated radiation therapy* (IMRT) for head and neck (Molineu et al. 2013).

In these instances researchers provided recommendations, and task group reports were published to improve the consistency of patient care through standardization.

Currently, the field of radiation oncology continues efforts to standardize practice across the field through task groups such as AAPM TG-263 that aim to improve reporting by standardizing naming conventions and dose reporting, thereby facilitating big data analysis of treatment efficacy and defining dose-response relationships (Mayo et al. 2018). Nuclear medicine and radiopharmaceutical therapy are currently positioned similarly to early external beam therapy, wherein technological advancements have made radiation dosimetry possible, but practitioners need to collectively support consistent, standardized practices. In this way, in addition to successful treatment delivery, reliable dosimetric estimates can be obtained and contribute to large scale dose-response data (Sgouros 2005).

Development of population-based predictive models of treatment efficacy and understanding organ/dose toxicity relationships requires accurate radiopharmaceutical dosimetry. Results from dosimetry studies are dependent upon the data with which the results were obtained. Organizations have developed freely available biodistribution templates to facilitate standardizing data organization and reporting (Kesner et al. 2017; Lassmann et al. 2011).

17.2 Dosimetry Workflow

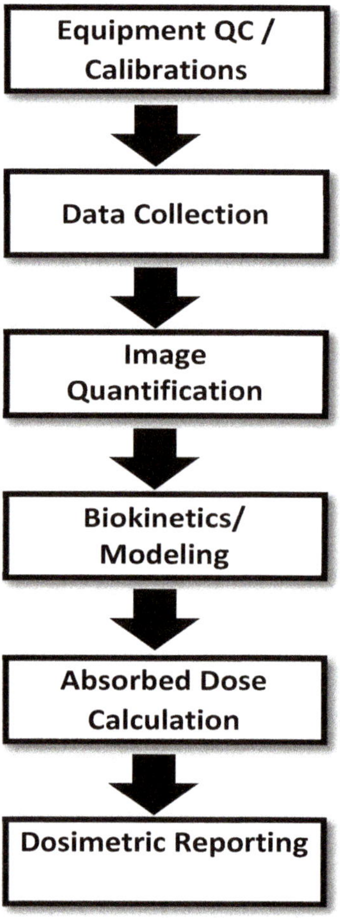

Figure 17.1 Typical radiopharmaceutical dosimetry workflow.

Dosimetry of internally administered radiopharmaceuticals can be generalized into the workflow shown in Figure 17.1. Each step in the acquisition and estimate of the absorbed dose builds upon the last. Having consistent, standardized processes can eliminate potential sources of variation in each workflow step, resulting in consistent and reproducible dose estimates. In addition, *quality control* (QC) checks within the workflow are useful tools to validate the accuracy of the data as it is manipulated during dosimetric calculation.

Performing dosimetry for RPT is not limited to the calculation; the entire clinical team plays a role, from the technologist accurately assaying and recording the administered activity to the physician assessing patient response. Multiple members of the clinical team will have a role in the dosimetry workflow. Professional organizations are working to develop consensus-based recommendations for facilities wishing to implement theranostic programs and administer RPTs (Herrmann et al. 2022; Bodei et al. 2013; Hurwitz et al. 2020; Love et al. 2022; ACR 2022). However, these recommendations can take time, whereas new RPTs are rapidly entering the clinic. As the community awaits these future recommendations, practices should be implemented for best patient care and the ability to obtain accurate and reproducible dose estimates.

17.3 Quality and Safety Management

Theranostic programs are complex in terms of departmental structure, personnel, and technical processes. Quality and safety initiatives are integral to the safe administration of RPT and the implementation of theranostic programs. The *Safety Is No Accident* publication, although specific to radiation oncology, is a valuable reference for institutions providing medical care involving multiple departments (Zietman et al. 2012). Many of the safety initiatives outlined are also applicable to radionuclide therapies. Safety and quality management is the cornerstone of safe and effective treatment using radiation.

17.3.1 Safety Culture

Radiopharmaceutical therapies are complex and evolving quickly. Facilities should establish multi-disciplinary safety committees responsible for the introduction of new RPTs and the review of existing therapies. Safety committees can be tasked with ensuring that processes, mitigation strategies, roles, responsibilities, and training are implemented prior to initiation and are routinely reviewed for continuous quality improvement. These committees should promote incident or event reporting in order to identify processes or areas for improvement. Safety committees may set departmental safety or compliance goals to improve identified areas of noncompliance.

Standardization and efficiency will be necessary as the demands of clinical staff are increased or modified. New roles and responsibilities will need to be handled by either training existing personnel or requiring the introduction of new personnel in the event specialized, domain-specific knowledge is required. Deviation from the routine is often met with resistance, but following a carefully developed plan is necessary to the success of this treatment modality. The initial implementation of a theranostic program may need to be modified to accommodate this evolving treatment modality or to find the optimal approach for the safe and efficient implementation of this therapy. A safe clinical environment and efficient workflow requires a coordinated effort by all staff involved in the care of patients administered radiopharmaceuticals for therapy, and this can only be accomplished in a safety-minded culture (Zietman et al. 2012).

In response to the tens of thousands of people harmed by healthcare every year, the United States Congress passed the “Quality Improvement Act of 2005” to improve patient safety. Departments must take a proactive approach to implementing a culture of safety, rather than a reactive approach, to ensure processes are in place for the administration of safe and efficient treatments while continuously improving the delivery paradigm (Helmreich 2000; Zietman et al. 2012).

17.3.2 Education and Training

The use of radiopharmaceuticals and radionuclides in therapy is a multidisciplinary approach, involving not only physicians and physicists, but also nurses, nurse practitioners, technologists, administrators, environmental services, billing, and radiation safety. At the most basic level, departments introducing theranostic programs need to ensure all divisions are aware of the restrictions, requirements, and safety concerns for each radiopharmaceutical and administered patient. Beyond awareness of the regulations is training; members of the clinical team need to be trained in their responsibilities as they pertain to the treatment of the patient. It is a lack of training and education that is often cited as a causative factor in medical events in the medical community (Pronovost et al. 2006; Richardson 2012; Thomadsen et al. 2003).

Board certification is traditionally the basis for establishing competency for clinical personnel. However, formal training in dosimetry for RPT is currently lacking (Graham and Buatti 2019). Recognizing this deficiency, multiple solutions have been proposed, such as training within the current educational paradigm (Kiess et al. 2022) and creating a new specialty or subspecialty that combines training in nuclear medicine, radiation oncology, and general oncology (Sgouros et al. 2020).

While the community awaits consensus recommendations, it is essential that each facility's theranostic program ensures appropriate training and education.

17.3.3 Communication and Coordination

A clinical team is required to provide the medical services associated with a theranostic program and the administration of RPTs. Clinical team members beyond the physician involved in the process of care include the medical physicist, NM technologist, radiation safety, nursing staff, and ancillary services. Clinical staff must function harmoniously as a close collaboration between a team of highly qualified individuals is necessary to provide the care required for patients administered radiopharmaceuticals for therapy. Effective communication of patient needs—particularly if those needs go beyond the SOP (e.g., patients who have catheters)—among the entire team ensures the safety of the patient, staff, and general population. Furthermore, these details will need to be communicated to the technical staff performing dosimetry calculations. Such details may alter the biokinetics of the radiopharmaceutical and, thus, the dose calculation.

Routine multidisciplinary "huddles" have been shown to improve care coordination, acting as an effective tool to ensure team members are aware of upcoming cases and potential or ongoing issues, and flag concerns proactively (Scoville al. 2016). Many RPTs require multiple administrations, and dosimetry typically requires multiple-time-point imaging. In delivering the treatment or performing dosimetry, additional tests, studies, or procedures may be deemed necessary. Communication is key in a theranostic program, and an effective method for communicating within a multidisciplinary team should be established.

Coordinating the schedules of the patient, equipment, and the technical staff needed for RPT and dosimetry will require additional consideration. Dosimetry for the purpose of determining a *maximum tolerated activity* (MTA) or ensuring that EQDXs to *organs at risk* (OARs) are below toxic thresholds will require special expertise held by a limited number of individuals participating in the care of patients administered RPTs. Repeatability of quantitative imaging may require that only specific equipment be utilized. Special coordination may be required to ensure the necessary equipment and expertise are available.

17.4 Safety Tools

As much of RPT dosimetry is currently nonstandard, clinical staff involved in the dosimetry workflow may use varied approaches in the acquisition and processing of data. Having multiple approaches may lead to increased variability of results, which can be further complicated if multiple machines, rotating staff, or different locations are utilized. Facilities should establish standardized practices and consistent expectations to ensure their respective programs have coherent processes leading to reduced variability and reproducible results.

17.4.1 Checklists

Checklists contain a listing of actions to be performed in a given process to help ensure no step is forgotten prior to moving on to the next process in the workflow. Patients receiving RPT will have multiple staff involved in their care between the time of consultation and administration of therapy, resulting in the patients and their information being "handed off" at multiple points as the patients move through this process. Communication errors between groups regarding completed/remaining tasks or patient status (the hand-off of information) have been identified as a factor contributing to medical events in healthcare (Burns et al. 2021).

Checklists can help ensure the continuity and consistency of care in the hand-off of patients between staff members by ensuring that necessary tasks have been completed and necessary information collected, thus preventing the loss of key information or critical missed steps. Effective checklists should utilize standard content and language that reflects the underlying task and is understood by the responsible clinical groups. Hand-off standardization must balance completeness with efficiency and intelligibility (Burns et al. 2021; Clay-Williams and Colligan 2015; Jewell et al. 2016).

Checklists help ensure every step in the process of treatment delivery has been met and reduces cognitive workload by avoiding task/process memorization so that the operator can instead focus on performing the task. Distractions and interruptions are common occurrences in healthcare. The verbal exchange of information can be forgotten or misinterpreted; having a standard document to ensure each task has been completed can ensure best patient treatment care (Zietman et al. 2012; Gawande 2009).

17.4.2 Clinical Workflow

In diagnostic NM departments, patients are typically scheduled for a single diagnostic session based on machine availability; however, RPT requires that special attention be paid to radiopharmaceutical availability as well as patient and machine scheduling. Importantly, depending on the particular RPT, dosimetry patients must receive imaging at specific time points after the administration of a diagnostic dosage or the initial cycle of therapy. Department schedulers must be able to identify these patients and take this into account by ensuring patients have appointment times that permit imaging at necessary time points. Additionally, whenever possible, scheduling must ensure repeat measurements are made on the same machine for the sake of consistency and reduced variability. Using different equipment during sequential imaging may potentially disrupt the quantitative accuracy of the dosimetric assessment of treatment. Ideally, all scanners should be harmonized or well characterized such that results are comparable or can be corrected with minimal difference across scanners.

Process maps are useful tools in developing efficient clinical workflows. These provide pertinent staff an opportunity to identify value-added and unnecessary steps in refining the workflow. Removing nonessential steps leads to a more streamlined workflow (Zietman et al. 2012; IHI 2005).

17.4.3 Policies and Procedures

Each facility must develop and implement detailed SOPs for the process of obtaining dosimetric data and delivering RPT. Standard operating policies and procedures should be specific to each treatment and should not only document the process, but also the roles and responsibilities of each team member with associated checklists. The NRC published a summary of misadministration event investigations from1991–1992 involving medical radionuclides. It found these events occurred primarily because SOPs did not exist, or that whatever procedures did exist were not sufficiently detailed, comprehensive, specific, or clearly written (Ostrom et al. 1994). Policies and procedures should be reviewed regularly and be accessible to all clinical staff.

17.5 Quality Assurance

17.5.1 Quantification

The first step in the calculation of the absorbed dose to a patient is the measurement of activity. Any quantitative measurement used in the calculation of absorbed dose—from administered activity to calibration of image data—needs to be known and traceable to an international standard (Lassmann and Eberlein 2018).

Each piece of equipment requires rigorous QC and careful calibration, as discussed in chapters 7, 8, and 1. It must be understood that quantitative images are dependent upon the equipment used, the acquisition settings, and the reconstruction algorithms employed. If the intention is to create quantitative images, image quantification will require careful calibration procedures and strict tolerances. Standard practice should include verification of the quantification with a follow-up phantom acquisition. The frequency of these calibrations is not currently mandated by accrediting bodies. Until recommendations and guidelines are available, it is incumbent upon the users to understand and account for the variability (drift) in the calibration factors for their individual pieces of equipment. If the calibration is found to drift rapidly, then frequent calibrations may be necessary. If the calibration factors are found to be stable, less frequent calibrations may be warranted.

A method to document these parameters for each patient must be implemented, which becomes especially crucial for retrospective studies. Community templates are available to help facilities document these parameters in an effort to standardize data organization and reporting (Kesner et al. 2017; Siegel et al. 1999).

17.5.2 Equipment Quality Control

Each facility must have an established quality assurance program. However, with the implementation of theranostics, additional QC procedures and tighter tolerances than those required for diagnostic imaging or by regulatory and accreditation bodies may be necessary. Professional recommendations on equipment QC for theranostics is forthcoming, however these recommendations can take time to develop. For example, specialized radiation therapy treatment machines like CyberKnife and Tomotherapy were all in clinical use years before the associated dedicated equipment QA task group reports on them were published. Thus, it falls to the facility to establish rigorous QC procedures to ensure accurate quantitative data (see Chapter 11).

All systems must also undergo system connectivity and acceptance testing before release to the clinic. Testing the interconnections and interoperability of systems must be part of the facility QA program, as any flaws in communication of systems can allow errors to occur. Acceptance testing of systems must be overseen by qualified personnel. These tests ensure specifications are met and provide a benchmark against which future testing can be compared. After the system has been released for clinical use, routine QC as part of the QA program must be performed.

17.5.3 Data Collection

Dosimetry estimates are derived by projecting biodistribution measurements into dosimetry calculations, and generally require multiple time-point measurements. Temporal sampling, acquired for either probe measurements or image data, must be appropriate for the dosimetry study and carefully documented. Prescriptive guidelines for performing dosimetry studies can be referenced from professional societies, such as the SNMMI, EANM, and IAEA, for common clinical radiotherapeutic compounds.

Data collection will include participation from multiple members of the clinical team. Training, SOPs, and checklists/templates will need to be in place to ensure accurate, reproducible, and appropriate data collection. Standardized and streamlined checklists and templates can ensure each task has been completed and the correct data recorded. A comprehensive list of key details to record can be referenced from the EANM guidance document on good practice for clinical dosimetry reporting. The EANM recommendations are intended to define and standardize data acquisition and processing reporting, thereby ensuring important details are not omitted and the study methodology can be evaluated and reproduced (Lassmann et al. 2011). The collection of the biodistribution data is perhaps more informative than the absorbed dose estimate itself, as this can be used to support large-scale data sets and the development of population-based estimates, ultimately leading to improved treatments. The EANM has provided a freely available template, a sequel to the guidance on good dosimetry reporting, to support this effort (Kesner et al. 2017).

However, each manual data entry carries with it an associated risk of error. Solutions to manual data entry errors include:

- automating data entry, whereby information is directly populated from a patient's medical data,
- creating templates that also include expected ranges that can be used to flag an entry that falls outside of the expectation, and

- creating a system of double checks/peer review.

Imaging data not acquired correctly or transferred accurately may have detrimental effects downstream. A brief QC check implemented at this point can pay dividends with respect to error reduction. One might review the *standardized uptake value* (SUV) in a normal region of the image data, or if a standard is included in the image data acquisition, the time activity data of the standard can be used to ensure that the standard exhibits only physical decay.

17.5.4 Multiple Time-Point Imaging

Radiopharmaceutical therapy dosimetry requires assessment of the distribution of the radiopharmaceutical in the patient over time, often at multiple time points separated by long intervals. Single-session scans over an extended period of time often results in intra-session motion, which impacts the quantitative accuracy of the resultant image, as well as activity measurement within the contoured target when performing dosimetry. Similarly, images of patients returning hours or days later will show deviation in positioning relative to their initial scan, requiring either rigid or deformable, registration if initially drawn targets are to be compared.

To facilitate intra-session and inter-session reproducibility and consistency, immobilization or support devices could be used to return patients to their original positioning. Use of these devices is common practice in external beam radiation therapy, and the accuracy of repeat measurements is no less important in RPT when quantifying uptake from imaging scans. At a minimum, every effort should be made to image patients reproducibly.

17.5.5 Organ Delineation

Perhaps the largest source of variation in organ-level dosimetry calculations is organ delineation (contouring) for mass estimates or organ-level time-integrated activity estimates (Uribe et al. 2021). Delineation can be made on either morphologic or functional image data. Organ contouring being the largest source of variation suggests that this would be an optimal point in the process for automation. Many software packages exist to facilitate this process and offer a standardized means of estimating target volume. However, it is prudent to evaluate the accuracy of the automated segmentation.

Presentation as part of a departmental committee review, such as a tumor board or chart rounds, are another option to review accuracy of contours. These are commonly conducted among colleagues in other areas of medicine and are a natural way to review data that is common knowledge to a community of peers.

17.5.6 Biokinetics/Modeling

The absorbed dose calculation requires an estimate of area under the curve, either for source region time-integrated activity coefficients calculated as areas under the time-activity curves, or as areas under the dose rate curve. The accuracy of the estimate is dependent upon the sampling of these data points (both the number of data points and the interval of sampling). The time-integrated activity or dose coefficients are estimated from the time-activity data or dose rate data, respectively, and small variabilities can be further compounded by the chosen analytical method for determining the area under the curve (compartmental modeling, sum of exponentials, trapezoidal integration) and the chosen method to account for the terminal contribution. Significant variability has been observed in the calculation of these areas under the curve. Institutionally, for a given radiopharmaceutical therapy, efforts should be made to standardize the analytical methods used in the dose calculation.

Moving through the flow chart of a dosimetric calculation, it becomes clear that manual intervention is necessary at multiple points. A relatively easy interventional QC check can be implemented here, which might include a brief review to ensure activity of no more than 100% is applied in the absorbed dose calculation.

17.5.7 Tumor Dosimetry/Absorbed Dose

Calculations of absorbed dose vary widely, the three primary approaches being S factor calculations based on the MIRD formalism, dose point kernel convolution, and Monte Carlo radiation transport simulation. These can be applied at the organ, suborgan, or voxel level. Currently, there is no universally accepted standard practice for evaluating the dose from internally administered radionuclides, contributing to the lack of dose-response data in RPTs. Furthermore, no standard platform exists that is universally used, resulting in many facilities developing and using in-house applications to perform these calculations. Due to potential intellectual property claims and concerns regarding liability, institutional sharing of such applications is challenging. The community will need to establish practice guidelines to standardize the dosimetry for radionuclide therapy such that dose-response data can be evaluated with large-scale data sets.

17.5.8 Recordkeeping

Detailed recordkeeping of each step in the dosimetry calculation should be strictly adhered to. In addition to the details of the patient exam, it is recommended that each patient study have the ability to cross-reference the details of the quantification process.

Professional societies have provided guidance on best practice of clinical dosimetry reporting, and this should be incorporated into the institutional practice of RPT (Kesner et al. 2017, 2018; Lassmann et al. 2011).

17.5.9 Commercial Platforms

Commercial platforms offer the potential to improve the efficiency of dosimetric calculations by standardizing the inputs to the dose calculation and the dose calculation steps themselves, while at the same time simplifying the process. Additionally, platforms offer a standardized selection of dose prescriptions, as well as dose reporting using universal templates.

Commercial products and automation will become integral to RPT and theranostic programs. Prior to clinical implementation, users should familiarize themselves with the capabilities of commercial products through vendor training and commissioning. A partnership between the vendor and the user should be established, whereby the two work cooperatively toward the common goal of safe and useful products used in RPT as vendor dosimetry solutions continue to mature. Users must validate the products before clinical use and share their concerns with the vendors to improve products (Zeitman et al. 2012; Frass et al. 1998).

As with verification of machine calibration and accuracy of segmentation, verification or acceptance testing and commissioning of any dosimetry platform, commercial or in-house, should become a standard of practice. Institutions are not permitted to perform patient exams without the equipment undergoing a rigorous set of tests to ensure specifications are met. There is an unfortunate history of serious patient harm from computer codes being used without their undergoing rigorous testing (Leveson 2017). No platform, even commercial, should be implemented clinically without proper validation, nor should they be trusted implicitly. Full end-to-end testing for representative treatments to confirm accurate estimates of dose assessment components must be undertaken to confirm accurate outputs. Furthermore, periodic constancy checks should be included as a routine QC process to ensure continued quantitative accuracy. Recognizing the importance of software QC, the American Association of Physicists in Medicine (AAPM) tasked professionals in the field with providing guidance on software QA for radiotherapy treatment planning systems (Fraass et al. 1998; Zeitman et al. 2012). Given the significant risks associated with inaccurate dose estimates for RPTs, rigorous software QC must be implemented.

17.6 Risk Management

17.6.1 Peer Review (Double Checks)

For institutions performing patient-specific dosimetry calculations, double checks should be implemented in the workflow of patient treatment. These checks should not be made by the same individual, as that would defeat the purpose of having an independent review of treatment calculations. Peer review is a means to improve quality and uphold standards. Receiving input and feedback from colleagues benefits all members of the clinical team.

Beyond review of individual patient dosimetry calculations, tumor boards or chart rounds are useful interdisciplinary peer review opportunities. Treatment details may be presented in these settings that offer an opportunity for a more thorough review of the entire patient treatment from a multidisciplinary team.

17.6.2 Reporting of Incidents

Implementing a system of reporting and reviewing errors should be considered. Institutions report medical events per regulations outlined by the NRC or their Agreement State. The implementation of a reporting and review system for more than reportable medical events—for example, near miss events—will give the facility an opportunity to review these events and potentially implement corrective action to prevent similar errors that could result in a medical event from occurring in the future. In addition, such review systems may identify areas that would benefit from additional training or operational improvement. (Zeitman et al. 2012).

17.6.3 Risk Analysis (Failure Mode Effects Analysis)

Finally, as already stated, radiopharmaceutical therapy is growing, outpacing regulatory requirements and consensus recommendations by professional societies. Regardless, it is incumbent upon the theranostic program to ensure safe and effective patient care. Traditionally, radiopharmaceutical therapies have been a small fraction of NM procedures performed and, thus, the rigorous quality management programs that are implemented for modalities such as EBRT have not been developed for this newer treatment modality. To this end, departments should consider implementing an internal assessment of risk proactively rather than waiting for errors to occur and evaluate the cause with a *root cause analysis* (RCA).

Given this is a developing treatment modality with many departments being integrated, some of which are new to the process, having an internal risk analysis that includes all those in the process of caregiving helps identify the critical factors in the treatment workflow. An effective process requires active collaboration among all members, including physicians, nurses, technologists, physicists, and administrators. The important takeaway is that this process is specific to an individual facility based on the facility's own processes and procedures. Each facility can direct its own resources or identify its own areas needing improvement to achieve patient safety and treatment quality.

The AAPM has a guidance document (Huq et al. 2016) devoted to *failure mode effects analysis* (FMEA). While that document is applicable to radiation therapy, it can be effectively utilized in RPT and theranostic programs. The goal of the FMEA is to identify the potential failure modes and how they might occur and propagate, allowing the team an opportunity to discuss where to place QA and QC interventions to prevent these failures from occurring.

Today, FMEA is a widely adopted risk analysis model due to its simplicity and utility in the reduction of risk (Liu et al. 2020). Failure mode effects analysis is particularly useful for healthcare quality improvement and error reduction, having been introduced in the hospital setting in the 1990s where proactive measures to avoid or minimize problems before they occurred became necessary with the increase in complexity and sophistication of the equipment and procedures.

17.7 Conclusion

Radiopharmaceutical therapies and theranostic programs are rapidly growing, and safe implementation requires a coordinated effort. Good clinical practices usually evolve over many years, however the growth of RPTs is outpacing professional recommendations and regulatory guidelines. It is, therefore, incumbent upon the institution/facility to implement a culture of safety and develop programs and procedures to ensure safe performance of this growing treatment modality while awaiting published guidance documents.

To create robust large-scale dose-response data, the community will need to harmonize calibration procedures, acquisition protocols, reconstruction techniques, and dose calculations. These recommendations will take time and, thus, it falls to the users to ensure best practices are in place and utilized.

References

ACR (2022). "ACR–AAPM–ACNM–SNMMI–SPR Technical Standard for Therapeutic Procedures using Radiopharmaceuticals." American College of Radiology. Approved October 11, 2022.

Avram, A. M., L. Giovanella, B. Greenspan, S. A. Lawson, M. Luster, D. Van Nostrand, J. G. Peacock, P. P. Ovčariček, E. Silberstein, M. Tulchinsky, F. A. Verburg, and A. Vrachimis (2022). SNMMI Procedure Standard. "EANM Practice Guideline for Nuclear Medicine Evaluation and Therapy of Differentiated Thyroid Cancer." https://www.snmmi.org/ClinicalPractice/content.aspx?ItemNumber=6414#approved.

Bodei, L., J. Mueller-Brand, R. P. Baum, M. E. Pavel, D. Hörsch, M. S. O'Dorisio, T. M. O'Dorisio, J. R. Howe, M. Cremonesi, D. J. Kwekkeboom, and J. J. Zaknun (2013). "The joint IAEA, EANM, and SNMMI practical guidance on peptide receptor radionuclide therapy (PRRNT) in neuroendocrine tumours." *Eur. J. Nucl. Med. Mol. Imaging* 40(5):800–16. DOI: 10.1007/s00259-012-2330-6.

Burns, J., S. Ciccarelli, E. Mardakhaev, A. Erdfarb, S. Goldberg-Stein, and J. A. Bello (2021). "Handoffs in Radiology: Minimizing Communication Errors and Improving Care Transitions." *J. A. Coll. Radiol.* 18(9):1297–1309. https://doi.org/10.1016/j.jacr.2021.04.007.

Clay-Williams, R. and L. Colligan (2015). "Back to basics: checklists in aviation and healthcare." *BMJ Qual. Saf.* 24(7):428–31. https://doi.org/10.1136/bmjqs-2015-003957.

Cooper, D. S., G. M. Doherty, B. R. Haugen, R. T. Kloos, S. L. Lee, S. J. Mandel, E. L. Mazzaferri, B. Mciver, S. I Sherman, and R. M. Tuttle (2006). "Management Guidelines for Patients with Thyroid Nodules and Differentiated Thyroid Cancer The American Thyroid Association Guidelines Taskforce." *Thyroid* 19(11):1167–214.

Ding, G. X., D. M. Duggan, B. Lu, D. E. Hallahan, A. Cmelak, A. Malcolm, J. Newton, M. Deeley, and C. W. Coffey (2007). "Impact of inhomogeneity corrections on dose coverage in the treatment of lung cancer using stereotactic body radiation therapy." *Med. Phys.* 34(7):2985–94.

Eiber, M., K. Herrmann, J. Calais, B. Hadaschik, F. L. Giesel, M. Hartenbach, T. Hope, R. Reiter, T. Maurer, W. A. Weber, and W. P. Fendler (2018). "Prostate Cancer Molecular Imaging Standardized Evaluation (PROMISE): Proposed miTNM Classification for the Interpretation of PSMA-Ligand PET/CT." *J. Nucl. Med.* 59(3):469–78.

Flux, G. D., K. S. Gleisner, C. Chiesa, M. Lassmann, N. Chouin, J. Gear, M. Bardiès, S. Walrand, K. Bacher, U. Eberlein, M. Ljungberg, L. Strigari, E. Visser, and M. W. Konijnenberg (2018). "From fixed activities to personalized treatments in radionuclide therapy: lost in translation?" *Eur. J. Nucl. Med. Mol. Imaging* 45(1):152–54. https://doi.org/10.1007/s00259-017-3859-1.

Fraass, B., K. Doppke, M. Hunt, G. Kutcher, G. Starkschall, R. Stern, and J. van Dyke (1998). "Quality Assurance for Clinical Radiotherapy Treatment Planning." Report No. 62 of American Association of Physicists in Medicine Radiation Therapy Committee Task Group 53. *Med. Phys.* 25(10):1773–829. https://doi.org/10.1118/1.598373.

Gawande, A. *The Checklist Manifesto: How to Get Things Right.* New York: Metropolitan Books, 2009.

Graham, M. M. and J. M. Buatti (2019). "Training Requirements for Theranostics: A Unique Opportunity for Collaboration." *J. Nucl. Med.* 60(9):1205–06.

Helmreich, R. L. (2000). "On error management: lessons from aviation." *BMJ* 320(7237):781–85. https://doi.org/10.1136/bmj.320.7237.781.

Herrmann, K., L. Giovanella, A. Santos, J. Gear, P. O. Kiratli, J. Kurth, A. M. Denis-Bacelar, R. Hustinx, M. Patt, R. L. Wahl, D. Paez, F. Giammarile, H. Jadvar, N. Pandit-Taskar, M. Ghesani, and J. Kunikowska (2022). "Joint EANM, SNMMI and IAEA enabling guide: how to set up a theranostics centre." *Eur. J. Nucl. Med. Mol. Imaging* 49(7):2300–09. https://doi.org/10.1007/s00259-022-05785-x.

Huq, M. S., B. A. Fraass, P. B. Dunscombe, J. P. Gibbons, G. S. Ibbott, A. J. Mundt, S. Mutic, J. R. Palta, F. Rath, B. R. Thomadsen, J. F. Williamson, and E. D. Yorke (2016). "The report of Task Group 100 of the AAPM: Application of risk analysis methods to radiation therapy quality management." *Med. Phys.* 43(7):4209–62. https://doi.org/10.1118/1.4947547.

Hurwitz, M., J. R. Buscombe, H. A. Jacene, A. K. Klitzke, D. Lamonica, Y. Lu, D. A. Pryma, E. M. Rohren, T. W. Speer, R. M. Subramaniam, H. M. Thompson, M. B. Tomblyn, W. W. Wong, Y. Xiao, K. P. Banks, and R. K. J. Brown (2020). "ACR-ACNM-ASTRO-SNMMI Practice Parameter for the Performance of Therapy With Radium-223." *Am J. Clin. Oncol.* 43(8):539–44.

IHI (2005). "Going Lean in Health Care." IHI Innovation Series white paper. Cambridge, MA: Institute for Healthcare Improvement, 2005.

Jewell, J. A., J. M. Percelay, V. L. Hill, C. K. Preuschoff, D. A. Rauch, and R. A. Salerno (2016). "Standardization of Inpatient Handoff Communication." *Pediatrics* 138(5)e20162681. https://doi.org/10.1542/peds.2016-2681.

Kesner, A. L., G. L. Poli, S. Beykan, and M. Lassmann (2017). "The IAEA Radiotracer Biodistribution Template–A community resource for supporting the standardization and reporting of radionuclide pre-dosimetry data." *Phys. Med.* 44:83–85. https://doi.org/10.1016/j.ejmp.2017.07.022.

Kesner, A. and L. Bodei (2018). "Modern Radiopharmaceutical Dosimetry Should Include Robust Biodistribution Reporting." *J. Nucl. Med.* 59(10):1507–09.

Kiess, A. P., R. F. Hobbs, B. Bednarz, S. J. Knox, R. Meredith, and F. E. Escorcia (2022). "ASTRO's Framework for Radiopharmaceutical Therapy Curriculum Development for Trainees." *Int. J. Radiat. Oncol. Biol. Phys.* 113(4):719–26. https://doi.org/10.1016/j.ijrobp.2022.03.018.

Lassmann, M., C. Chiesa, G. Flux, and M. Bardiès (2011). "EANM Dosimetry Committee guidance document: Good practice of clinical dosimetry reporting." *Eur. J. Nucl. Med. Mol. Imaging* 38(1):192–200. https://doi.org/10.1007/s00259-010-1549-3.

Lassmann, M. and U. Eberlein (2018). "The Relevance of Dosimetry in Precision Medicine." *J. Nucl. Med.* 59(10):1494–99. https://doi.org/10.2967/jnumed.117.206649.

Lassmann, M., H. Hänscheid, C. Chiesa, C. Hindorf, G. Flux, and M. Luster (2008). "EANM Dosimetry Committee series on standard operational procedures for pre-therapeutic dosimetry I: blood and bone marrow dosimetry in differentiated thyroid cancer therapy." *Eur. J. Nucl. Med. Mol. Imaging* 35(7):1405–12. https://doi.org/10.1007/s00259-008-0761-x.

Leveson, N. G. (2017). "The Therac-25: 30 Years Later." *Computer* 50(11):8–11. https://doi.org/10.1109/MC.2017.4041349.

Liu, H., L. Zhang, Y. Ping, and L. Wang (2020). "Failure mode and effects analysis for proactive healthcare risk evaluation: A systematic literature review." *J. Eval. Clin. Pract.* 26(4):1320–37. https://doi.org/10.1111/jep.13317.

Love, C., N. B. Desai, T. Abraham, K. P. Banks, L. Bodei, T. Boike, R. K. J. Brown, D. L. Bushnell, L. E. DeBlanche, M. M. Dominello, T. Francis, E. C. Grady, R. F. Hobbs, T. A. Hope, J. S. Kempf, D. A. Pryma, W. Rule, B. Savir-Baruch, I. Sethi, R. M. Subramaniam, Y. Xiao, and N. R. Schechter (2022). "ACR-ACNM-ASTRO-SNMMI Practice Parameter for Lutetium-177 (Lu-177) DOTATATE Therapy." *Am. J. Clin. Oncol.* 45(6):233–42.

Mayo, C. S., J. M. Moran, W. Bosch, Y. Xiao, T. McNutt, R. Popple, J. Michalski, M.Feng, L. B. Marks, C. D. Fuller, E. Yorke, J. Palta, P. E. Gabriel, A. Molineu, M. M. Matuszak, E. Covington, K. Masi, S. L. Richardson, T. Ritter, et al. (2018). "American Association of Physicists in Medicine Task Group 263: Standardizing Nomenclatures in Radiation Oncology." *Int. J. Radiat. Oncol. Biol. Phys.* 100(4):1057–66. https://doi.org/10.1016/j.ijrobp.2017.12.013.

McEwen, M., L. DeWerd, G. Ibbott, D. Followill, D. W. O. Rogers, S. Seltzer, and J. Seuntjens (2014). "Addendum to the AAPM's TG-51 protocol for clinical reference dosimetry of high-energy photon beams." *Med. Phys.* 41(4):041501. https://doi.org/10.1118/1.4866223.

Molineu, A., N. Hernandez, T. Nguyen, G. Ibbott, and D. Followill (2013). "Credentialing results from IMRT irradiations of an anthropomorphic head and neck phantom." *Med. Phys.* 40(2):022101. https://doi.org/10.1118/1.4773309.

O'Donoghue, J., P. Zanzonico, J. Humm, and A. Kesner (2022). "Dosimetry in Radiopharmaceutical Therapy." *J. Nucl. Med.* 63(10):1467–74. https://doi.org/10.2967/jnumed.121.262305.

Ostrom, L. T., T. J. Leahy, and S. D. Novack. *Summary of 1991–1992 Misadministration Event Investigations.* NUREG/CR-6088 EGG-2707. Washington, D.C.: U.S. Nuclear Regulatory Commission, 1994. https://doi.org/10.2172/10134993.

Ovčariček, P. P., M. C. Kreissl, A. Campenni, et al. (2022). "SNMMI/EANM practice guideline vs. ETA Consensus Statement: differences and similarities in approaching differentiated thyroid cancer management—the EANM perspective." *Eur. J. Nucl. Med. Mol. Imaging* 49:3959–63. https://doi.org/10.1007/s00259-022-05935-1.

Pronovost, P. J., D. A. Thompson, C. G. Holzmueller, L. H. Lubomski, T. Dorman, F. Dickman, M. Fahey, D. M. Steinwachs, L. Engineer, J. B. Sexton, A. W. Wu, and L. L. Morlock (2006). "Toward learning from patient safety reporting systems." *J. Crit. Care* 21(4):305–15. https://doi.org/10.1016/j.jcrc.2006.07.001.

Richardson, S. (2012). "A 2-year review of recent Nuclear Regulatory Commission events: What errors occur in the modern brachytherapy era?" *Pract. Radiat. Oncol.* 2(3):157–63. https://doi.org/10.1016/j.prro.2011.08.004.

Rozich, J. D., R. J. Howard, J. M. Justeson, P. D. Macken, M. E. Lindsay, and R. K. Resar (2004). "Standardization as a Mechanism to Improve Safety in Health Care." *Jt. Comm. J. Qual. Saf.* 30(1):5–14. doi: 10.1016/s1549-3741(04)30001-8.

Scoville, R., K. Little, J. Radover, K. Luther, and K. Mat (2016). "Sustaining Improvement." IHI White Paper. Cambridge, Massachusetts: Institute for Healthcare Improvement.

Sgouros, G., G. Barest, J. Thekkumthala, C. Chui, R. Mohan, R. E. Bigler, and P. B. Zanzonico (1990). "Treatment planning for internal radionuclide therapy: three-dimensional dosimetry for nonuniformly distributed radionuclides." *J. Nucl. Med.* 31(11):1884–91.

Sgouros, G. (2005). "Dosimetry of internal emitters." *J. Nucl. Med.* 46 Suppl 1:18S–27S. doi.org/10.1007/s00259-014-2824-5.

Sgouros, G., L. Bodei, M. R. McDevitt, and J. R. Nedrow (2020). "Radiopharmaceutical therapy in cancer: clinical advances and challenges." *Nat. Rev. Drug Discov.* 19(9):589–608. https://doi.org/10.1038/s41573-020-0073-9.

Siegel, J. A., S. R. Thomas, J. B. Stubbs, M. G. Stabin, M. T. Hays, K. F. Koral, J. S. Robertson, R. W. Howell, B. W. Wessels, D. R. Fisher, D. A. Weber, and A. B. Brill (1999). "MIRD pamphlet no. 16: Techniques for quantitative radiopharmaceutical biodistribution data acquisition and analysis for use in human radiation dose estimates." *J. Nucl. Med.* 40(2):37S–61S.

Strand, S.-E., B.-A. Jönsson, M. Ljungberg, and J. Tennvall (1993). "Radioimmunotherapy Dosimetry—A Review." Acta *Oncol.* 32(7–8):807–17. https://doi.org/10.3109/02841869309096140.

Tailor, R. C., W. F. Hanson, and G. S. Ibbott (2003). "TG-51: Experience from 150 institutions, common errors, and helpful hints." *J. Appl. Clin. Med. Phys.* 4(2):102–11.

Thomadsen, B., S.-W. Lin, P. Laemmrich, T. Waller, A. Cheng, B. Caldwell, R.Rankin, and J. Stitt (2003). "Analysis of treatment delivery errors in brachytherapy using formal risk analysis techniques. *Int. J. Radiat. Oncol. Biol. Phys.* 57(5):1492–1508. https://doi.org/10.1016/S0360-3016(03)01622-5.

Thomas, S. R., R. C. Samaratunga, M. Sperling, and H. R. Maxon (1993). "Predictive estimate of blood dose from external counting data preceding radioiodine therapy for thyroid cancer." *Nucl. Med. Biol.* 20(2):157–62. https://doi.org/10.1016/0969-8051(93)90108-7.

Uribe, C., A. Peterson, B. Van, R. Fedrigo, J. Carlson, J. Sunderland, E. Frey, and Y. K. Dewaraja (2021). "An International Study of Factors Affecting Variability of Dosimetry Calculations, Part 1: Design and Early Results of the SNMMI Dosimetry Challenge." *J. Nucl. Med.* 62:36S–47. https://doi.org/10.2967/jnumed.121.262748.

Xiao, Y., L. Papiez, R. Paulus, R. Timmerman, W. L. Straube, W. R. Bosch, J. Michalski, and J. M. Galvin (2009). "Dosimetric Evaluation of Heterogeneity Corrections for RTOG 0236: Stereotactic Body Radiotherapy of Inoperable Stage I-II Non–Small-Cell Lung Cancer." *Int. J. Radiat. Oncol. Biol. Phys.* 73(4):1235–42.

Zietman, A. L., J. R. Palta, M. L. Steinberg, eds. *Safety is No Accident.* Fairfax, VA: American Society for Radiation Oncology, 2012. Available online at https://www.astro.org/uploadedFiles/Main_Site/Clinical_Practice/Patient_Safety/Blue_Book/SafetyisnoAccident.pdf.

Zhuang, T., T. Djemil, P. Qi, A. Magnelli, K. Stephans, G. Videtic, and P. Xia (2013). "Dose calculation differences between Monte Carlo and pencil beam depend on the tumor locations and volumes for lung stereotactic body radiation therapy." *J. Appl. Clin. Med. Phys.* 14(2):38–51.

Chapter 18

Principles of Voxelized Dosimetry

Stephen A. Graves, Ph.D., DABR

Assistant Professor of Radiology
Division of Nuclear Medicine
University of Iowa
Iowa City, IA

18.1 Introduction

Voxelwise or *voxelized dosimetry* refers to the specification of absorbed dose at the voxel level. The end result is analogous to conventional *external beam radiotherapy* (EBRT) and brachytherapy treatment planning and dosimetry in that these techniques provide a patient-specific dose map superimposed on anatomic imaging (CT, MRI, or ultrasound). Advantages of this approach over organ-level dosimetry methods include:

1. It potentially provides improved whole-organ mean dose estimates, by way of transporting energy within the patient's own anatomy (rather than anthropomorphic phantom-derived energy transfer estimates).
2. It provides dose heterogeneity information within a given tissue structure.

Potential drawbacks of voxelwise dosimetry depend on the implementation, but notable disadvantages include:

1. There are difficulties correcting for partial volume effects and image noise, which can lead to underestimation of absorbed dose in smaller structures.

2. There can be overestimation of dose heterogeneity in a noise-dependent fashion.
3. Furthermore, other image-degrading factors (e.g., scatter, attenuation correction inaccuracies, patient motion) may be more likely to adversely impact the result of voxelwise dosimetry in comparison to whole-organ-based techniques.

Although the end result (absorbed dose per voxel) is similar to conventional 3D EBRT dosimetry, a key difference is that dose calculation for radiopharmaceuticals is fundamentally rooted in quantitative nuclear imaging for determination of the activity distribution, and thereby the distribution of energy being released in matter. This differs from external beam or sealed source brachytherapy-based dosimetry, whereby the radiation source is carefully characterized by nonimaging-derived means, such as experimental ionization measurements or Monte Carlo-derived estimates of beam quality and dose distribution relative to the radiation source. This difference (image-based dosimetry vs. source-model-based dosimetry) is critical to recognize because it demonstrates the fact that the image-generation process has the potential to significantly alter the estimated dose distribution for radiopharmaceuticals. Therefore, in addition to discussing typical dosimetry concepts (e.g., time-activity curve integration, radiation transport, dose map interpretation), we first must consider the impact of image generation on subsequent steps in the dosimetry process.

18.2 Image Formation

Voxelwise dosimetry relies on calculations from a quantitative nuclear image or series of time-ordered images for a particular radiopharmaceutical in a particular patient. The two primary imaging modalities (PET or SPECT) differ in several ways other than the typical isotopes being imaged (i.e., isotopes emitting positrons vs. low- to intermediate-energy gammas). In PET, coincident events are acquired with very high count rate and total efficiency, leading to relatively high "count density" in structures of interest. By comparison, SPECT imaging typically operates in the range of 2,000 to 20,000 counts per second, with low overall detection efficiency due to fewer than 1 in 10,000 patient-emitted photons successfully traversing the collimator.

As a result, SPECT images have comparatively low count density in structures of interest. This can be partially offset in some cases by substantially elevated administered activity. However, in general it is reasonable to expect fewer than 5 million events in a SPECT field of view acquisition, and often fewer than 1 million events. For a typical reconstruction technique (128 × 128, 4 mm slices, 40 cm axial FoV), this equates to roughly 0.5 to 3.0 acquired photon events per reconstructed voxel, on average. Although this would seem to indicate an unacceptably high level of statistical uncertainty within a given voxel, these photon events are typically quite concentrated in subregions of the reconstructed volume, such as radiopharmaceutical-avid tumors and clearance tissues (blood, liver, kidneys, etc.). For this reason, higher count densities are observed in regions of interest, although they are still substantially lower than in PET imaging and, therefore, statistical precision and associated image noise can still be quite problematic depending on the reconstruction technique and desired spatial resolution.

SPECT image reconstruction is typically performed using iterative methods, such as *ordered subsets expectation maximization* (OSEM) or *ordered subset conjugate gradient maximization* (OSCGM). Similarly, PET image reconstruction is typically performed using OSEM or noise-penalized iterative techniques. In all cases, the use of iterative image reconstruction allows for incorporation of image-degrading factors (such as attenuation, scatter, and spatial resolution) within the forward and/or backward projections steps of the iterative process, thereby effectively correcting for these degrading factors in the resulting reconstruction. Although theoretically these factors can be fully corrected for, in practice it is typical to avoid iterating to convergence to avoid excessive noise amplification and because estimates of these effects are subject to methodological limitations. Additionally, in some cases it is beneficial to intentionally filter/smooth projection data or the final reconstructed image to affect the reconstructed image characteristics. Although this practice degrades the reconstructed spatial resolution, in some cases the noise reduction trade-off may be favorable. The interplay between the number of iterative updates, smoothing, spatial resolution, and noise is illustrated by the Lu-177 SPECT/CT reconstructions in Figure 18.1.

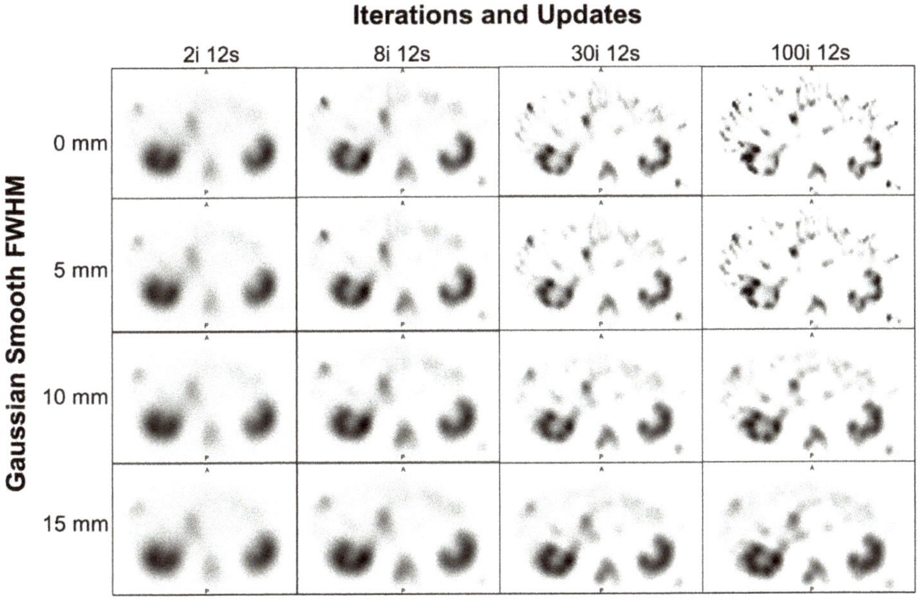

Figure 18.1 Lutetium-177 SPECT/CT reconstructions as a function of iterative updates and post-recon smoothing. Patient received 7.4 GBq of ^{177}Lu-DOTATATE and was scanned at 24 hours post-administration (120 views, 128 x 128 matrix, 1.0 zoom, 15 s/view, noncircular orbit, two FOV acquisition). All images are all shown on a linear grayscale, with a window of 0 to 80 counts per voxel.

Notably, resolution/noise trade-off considerations for dosimetry applications differ from those associated with human observer tasks, such as image interpretation and pathology detection. For detection tasks, reconstructions with high spatial resolution and high noise are typically associated with higher sensitivity, but greatly reduced specificity and positive predictive value. In a dosimetry context, however, the primary objective is to obtain a quantitative estimate of absorbed dose to critical organs and tumors. Therefore, the image reconstruction parameters used for dosimetry will typically differ from those commonly used for clinical image interpretation, with more reconstruction iteration and less smoothing being preferred. With that said, noise amplification remains undesirable in the context of voxelwise dosimetry, as this noise translates into noise in the voxelwise dose estimate.

18.3 Time-Activity Integration

Time-activity integration is the process of determining the total number of decays occurring in a region or, in the case of voxelwise dosimetry, within a particular patient voxel. If one assumes only physical decay (or other mono-exponential elimination behavior), as is often the case with ^{90}Y-microspheres, the process of voxel-level time-activity integration is straightforward, and given by:

$$\tilde{\tau} = \int_0^\infty A(t)\,dt = \int_0^\infty A_0 e^{-\lambda t}\,dt = \frac{A_0}{\lambda}. \tag{18.1}$$

If the activity in a voxel is measured (A_m) at some time (t_m), then time-integrated activity can be expressed as

$$\tilde{\tau} = \frac{1}{\lambda}\left(\frac{A_m}{e^{-\lambda t_m}}\right). \tag{18.2}$$

Typically this time-integrated activity coefficient will carry units of mCi • h (1 mCi • h = 1.332 • 10^{11} decays), or it can be normalized per administered activity (mCi • h/mCi or MBq • h/MBq).

More complex time-activity curve-fitting functions, or *piecewise fitting*, can be utilized if multiple image timepoints are obtained, or if clearance behavior has been characterized for a particular patient population (Madsen et al. 2018). Typical fitting functions and approaches to fitting are discussed elsewhere in this book (J. O'Donoghue), but two unique considerations are needed in the context of voxelwise dosimetry. These considerations are as follows:

1. How do we identify the same voxel in sequential imaging timepoints?
2. How do we fit a "noisy" time-activity curve?

18.3.1 Voxel Alignment

The first consideration, aligning voxels between images, is not trivial. This is true for several reasons, such as differences in patient positioning, differences in breathing phase, changes in patient anatomy—including varying organ positions within the body—as well as variation in local SPECT-CT misalignment.

The alignment challenge can be addressed by global deformable registration of the sequential SPECT/CT images to a common "gold-standard" imaging timepoint using the associated CT scans for intensity-based or mutual information-based deformation. Although commonly utilized, this approach can often yield unsatisfactory results, requiring manual adjustment or compromised voxel-level dosimetric accuracy. Additionally, care must be taken to ensure that the SPECT deformation process is activity-preserving.

An alternative to CT-CT deformation-based voxel alignment is to utilize rigid SPECT-SPECT intensity-based registration on an ROI-specific basis (Mirando et al. 2020). This approach has the benefit of not relying on the SPECT-CT alignment at each timepoint, which can be significantly impacted by respiratory and patient motion. An example of how contour-based SPECT-SPECT registration can improve voxel alignment can be found in Dewaraja et al. (2022). Practically speaking, if tissue regions of interest are drawn on CT imaging, an expansion to account for partial volume effects (~5 mm for PET or ~10 mm for SPECT) should be utilized when masking the nuclear medicine activity distribution for intensity-based alignment.

18.3.2 Curve Fitting and Regularization

Following voxel alignment, a time-activity curve must be generated for each voxel. For multiple imaging timepoints, an estimate of the voxel activity at each timepoint is provided for fitting. However, due to the generally low counts per voxel encountered in SPECT, the voxel-level data is often not well-behaved. This can be easily demonstrated by assuming Poisson-distributed counting statistics in each voxel—which is not strictly true, but illustrative for the sake of argument and reasonable to first order. Assuming Lu-177 imaging timepoints of 4 hours and 72 hours and a "true effective half-life" of 70 hours, one can estimate the percentage of voxels with non-physical time-activity curves (a voxel where the activity appears to be greater at 72 hours than at 4 hours) as a function of mean counts per voxel (exact calculation methods are left as an exercise to the reader). Results for this situation are shown in Figure 18.2. Purely as a result of statistical fluctuation in the measured projection data and the corresponding reconstructed voxels, a significant fraction of voxels are inevitably observed to have an apparent increase in activity concentration or clearance that is less than physical decay, despite a true two-fold decrease in activity concentration. This effect is particularly pronounced for count densities below ~50 counts per voxel, as would be encountered in most patient anatomy, other than the most avid tissue regions.

For weakly constrained fit functions (i.e., more fit parameters than data points), or in the case of fitting that leads to divergence or nonphysical convergence, assumptions must be made. Some of the possible assumptions are as follows:

- If the TAC slope is positive, use trapezoidal integration, followed by physical decay following the last timepoint.

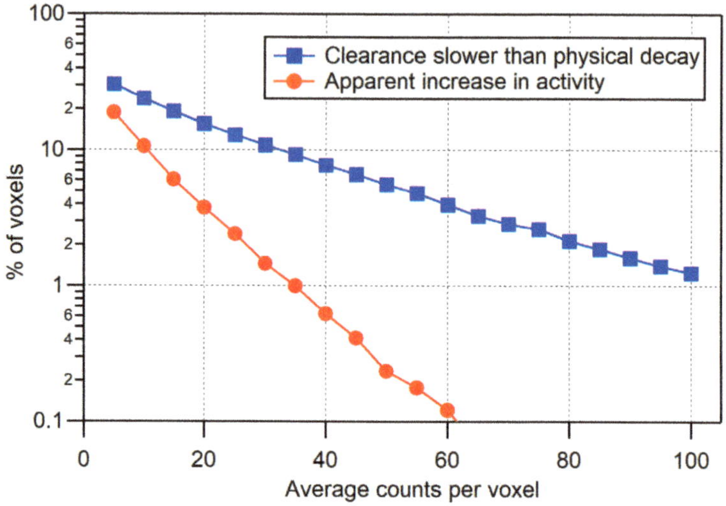

Figure 18.2 Percentage of voxels with nonphysical time activity curves as a function of the average number of counts per voxel in the t = 4-hour image. Results indicate that a significant fraction of voxels are not well described by a monoexponential fit and, therefore, rate-constant regularization is needed.

- If the TAC slope is positive or nonphysical, use a mono-exponential fit that matches the aggregate rate constant of neighboring voxels, or of the entire organ/tissue ROI, scaled by the measured activity in the particular voxel.
- Enforce a minimum rate constant, often chosen to be the physical decay constant. For multi-exponential fitting, the dominant/terminal term should have a rate constant that is greater than or equal to the decay constant.
- Utilize "prior information," such as the average rate constant for this patient population, or the organ average rate constant from a particular patient's prior course of therapy.

The second of these assumptions relies on the presence of a preexisting region of interest, similar to the SPECT-SPECT local rigid registration technique described in Section 18.3.1. The approach of using a whole organ, or whole tumor region of interest, for determination of certain fitting parameters, i.e., exponential rate constant(s), has been implemented in several dosimetry software packages. For regions of the image that are not associated with a particular ROI, one can either rely on local regularization (which may still be prone to nonphysical time-activity curves in areas of low information density), or one can assume that these regions behave similarly to the whole-body clearance kinetics.

In addition to the regularization technique applied to voxels with nonphysical time-activity data, a decision also must be made about how the fitting will be performed for voxels with "well-behaved" time-activity curves. It is straightforward to treat these voxels the same as voxels with nonphysical behavior, however one may also decide to use a different approach, such as allowing the rate constant(s) to vary based on voxel-level time-activity curve behavior.

18.3.3 Time-Activity Integration vs. Time-Dose-Rate Integration

Following voxel alignment, there are generally two approaches to voxelwise dosimetry, based on whether absorbed dose calculation is performed before or after the time-integration step. Thus far in this section, we have discussed time-activity integration, but there are theoretical advantages to the alternative approach: calculating a dose rate map from each aligned voxel image and performing a time-dose-rate integration at the voxel level. This approach essentially "smooths" the activity images according to their appropriate energy deposition pattern. Some energy will be deposited locally, with small to negligible smoothing (i.e., alpha radiation or low-energy beta particles), but absorbed dose from photon emissions will generally be spread over a larger region of tissue. This can greatly improve the noise characteristics of background regions of the image, thus providing a significant level of regularization prior to the integration step and reducing the need for prior information, regional/ROI regularization, or numerical integration approaches. The only drawback to computing dose maps prior to integration is increased computation burden by a factor equal to the number of imaging timepoints.

18.4 Radiation Transport and Dose Calculation

Three primary approaches exist for voxelwise radiopharmaceutical dose calculation: local energy deposition, kernel convolution, and Monte Carlo. These approaches to radiation transport, energy deposition, and absorbed dose calculation are described below.

18.4.1 Monte Carlo

The most accurate, as well as most computationally expensive, approach to radiopharmaceutical dosimetry involves directly transporting individual ionizing radiation particles through the patient anatomy. This approach uses random number generators, in combination with known physical effect probabilities, to determine various transport parameters. For photons, this would include the interaction-free pathlength along a certain trajectory, the type of interaction (e.g., photoelectric, Compton scattering, Rayleigh scatter, pair/triplet production), as well as the scattering angle and trajectory in the event of a scatter-inducing interaction. Charged particle transport, in most cases electrons, will simulate small- and large-angle scattering, as well as energy loss during transport (collisional and radiative stopping power). In addition to transporting primary decay emissions, Monte Carlo methods can also involve transporting secondary radiations, such as Compton-scattered electrons, bremsstrahlung x-rays, or other secondary ionizing radiation modes.

Specific assumptions regarding the radiation transport can have a large impact on the computation time, as well as a negligible to large impact on the resulting dosimetric accuracy. For example, most Monte Carlo dosimetry implementations have a lower-energy cutoff on photon and charged particle transport, corresponding to a pathlength of less than one voxel (or some fraction of a voxel). Additionally, look-up tables for cross sections and probability distributions can be optimized by modifying the data binning width and data interpolation methods.

Perhaps most importantly, a particular implementation can leverage acceleration techniques, analogous to condensed history methods (Bhan et al. 2007), but typically involving non-uniform spatial- and emission-sampling of radioactive decays within the simulation volume. The number of simulation histories per voxel may be chosen to be proportional to the activity in that voxel, as opposed to simulating the same number of histories in each voxel. This has the effect of improving statistical precision much more quickly (relative to the total number of simulation histories) in regions of higher activity accumulation, which tend to be of clinical interest.

Similarly, for a complex decay with many emission energies and modes, one may choose to sample emissions probabilistically, rather than simulating each emission individually, and weighting the resulting energy deposition by the emission probability to complete each "decay history." Although not related to the radiation transport method directly, the computational problem may be accelerated by "parallelizing" the task into multiple independent simulation "jobs," such as separately simulating decays within each axial slice, followed by generating a composite energy deposition map from the parallel simulations (Besemer et al. 2018).

Although Monte Carlo-based RPT dosimetry has historically been considered to be too time consuming for clinical use, there are now at least two FDA-cleared software platforms for radiopharmaceutical dosimetry that advertise calculation times of less than 30 minutes for typical radioisotopes and calculation matrix sizes.

18.4.2 Kernel Convolution

Pre-tabulated energy deposition patterns surrounding a point or voxel source of a radionuclide are often referred to as "kernels" in the context of radiation dosimetry. These kernels can take the form of *dose point kernels*, which describe the deposition of energy or absorbed dose surrounding a point source, or voxel *S value kernels*, which describe the dose absorbed by voxels surrounding, and including, a source voxel of radioactivity. The generation and use of these kernels for 3D voxelwise dosimetry has been quite popular, owing to their availability, ease of implementation, and generally minimal computation time.

18.4.2.1 Kernel Generation and Convolution

Dose kernels are generated using Monte Carlo transport simulation in a uniform media, either tissue or water. The source can be defined as a point, or as having activity uniformly distributed in a source voxel. As discussed in the literature (Graves et al. 2019), these kernels should generally have dimensions that are large enough to encapsulate a majority (~99%) of energy released during each decay. For some isotopes with few gamma emissions (such as ^{32}P, ^{90}Y, and ^{89}Sr) a relatively small kernel (<1 cm radius) is sufficient to convey a complete picture of the energy deposition pattern. For others, such as ^{131}I, a relatively large kernel (>60 cm) is required to adequately describe the energy deposition pattern. A summary of the R_{50}, R_{90}, and R_{99} (the radii

Table 18.1 Energy per decay for common medical radioisotopes and the radii at which 50%, 90%, and 99% of decay energy is absorbed in water (adapted from Graves et al. 2019)

Isotope	E_{ave} per Decay (MeV)	$R_{50\%}$ (cm)	$R_{90\%}$ (cm)	$R_{99\%}$ (cm)
^{11}C	1.40	9.8	38.9	72.0
^{18}F	1.23	12.0	40.5	73.3
32P	0.70	0.16	0.37	0.55
^{52g}Mn	3.53	18.8	53.6	98.1
^{64}Cu	0.31	5.1	36.2	70.5
^{67}Cu	0.27	0.05	24.4	49.6
^{67}Ga	0.19	10.7	33.3	60.2
^{68}Ga	1.69	3.5	35.2	69.5
^{89}Sr	0.59	0.13	0.32	0.48
^{89}Zr	1.26	16.1	48.3	87.8
^{90}Y	0.93	0.23	0.54	0.81
^{99m}Tc	0.14	12.1	31.9	54.1
^{111}In	0.44	12.7	34.9	60.2
^{117m}Sn	0.31	0.02	25.0	49.3
^{123}I	0.20	9.8	31.8	56.9
^{124}I	1.31	14.4	48.7	94.9
^{125}I	0.06	1.5	7.1	14.6
^{131}I	0.58	7.4	35.5	66.6
^{153}Sm	0.33	0.04	8.8	31.6
^{177}Lu	0.18	0.02	12.4	40.9
^{186}Re	0.36	0.07	0.22	25.1
^{188}Re	0.84	0.20	0.58	37.8
^{211}At	7.68	<0.01	2.3	47.7
^{212}Pb	10.2	<0.01	13.1	76.7
^{213}Bi	9.01	<0.01	<0.01	15.8
^{223}Ra	27.6	<0.01	<0.01	2.81
^{225}Ac	28.3	<0.01	<0.01	0.21

at which 50%, 90%, and 99% of decay energy has been absorbed in water) for common medical radioisotopes is included in Table 18.1.

Although *dose point kernels* (dose as a function of radius from a point source) are relatively straightforward to generate, they are not immediately useful for radiopharmaceutical dosimetry. Compatibility with the nuclear medicine image, or time-integrated activity map, requires that the kernels be formatted in a voxel matrix. These kernels can be generated directly using Monte Carlo simulations, or radial dose point kernels can be reformatted by point-wise sampling onto a high-resolution matrix, convolving the high-res matrix with a high-res 3D rectangular function representing a source voxel, down-sampling to the result to the clinical image voxel size, and scaling the result to ensure conservation of energy. In the case of pre-generated voxel s value

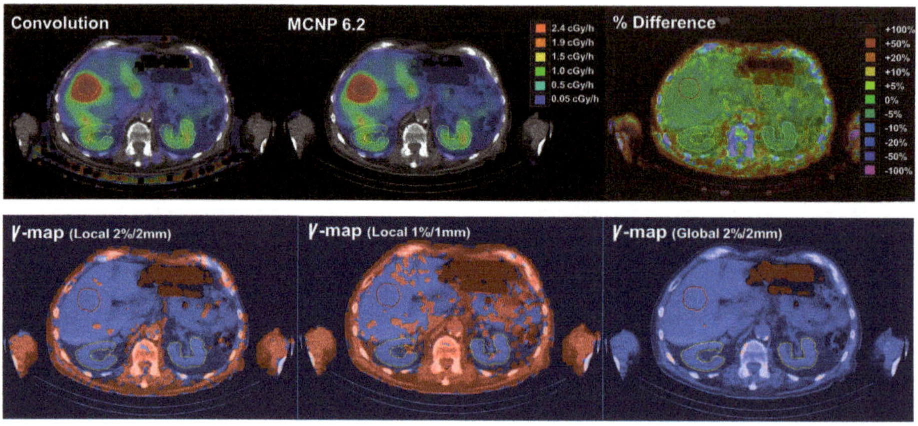

Figure 18.3 Comparison of absorbed dose maps generated by kernel convolution and Monte Carlo simulation. (Figure from Graves et al. 2020a.)

kernels, the kernel must be resampled to the clinical image resolution, or vice versa. Throughout this process, it is critical that activity, dose rate, or energy be conserved.

In practice, convolution dose calculation is typically performed using the *fast Fourier transform* (FFT) method, as this process has computational efficiency of $O(N^3\log(N^3))$, as opposed to the general convolution algorithm with efficiency of $O(N^6)$. Adequate zero-digit padding of the clinical image and kernel are required for accurate completion of the FFT-based 3D convolution, which can temporarily increase computer memory requirements.

The accuracy of convolution-based dosimetry has been shown to be quite good in situations where structures of interest are comprised of soft tissue, and a majority of energy deposited in these structures comes from the "self-dose" component, i.e., energy released from decays within the structure itself (Graves et al. 2020a). This is illustrated in the comparison between kernel convolution and Monte Carlo (MCNP 6.2) in Figure 18.3.

Gamma analysis reveals excellent agreement in kidneys and tumors, whereas kernel convolution tends to overestimate energy deposition (and, therefore, absorbed dose) in low-density regions (lungs, intestinal voids) as well as underestimating absorbed dose in high-density regions (bone and bone marrow). Convolution dosimetry's tendency of overestimating absorbed dose in low-density regions can be explained by the lack of effective pathlength scaling, i.e., a direct convolution assumes equivalent radiation interaction probability throughout the entire volume, whereas Monte Carlo simulations accurately describe the longer interaction-free pathlength in low-density regions. Similarly, convolution methods underestimate the effective pathlength and relative increase in photoelectric absorption in bone, resulting in absorbed dose underestimation. The clinical relevance of these inaccuracies is situationally dependent; in many cases absorbed dose to soft tissue sources (kidneys, liver, tumors, salivary glands) is the primary endpoint, however some situations involving bone (metastatic disease, bone marrow) or lungs (lung metastases) may benefit from more accurate techniques.

18.4.2.2 Collapsed-Cone Convolution Superposition

As mentioned above, standard convolution methods will overestimate effective pathlength in low-density regions and underestimate effective pathlength in high-density regions. Correction-based methods for convolutional dosimetry have long been a mainstay of external beam radiotherapy dosimetry. In particular, the *collapsed cone convolution* method employs kernel "stretching" along a fixed number of rays originating from each source voxel (Figure 18.4) (Papanikolaou et al. 2009). In this method, energy from a source voxel is assumed to only be deposed within voxels encountered by each ray, weighted by the effective pathlength in each voxel. Perhaps surprisingly, this approximation results in acceptable accuracy due to the effect of averaging over many regional source voxels, thus ensuring that "target voxels" are adequately covered by energy deposition rays.

As shown in Figure 18.5, a relatively small number of energy-deposition rays ("cones") are required for electron energy deposition. A larger number of cones are required for radiopharmaceutical energy transport, due to the gamma emissions encountered from most isotopes (i.e., ^{177}Lu and ^{131}I) having substantially longer pathlengths in tissue. Furthermore, unlike megavoltage external beam radiotherapy, the sub-MeV energy of photons emitted following radioactive decay require special consideration due to the physical effective pathlength not fully compensating for the increased photoelectric absorption in higher-Z (higher-density) structures, such as bone (Graves et al. 2020b). For this reason, empiric nonlinear pathlength scaling is necessary for accurate energy deposition (as determined by agreement with Monte Carlo methods) in tissue structures with densities greater than ~1.1 g/cm^3.

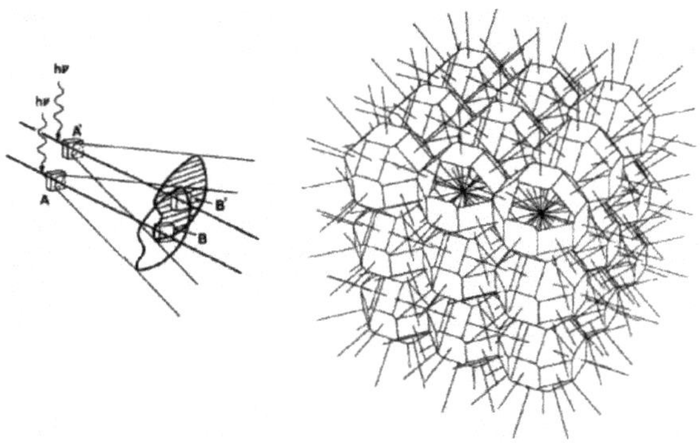

Figure 18.4 Illustration of the collapsed cone convolution method for dose calculation. Radiation energy is assumed to be deposited along a finite number of rays originating from each source voxel. (Figure obtained from Papanikolau and Stathakis 2009.)

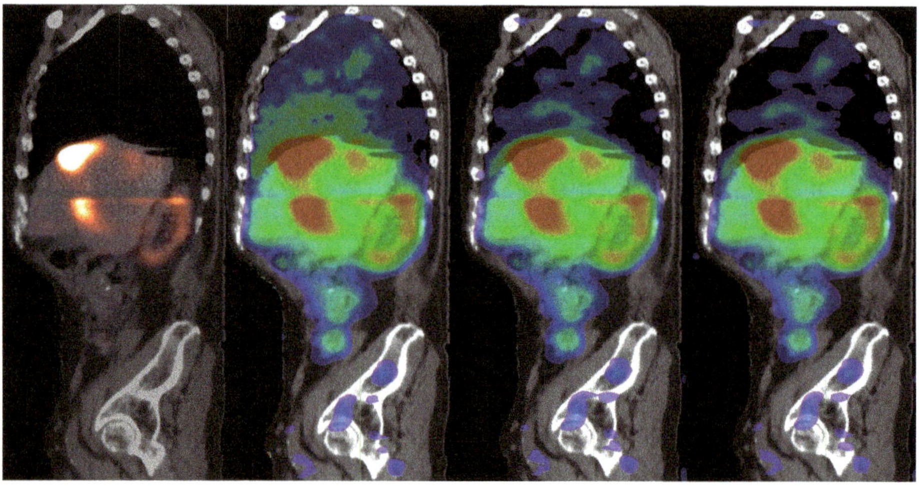

Figure 18.5 Comparison of simple convolution, collapsed cone convolution superposition (CCCS), and Monte Carlo (MCNP 6.2) for dose rate calculation from a ^{177}Lu SPECT/CT image (Graves et al. 2020b). Lung dose is overestimated in the convolution-based dose calculation, due to overestimation of effective pathlength. Good agreement is observed when comparing the CCCS and MCNP6.2 calculation results.

18.4.3 Local Energy Deposition

Local energy deposition assumes some portion, or all, of the emitted radiation is absorbed only by the source voxel. Often this assumption is applied for low-energy beta particles and alpha particles where the expected range is less than the size of one voxel. However, even in situations where the beta range is longer (i.e., ^{90}Y), this assumption of local energy deposition can be considered optimal (Pasciak et al. 2009; Pasciak et al. 2014; Pacilio et al. 2015). For pure beta emitters, if the reconstructed SPECT or PET image has a spatial resolution FWHM that is less than twice the $R_{50\%}$ value listed in Table 18.1, it is likely most appropriate to assume local energy deposition. This is due to the effect of spatial resolution blurring of the true underlying activity distribution being of greater magnitude than the energy transport "blurring" that would be applied from Monte Carlo or convolution dosimetry methods. Similarly, $R_{50\%}$ values calculated from only the beta and alpha emissions of other isotopes with mixed emissions could be used to evaluate the applicability of local energy deposition. Exceptions to this generality may include energy transport in the lungs, where pathlengths can be greater, as well as situations where recovery coefficients are already utilized for partial volume correction, in which case local deposition may slightly overestimate dose in source voxels.

In routine practice, local energy deposition is appropriate for ^{90}Y microsphere voxelwise absorbed dose calculations, whereas ^{131}I and ^{177}Lu systemic agents are better suited to convolution or Monte Carlo methods. One could utilize local energy deposition for the beta component of ^{131}I or ^{177}Lu, but this does not practically simplify the dosimetry calculations, due to the need for photon energy transport.

18.5 Dose Map Interpretation

Evaluation of each voxelwise dose map should be performed keeping three things in consideration:

1. Misregistration between the SPECT and CT scans at the reference timepoint will be propagated to the voxelwise dose map.
2. Noise and spatial resolution effects are largely propagated to the final image, with some benefit to noise from regularization and absorbed dose calculation.
3. Dose rate information and corresponding radiobiological effects are "lost" when presenting information only as absorbed dose (Gy).

Unlike external beam radiotherapy, where a physical absorbed dose is designed according to a planned dose/fractionation schedule, the time-dose relationship for RPT is lost during the time-integration step. In the situation where one computes dose rate for each imaging timepoint prior to time-integration, it is possible to produce a voxelwise map of *biologic effective dose* (BED) rather than physical absorbed dose (Besemer et al. 2018). The benefit of precomputing dose-rate maps individually is additive with the benefits described in Section 18.3.3 of this chapter.

Following computation of a voxelwise dose map, and given a CT-based organ contour, it is relatively straightforward to determine a whole-organ mean absorbed dose. With that said, the first two considerations above (potential for misregistration, as well as spatial resolution/partial volume effects) do require some special treatment. In some situations, the dose map (and corresponding reference NM image) may need to be locally aligned with the CT-based contour. For medium- to small-sized objects (~300 cm^3 and below), partial volume correction is recommended. Mean concentration activity recovery coefficients (RC_{mean}) for the scanner, isotope, collimator, and reconstruction profile as a function of object size can be utilized, but direct application of RC_{mean} values to raw organ absorbed dose results will generally produce an overestimation of true absorbed dose. This is due to background spill-in high background regions, as well as "background" photon absorbed dose from other regions of the body in the case of gamma-emitting radionuclides. Therefore, for isotopes with beta/alpha particle energies that are conducive to local energy deposition assumption, an expression for partial volume correction in the voxelwise dose domain can be expressed as:

$$D = \frac{(D_0 - D_{bkgd})}{RC_{mean}} + D_{bkgd}. \tag{18.3}$$

where D is the structure mean absorbed dose corrected for PVE, D_0 is the raw dose estimate, and D_{bkgd} is the dose that would be calculated from a time-integrated activity map that has been modified to set the number of source region decays to zero. In practice, accurate estimation of D_{bkgd} is challenging, although a rough approximation can be made by drawing a background region near the tissue of interest in an area with minimal activity concentration. These considerations—as well as additional nuances associated with longer-range charged particles (^{90}Y)—were recently discussed by Marsh et al. (2022) at the 2022 AAPM annual meeting.

18.6 Conclusions

In comparison to organ-level dosimetry, voxelwise dose calculation has the advantage of providing insight into the nonuniformity of organ and tumor dose, however careful consideration of all dose map generation steps is necessary: image reconstruction, registration, voxel alignment, time-activity integration and regularization, dose calculation, and correction for image-degrading factors. In general, one expects whole-organ mean absorbed dose estimates determined from voxelwise approaches to converge to results from conventional absorbed fraction whole-organ techniques, but differences in the energy transport method—as well as the patient/phantom geometry used for dose calculation—may lead to differences.

References

Besemer, A. E., Y. M. Yang, J. J. Grudzinski, L. T. Hall, and B. P. Bednarz (2018). "Development and validation of RAPID: a patient-specific Monte Carlo three-dimensional internal dosimetry platform." *Cancer Biother. Radiopharm.* 33(4):155–65.

Bhan, K. and J. Spanier (2007). "Condensed history Monte Carlo methods for photon transport problems." *Computat. Phys.* 225(2)1673–94.

Dewaraja, Y. K., D. M. Mirando, A. B. Peterson, J. Niedbala, J. D. Millet, J. K. Mikell, K. A. Frey, K. K. Wong, S. J. Wilderman, and A. S. Nelson (2022). "A pipeline for automated voxel dosimetry: application in patients with multi-SPECT/CT imaging following 177Lu peptide receptor radionuclide therapy." *J. Nucl. Med.* 63(11):1665–72.

Graves, S. A., R. T. Flynn, and D. E. Hyer (2019). "Dose point kernels for 2,174 radionuclides." *Med. Phys.* 46(11):5284–93.

Graves, S., A. Tiwari, A. Kruzer, A. Nelson, D. Mirando, Y. Dewaraja, and J. Sunderland (2020a). "Monte Carlo validation of convolution-based voxelwise dosimetry." *J. Nucl. Med.* 61(supplement):1019.

Graves, S., A. Tiwari, and J. Sunderland (2020b). "Collapsed-cone convolution superposition for improved accuracy of voxelwise dosimetry." *J. Nucl. Med.* 61(supplement 1):535.

Madsen, M. T., Y. Menda, T. M. O'Dorisio, and M. S. O'Dorisio (2018). "Single time point dose estimate for exponential clearance." *Med. Phys.* 45(5):2318–24.

Marsh, I. R., T. J. Bradshaw, J. J. Grudzinski, R. Hernandez, E. Aluicio-Sarduy, and B. P. Bednarz (2022). "A Method for Partial Volume Correction of Dose Distributions Derived From PET-Based Theranostic Dosimetry." Presentation at the 2022 American Association of Physicists in Medicine annual conference, June 10, 2022.

Mirando, D., Y. K. Dewaraja, N. Cole, and A. S. Nelson (2020). "In pursuit of fully automated dosimetry: evaluation of an automatic VOI propagation algorithm using contour intensity-based SPECT alignments." Eur. J. Nucl. Med. Mol. Imaging 47(suppl 1):S236.

Pacilio, M., E. Amato, N. Lanconelli, C. Basile, L. A. Torres, F. Botta, M. Ferrari, N. C. Diaz, M. C. Perez, M. Fernández, M. Lassmann, A. V. Gil, and M. Cremonesi (2015). "Differences in 3D dose distributions due to calculation method of voxel S-values and the influence of image blurring in SPECT." *Phys. Med. Biol.* 60(5):1945.

Papanikolaou, N. and S. Stathakis (2009). "Dose-calculation algorithms in the context of inhomogeneity corrections for high energy photon beams." *Med. Phys.* 36(10):4765–75.

Pasciak, A. S. and W. D. Erwin (2009). "Effect of voxel size and computation method on Tc-99m MAA SPECT/CT-based dose estimation for Y-90 microsphere therapy." *IEEE Trans. Med. Imaging* 28(11):1754–58.

Pasciak, A. S., A. C. Bourgeois, and Y. C. Bradley (2014). "A comparison of techniques for 90Y PET/CT image-based dosimetry following radioembolization with resin microspheres." *Front. Oncol.* 4:121.

Chapter 19

Multiple Timepoint Image Registration

Bryan Bednarz, Ph.D.

Department of Medical Physics, Human Oncology,
Radiology, and Engineering Physics
University of Wisconsin–Madison
Madison, WI

Radiopharmaceutical therapy (RPT) dosimetry calculations are often performed by utilizing serial measurements of SPECT/CT or PET/CT images taken days apart to adequately characterize the pharmacokinetic behavior of the RPT drug in patients (St. James et al. 2021; Sgouros et al. 2010, 2020; ICRU 2021). Patient motion, inconsistent patient positioning on the imaging couch, and organ deformation will likely occur between or during imaging sessions causing misalignment in the image data sets between timepoints. Correct registration of these serial images is important to support accurate and precise dosimetry calculations in tumors and organs of interest. Misalignment of the serial SPECT images with a translational error of less than 9.5 mm and rotation error of less than 6° could lead to absorbed dose errors up to 90% (Papavasileiou et al. 2007).

Multi-timepoint dosimetry should utilize a combination of either whole-body or piece-wise rigid registration followed by deformable registration. While there are many commonalities between registration problems encountered in *external beam radiation therapy* (EBRT) and RPT, there are also some key differences. RPT dosimetry requires the registration of large-field-of-view images that are acquired hours or even days apart, whereas EBRT dosimetry typically involves the registration of limited-field-of-view images (e.g., adaptive planning and patient positioning) that include the primary tumor and surrounding healthy tissues. Also, RPT dosimetry may require absorbed dose calculations done on several index lesions that are typically much smaller than tumors treated with EBRT. Therefore, RPT dosimetry has different challenges when it comes to image registration compared to EBRT.

This chapter gives an overview of image registration used for RPT dosimetry. It will discuss the unique challenges associated with the registration of multi-timepoint imaging data.

19.1 How Image Registration Fits within the RPT Dosimetry Workflow

The general procedure for RPT dosimetry is highlighted in Figure 19.1. This chapter primarily covers three-dimensional, patient-specific dosimetry. Other techniques that can be used include two dimensional planar images or point-wise measurements, such as blood draws or whole-body survey meter readings, but those techniques are not covered here. The first step in the workflow is to administer the radiopharmaceutical that is going to be imaged. This is often the drug that is used for therapy which is imaged after a treatment cycle is administered, but may be a surrogate. This approach toward patient-specific dosimetry is called *pretreatment dosimetry.*

Imaging is performed using nuclear medicine scanners. In the case of pretreatment dosimetry, either SPECT/CT or PET/CT imaging can be used depending on the tracer or surrogate being imaged to acquire three-dimensional images of the activity of the drug in the body. To better quantify the pharmacokinetic distribution of the drug, patients are generally imaged at multiple timepoints that span multiple days post injection. The ideal number of timepoints depends on the physical and biological properties of the drug, but most studies have used 3–4 timepoints as a compromise between accuracy and logistical challenges related to patient scheduling. There has been a recent push to continue to reduce the number of imaging timepoints used in the dosimetry workflow—even down to a single timepoint. These approaches replace patient-specific imaging data with population data to model the pharmacokinetics of the drug. Reducing imaging timepoints can impact the overall accuracy of the dose calculation and make dosimetry less amenable to identifying outliers in the patient population.

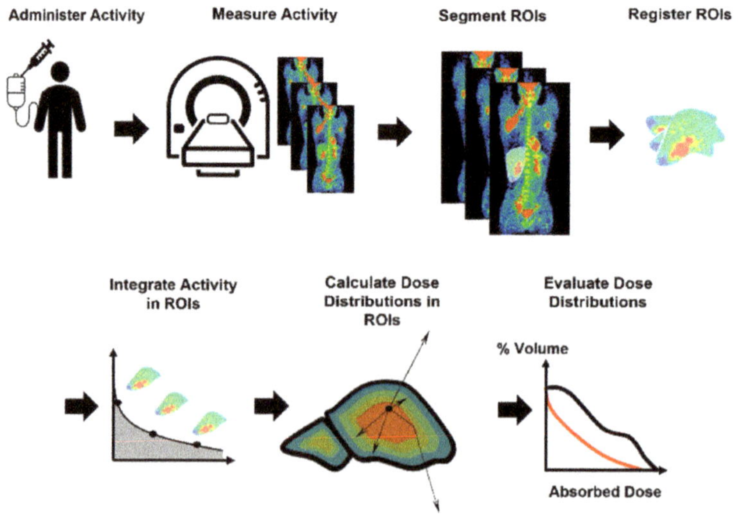

Figure 19.1 Voxel-level, patient-specific dosimetry workflow.

Segmentation and registration are important components of the dosimetry workflow. The absorbed dose distribution in RPT is dependent on the biological and physical properties of a drug in an ROI, so the ROI contours serve to quantify the existing absorbed dose distribution. In general, two main segmentation techniques have been used for defining ROIs:

- ***Manual anatomical or functional segmentation*** using either the CT image or the SPECT image is typically done for segmenting normal organ/tissue ROIs.
- ***Segmentation using universal or regional thresholding techniques*** is often used to define tumor ROIs.

In the case of segmentation, delineation of an ROI can impact the quantification accuracy of SPECT/CT or PET/CT significantly (Uribe et al. 2017; Hobbs et al. 2009).

Several variables can impact the accuracy of ROI segmentation and, ultimately, the predicted absorbed dose in the ROI including:

1. inter-observer variability,
2. temporal variability over a series of images, and
3. the choice of a segmentation threshold in the case of segmenting using SPECT thresholding techniques.

Image registration, often multi-modal, is needed to calculate the total absorbed dose burden to the patient using serial SPECT/CT or PET/CT images at multiple timepoints. Most often, rigid registration is used for initial alignment of images followed by a deformable image registration step. An image at a single timepoint is chosen as the "stationary image" to guide the registration process. Every other "moving image" is registered to the anchor image. Typically the stationary image is the first or last timepoint in the image series. Registration accuracy can greatly impact the accuracy of the overall dosimetry process. Voxel-level dosimetry is more impacted by registration accuracy than ROI-level or organ-level dosimetry.

The time-integrated activity in an ROI (or voxels within an ROI) is calculated by fitting the activity over multiple timepoints and then integrating the fit over the time post-administration. Different fitting methods can be used depending on the behavior of the drug and can be utilized to calculate absorbed dose distributions. In most cases, the pharmacokinetic behavior of an RPT drug in an organ or tumor can be modeled by the combination of exponential functions representing the uptake or elimination of the drug over time. Several different exponential models can be used to fit activity concentrations or dose rates as a function of time. Another approach is to use the trapezoidal rule for all but the last timepoint and adjusting the integration to reflect radiological decay following the last timepoint. It is also possible to calculate the dose rates to an ROI or voxels within an ROI and then integrate these rates to get the total absorbed dose. This approach has been used in order to calculate radiobiologically standardized dosimetric quantities such as *biological effective dose* (BED) (Hobbs et al. 2009; Fowler 2010).

The final step in the dosimetry workflow is the dose calculation step. Historically, dose calculation was performed at the organ level using 3D imaging data and pre-calculated dose factors. These dose factors, which are known as *S values*, are derived from Monte Carlo simulations in computational reference phantoms. They are calculated assuming homogeneous activity in each source organ, and the dosimetric contribution from these source organs is determined to itself (i.e., self-dose) or to neighboring organs (i.e., cross-fire). The absorbed dose to a target organ from a source organ is calculated by taking the product of the S value and the cumulative activity in the source organ determined by methods described previously. The total absorbed dose to each target organ is the sum of the absorbed dose contributions resulting from all source organs.

There are two main limitations of organ-level approaches:

1. ***The anatomical specificity of the patient is not considered since S values are calculated in computational phantoms***. While S values can be scaled by patient-specific organ masses to improve dosimetric accuracy, the relative positions of the organs cannot be adjusted. In addition, tumors are not present in these phantoms. Therefore, tumor dosimetry is often done by using specific absorbed fractions calculated in spheres, which often misrepresents the actual shape of the tumor and neglects cross-fire dose contributions from source organs adjacent to and at larger distances from the tumor.
2. ***Nonuniform uptake of activity in the source organ cannot be modeled***. Uniform activity is assumed in all regions or organs and tumors, which will often misrepresent the nonuniform activity distribution in relevant organ and tumor ROIs.

A more accurate alternative to organ-level dosimetry is to perform voxel-level dosimetry. Voxel-level dosimetry methods utilize patient-specific 3D anatomical information acquired from CT or MR scans to model the patient anatomy and 3D SPECT or PET scans to determine the time-integrated activity in each voxel. The absorbed dose is calculated at the voxel level by simulating radiation transport in the patient anatomy (see Chapter 18).

Finally, information regarding the patient-specific absorbed dose distribution in ROIs is often calculated. Absorbed dose statistics can be determined in ROIs in the form of the mean dose, maximum/minimum dose, or any other quantity of interest. If more granularity is desired, then the dose-volume histogram in ROIs can be calculated.

19.2 Overview of Image Registration

Image registration is the determination of one-to-one mapping between the coordinates in one space and those in another, such that points in the two spaces that correspond to the same anatomical or functional point are mapped to each other. Four components are required for image registration:

1. a geometric transformation,
2. a registration metric,
3. an optimizer that is used to find the best match between two images, and
4. image resampling and application of the transformation.

The general procedure for image registration is outlined in Figure 19.2. More details on image registration can be found in the following references (Hill et al. 2001; Oliveira et al. 2013; Brock et al. 2017).

The primary task of image registration is to find an appropriate transformation to best match points on a moving dataset to a stationary dataset. Registration can either be done by *extrinsic methods* or *intrinsic methods*.

- ***Extrinsic methods*** are based on foreign objects that are introduced into the image space.
- ***Intrinsic methods*** are based on information generated by the patient. For intrinsic methods, registration can be based upon a set of salient points (i.e., landmarks), on segmented binary structures (i.e., segmentation based), or directly on measures computed from intensity values in the image (i.e., voxel-based).

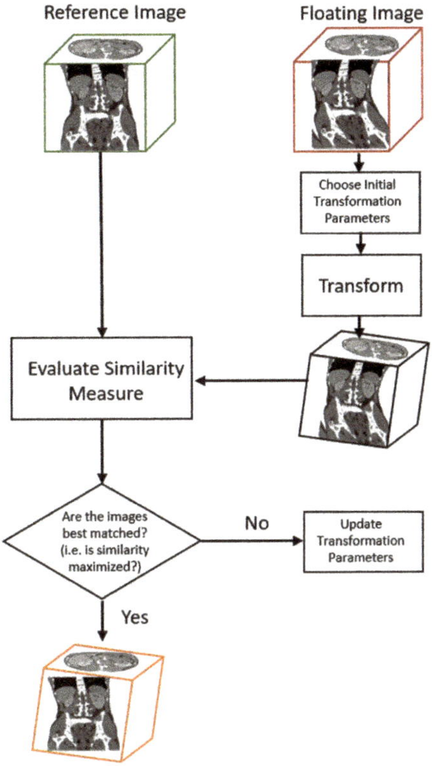

Figure 19.2 General procedure for image registration.

The most common transformations performed in nuclear medicine are based on voxel intensity values. The simplest similarity measures are those that directly compare intensity measures between two images on a voxel-by-voxel basis. Obviously, this approach is only useful for intra-modality registration, where the voxel values are of the same type (e.g., HU versus HU). Inter-modality image registration requires other similarity measures, including principle axes, minimum intensity difference, cross-correlation, minimum variance, or the use of voxel intensity histograms. Table 19.1 summarizes each of these similarity measures.

In general, transformations can be global or local. *Global transformation* implies that the transformation applies to the entire dataset, whereas *local transformation* implies the transformation is only applied to a subset of the image dataset. *Rigid registration* uses the simplest type of transformation. Here the transformation represents a rigid body and consist of three translation and three rotation parameters. *Affine transformations* include transformations from rigid registrations and adds additional transformations, including scaling, shearing, and plane reflection. A more complex transformation is used in *deformable registration*. In this case, the transformation is spatially variant, and the number of degrees of freedom can be as large as three times

Table 19.1 Similarity metrics used for image registration

Similarity Measure	Description	References
Principle Axes	Based on the center of mass and its orientation with respect to its center of mass (i.e. principle axis). Registration is achieved by aligning center of mass and principle axis.	Alpert et al. 1989
Minimum Intensity Difference	Based on calculating the absolute or squared intensity differences between voxels in two images. Commonly, the sum of squared differences is often used.	Papavasileiou et al. 2007
Cross-Correlation	Based on calculating the correlation between voxels in two images.	Flynn et al. 1999; McParland et al. 1995
Minimum Variance	Assumes image consists of mostly homogenous regions (i.e. small variance) that differ substantially from one another (i.e. large variance). If two images are correctly registered a homogeneous region from one region will correctly align with the matching homogeneous region for the other image. The intensity values in each image need not be related.	Ardekani et al. 1995
Intensity Voxel Histogram	Utilize information theoretic approaches. These include, joint entropy, mutual information, or the correlation ratio.	Pluim et al. 2003

the number of voxels in the dataset. Most commonly, the so-called vector fields that represent these transformations are constrained by a regularization function to ensure unrealistic changes are not made to anatomy or physiology. A list of different types of transformations are provided in Table 19.2.

Parameters that make up the registration transformations mentioned above are searched for by finding an optimum of some function defined on the parameter space. The function attempts to measure the similarity as defined by a metric between two different image sets. If the similarity function is well behaved, then several different optimizations techniques can be used. The primary goal of the optimizer is to find search directions intelligently and efficiently through the parameter space, while also searching the entire parameter space to find the most accurate solutions as defined by the similarity metric. Most often optimization algorithms employed for registration tasks use gradient-based methods or stochastic methods. *Gradient-based methods* take repeated steps, typically in the opposite direction of the gradient of the similarity function because this is the direction of the steepest descent (i.e., gradient-descent), to best choose the search directions that tend toward a minimum or maximum value. *Stochastic methods* randomly sample the search space in a confined fashion to find the maximum or minimum values of the similarity function. Common stochastic methods include *simulated annealing* and *genetic* methods.

The final general step in image registration is resampling. *Resampling* is the process of matching the image resolution of the moving image to the stationary while trying to minimize any loss of information during the process. There will not always be an exact match of intensity values of a voxel in the moving image matched to the sta-

Table 19.2 Types of transformations used for image registration

Transformation	Description
Rigid	Permits translation in 3 directions and rotations about 3 axes
Non-rigid	Local and voxel-based deformation often regularized by a smoothing parameter
Affine	Rigid-transformation as well as three scale and three shear parameters
Spline-based methods	Uses splines to define a vector field that maps each voxel in a moving image to a voxel in a stationary image
Viscous/elastic/optical flow	Spatially variant voxel displacement by a vector field governed by intensity gradients
Finite element methods (FEM)	Spatially variant voxel displacement by a vector field governed by biomechanical tissue properties

tionary image. Therefore, *interpolation* between the image values has to be performed. There are several methods for interpolation including *nearest neighbor*, *bilinear interpolation*, and *cubic convolution*. If resampling is not done correctly it can be a significant source of error in the overall registration process.

19.3 Image Fusion and Registration in Nuclear Medicine

Image registration has played and continues to play an important role in nuclear medicine. Historically, there have been three main clinical motivations for registering nuclear medicine data:

1. to combine functional and anatomical images to aid in diagnosis of disease,
2. to aid in the combination of images from serial imaging studies, and
3. to aid in treatment planning of radiotherapy or surgery.

Registration was a critical, yet difficult, task to combine anatomical and functional images prior to the integration of CT or MR scanners with SPET or PET scanners. The most common registration task was fusing SPET or PET data with CT or MR data in the brain, but other non-brain studies were performed as well. Many of the challenges associated inter-modality registrations were eliminated when PET/CT and SPECT/CT scans were introduced into the clinic. Registration was also used to combine serial PET or SPECT images in patient brains over multiple image timepoints in intra-modality, intra-subject studies. In addition, PET and SPECT registered to high-resolution CT continues to assist in the treatment planning process of several tumors in radiotherapy and surgery.

Image registration in the context of patient-specific dosimetry has been applied by several groups. A few methods were used for fusing SPECT and CT data for dosimetry prior to SPECT/CT imaging, including a point-marker-based registration using rigid-body transformation and a mutual information registration, also employing a rigid body transformation. Landmark-based methods (Brown et al. 1993) and

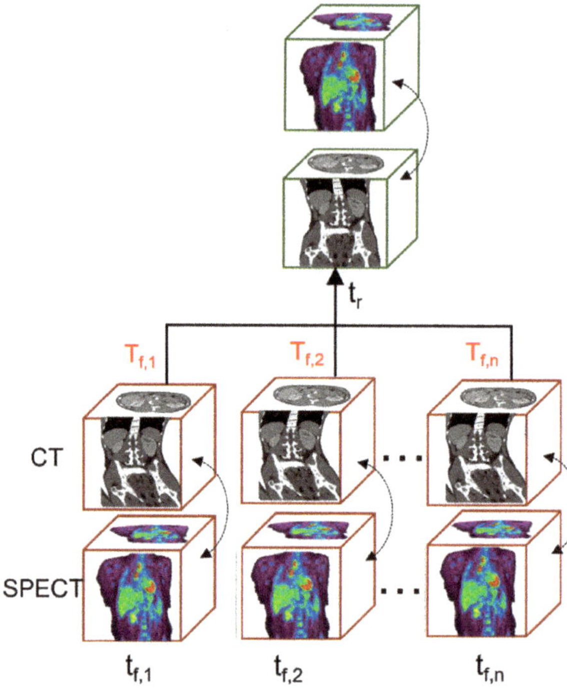

Figure 19.3 Graphic showing CT-CT image registration for RPT dosimetry. Floating CT images (f,1–n) are registered to the reference CT image (r). Rigid transformations, deformable transformations or a combination of both can be used.($T_{f,1-n}$). The transformation results are subsequently applied to the corresponding SPECT data.

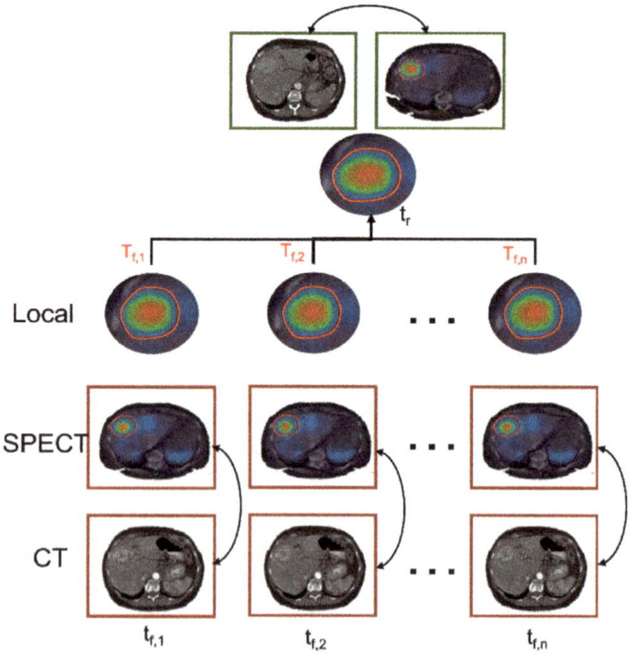

Figure 19.4 Graphic showing CT-CT image registration for RPT dosimetry.

contour-based methods (Li et al. 1995) have also been used with affine transformation for this problem. The latest motivation for registration in nuclear medicine is combining images from serial imaging studies to perform patient-specific dosimetry calculations. Papavasileiou et al. (2001; 2007) developed a method for simultaneous fitting of time-activity curves and registration of serial SPECT scans using affine registrations and voxel-intensity similarity measures. Sjögreen-Gleisner et al. (2009) used rigid transformations and mutual information as the similarity measure to register SPECT/CT scans and serial SPECT scans.

When hybrid SPECT/CT imaging is performed, CT-CT registration is often utilized for patient-specific dosimetry. The advantages of CT-CT registration include superior spatial resolution of CT and avoiding possible registration errors caused by the time-varying activity distribution in the patient. CT-CT registration is especially useful for dosimetry in normal organs. The most common transformation used for serial CT-CT registration is B-spline-based, and mutual information is often used as the similarity metric (Grassi et al. 2018; Jackson et al. 2013).

Either a rigid registration or deformable registration, or a combination of both, is performed to find a reasonable match between the floating CT images and the reference CT image. An illustration demonstrating the CT-CT registration approach is provided in Figure 19.3.

For registration of tumor volumes from multiple timepoint scans, local contour-based SPECT to SPECT registration could offer advantages. Typically, this approach uses tumor contours plus a registration margin and intensity-based similarity metrics to register SPECT scans acquired at different timepoints. A rigid registration is often performed, but deformable registration could also be possible. Registration errors of less than 2% have been reported in tumors treated with ^{177}Lu-DOTATATE using this approach (Dewaraja et al. 2022). An illustration of this approach is provided in Figure 19.4.

19.4 Impact of Registration on Dosimetry

Despite the impact of registration accuracy on RPT dosimetry there is a limited amount of data on the subject in the literature. Papavasileiou et al created three simulated SPECT datasets using patient studies of two ^{131}I-MIBG patients and a ^{186}Re-HEDP patient(5). The group purposely mis-registered these multi-timepoint datasets to evaluate the impact of registration accuracy on the final dosimetry results. A total of 100 mis-registrations were generated using random sampling. A simple voxel S-value approach was used for the absorbed dose calculation assuming energy deposition in water. The group showed that even small mis-registrations (i.e., translation of less than 9 mm and a rotation of less than 5 degrees) might cause differences in the absorbed dose of up to 90% in tumor volumes. These errors became even more significant when the volumes of interest were further away from the center of the SPECT image. It is important to note that only SPECT data was used in this study.

Regarding SPECT/CT datasets a few groups have quantified the impact of performing different types of registration on the dosimetry. Sjögreen-Gleisner et al.

(2009) evaluated the impact of rigid and nonrigid methods on the absorbed dose calculation process using multi-timepoint SPECT/CT data of an ^{111}I-labeled mAb. The registrations were performed using the CT data and modifying the SPECT data accordingly. The CT-based registrations were superior to the SPECT-based approaches when evaluating activity consistency. For these datasets, the differences between using rigid and nonrigid registrations were small, but nonrigid methods ultimately were more robust than rigid methods. In this study the registration process had the largest dosimetric impact on tumor volumes. Similar results were reported by Ao et al. (2015) using simulated SPECT/CT data in the XCAT phantom. Using the phantom data, they reported a substantial reduction in misalignment errors, going from 15.5% to 2.1% in the spleen and from 7.3% to 0.2% in the liver when deformable registration was used instead of rigid registration.

Grassi et al. (2018) evaluated the impact of multi-timepoint SPECT/CT registration in ^{177}Lu-DOTATAC patients. Both a rigid and deformable registration approach were evaluated. The group showed that the rigid registration approach led to absorbed doses in the kidneys that were consistently lower than absorbed doses calculated using the deformable registration approach. The percent difference between the kidney absorbed doses reported ranged from –19% to 4% in a total of 11 patients. Tumor absorbed doses were also consistently lower when rigid registration was used. Similar to earlier results, the choice of registration technique had the greatest impact on the tumor absorbed doses. The percent differences between the tumor absorbed doses ranged from –67.2% to 100.7%.

19.5 Concluding Remarks

There is increasing evidence that supports the use of patient-specific dosimetry to help guide RPT therapies. Obstacles that once hampered dosimetry efforts in the clinic are being removed, and it is not unreasonable to suggest that dosimetry will play a pivotal role in the RPT workflow in the next few years. Given the likely clinical impact of dosimetry in the near future, it is important to attempt to reduce uncertainties in the dosimetry process to the greatest extent possible. Previous investigations have indicated that image registration has a significant impact on the accuracy of absorbed dose calculations. Based on a limited amount of data, CT-CT rigid registration followed by deformable registration is likely the most robust registration approach for absorbed dose calculations in normal organs. Local-regional rigid or deformable registration of SPECT data is likely the most robust registration approach for tumor d

osimetry. However, it is important to keep in mind that these general conclusions are based upon limited data generated for only a few types of RPTs. Different registration approaches might be warranted for different types of therapies.

There is a clear need for more investigation into the impact of registration on the dosimetry process, and perhaps novel approaches pertaining to registration of multi-timepoint and multi-modal images. One obvious approach could be to integrate machine learning and artificial intelligence in the registration process (Brosch-Lenz et al. 2021). Several studies outside of the RPT domain have shown improvements in image registration errors when AI techniques are used over conventional techniques. Similar improvements would be expected for registration tasks in RPT. Furthermore, AI-based registration can help reduce image artifacts that sometimes appear when conventional approaches are used.

The present excitement pertaining to RPT is palpable. RPT drugs are being introduced into the clinic at an unprecedented rate. Centers devoted to RPT are springing up across the United States. There are well over 200 clinical trials in the United States alone that involve RPT agents. Fueling the RPT renaissance is the significant amount of money being invested by companies in the development and translation of novel RPT agents. It is becoming more transparent that patient-specific dosimetry improves clinical RPT outcomes (Wahl et al. 2003, 2021; Sandström et al. 2020; Ilan et al. 2015; Del Prete et al. 2019; Sundlöv et al. 2022). Therefore, patient-specific dosimetry that relies on robust image registration will become an integral part of patient planning and management in the near future.

References

Alpert, N. M., J. F. Bradshaw, D. Kennedy, and J. A. Correia (1989). "The Principal Axes Transformation—A Method for Image Registration." *J. Nucl. Med.* 31(10):1717–22.

Ao, E. C. I., N. Y. Wu, S. J. Wang, N. Song, and G. S. P. Mok (2015). "Improved dosimetry for targeted radionuclide therapy using nonrigid registration on sequential SPECT images." *Med Phys*. 42:1060–70.

Ardekani, B. A., M. Braun, B. F. Hutton, I. Kanno, and H. Iida (1995). "A fully automatic multimodality image registration algorithm." *J. Comput. Assist. Tomogr.* 19:615–23.

Besemer, A. E., B. Titz, J. J. Grudzinski, et al. (2017). "Impact of PET and MRI threshold-based tumor volume segmentation on patient-specific targeted radionuclide therapy dosimetry using CLR1404." *Phys. Med. Biol.* 62:6008–25.

Brock, K. K., S. Mutic, T. R. McNutt, H. Li, and M. L. Kessler (2017). "Use of image registration and fusion algorithms and techniques in radiotherapy: Report of the AAPM Radiation Therapy Committee Task Group No. 132." *Med. Phys.* 44:e43–76.

Brosch-Lenz, J., F. Yousefirizi, K. Zukotynski, J.-M. Beauregard, V. Gaudet, B. Saboury, A. Rahmim, and C. Uribe (2021). "Role of Artificial Intelligence in Theranostics: Toward Routine Personalized Radiopharmaceutical Therapies." *PET Clin.* 16:627–41.

Brown, L. G., G. Q. Maguire, Jr., and M. E. Noz (1993). "Landmark-based 3D fusion of SPECT and CT images." *SPIE International Society for Optical Engineering* 2059:166–74. https://doi.org/101117/12150244.

Del Prete, M., F.-A. Buteau, F. Arsenault, N. Saighi, L.-O. Bouchard, A. Beaulieu, and J.-M. Beauregard (2019). "Personalized ^{177}Lu-octreotate peptide receptor radionuclide therapy of neuroendocrine tumours: initial results from the P-PRRT trial." *Eur. J. Nucl. Med. Mol. Imaging* 46:728–42.

Flynn, A. A., A. J. Green, G. Boxer, R. B. Pedley, and R. H. J. Begent (1999). "A comparison of image registration techniques for the correlation of radiolabelled antibody distribution with tumour morphology." *Phys. Med. Biol.* 44(7):N151–59. doi: 10.1088/0031-9155/44/7/402.

Fowler, J. F. (2010). "21 years of biologically effective dose." *Br. J. Radiol.* 83:554–68.

Frey, K. A., K. K. Wong, S. J. Wilderman, and A. S. Nelson (2022). "A pipeline for automated voxel dosimetry: application in patients with multi-SPECT/CT imaging following ^{177}Lu peptide receptor radionuclide therapy." *J. Nucl. Med.* 63(11):1665–72.

Grassi, E., F. Fioroni, S. Berenato, N. Patterson, V. Ferri, L. Braglia, A. Filice, A.Versari, M. Iori, and E. Spezi (2018). "Effect of image registration on 3D absorbed dose calculations in ^{177}Lu-DOTATOC peptide receptor radionuclide therapy." *Phys. Med.* 45:177–85.

Hill, D. L. G., P. G. Batchelor, M. Holden, and D. J. Hawkes (2001). "Medical image registration." *Phys. Med. Biol.* 46(3):R1–45.
Hobbs, R. F. and G. Sgouros (2009). "Calculation of the biological effective dose for piecewise defined dose-rate fits." *Med. Phys.* 36:904–07.
ICRU (2021). ICRU Report 96. "Dosimetry-Guided Radiopharmaceutical Therapy." *J. ICRU* 21(1):1–212. https://www.icru.org/report/icru-report-96-dosimetry-guided-radiopharmaceutical-therapy/.
Ilan, E., M. Sandström, C. Wassberg, A. Sundin, U. Garske-Román, B. Eriksson, D. Granberg, and M. Lubberink (2015). "Dose response of pancreatic neuroendocrine tumors treated with peptide receptor radionuclide therapy using 177Lu-DOTATATE." *J. Nucl. Med.* 56(2):177–82.
Jackson, P. A., J. M. Beauregard, M. S. Hofman, T. Kron, A. Hogg, and R. J. Hicks (2013). "An automated voxelized dosimetry tool for radionuclide therapy based on serial quantitative SPECT/CT imaging." *Med. Phys.* 40:112503.
Li, H., B. S. Manjunath, and S. K. Mitra (1995). "A Contour-Based Approach to Multisensor Image Registration." *IEEE Trans. Image. Process.* 4:320–34.
McParland, B. J. and J. C. Kumaradas (1995). "Digital portal image registration by sequential anatomical matchpoint and image correlations for real-time continuous field alignment verification." *Med Phys.* 22:1063–75.
Papavasileiou, P., G. D. Flux, M. A. Flower, and M. J. Guy (2001). "An automated technique for SPECT marker-based image registration in radionuclide therapy." *Phys. Med. Biol.* 46:2085–97.
Papavasileiou, P., A. Divoli, K. Hatziioannou, and G. D. Flux (2007). "The importance of the accuracy of image registration of SPECT images for 3D targeted radionuclide therapy dosimetry." *Phys. Med. Biol.* 52(24):N539–48.
Pluim, J. P. W., J. B. A. Maintz, and M. A. Viergever (2003). "Mutual-Information-Based Registration of Medical Images: A Survey." *IEEE Trans. Med. Imaging* 22(8):986–1004.
Sandström, M., N. Freedman, K. Fröss-Baron, T. Kahn, and A. Sundin (2020). "Kidney dosimetry in 777 patients during ^{177}Lu-DOTATATE therapy: aspects on extrapolations and measurement time points." *EJNMMI Phys.* 7(1):73.
Sgouros, G., J. C. Roeske, M. R. McDevitt, et al. (2010). "MIRD pamphlet No. 22 (Abridged): Radiobiology and dosimetry of α-particle emitters for targeted radionuclide therapy." *J. Nucl. Med.* 51:311–28.
Sgouros, G., L. Bodei, M. R. McDevitt, and J. R. Nedrow (2020). "Radiopharmaceutical therapy in cancer: clinical advances and challenges. *Nat. Rev. Drug Discov.* 19(9):589–608.
Sjögreen-Gleisner, K., D. Rueckert, and M. Ljungberg (2009). "Registration of serial SPECT/CT images for three-dimensional dosimetry in radionuclide therapy." *Phys. Med. Biol.* 54:6181.
St. James, S., B. Bednarz, S. Benedict S, et al. (2021). "Current Status of Radiopharmaceutical Therapy." *Int. J. Radiat. Oncol. Biol. Phys.* 109:891–901.
Sundlöv, A., K. S. Gleisner, J. Tennvall, M. Ljungberg, C. F. Warfvinge, K. Holgersson, A. Hallqvist, and P. Bernhardt (2022). "Phase II trial demonstrates the efficacy and safety of individualized, dosimetry-based ^{177}Lu-DOTATATE treatment of NET patients." *Eur. J. Nucl. Med. Mol. Imaging* 49(11):3830–40.
Uribe, C. F., P. L. Esquinas, J. Tanguay, M. Gonzalez, E. Gaudin, J.-M. Beauregard, and A. Celler (2017). "Accuracy of ^{177}Lu activity quantification in SPECT imaging: a phantom study." *EJNMMI Phys.* 4(1):2.
Wahl, R. L. (2003). "The clinical importance of dosimetry in radioimmunotherapy with tositumomab and iodine I 131 tositumomab." *Semin Oncol.* 30(2 Suppl 4):31–38.
Wahl, R. L., E. C. Frey, H. A. Jacene, B. S. Kahl, S. Piantadosi, J. A. Bianco, R. J. Hammes, M. Jung, W. Kasecamp, H. Bin, G. Sgouros, I. W. Flinn, and L. J. Swinnen (2021). "Prospective SPECT-CT Organ Dosimetry-Driven Radiation-Absorbed Dose Escalation Using the In-111 (^{111}In)/Yttrium 90 (^{90}Y) Ibritumomab Tiuxetan (Zevalin®) Theranostic Pair in Patients with Lymphoma at Myeloablative Dose Levels." *Cancers (Basel)* 13(11):2828.

Chapter 20

Uncertainties and Relative Merits of Voxelized vs. Organ-Level Dosimetry

Jonathan I. Gear, Ph.D.

Joint Department of Physics
Institute of Cancer Research and
Royal Marsden NHS Foundation Trust
United Kingdom

20.1 Introduction

20.1.1 Recap of Organ-Level Dosimetry

The methodology for organ-level dosimetry has been extensively covered in Chapter 13, and the fundamental concepts and different sources of uncertainty were presented in Chapter 16. To summarize the main aspects, organ-level dosimetry generally requires the operator to identify and outline the target volume at the beginning of the dosimetry process. In most cases, this is achieved using some anatomical scan data, such as CT or MRI. For radiopharmaceutical therapy, the activity contained within the target organ is the primary source of irradiation and, therefore, cross-dose from one organ to another is generally ignored to simplify the calculation process.

Having identified the target/source volume, the activity in the volume is determined. For a non-quantified SPECT scan this requires measuring the total counts with the *volume of interest* (VOI) and multiplying by some predetermined sensitivity factor. If the SPECT scan has been internally quantified within the reconstruction software, then voxel units are usually given in Bq/ml. Therefore, total activity within the VOI is the average voxel concentration multiplied by the VOI volume.

Provided similar values for system sensitivity are used, these two methods will produce identical results. An important secondary step at this stage is to account for partial volume effects, whereby events that originated within the source appear outside the VOI on the SPECT image. A recovery correction is then applied based on the "spill-out" from similarly sized phantom measurements. An alternative approach is to use a secondary VOI larger than the initial anatomical VOI which is believed to encompass all corresponding events. The difficulty here is that it is often unclear how large to make the secondary VOI or at what point neighboring sources can "spill-in."

Whichever approach to quantification is used, it is often repeated at multiple time points, and the measured activity is plotted in the form of a time-activity curve. A function can then be fitted to this data and integrated to define the time-integrated activity. *S values* are generally determined based on the preestablished reference models. These can either be taken from organ models or, as is of often the need for lesions, unit density sphere models. The S value should be chosen to match the mass of the identified target, and it is recommended to scale the reference S value according to the ratio of reference and patient organ mass.

20.1.2 Recap of the Voxel Method

The methodology for voxel-level dosimetry is ultimately very similar to that at the organ level, just at a smaller scale. The operator does not necessarily need to identify the target at the initial stage, and the methodology allows the absorbed dose to be calculated to multiple targets at the same time. More correct is to say that each voxel in the SPECT image is defined as a source and target. The necessity to include cross-dose from one voxel to another is arguably higher than when working at the organ level (depending on particle range), and the choice of how this is performed is decided later. As with the organ level, quantification can be performed during or after reconstruction, although it is necessary to generate a quantified image where voxels are in units of activity or activity concentration. As already stated, this is standard for quantification performed during reconstruction. Post-reconstruction quantification will require scaler multiplication at the voxel level using an appropriate sensitivity factor.

Once the sequential SPECT images have been quantified, it is necessary to register them to ensure the respective voxels are spatially aligned (see Chapter 19). Once a set of registered and quantified SPECT images have been obtained, the methodology can take one of two routes:

- A time-activity image can be generated by integrating the activity images at the voxel level, or
- The activity images can be converted to absorbed dose rate images at a dose calculation step prior to integration.

The two approached ultimately have the same result. As we will be making comparisons to the organ-level calculation in this instance, we will concentrate on the former approach and consider the generation of the *time-integrated activity* (TIA) image.

Table 20.1 A comparison of organ and level dosimetry

	Organ Level	Voxel Level
Image acquisition	Planar, SPECT or Hybrid	SPECT only
Quantification	During reconstruction at the voxel level or post reconstruction at the organ level.	At the voxel level. Generally available in most image processing software
Partial Volume correction	Correction of the image data is and post reconstruction using recovery data or oversized VOIs. Performed outside of the image processing software	Corrected on the image data, either during reconstruction or using some post-image processing. Generally available in most image processing software
Registration	Not required but VOIs on sequential images must be consistent.	At the voxel level. Generally available in most image processing software
TAC fitting	Generally performed outside of the image processing software.	At the voxel level. Only available with dedicated dosimetry software,
TAC integration	Generally performed outside of the image processing software.	At the voxel level. Only available with dedicated dosimetry software,
Absorbed dose calculation	Using mass scaled S values, only considering self-dose to lesions.	Convolution or Monte Carlo simulations taking into account cross dose
Absorbed dose reporting	Can only report mean absorbed dose to VOI	Dose reported at the voxel level allows other absorbed dose metrics to be reported.

Conversion of time-integrated activity to absorbed dose is performed using one of a number of options. Either convolution with a dose point kernel, which is essentially an array of voxel S values, or using a Monte Carlo simulation with the TIA array and CT data as input. The merits and details of each technique are beyond the scope of this chapter, but they are discussed in detail in Chapter 18. The key point here is that the final output (whichever method is used) is that of an absorbed dose map that an operator can interrogate to report the necessary data regarding any desired target or organ.

20.1.3 Fundamental Differences

As described briefly above, there are a number of key differences between organ- and voxel-level dosimetry, with each of the two methods having unique advantages and disadvantages. In addition, there are a number of sources of uncertainty that can be identified in each step in the dosimetry processes. In the following section we will discuss many of the aspects and consider the merits of the two techniques. Table 20.1 summarizes some of the fundamental differences that will be discussed in the latter sections with consideration of the associated uncertainties.

20.2 Quantification

When considering the process of quantification, the application of a system sensitivity factor is largely similar between the two dosimetry methodologies. No matter at which point the scaling from counts to activity occurs—or if it happens at a voxel or VOI level—it is simply a scalar multiplication and, as such, the fractional uncertainty in the sensitivity factor is propagated into that of activity. Furthermore, this can be considered as a systematic uncertainty that is consistent at all time points and, therefore, can be propagated directly into the uncertainty in absorbed dose. The consequence of this is that if there is, say, a 5% uncertainty in the defined sensitivity factor, then there is a minimum 5% uncertainty in the reported absorbed dose, whether that absorbed dose is calculated at the organ or voxel level.

20.2.1 Image Noise

While the effect of calibration uncertainties may be consistent between organ or voxel dosimetry, there are some other aspects that can have a different impact. Of consequence is the presence of noise in the SPECT data. When working at the organ level, the total counts across all voxels in the VOI are usually sufficiently high that random errors due to image noise are largely inconsequential. Where images are affected by noise, the causality will propagate into random errors in the TAC data and, as discussed in Chapter 16, are encompassed within the uncertainties estimated within the TAC fit parameters. A good example of this is that observed in an organ such as the liver, as shown in Figure 20.1 for a ^{177}Lu DOTATATE patient. An initial observation would indicate uniform uptake of tracer in the liver, and without evidence of disease, there is no indication to believe otherwise.

Two reconstructions have been performed in this instance: one is a standard OSEM algorithm without any noise suppression or image filtering, and the other uses a Bayesian smoothing prior to help reduce the effect of noise. Even with noise suppression, there is variation in image intensity at the voxel level (range 290 to 390 counts per voxel in the indicated VOI) that would result in a 30% difference in absorbed dose. The effect is even more exaggerated in the OSEM reconstruction (count range 4 to 1400), where voxel doses in the same organ could vary by a factor of over 300. While one could argue that the OSEM reconstruction is potentially unrealis-

Figure 20.1 ^{177}Lu DOTATATE data fused with CT (left), reconstructed with a Bayesian smoothing prior (middle), and with a non-filtered OSEM algorithm (right).

tic and would not likely be used for an absorbed dose calculation, the important point to recognize is that when calculating based on an organ VOI, the effect of noise is less pronounced. Indeed, the activity measured based on the total counts in the indicated VOIs vary by less than 1% between the OSEM and Bayesian reconstruction.

20.2.2 Partial Volume Effects

A downside to suppressing noise—particularly if achieved through the use of post-reconstruction filtering—is that it smooths the image data and further reduces the resolution of the already limited SPECT image. When performing dosimetry at the VOI level, the impact of this can be negated using a *recovery factor*. With such a technique, even with highly filtered and smooth data, the total counts should still be resolvable, provided the recovery phantoms are representative of the structure under investigation. In addition, uncertainty estimates can be calculated if using a *recovery curve* (see Chapter 16). This recovery correction is, for obvious reasons, not a viable option if reporting doses at the voxel level, and one must instead rely on improving the resolution of the image data. However, these algorithms are never perfect and are prone to additional sources of error.

We can examine the impact of resolution and partial volume effects by considering the case of small lesions or organs. Figure 20.2 indicates a small 18-mm lung lesion of a patient being treated with ^{177}Lu-DOTATATE. Figures 20.2(b)–(d) are corresponding SPECT images reconstructed using different algorithms. It is clear that the

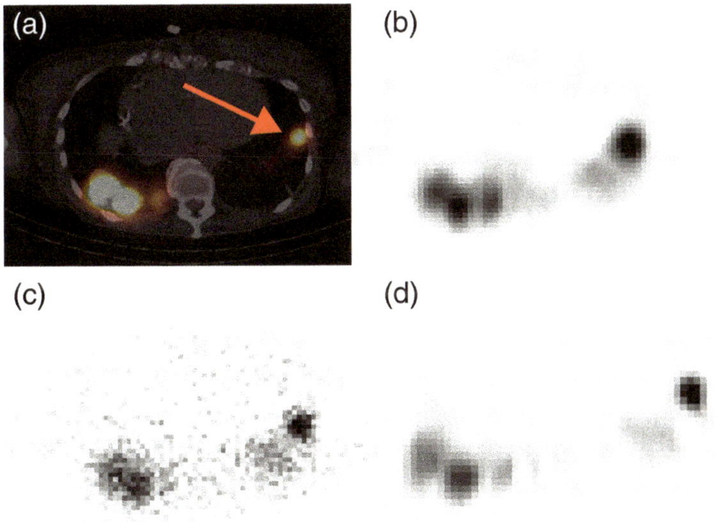

Figure 20.2 ^{177}Lu DOTATATE data (a) fused with CT, (b) reconstructed with a Bayesian smoothing prior, (c) with a non-filtered OSEM algorithm, and (d) with an OSEM-PSF algorithm employing collimator detector response modeling.

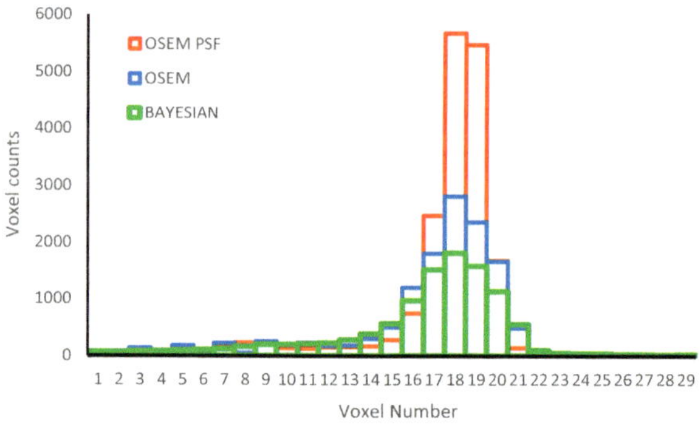

Figure 20.3 Count profiles through a lesion when reconstructed using different algorithms.

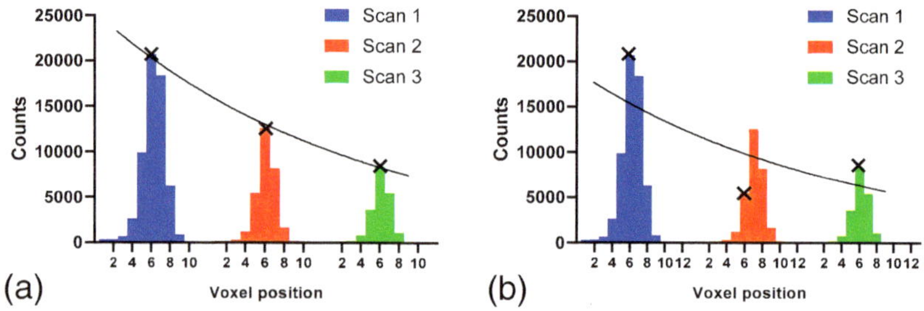

Figure 20.4 Count profiles through a lesion at different time points: (a) when spatially aligned and (b) when misaligned.

count distribution varies between the three SPECT images. The absorbed dose distribution will, therefore, also be similarly different.

The extent of this variation can be further examined by looking at the profiles of voxel counts through the lesion (Figure 20.3). The lesion in question is equivalent to four SPECT voxels in diameter and, therefore, a dose map is going to attribute much of this activity outside of the volume. For this example, the absorbed dose to the lesion is also observed to vary by a factor of 2 depending on how the image is reconstructed.

20.3 Time-Integrated Activity

20.3.1 Voxel Registration

When fitting a time activity curve at the voxel level, it is necessary to ensure the appropriate voxels are correctly assigned for each time point. This is particularly important when the site of interest is small or if there is a high activity gradient. Consider the lesion from the previous example. A profile through the lesion at each time point is shown in Figure 20.4. The count data corresponding to voxel 6 (marked with a cross) demonstrates a clear decrease at the different time points. However, if a slight misalignment is present (in this case by a single voxel at scan 2) the data now follows a very different trend. Indeed, the data for the second time point is now less than that at the third.

20.3.2 TAC Fitting vs Trapezoid

The effect of misalignment will, in most cases, result in an error in the estimate of time-integrated activity. The extent of this error will, in turn, depend on the method used to fit and integrate the TAC data.

Similar to the organ level, if fitting a function to the TAC data, estimates of uncertainty in the fit parameters can be made based on how well the chosen function fits to the data. For the two cases above, exponential fits to the TACs are shown in Figure 20.5 where uncertainty bounds are indicated by dotted lines. Estimates of time-integrated activity vary by 60% in these two scenarios, with associated uncertainties increasing from 7% to 39%.

These estimates of uncertainly are particularly important at the voxel level, as it is unlikely that the TAC for each voxel in the TIA matrix can be visually interrogated to determine the appropriateness of the fit. However, if the software allows an uncertainty in the fit parameters, the TIA estimate can be visualized in the form of an uncertainty map.

There is scope—particularly in areas of low count density and high noise—for a fitted function to result in a positive slope. This needs to be accounted for in the software to avoid the calculation of an infinitely large absorbed dose. Software may avoid this situation by setting such cell values to zero or performing an integration method that will not assume an ever-increasing activity. One method used frequently when fitting at a voxel level, particularly due to its quick computation time, is the *trapezoid approach*. This method performs linear interpolation between data points, representing the area under each section of the curve as a trapeze. This method works ade-

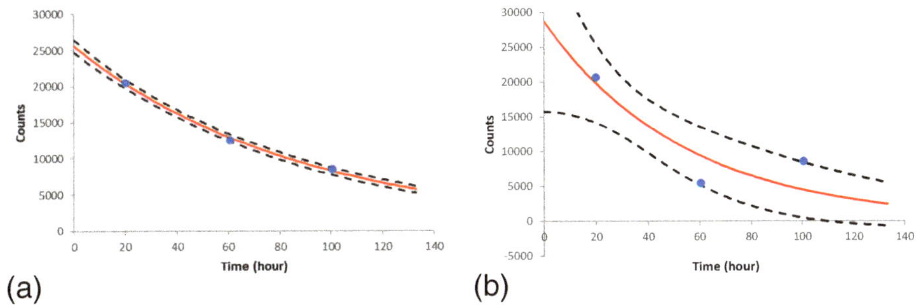

Figure 20.5 Single voxel time activity curve (a) when spatially aligned and (b) when misaligned by a single voxel.

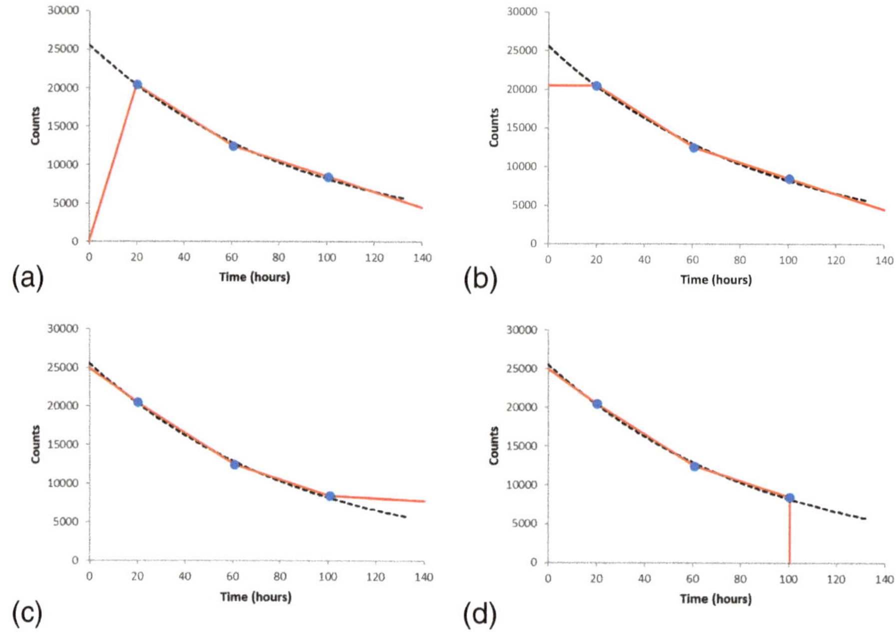

Figure 20.6 Examples of different approaches to extrapolation when using the trapezoid method and the consequence on the estimate of time-integrated activity.

quately for well-sampled TAC data, but it can introduce large errors when performed with only a few data points. There is the added difficulty in deciding how to extrapolate the TAC after the last data point and before the first. Some software may interpolate from the period between the last two data points, while others opt to assume physical decay, and some may even ignore this period completely.

Figure 20.6 demonstrates how the estimation of TIA can vary depending on which of these approaches is adopted. In Figure 20.6(a), data is extrapolated from the origin to the first point, and last two points used to extrapolate beyond 100 hours post administration. This method results in a TIA 73% of that when fitting an exponential function. In Figure 20.6(b), activity at time t = 0 is set equal to that at eth first measurement point, resulting in a TIA 82% that of the fitted function. In Figure 20.6(c) data is extrapolated back to t = 0 using the first two data points, but physical decay is assumed after the last, resulting in a TIA 155% that of the fitted function. Finally in Figure 6(d), TIA is not determined after the last point, resulting in a value 68% of the fitting method. These data are calculated for a relatively well-fitted TAC, and it is not difficult to see how such variation will further exacerbate in scenarios such as that presented in Figure 20.5(b).

20.3.3 Covariant Voxels

The concept of *covariance* and the need for recognizing when parameters have some inter-dependencies was introduced in Chapter 16. This concept holds true for voxel dosimetry, but in some respect has a greater and more complex relationship.

Traditionally, time-integrated activity in each voxel is calculated independently and treated as separate sources of activity. However, we have previously discussed that due to the resolution of imaging systems, activity contained in one part of the body can appear in the image at a location it did not originate. Consider two sources of activity with different effective half-lives. If a proportion of the activity from source A is observed to "spill in" to source B, the observed effective half-life measured in B will be a combination of the biokinetic of the two sources. At the voxel level there is then a situation where the observed TAC is a combination of multiple biokinetic functions of the surrounding voxels. This situation becomes far too complex to model, particularly given the limited available temporal data. However, the concept can be used to recognize where neighboring voxels are likely to share similar kinetics—such as voxels located in the same organ or tissue. The methodology employed here is to identify volumes within the image that likely share the same kinetics. A TAC fit is performed using the data from all the voxels in the VOI, and the respective decay constant(s) derived. Some semblance of independence is kept by then fitting the same function to the individual voxels, but fixing the decay values.

Putting this method into practice with the example above, our TAC fits for the two scenarios now look like that of Figure 20.7. It can be seen that for the first example with its well-registered voxels, the fit is now slightly poorer as it is being influenced by the surrounding pixels. However, the example that was misaligned is much more plausible given what we observed in our ideal case. The obvious disadvantage of this approach is the need and reliance on an operator-derived VOI to ascertain which voxels will share an effective decay. Its accuracy also relies on this assumption being true.

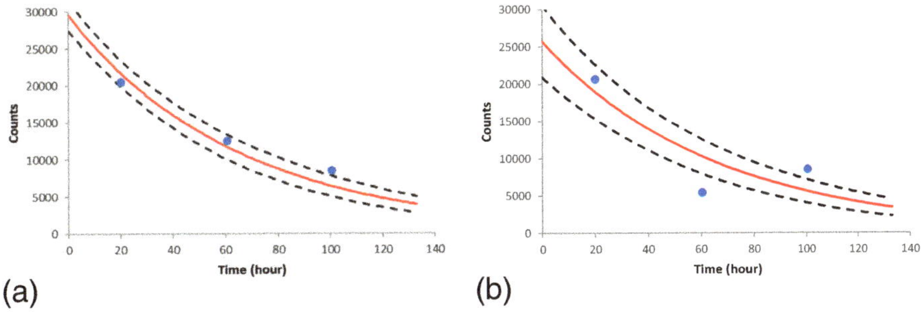

Figure 20.7 Single-voxel time activity curves, fitting using (a) a VOI-defined half-life when spatially aligned and (b) when misaligned by a single voxel.

20.4 Reporting Absorbed Dose

When calculating an absorbed dose, the operator should give consideration to how that absorbed dose is going to be reported. An organ-level dose calculation will be limited by the number of source organs the operator is willing and able to outline, and generally only a few dosimetric parameters can be reported. What is more, as outlining is the first step of the calculation, this decision needs to be made at an early stage. The output from a voxel dose calculation is a dose map and, therefore, this definition of a VOI is not necessary. However, as with any imaging modality, the examination does not complete at the creation of the image. At some point that image needs to be interpreted and the clinical relevance reported. This will inevitably require the interrogation of dose values at particular sites within the body and, therefore, requires some definition of what and where that site is.

With organ-level dosimetry, we have no knowledge of the nonuniformity of the dose distribution within the organ. Therefore, mean absorbed dose is the most commonly reported parameter. At the voxel level, nonuniformity can be reported in the form of a dose volume histogram or equivalent uniform dose (O'Donoghue et al. 1999). The dose volume histogram (and associated parameters such as D_{50}) are common in radiotherapy, but they are not widely expressed for radiopharmaceuticals, meaning clinical relevance of such data is not yet clear. Misinterpretation of DVH data has also raised concerns. One example is the reporting of the area under the DVH, without recognizing this is mathematically equivalent to mean dose (Walrand et al. 2019).

Where clinical relevance of dosimetry data is most likely to have an impact is when direct comparisons to similar data can be made. This comparison may be between different sites of the body, between treatment cycles, or against a population distribution. However, such comparisons still require the dosimetry data to be condensed into a single parameter and, as such, mean absorbed dose is still likely to be the most important output from any dose calculation—whether it be performed at the organ or voxel level.

20.5 Example: ^{177}Lu DOTATATE

An example to demonstrate the implementation of the two dosimetry approaches is given in this section. For the organ-level calculation, this is complemented with a formal analysis of uncertainty, as described in Chapter 16. As there is no formal method for uncertainty analysis for voxel dosimetry, sources of error are demonstrated and comparisons made to the organ-level doses. The example given is that of a 79-year-old female patient with a history of metastatic neuroendocrine disease of unknown origin treated with 7462 MBq of ^{177}Lu DOTATATE. Pretherapy ^{68}Ga DOTATATE PET-CT showed moderate to intense activity at most sites of disease, with minimal uptake on ^{18}F FDG PET. The patient also presented with borderline impaired renal function and mild anemia, so a comparison of lesion and kidney doses was of interest. For the example, two sites of disease and the kidneys will be the focus of the calculations. The first lesion is an 18-mm mass in the lungs (already shown in Figure 20.2) and a larger 50-mm mass in the left gluteal muscle.

Table 20.2 Reconstruction parameters of the SPECT data used for the organ and voxel absorbed dose calculation

	Algorithm	Attenuation Correction	Scatter Correction	Partial Volume Correction	Iterations	Subsets
Organ Level	Bayesian	CT	Monte Carlo	None	8	4
Voxel Level	OSEM	CT	Monte Carlo	PSF modeled	20	6

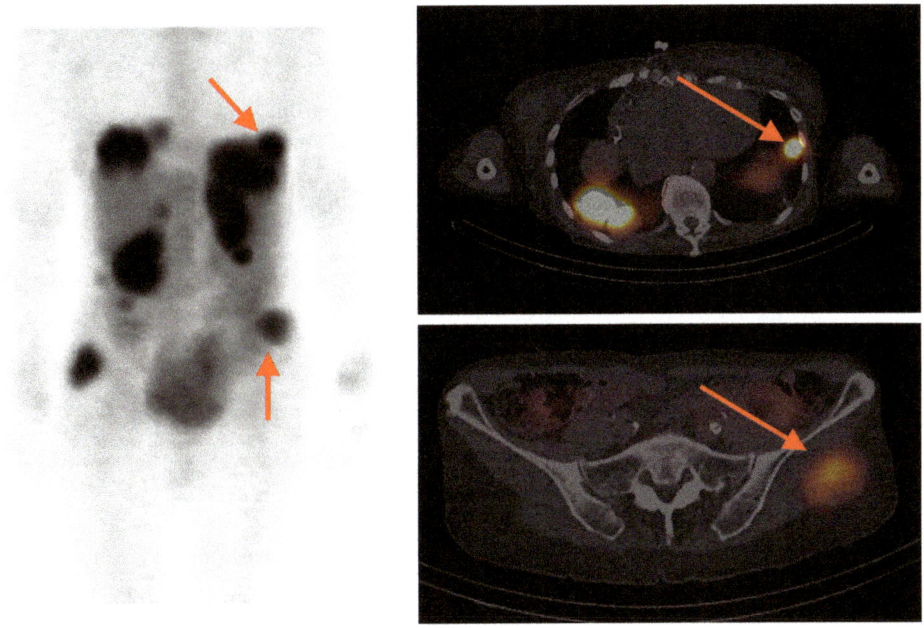

Figure 20.8 Maximum intensity projection and transaxial images of Lu-177 scan indicating lesions that will be the focus of the absorbed dose calculations.

Image Acquisition

Absorbed doses for the lesions were calculated using sequential SPECT acquisitions performed at 13.1 h, 94.3 h, and 118.5 h post administration. Sixty-four projections were acquired in a 128 matrix for 40 s per view, each SPECT covered 2 fields of view using both the 113-keV and 208-keV photopeaks. Two different reconstructions were applied for the organ and voxel dosimetry, as summarized in Table 20.2.

The reconstructed SPECT voxel size was 4.67 mm. ^{177}Lu-DOTATATE SPECT/CT images are shown in Figure 20.8 with the lesions of interest identified.

20.5.1 Approach #1: Organ Dosimetry

Volume Outlining

Volumes of interest of the two lesions and kidneys were defined manually using the low-dose CT acquired with the SPECT data. As the VOIs were drawn directly on the CT data, contributions of voxelization and spatial resolution on the estimate of uncertainty were considered negligible compared to the random component of uncertainty attributed to intra-operator variability. Instead, three operators were each asked to outline the volumes, and the average volume was used for the computation of absorbed dose. Uncertainty was defined as the standard deviation of these values divided by root 3. Volumes and uncertainties are summarized in Table 20.3.

Calibration

Calibration of the system was performed by imaging a uniform cylindrical phantom filled with 0.034 MBq/ml of ^{177}Lu. The phantom was imaged with the same acquisition and reconstruction parameters as for the patient scans. A spherical VOI with a diameter of 10 cm was placed centrally within the phantom, ensuring uniform count distribution and unaffected by partial volume effects.

The fractional standard uncertainty in activity was taken as 10%, the typical uncertainty attributed to field dose calibrators. Given the high number of counts measured within the VOI, statistical uncertainties on this value are considered negligible: Q = 16.8 cps/MBq, with a standard uncertainty of 1.6 cps/MBq.

Recovery Coefficient

Spherical inserts used for the recovery data ranged from 1.0 to 6.5 cm in diameter filled with 0.23 MBq/ml of ^{177}Lu. Image acquisition and reconstruction were similar to that above, excerpt in this case the sensitivity value was used during reconstruction to convert voxel values to units of MB/ml. VOIs on the image were generated by selecting the appropriate percentage threshold to give a volume that matched the known physical volume of the insert. A recovery coefficient for each insert was then calculated as the mean voxel value over the actual insert concentration. The generated recovery curve is given in Figure 20.9. The empirical function fitted to the example data takes the form of a two-parameter function, with respect to volume v,

$$R(v) = 1 - \frac{1}{1 + \left(\frac{v}{p_1}\right)^{p_2}}. \tag{20.1}$$

Fit parameters of the curve with associated uncertainties are given in Table 20.4. The covariance between the parameters was calculated to be –0.0008.

Recovery coefficients for the specified volume of Table 20.5 with associated uncertainty VOIs are summarized in Table 20.3.

Table 20.3 Volume and associated standard uncertainty for lesions and kidneys

Volume	Volume (cm^3)	Standard Uncertainty (cm^3)	Standard Uncertainty (%)
Lung lesion	7.17	0.55	7.7
Gluteal lesion	44.0	1.1	2.5
Left kidney	101.4	6.1	6.0
Right kidney	90.14	5.4	6.0

Table 20.4 Recovery curve fit parameters and associated standard uncertainties

Parameter	Value	Standard Uncertainty	Fractional Standard Uncertainty (%)
p_1	6.83	0.49	7.1
p_2	0.78	0.04	5.4

Table 20.5 Recovery coefficients and associated standard uncertainties for lesions and kidneys

	R	$u(R)$	$u(R)\,/\,R$
Lung lesion	0.51	0.02	4.00%
Gluteal lesion	0.81	0.02	1.95%
Left kidney	0.89	0.01	1.42%
Right Kidney	0.88	0.01	149%

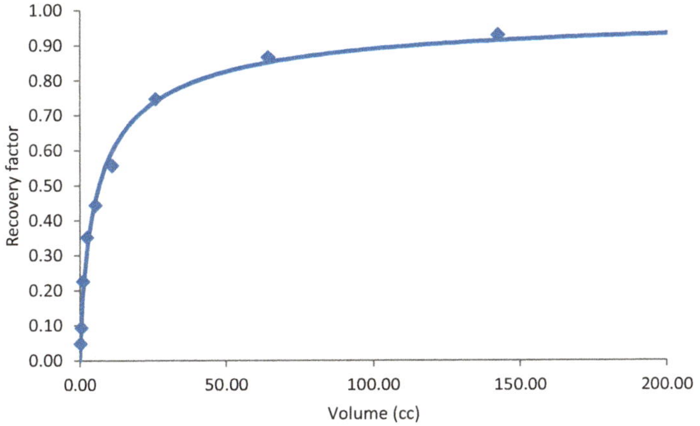

Figure 20.9 Recovery curve used to correct for partial volume effects in organ-level absorbed dose calculation.

Table 20.6 Count rate data and associated standard uncertainties for lesions and kidneys at each time point

Volume	Scan	Count Rate (cps)	Standard Uncertainty (cps)	Standard Uncertainty (%)
Lung lesion	1	103	6.2	6.0
	2	44	2.5	5.7
	3	38	4.2	11.0
Gluteal lesion	1	226	4.6	2.0
	2	100	2.1	2.1
	3	79	1.5	1.9
Left kidney	1	1107	23.7	1.9
	2	338	8.1	2.1
	3	230	5.6	2.4
Right kidney	1	1281	30.8	2.4
	2	362	8.3	2.3
	3	251	6.0	2.4

Table 20.7 ^{177}Lu activities and associated fractional standard uncertainties.

VOI	Scan 1 (MBq)	Scan 2 (MBq)	Scan 3 (MBq)	Fractional Std Uncertainty (%)
Lung lesion	12.00	5.13	4.42	11.0
Gluteal lesion	16.65	7.42	5.80	10.3
Left kidney	74.08	22.58	15.36	10.8
Right Kidney	86.6	24.5	16.9	10.8

Table 20.8 TAC parameters and associated standard uncertainties

	Parameter	Fitted Value	Uncertainty	Fractional Uncertainty
Lung lesion	A_0 (MBq)	15.3	0.75	4.9%
	λ (h^{-1}	11×10^{-3}	0.9×10^{-3}	8.2%
Gluteal lesion	A_0 (MBq)	21.4	0.21	1.0%
	λ (h^{-1})	11×10^{-3}	0.2×10^{-3}	1.8%
Left kidney	A_0 (MBq)	108.2	0.57	0.5%
	λ (h^{-1})	17×10^{-3}	0.1×10^{-3}	0.6%
Right kidney	A_0 (MBq)	127.9	2.23	1.7%
	λ (h^{-1})	17×10^{-3}	0.4×10^{-3}	2.4%

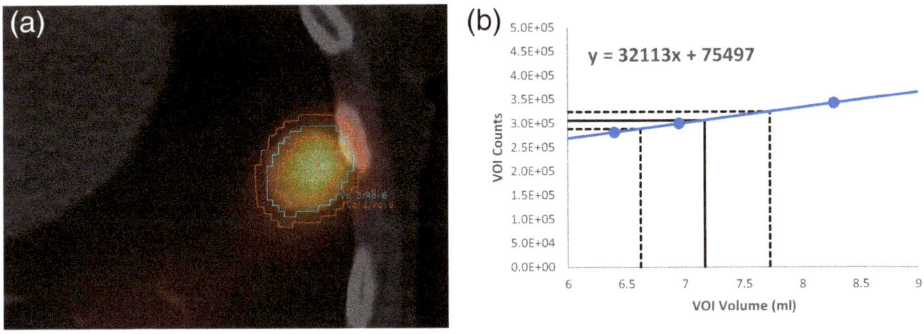

Figure 20.10 (a) SPECT/CT image of lung lesion with three VOIs used to measure count data. (b) Plots of VOI, with v = counts versus volume used to derive volume-corrected count data and associated uncertainty.

Count Rate

Count rates for each lesion and kidney were obtained by drawing three VOIs for each respective volume. Rather than try to accurately match these volumes to that of Table 20.3, the 3 VOIs were chosen about the CT volume. Linear interpolation of these data could then be used to determine an estimate of C at the specified volume, and an estimate of the sensitivity coefficient (dC/dv) with which to determine uncertainty. An example of these outlines is given in Figure 20.10(a), and a graphical representation of the interpolation is given in Figure 20.10(b).

Estimated counts are essentially gained by reading off the graph at the specified volume, such that at 7.17 cc the counts would be $32{,}113 \times 7.17 + 75{,}497 = 305{,}747$. As the volume uncertainty has already been estimated as 0.55 cc, the count uncertainty is simply taken from the gradient of the graph, $32{,}113 \times 0.55 = 17{,}662$ counts. A summary of these data for all outlines is given in Table 20.6.

Activity

Conversion of counts to activity is now simply a matter of dividing the recorded counts by both the sensitivity and recovery factor. Uncertainty in activity is calculated by adding the fractional uncertainties of these input parameters. Concerning count rate, as we have a slightly different uncertainty associated with each time point, for convenience (and due to the small variation) we use the average percentage uncertainty across all time points (see Table 20.7).

TAC Fitting

The Gauss-Newton algorithm was used to minimize the sum of square difference between the measured and fitted data. A single exponential function was used to describe the data, and the uncertainties in the fit parameters A_0 and λ were determined using Equation (16.30) in Chapter 16. Time-activity curves for the two lesions are given in Figure 20.11(a) and kidneys in Figure 20.11(b) with the fitted exponential functions. Error bars on the data points are equal to the estimated standard uncertainty in activity.

Solution parameters for the fits, with associated random and systematic components of uncertainty, are summarized in Table 20.8.

Time-Integrated Activity

Uncertainties in the estimation of time-integrated activity are separated into components of random (from the fitting) and systematic (from calibration and VOI outlining) components of uncertainty and are summarized in Table 20.9. It can be observed that data to the TAC are well fitted in this example, and the major source of uncertainty comes from the outlining and the 10% uncertainty associated with system calibration.

S Value and Absorbed Dose

The S values for lesions were determined by scaling the relevant reference S value data with the delineated organ volume. Lesion data were scaled using standard unit density sphere models (Stabin et al. 2000). As discussed in Chapter 16, uncertainties in the S values are largely a result of the uncertainty associated with the volume definition and, therefore, fractional uncertainties are equivalent to that defined in Table 20.10. A summary of S value data are given in Table 20.10.

The uncertainty in the absorbed dose is determined from Equation (16.8) of Chapter 16, for which the covariance is required using Equation (16.4). Solutions for $u(\tilde{A},S)$ are summarized in Table 20.10. It can be seen from the correlation coefficients, $r(\tilde{A},S)$, that the covariance between $\tilde{A}$ and S is fairly significant. In addition, the negative nature of the correlation results in a reduction in calculated uncertainty.

Table 20.9 TIA and components of uncertainty

	TIA (MBq.h)	$u(\tilde{A})$ Fitting	$u(\tilde{A})$ Systematic	$u(\tilde{A})$ Total
Lung lesion	1383	5.23%	11.0%	12.8%
Gluteal lesion	1980	1.06%	10.3%	10.3%
Left kidney	6549	0.48%	10.8%	10.8%
Right kidney	7419	1.57%	10.8%	10.8%

Table 20.10 Summary of VOI S value and absorbed dose data with standard uncertainties

	S (Gy/MBq.h)	$u(S)$ (Gy/MBq.h)	$u(\tilde{A},S)$	$r(\tilde{A},S)$	D	$u(D)$
Lung lesion	1.18E–2	9.12E–4	–0.039	–0.24	16.4	2.2
Gluteal lesion	1.89E–3	7.88E–5	–0.022	–0.14	3.7	0.4
Left kidney	8.12E–4	4.87E–5	–0.011	–0.32	5.3	0.6
Right kidney	9.13E–4	5.48E–5	–0.015	–0.35	6.8	0.7

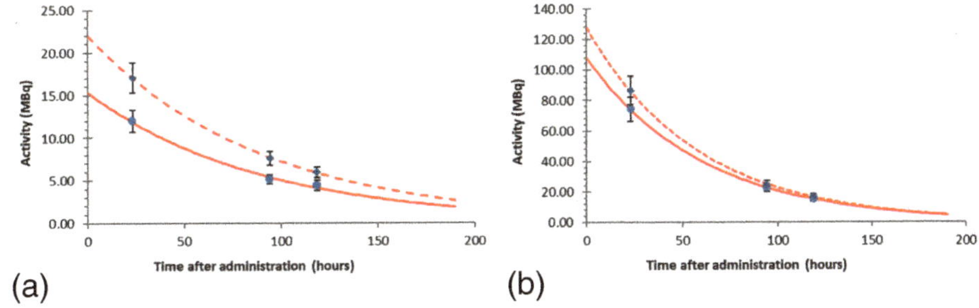

Figure 20.11 (a) Time activity curves for the lesions (lung lesion solid line, gluteal dotted) and (b) the kidneys (left solid line, right dotted).

20.5.2 Approach #2: Voxel Dosimetry

Voxel dosimetry was performed using the same data presented in the organ dosimetry example, except the reconstruction has been adjusted to include a PSF corrections for reasons previously described. Calibration was also performed in a similar manor, but this time the sensitivity factor was included in the reconstruction as the dosimetry software requires quantitative SPECT images. Voxel dosimetry calculations were performed using a locally developed software (Rushforth et al. 2019), developed as a scripted loadable module for the 3D Slicer platform in the python programming language. The code incorporates all aspects of the dosimetry chain, from image registration to time activity fitting to dose calculation and segmentation. The code allows for the creation of 3D dose maps following administration of therapeutics and also outputs some aspects of uncertainty (such as that associated with TAC fitting).

Image Registration

Difference maps of scan 1 and 2 are shown in Figure 20.12. Where a similar distribution of activity is observed, voxels appear gray, while blue or red voxels indicate a difference between normalized voxel values. The importance of proper registration is clearly evident. A nonrigid registration was then applied, and the resulting equivalent difference maps are shown in Figure 20.13. Note that there is still some color at the periphery of the lesion and organs, but otherwise the images appear well registered.

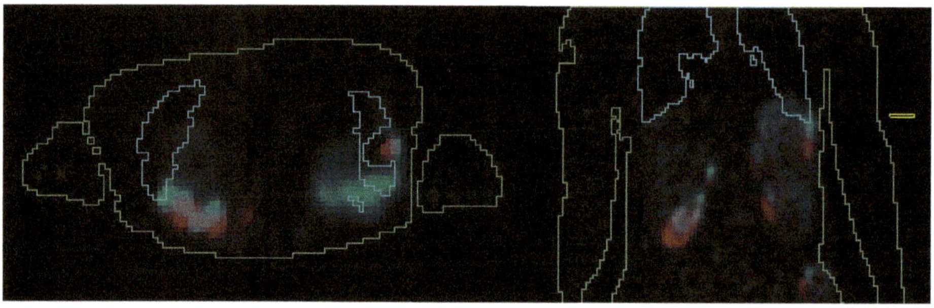

Figure 20.12 Difference maps of scan 1 and scan 2 before performing image registration.

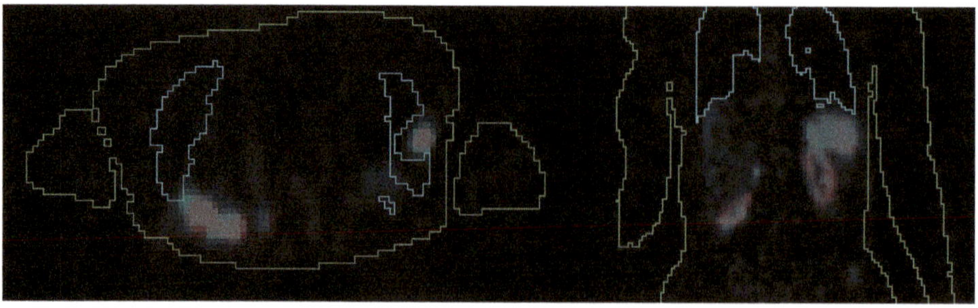

Figure 20.13 Difference maps of scan 1 and scan 2 after performing image registration.

Time Activity Fitting and Time-Integrated Activity

TAC fitting was performed at the voxel level, fitting a single exponential function. Time-integrated activity maps are shown in Figure 20.14. Corresponding uncertainty maps for TIA, based on the fitting parameters, are given in Figure 20.15. The largest uncertainties are generally observed at the periphery of areas of uptake, which similarly corresponds to where the registration was observed to be poorest. Large uncertainties are also observed in areas of the bowel, where it is inappropriate to model transit as an exponential function.

Typical time activity curves for voxels located in the lung lesion are given in Figure 20.16. A central voxel with good fitting is shown in Figure 20.16(a), and an edge voxel is given in Figure 20.16(b). These cures follow a very similar pattern to that presented earlier in the chapter.

Dose Calculation

Voxel absorbed doses were calculated using both a convolution technique and Monte Carlo simulation. $10 \times 10 \times 10$ voxel convolution kernels were generated using the EGSnrc Monte Carlo assuming a unity density tissue media. The full Monte Carlo

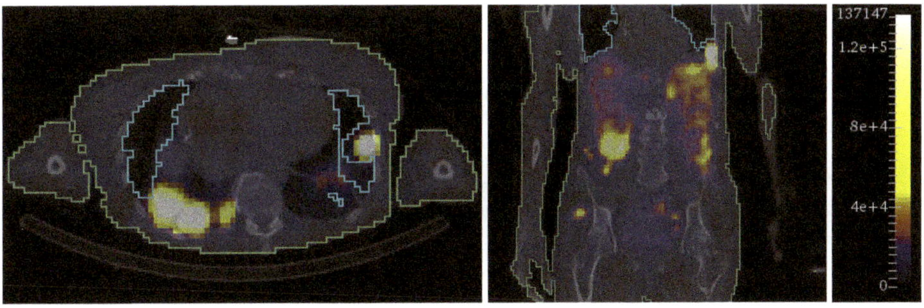

Figure 20.14 Time-integrated activity map.

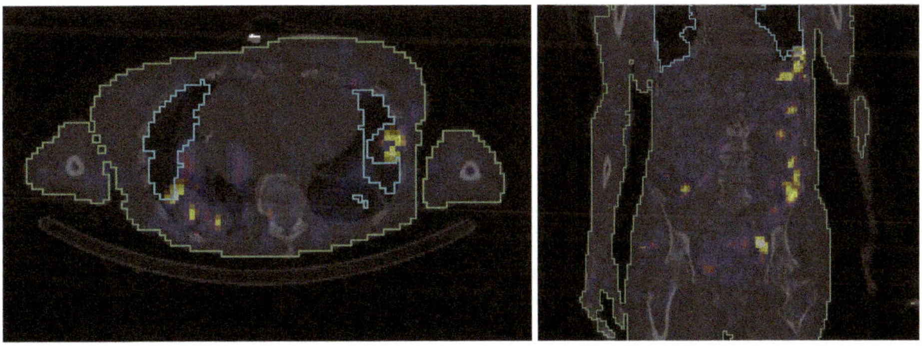

Figure 20.15 Time-integrated activity uncertainty map.

simulation was also performed using the EGSnrc code. The time-integrated activity map along with the CT scan formed the inputs to the Monte Carlo code. One million particle histories were simulated using the ^{177}Lu decay scheme as input. Photons were tracked down to 10 keV, and electrons down to 521 keV.

Immediate differences and similarities are observed between the two dose maps (Figure 20.17). Absorbed doses to the lesions within the lungs are similar, although the dose to background tissue (such as the heart) is higher in the Monte Carlo simula-

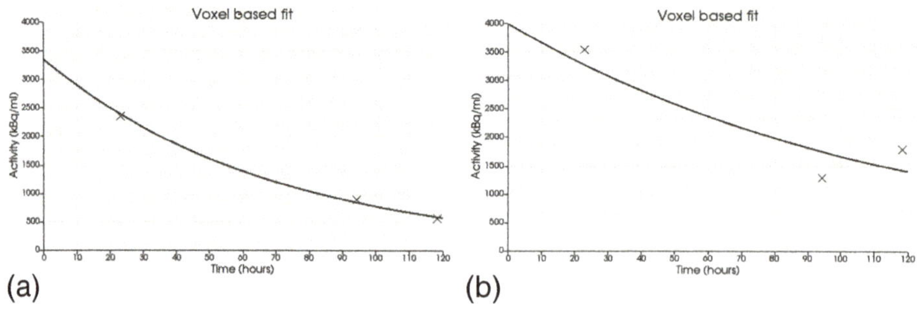

Figure 20.16 Time-activity curves for a single voxel located (a) in the center and (b) the edge of the lesion.

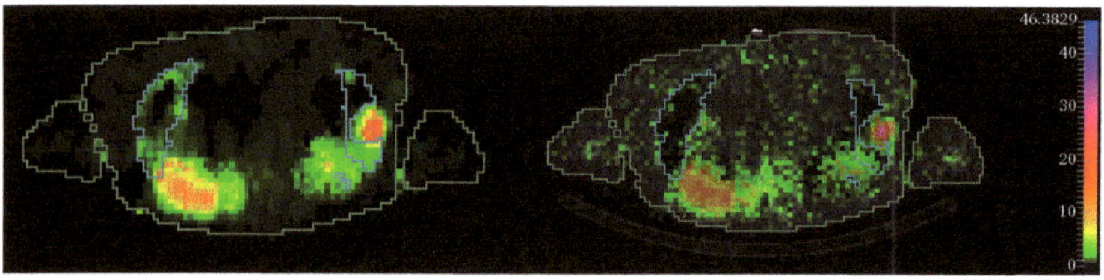

Figure 20.17 Absorbed dose maps generated using a convolution method (left) and full Monte Carlo simulation (right).

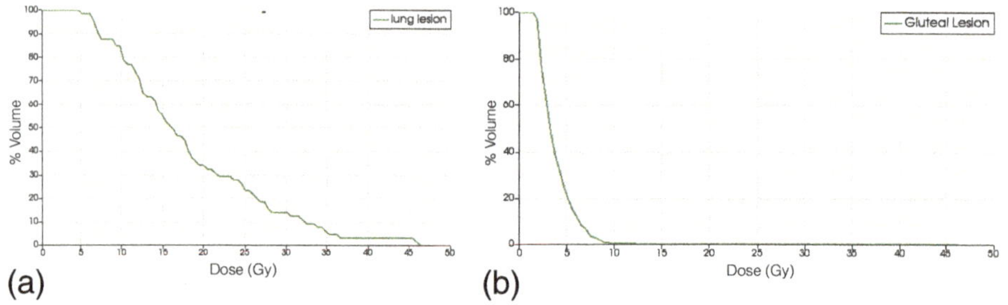

Figure 20.18 Cumulative dose volume histogram within a VOI defining (a) the lung lesion and (b) the gluteal lesion.

Table 20.11 Summary of absorbed dose data calculated using the organ and voxel methods

	Organ Level		Voxel Level	
	Absorbed Dose (Gy)	Standard Uncertainty (Gy)	Absorbed Dose (Gy)	Standard Deviation (Gy)
Lung lesion	16.4	2.2	18.3	9.5
Gluteal lesion	3.8	0.4	3.8	1.8
Left kidney	5.3	0.6	4.8	2.7
Right kidney	6.8	0.7	5.8	2.7

tion, presumably due to photon emissions not adequately modeled in the convolution approach. The convolution technique also results in higher doses to the lung tissue, presumably due to a mismatch in tissue density between the dose kernel and the lung tissue. The Monte Carlo simulation results in a significantly noisier image than the convolution technique, and it also shows higher doses to body areas where tissue density is more accurately modeled. Noise in the Monte Carlo-derived dose map could be reduced by simulating a higher number of particle histories, but this would obviously increase the computation time.

Cumulative dose volume histograms (cDVH) derived using the Monte Carlo dose map are shown in Figures 20.18(a) and 20.18(b) for the lung and gluteal lesion. Of interest is the shape of the cDVH of the lung lesion, which indicates a very nonuniform distribution of dose within the lesion (range 7 to 43 Gy). The solution to interpreting this nonuniformitity remains unanswered, as it is unclear whether it is a true representation of real differences in absorbed doses or (more likely) a result of partial volume effects, PSF correction, and noise in the Monte Carlo simulation.

20.5.3 Comparison and Conclusions

The production of cDVHs inevitably requires some selection of volume within which to report doses. In this example, volumes were selected that matched the identified source volumes given in the organ-level calculation, and that now enables a direct comparison of the two sets of results. Table 20.11 summarizes these results, reporting a mean and standard error for the organ-level calculation. Although we were able to discuss and highlight sources of uncertainty in the voxel approach, these have as yet not been fully propagated into an estimate of uncertainty in absorbed dose. Instead, for the voxel approach, we report the mean voxel dose in the VOI and associated standard deviation.

What is perhaps most apparent is the similarity of the obtained results for the two techniques. Despite using different reconstructions and given many of the difficulties discussed with the voxel method, the mean doses are indeed in good agreement. Both techniques highlight absorbed doses to the gluteal lesion that are less than that of the kidneys. The variation between the left and right kidney is also evident in both techniques which, interestingly, matched a similar observation in an earlier split renal function test.

For all four volumes, dose values for the two techniques mostly fall within the uncertainty bound calculated at the organ level, and although a slightly larger difference in absorbed dose for the lung lesion is apparent, this is unlikely to have a significant clinical impact.

The choice of performing a dose calculation at the organ or voxel level will inevitably come down to operator preference. Organ-level dosimetry can be performed with simple nuclear medicine image processing software and calculation spreadsheets. This approach will require more operator input, and when uncertainty estimates are included, this can quickly become very user intensive. The advantage here is that operators have more control over the choices they make and how to deal with specific situations. Voxel dosimetry tends to be more automated, requiring specialized software to undertake the separate voxel calculations. This somewhat limits the reporting of the calculation to graphical or image representations, rendering a rigorous interrogation of the dose calculation difficult. Sources of error can be highlighted or observed, but there is as yet no clear way of dealing with them. It remains the opinion of the author that in many cases the additional data obtained from a voxel approach will not necessarily offer a clinical benefit over the reporting of a mean dose. Variation in absorbed doses within an organ or lesion should also be considered carefully as to whether this is a "real" effect or the result of some reconstruction or simulation artifact.

References

O'Donoghue, J. A. (1999). "Implications of Nonuniform Tumor Doses for Radioimmunotherapy." *J. Nucl. Med.* 40(8):1337–41.

Rushforth, D. P., et al. (2019). "Validation of a voxel based MRT dosimetry software module for 3D slicer." *Eur. J. Nucl. Med. Mol. Imaging* 46(Suppl 1):S364–65.

Stabin, M. G. and M. W. Konijnenberg (2000). "Re-evaluation of absorbed fractions for photons and electrons in spheres of various sizes." *J. Nucl. Med.* 41(1):149–60.

Walrand, S., C. Chiesa, P. M. Gabina, N. Chouin, J. Gear, C. Stokke, P. Bernhardt, K. S. Gleisner, L. Strigari, M. Konijnenberg, and EANM Dosimetry Committee (2019). "Re: Tumor Targeting and Three-Dimensional Voxel-Based Dosimetry to Predict Tumor Response, Toxicity, and Survival after Yttrium-90 Resin Microsphere Radioembolization in Hepatocellular Carcinoma." *J. Vasc. Interv. Radiol.* 30(12):2047–48.

Chapter 21

Bio-Effect Modeling in Radiopharmaceutical Therapy

Robert F. Hobbs, Ph.D., DABR

Associate Professor
Department of Radiation Oncology
Johns Hopkins Medical Institute
Baltimore, MD

21.1 Introduction

This chapter will take a look at the more generally accepted and utilized bio-effect models in use in *radiopharmaceutical therapy* (RPT). The basic models and quantities used in RPT are the same as for *external beam radiation therapy* (EBRT), namely *biological effective dose* (BED), *equi-effective dose* (EQDX), *equivalent uniform dose* (EUD), etc. However, because of the different nature of the radiation that determines the absorbed dose distributions—namely beam arrangement and geometry for EBRT and physiologically driven uptake for RPT—these same quantities may be more or less useful in one modality compared to the other.

Interactions between radiation and living tissue results in biological effects, from DNA damage to cell kill, to tumor control and normal organ damage. All radiation therapy seeks to maximize the former and maintain the latter at safe levels. The *absorbed dose*, equal to energy deposited in tissue per unit mass, is the primary physical quantity that translates this amount of damage done to cells and tissue, and it is used to predict both therapeutic effects and potential adverse tissue reactions.

However, in addition to absorbed dose, the biological response to ionizing radiation depends on numerous physical and biological factors.

Physical factors include, but are not limited to:

- particle types and energies,
- *linear energy transfer* (LET) of the particles,
- placement of the isotope within the cell,
- absorbed dose rate, and
- nonuniformity of the dose distribution.

Biological factors include:

- the rate of cell proliferation,
- the degree of cell differentiation,
- the intrinsic radiosensitivity of the irradiated cells,
- the phase of the cell cycle at the time of irradiation, and
- the oxygen concentration.

21.2 Radiobiology vs. Bio-Effect Modeling

Bio-effect modeling is rooted in the discipline of radiobiology, which has long studied the relationship between radiation and the biological response, as well as the factors influencing the responses, including the elements listed above. Classically, many of these studies have been done *in vitro* or in preclinical models, and the field is much more diverse and advanced than we can relay in the simple clinical bio-effect models we will present here. This is true also in EBRT, but the lack of more accepted bio-effect models is exacerbated in RPT by a lack of dosimetric data from clinical data. Most administrations of RPT are performed with fixed activity or mass-based paradigms. Correlations with absorbed dose or other dosimetric quantities or application of bio-effect models to determine treatment are rare in RPT, with a few exceptions that will be discussed. For a more in-depth discussion of radiobiology and its relationship to bio-effect modeling, the reader is referred to Chapter 5 of the MIRD Primer 2020 (Bartlett, et al. 2022) or Chapter 4 of ICRU Report 96 (Sgouros et al. 2021), while some basic concepts are outlined here.

21.2.1 Basic Radiobiology

Classically, radiobiology includes the four Rs: repair, reoxygenation, repopulation, and redistribution.

The concept of repair is straightforward: cells that are damaged by radiation are not necessarily killed, but have repair mechanisms that allow for continued normal life cycles. Cell nuclear DNA damage is the primary cause of radiation-induced cell death. Ionizing radiation can deliver a range of structural DNA lesions, including base damage (point mutations), nucleotide damage, single-strand breaks (SSB), double strand breaks (DSB), protein-DNA cross-links, and protein-protein cross-links. Typically, the vast majority of these are repaired: the number of lesions induced in the DNA of a cell nucleus at an absorbed dose of 1 to 2 Gy includes more than 1000 point mutations, approximately 1000 SSB, and approximately 40 DSB (Goodhead 1989). Human cells are efficient at DNA repair, such that the proportion of lesions that lead to cell death is relatively small. DSBs are most difficult to repair and are often considered the main factor in cell kill. Repair can also exist at a tissue level, where whole cells damaged or killed within an organ may be replaced over time, the effect generally classified as repopulation.

Repopulation is also a relatively basic concept. Cells can multiply through mitosis, which in healthy adult tissue is generally only used to replace dead cells. However, tumor cells undergo cell division on a more ad hoc basis and proliferate, which is the mechanism of tumor growth. In RPT, since the absorbed dose is delivered over a more protracted period of time, cell damage and cell kill compete against tumor repopulation for overall effect.

Biological response to ionizing radiation is also affected by the oxygen concentration in the tissue, with cells and tissues being markedly less affected when they are hypoxic than when they are normoxic, particularly for low-LET radiation (beta electrons and photons). This effect is less pronounced in high-LET radiation (alpha particles and Auger electrons).

The susceptibility of cells to permanent damage and death depends on their phase of the cell cycle, with G2/M phase being the most sensitive. This correlates well with the notion that rapidly proliferating cell populations are generally more radiosensitive. Therefore, in a cell population that is irradiated with regular intermittence, proliferating cells will progressively tend to be in the same phase because those on the more sensitive phases at the time of irradiation will die out. This is known as redistribution.

Ideally, the biological effects of ionizing radiation would be correlated with a single quantity that accounts for all of these factors, but the complexities involved usually preclude such a simple approach. The bio-effect models in this chapter are more centered on the concepts of repair and (to a lesser extent) repopulation.

Cell death is manifested in many distinct ways, including apoptosis, necrosis, autophagy, and senescence. *Apoptosis* is a genetically programmed cell death that may occur as a result of external factors (including radiation). *Necrosis* is a loss of cell membrane integrity and cessation of all cellular function. *Autophagy*, as its name indicates, is when the cell self-engulfs parts of its cytoplasmic components. *Senescence* is when the cell enters a permanent cell-cycle arrest, meaning it is no longer capable of cell division. *Cell reproductive failure* is probably the most clinically relevant of these in the context of RPT and EBRT (Bentzen et al. 2010).

21.2.2 Linear-Quadratic Model

Much of bio-effect modeling comes from the linear-quadratic (LQ) model of cell survival curves. This model derives from the observation of a quadratic fit to the surviv-

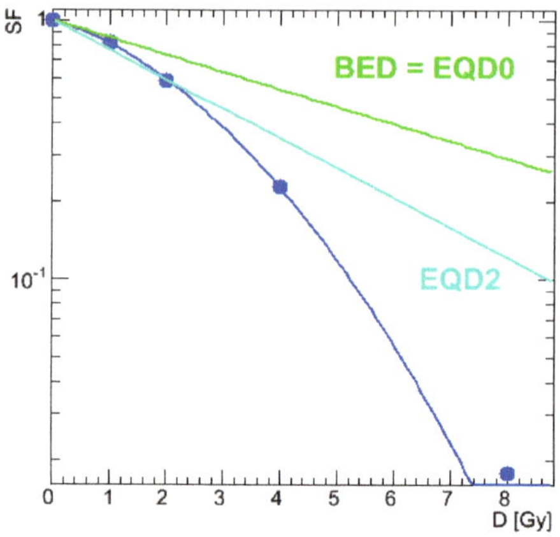

Figure 21.1 Surviving fraction of cells as measured by assay due to irradiation of cells from single boli of external beam radiation as a function of bolus absorbed dose in a log-linear plot. The fit to the response is a quadratic equation with linear and quadratic terms (dark blue). The BED (or EQD0) is the tangent to the curve at D = 0 (green line). EQD2 is the line from D = 0 that passes through the curve at D = 2 (light blue line).

ing fraction of cells in *in vitro* studies as a function of single bolus of absorbed dose plotted on a log-linear graph (see Figure 21.1).

The surviving fraction of cells is fit by a curve of the form:

$$SF = e^{-(\alpha D + \beta D^2)}, \tag{21.1}$$

where D is the absorbed dose of the single bolus, SF is the surviving fraction of cells, and α and β are known as the radiobiological parameters (cell line-specific or tissue type-specific). Originally, the interpretation of this response curve was that an unrepaired double strand break was necessary to kill a cell, with the linear term of the response curve representing a double-strand break "hit" and the quadratic term representing two successive unrepaired single-strand break "hits." This interpretation is no longer taken at face value, and the parameters do not reflect the underlying mechanisms of response. However, bio-effect modeling is still based on this linear-quadratic response, even though it is recognized as more of an empirical observation than a specific interpretation of biological mechanisms. It can be useful to think in terms of single and double "hits" in order to develop and manipulate the formalism.

Along these lines of thought, a slightly modified formalism for RPT was developed. Since it required two quadratic "hits" to create cell kill, and since the delivery of RPT is over a protracted period of time, there should be a modifier that accounts for single-strand break repair between the first and second hits. This factor, known as the *G-factor* or *Lea-Catcheside factor* was first introduced by Lea and Catcheside:

$$SF = e^{-(\alpha D + G(\infty)\beta D^2)}, \tag{21.2}$$

where,

$$G(\tau) = \frac{2}{D^2} \int_0^{\tau} \dot{D}(t)dt \int_0^{t} \dot{D}(w)e^{-\mu(t-w)}dw, \tag{21.3}$$

and μ is the repair constant, assuming exponential repair, (in theory) specific to each cell line and tissue type; t and w are integration variables that represent the two different times of single strand breaks. In the case where the dose rate as a function of time is best modeled by a mono-exponential, a simpler, more familiar form of the G-factor is obtained:

$$G(\infty) = \frac{\lambda}{\lambda + \mu} \tag{21.4}$$

where λ is the effective decay constant of the mono-exponential fit:

$$D(t) = \int_0^{\infty} \dot{D}(t)dt = \int_0^{\infty} \dot{D}_0 e^{-\lambda t}dt. \tag{21.5}$$

21.3 Absorbed Dose Rate and Standardization of Dose

According to the *linear-quadratic* (LQ) model, an increased amount of damage occurs if a greater absorbed dose per bolus (or fraction) is delivered for a same amount of total absorbed dose. This is consistent with the idea that more absorbed dose in a shorter period of time is less likely to allow for repair, resulting in an effective "hit." This is well understood in EBRT, and fractionation schemes are chosen deliberately by the treating physician to take advantage of this effect. In radiopharmaceutical therapy, the only point of control is the amount of activity administered, although the total activity is often fractionated as well. Individual patient pharmacokinetics determine how much activity is taken up in each region and what the retention is, such that the absorbed dose per unit time, the dose rate, can vary considerably from patient to patient for a same administered activity or a same absorbed dose to the target or nontarget region. This absorbed dose rate correlates to the absorbed dose bolus in the LQ model. For a same absorbed dose in RPT, if the dose rate is higher (and the delivery time shorter), the biological effect will be greater. Thus, knowledge of the absorbed dose alone does not give the ability to predict biological outcome as well as when the dose rate effect is folded into a dosimetry metric. Just as comparisons and equivalencies can be made between different fractionated regimens based on the LQ model in external beam, such comparisons and equivalencies are desirable in RPT, made more necessary by the lack of control of the treating physician of the individual patient's kinetics.

21.3.1 Biological Effective Dose (BED)

The need to account for dose rate or fraction size effect resulted in the introduction of the BED, originally called the *extrapolated tolerance dose*, ETD (Barendsen 1982; Dale 1996; Fowler 1990). The idea was to imagine a linear dose-response curve by taking advantage of the existing linear term of the LQ model. The BED would be the equivalent linear term dose necessary to deliver a same biological outcome as the absorbed dose (see Figure 21.1). In other words, since the quadratic term is considered a succession of two hits, the BED is the equivalent absorbed dose given at an infinitely slow dose rate. The definition of the BED from the cell survival curve is:

$$SF = e^{-(\alpha BED)}, \tag{21.6}$$

and by setting Equations (21.2) and (21.6) equal to each other:

$$BED = D\left(1 + D\frac{G(\infty)}{\alpha / \beta}\right). \tag{21.7}$$

For a simple mono-exponential decay of the dose rate, as is the case in brachytherapy, a more familiar formulation appears:

$$BED = D\left[1 + \left(\frac{\lambda}{\lambda + \mu} \times \frac{D}{\alpha / \beta}\right)\right]. \tag{21.8}$$

It is important to remember the condition of Equation (21.5), namely of a simple mono-exponential fit to the dose rate; a common error is to use Equation 21.8 even when the dose rate fit is not a mono-exponential, thereby creating a false BED value.

Equations of the BED have been derived for multicomponent exponential absorbed dose rate curves that are typically observed in nuclear medicine (Howell et al. 1994; Solanki et al. 2017). Fits to the dose rate curves that are not based on exponentials may require numeric integration (Hobbs et al. 2009) or discrete convolution (Gustafsson et al. 2013a; 2013b). The BED can be applied to any region, including voxels and cells. The radiobiological parameters α, β, and μ could also vary within a same given tissue or cell line spatial on an inter-voxel or inter-region level.

As the original mechanistic underpinning of the LQ model, and by extension the BED, have come into question, so too has the utility and validity of the BED. However, there is considerable evidence for the need to account for dose rate in RPT, and the BED formulation is mathematically aligned with the Wither's formula conversion of absorbed dose equivalency between different fractionated external beam regimens. Within RPT, confirmation of its importance to RPT—and arguably one of the single greatest successes of dosimetry in RPT—came when Barone et al. (2005) showed that kidney toxicity in somatostatin-receptor-targeted therapy correlated well with kidney BED, better than with absorbed dose. The MIRD Committee completed this analysis by comparing equivalent BED response curves for both external beam data and RPT peptide receptor data in MIRD Pamphlet 20 (Wessels et al. 2008).

The BED formalism is an extension of the LQ model. As such, the implicit understanding is that all cells under observation in a region where the BED is being calculated observe the same response. Clearly, this is generally not the case for normal organs, which are composed of different cell types. Response to the absorbed dose should also depend on the complexity of the different cell types within the region (organ) under consideration, but the "dose"-limiting response may also only be a marker from a single cell type response or an incomplete combination, as damage to many or all cells of a single cell type in an organ may be sufficient to create an organ-wide response. As long as the absorbed dose distribution is reasonably uniform within a given organ, the BED calculated from organ-level absorbed dose is a reasonable surrogate for cell type-level BED needed for correlations with response. An obvious exception, one that has been discussed in Chapter 14, is the case of ^{90}Y-microspheres. The distribution of microspheres is such that all activity is localized to the vasculature, and the absorbed dose distribution affects different anatomical components of the liver in dissimilar fashion, with the vasculature itself taking a particularly hard hit. In addition, the nontarget statistical distribution of the microspheres within the vasculature itself creates a much more complex tissue response. This is evident by the scale of absorbed doses used in ^{90}Y-microsphere TARE, where 120 Gy to the liver is a standard prescription, whereas such an absorbed dose value would be unthinkable in external beam. This discrepancy cannot not simply be remedied by applying Equation (21.8) and obtaining the BED; the conditions of validity of the BED formulation do not apply, and the resulting values are not BED values.

21.3.2 BED Uncertainty

Uncertainty in the BED calculation can be estimated using the GUM method presented in Chapter 16, starting with Equations (21.7) or (21.8) and considering the relative uncertainties from the absorbed dose D, the decay fit parameter(s) λ (or λ_i), and the radiobiological parameters. Here is where an issue arises; often, radiobiological parameter uncertainties are relatively large or unknown, especially for the repair rate parameter μ, where a common value is typically used for normal tissues (0.46 h^{-1}) (Bodey et al. 2003) and another for tumors (1.73 h^{-1}) (Brenner et al. 1991). This may also reflect the fuzziness in the G-factor, which may incorporate or hide a wider range of mechanisms. The radiobiological parameters are generally established in colony formation assays, and there are questions as to whether such values make sense in an *in vivo* context. As is true for most of RPT dosimetry, the uncertainties are comparatively large to other radiation modalities, and a relatively simple formulism is likely inadequate, yet its implementation is a significant improvement compared to its absence and represents an important first order step to incorporating radiobiological response to physical absorbed dose. Additionally, as shown by the MIRD Pamphlet 20 study (Wessels et al. 2008), to first order the BED values calculated from RPT are equivalent and translate to BED in EBRT.

21.3.3 BED in the MIRD Formalism

Calculation of the BED can, in theory, be accomplished with the MIRD S value formalism directly applied to the time-integrated activities of the source organs, which have been already obtained for the calculation of the absorbed dose. This can become quite complex, as the G-factor shown in Equation (21.3) depends on absorbed dose rate, not activity. The absorbed dose rate to a given target region, at a given time, is the sum of absorbed dose rate contributions from the different source regions at that time and the kinetics for each source organ are likely different. A full theoretical adaptation of the MIRD formalism using time-integrated activities (assuming the TIAs are fit by multi-exponential equations) for BED has been developed (Baechler et al. 2008); however, the formalism is quite complex, making it impractical and unwieldy to implement for more than two sources.

A more practical method for calculating the BED when the absorbed dose has been calculated using the MIRD method is to use the MIRD dose rate formula at each time point calculated from the activities in the sources regions at each time point, t_i:

$$\dot{D}(r_T,t_i) = \Sigma_{r_s} A(r_s,t_i) \times S(r_T \leftarrow r_s). \tag{21.9}$$

With the dose rates calculated, the absorbed dose can be calculated from the fits to the curve (see Chapter 15), as can the BED.

21.3.4 BED, Repopulation, and Time of Integration

One of the radiobiological quantities that is not taken into account in the above formalism is the repopulation of tumor cells or normal tissue repair; clearly these considerations should factor into the BED formulism. For tumor repopulation, Strigari et al. (2011) (in RPT) and Antipas et al. (2001) (in brachytherapy) have proposed adding a factor to the cell survival curve for tumor cell types:

$$SF = e^{-(\alpha D + G(\infty)\beta D^2 - \gamma\tau)}, \tag{21.10}$$

where γ is a parameter correlated with the doubling time of the clonogenic cells and τ is the time of treatment. This introduces another point of debate, namely, should integration of the BED (or even the absorbed dose) extend out to infinity, as that is clearly not clinically relevant. A better time cut-off might be once the dose rate is below a certain threshold or, in the case of the BED calculation, when the cell kill is outweighed by the tumor repopulation factor.

The BED is then expressed by:

$$BED = D\left(1 + D\frac{G(\infty)}{\alpha/\beta} - KT\right), \tag{21.11}$$

where K is a parameter to offset tumor repopulation, and T is the effective time of treatment. In principle, Equation (21.11) is a necessary and important adjustment to the BED formalism and should improve the correlation of BED with effect. So far, there has not been any major correlative study with the BED incorporating the repopulation factor. This certainly is due to a lack of data to attempt such correlations, as well as the strong possibility that this effect may already be incorporated or hidden underneath the existing parameters in the current repopulation-free formalism. An effective, proven formalism that incorporates repopulation would also better help understand the radiobiological effects of RPT fractionation and whether such fractionations are beneficial to overall tumor control. As it is now, this can still be done by considering tumor repopulation and BED separately and combining the competing effects.

Normal tissue repair is a complex issue and one that radiation oncologists are becoming more familiar with as patients returning for multiple rounds of EBRT are becoming more common. Maximum tolerated doses and BEDs are, in theory, for the patient's lifetime, but just as clearly there will be some tolerance build-up for additional radiation therapy several years or longer after the previous therapy. Multiple cycles of RPT are often only documented with potential MTA limits based on the potential for secondary cancer induction, rather than based on maximum organ absorbed dose or BED limits. The amount and rate of repair and renewed tolerance are unknown at this point. Safe practice is to document all radiation treatment and characterize absorbed dose and BED to different organs to inform future radiation therapy, whether RPT or EBRT, and possible dose constraints.

21.3.5 Equieffective Dose (EQDX)

In an effort to provide a standardized dosimetric quantity that incorporates dose rate effects, the ICRU is promulgating a quantity named *equieffective dose* (EQDX) (Bentzen 2012). The equieffective dose, like the BED, accounts for differences in absorbed dose rate. The general EQDX relates different absorbed dose delivery schedules to a reference (X-Gy) external beam absorbed dose fraction schedule and has the unit Gy. The ICRU recommends the use of EQD2, equivalent dose to a 2-Gy per fraction external beam absorbed dose based, in part, on the large body of published clinical response data for EBRT with 2-Gy fractions. This quantity will be discussed more in Chapter 22. The BED is equivalent to EQD0 in this formalism, and since both are linear in the LQ model, they are equally valid as standardized dosimetric quantities, neither having any intrinsic advantage over the other. Values and equations for one may be converted to the other by a simple multiplicative factor.

21.3.6 Standardization of Dose

Standardization of RPT dosimetry, practice, protocols, and results is a fundamentally important topic and one that is addressed in Chapters 17 and 22. The standardization of dose reporting and the incorporation of dose rate effect into biologically equivalent dose (EQDX) is just as critical. This problem is often exemplified when new drugs come to market and either the developers or the regulators refer to external beam absorbed dose limits as thresholds for toxicity from RPT drugs. Yet RPT absorbed dose varies and is rarely as potent as the external beam fractionation scheme (2 Gy) that provided the established threshold. More specifically, as an important example, the 23-Gy threshold for kidneys taken from external beam MUST be understood as an EQD2 value and not a generic absorbed dose threshold. Of course, given the complexity of RPT uptake and dosimetry from RPT to RPT, it could very well be that localized uptake of a drug changes the absorbed dose scale regardless (e.g., ^{90}Y-microspheres) for any specific RPT. However, to first order, as long as there is relatively uniform distribution of activity, the external beam values should provide a reasonable threshold for RPT, provided the correct conversions to a biologically standardized dose (EQDX) are made.

In a similar vein, repeat treatments are becoming more common as patients are living longer and longer with their disease. With the increase in use of RPT, there are more and more occurrences of sequential RPT-EBRT therapies. Unlike with chemotherapy, radiation toxicity thresholds are supposed to be valid for the lifetime of the patient. As patients are increasingly returning for additional radiation therapies, whether within the same modality or between different radiation modalities, this paradigm is being challenged. Certain repair is expected to occur, but quantitatively, there is no consensus as of yet as to how much repair and over what period of time this takes place. Presumably, these quantities would also depend on the organ; a formalism that would incorporate such parameters is clearly desirable, but here too there is a lack of data to inform potential new models. Physicians must judge based on their own experience and expertise and an individual patient's case. While providing absorbed dose maps to treating radiation oncologists instead of flat administered activity values is a significant step forward (ACR-NCI-ASTRO ^{177}Lu recommendations), providing relevant, relatable dosimetric values (EQDX, BED) for re-treatments is of even greater value and will help establish new dynamic thresholds that account for long-term tissue repair.

21.4 Application: Combination Therapies

Cancer therapy is moving increasingly toward combination therapies in efforts to create multi-pronged attacks upon resistant disease. This can occur as a result of sequential treatments or, increasingly, by design, where the different modalities are delivered within a contracted time period in order to achieve a synergistic effect.

Outside of radiation therapies, treatments are rarely tailored quantitatively to the individual patient, but are based on the category of the patient's disease type. Stratification and categorization of patients exists, depending on a range of factors, but ultimately patients are treated at best as part of a class or category of patients rather than with a personalized treatment. The rationale and ability for RPT to be more personalized, as well as the inherent disadvantages of a one-size-fits-all approach, have been outlined in Chapter 12. This rational dosimetry-based approach may also be implemented in radiation modality combinations.

Combinations with non-radiation modalities fall into two categories:

1. ***Sequential treatments***. Although not combinations per se, sequelae from the first treatment modality may well carry over to the next modality and affect the responses, particularly when adverse toxicities of each modality affect the same tissue, as with hematotoxicity from chemotherapy and RPT. Establishing new and different thresholds for certain organ toxicities specific to the sequential therapies may be necessary. Baechler et al. (2010) showed that hematotoxicity following a given absorbed dose to bone marrow from RPT with ^{131}I-tositumomab and ^{90}Y-ibritumomab tiuxetan was greater for patients that had recently undergone chemotherapy. Similarly, ^{177}Lu-DOTATE is contraindicated with chemotherapy, and a wash-out period is recommended before transitioning from one therapy to another due to the hematotoxicity that can arise when given simultaneously. Systemic chemotherapy or immunotherapy often damages the hematopoetic stem cells in bone marrow, resulting in a lower absorbed-dose threshold for adverse toxicity.

2. ***Qualitative combinations***. Here the combinations are planned using a multifaceted effort for greater efficacy, whether simply by benefiting from the additive effect from therapies with orthogonal toxicities or out of the belief in a synergistic effect. Here too the RPT MTDs may vary from the RPT alone case depending on a variety of factors. It is equally important to remember that synergy can be negative as well as positive, such that taking half of the dosage of each modality may result in toxicity greater than from the single modality treatment alone and may require dose escalation trials to determine the safe level of administered activity and drug dosage, independently of the monotherapy modality maximum tolerated dosages. Conversely, toxicities may be orthogonal, such that maximum tolerated combination dosages may be close to the individual maximum tolerated dosages. Fixed-activity or ad hoc methodologies are still the standard for RPT despite the push for greater personalization and dosimetry-based treatment planning. This approach, born from chemotherapy, is the go-to option when combination trials are proposed. More recently, several trials have been proposed combining RPT (particularly ^{223}Ra) with immunotherapy, where the activity of the RPT is determined not for any potential dosimetric impact but merely for stimulation of the immunotherapy itself. When setting up these trials, discussions around dosages for both the RPT and the chemotherapies or immunotherapies generally revolve around fractionation schedules and combination timing, rather than tolerated absorbed dose.

21.4.1 Quantitative Combinations: RPT–EBRT

EBRT and RPT are both radiation modalities that use ionizing radiation to eradicate diseased tissue; both can be quantified in terms of absorbed dose and equieffective dose. As seen previously, the equieffective dose (or the BED) is the same quantity in EBRT as it is in RPT. Consequently, RPT and EBRT combination therapy may be planned as a single therapy. The formalism for this approach was developed by Bodey et al., by showing how converting the RPT absorbed dose to BED would allow this option (Bodey et al. 2003; 2004). A practical implementation of this rational personalized dosimetry-based combination therapy was propose and implemented by Hobbs et al. (2011b). This method has been applied clinically for the case of ^{153}Sm-EDTMP combined with EBRT for treatment of patients with unresectable metastatic osteosarcoma. The reader is encouraged to read ICRU Report 96 (Sgouros et al. 2021) for further details. More recently, this approach is gaining interest, and a number of groups have implemented similar approaches (Kerr et al. 2022).

The rationale for these combinations is one of orthogonal toxicities. EBRT can deliver precise amounts of high EQDX to targets, but is generally limited by surrounding normal organs. RPT, on the other hand, has normal organ toxicities that are usually distant from the tumor site, thus allowing for an increase (boost) of absorbed dose to the target with a minimal increase in surrounding normal organ toxicity. At the same time, RPT alone may deliver less than the tumorical EQDX to the target, due to lack of tumor uptake or limitations on the amount of administered activity. Combining the two is ideal when the target EQDX is achieved in combined therapy scenarios where radiation from either of the modalities alone would not have achieved the target EQDX alone.

Often these methodologies are not planned exactly as combinations, but present as more of a sequential/combination hybrid; specifically, the RPT is generally administered first, either using a personalized MTD/MTA approach based on pre-therapeutic tracer dosimetry for the RPT, but sometimes even with fixed activity. The important step is to create a biologically relevant 3D voxelized dose map (i.e., in EQDX, where X is the dose per fraction of the external beam portion), preferably from the therapeutic administration of the RPT. This EQDX map is then imported into the external beam planning system and entered as a fixed EQDX, such that the planning system accounts for the EQDX distribution from RPT. This map requires the kinetics of the different regions to be known, as discussed in the previous section. Therefore, dose rate images must be produced at each time point and integrated, and then the biologically relevant dose map is created using the calculated G-factor from the kinetics.

Alternatively, the MIRD methodology may be implemented and mean EQDX values obtained for the targets and normal organs at risk. In that case, the target absorbed dose and dose constraints in the EBRT plan may be lowered according to the RPT values. A voxelized map is more appealing and may allow for more subtlety (dose painting) in the target region. However, when combining the modalities on the voxel level, the nature of the uncertainties in the two modalities must be understood. Voxel activity quantification (and consequently EQDX) is one of measurement: the activity is where it is, and it is a question of how well it can be measured, both quantitatively and spatially. EBRT voxel absorbed dose is very precisely delivered to the space that the patient has been imaged in. Reproducibility of that exact patient configuration on the treatment table is the major challenge of EBRT and has certain intrinsic uncertainties, with patient motion, inter- and intra-fraction variability playing roles. Thus, painting dose down to the specific voxel runs the risk of having a blurred EQDX interface between the two modalities.

A detail which can help in the process is to have the patient receive a CT simulation for the external beam portion of the therapy before initiating the RPT. The reason for this is that the patient-specific fixation devices used in external beam may then be taken to the nuclear medicine SPECT/CT and used to give the patient a more reproducible position in the SPECT. Nuclear medicine couches tend to be cots, to allow for patient comfort, but when the precision of the registration and dosimetry matter, as is the case here, using the fixation devices attached to a radio-opaque board on top of the couch will help in the registration process.

Finally, full implementation of a rationally combined RPT-EBRT therapy is slightly more involved, where the pre-therapeutic EQDX map is imported into the external beam planning system not as a fixed beam, but with the weighting of the beam an option for the optimizer. In practice, this doesn't work well, as the optimizer will generally set the contribution of the RPT beam to zero or close to zero. Instead, the different options are best explored using a manual weight of the RPT beam and considering the different options of RPT activity versus external beam arrangement, as discussed in ICRU Report 96 (Sgouros et al. 2021).

21.4.2 Quantitative Combinations: RPT– RPT

While combinations of chemotherapy are relatively commonplace, RPT combinations are more rare (Villard et al. 2012; Kunikowska et al. 2011) in spite of a number of theoretical studies (Kotzerke et al. 2005; Garin et al. 2006; Leichner et al. 2008; O'Donoghue et al. 1995). The rationale for combining different RPTs falls into two categories:

1. ***Normal organ toxicity optimization***. In this category, the two RPTs may have orthogonal toxicities. The first example for which a methodology was developed was for ^{131}I-MIBG and ^{90}Y-DOTATATE treatment of neuroendocrine tumors (Besse et al. 2009; Madsen et al. 2006). The ^{131}I-MIBG activity is limited by hematotoxicity and the ^{90}Y-DOTATATE activity by kidney toxicity. Of course, it is naïve to assume that there is no contribution to hemato- and renal toxicity from the other RPT. The system for determining the optimal therapeutic combination—defined as the amount of ^{90}Y-DOTATATE and ^{131}I-MIBG such that both the bone marrow (bm) and the kidneys (kid) were treated to MTD—was determined using a system of two equations:

$$\begin{aligned} MTD(kid) &= A_{I131}d_{I131}(kid) + A_{Y90}d_{Y90}(kid) \\ MTD(bm) &= A_{I131}d_{I131}(bm) + A_{Y90}d_{Y90}(bm) \end{aligned} \tag{21.12}$$

   where A is the administered activity (the two unknowns) of each radiopharmaceutical, d is the dose coefficient per RPT and normal organ at risk, and *MTD* is the maximum tolerated absorbed dose of each organ. To fully implement this methodology on a personalized basis would require two sets of pre-therapeutic administrations and measurements, one for each RPT, or, alternatively a robust dual isotope imaging of a tracer of ^{131}I and a ^{90}Y surrogate (^{111}In, perhaps), or a combination of PET and SPECT agents (^{124}I and ^{111}In or ^{131}I and ^{90}Y) using two different machines to quantify each isotope separately. In any case, the dosimetry would be rather involved. A simpler, but less efficient, alternative is to combine one isotope using population-based kinetics with patient-based kinetics from the more variable RPT.

   As discussed, combinations are ideally planned using a biologically relevant dosimetric quantity, since the effects of adding absorbed doses obtained with different kinetics may be unpredictable. Accordingly, a similar approach using a combination of ^{131}I-tositumomab and ^{90}Y-ibritumomab tiuxetan was proposed where the two equations in the system were quadratic and based on maximum tolerated BED (MTBED) for the lungs (the primary organ at risk for the ^{131}I-tositumomab) and the MTBED for the liver (the primary organ at risk for the ^{90}Y-ibritumomab tiuxetan) (Hobbs et al. 2013).

2. ***Tumor dose optimization***. For a single tumor, efficacy from RPT can theoretically be optimized by a radionuclide whose range of energy deposition is on the order of the size of the tumor (O'Donoghue et al. 1995; Howell et al. 1989). RPTs are systemic therapies, so for a distribution of tumors of differ-

ent sizes within a patient, it may be beneficial to use multiple isotopes for a same targeting agent, such as ^{177}Lu- (β energy Q-value 483 keV) and ^{90}Y-DOTATATE (β Q-value energy 2.28 MeV) for neuroendocrine disease. In most cases, the extent of micro-metastatic spread is unknown and could theoretically benefit from different isotopes than that used to treat the visible disease, including those with a larger photon contribution. Since the weight and distribution of disseminated disease is unknown, rigorous methodologies have not been developed, but they depend more on qualitative approaches. An example could be the use of a beta-emitter therapy for visible disease, followed alpha-particle emitter RPT for systemic control.

In the example of combined ^{131}I-tositumomab and ^{90}Y-ibritumomab tiuxetan methodology (Hobbs et al. 2013), tumor dosimetry was incorporated into the design by searching for the maximum EUD to the visible tumors along the line of maximum possible efficacy constrained by the quadratic equations; however, size distribution of the tumors was not considered.

21.5 Nonuniform Distribution of Dose: the EUD

21.5.1 EUD for Tumors

While tumor absorbed dose is generally uniform in EBRT as the variability of absorbed dose to each voxel with respect to another is kept to a minimum, the RPT uptake in the tumor can be highly nonuniform due to factors such as penetration, vascularization, and hypoxia. As has been discussed in previous chapters, voxelized dosimetry can provide a tumor DVH that provides more information than a mean absorbed dose to the tumor. The EUD is a single value that incorporates the intra-variability of absorbed dose within the tumor. As its name indicates, it was defined as the single absorbed dose value that each and every cell or voxel would have that would give the same cell survival probability for the nonuniform distribution (Niemierko 1997). Since bio-effect modeling is driving the definition of this quantity, the better definition would be to calculate the equivalent uniform BED rather than the equivalent uniform absorbed dose (O'Donoghue 1999). Most authors define the EUD in this manner, but the use of the acronym EUBED is also prevalent. Mathematically, this translates as:

$$SF = \frac{\sum_{i=1}^{N} e^{-\alpha BED_i}}{N}, \tag{21.13}$$

where the i iterates over each cell or voxel BED. And by setting Equations (21.13) and (21.6) equal to each other:

$$EUD = -\frac{1}{\alpha}\ln\left(\frac{\sum_{i=1}^{N} e^{-\alpha BED_i}}{N}\right). \tag{21.14}$$

The EUD has considerable theoretical appeal and has been used as such for single value calculation to represent the absorbed dose (or BED) distribution (Hobbs et al. 2012; Prideaux et al. 2007)), where the unit voxel or cell absorbed dose or BED is assumed to be known. However, its application to clinical dosimetry remains limited—the reason being that for low BED values, the EUD is nearly equal to the mean BED, and for high BED values the EUD tends toward the value of the lowest BED input value (where low and high are relative to α^{-1}). Clinically this means that for the EUD to be precise, the BED and every quantity it derives from upstream—specifically the absorbed dose and the activity—need to be known precisely in each and every voxel. Currently, as we have seen in Chapter 20, there is a significant uncertainty in individual voxels. In particular, the PVE effect, or spill-out effect, places activity from the edge voxels of a region and locates it outside the region. This effect is more pronounced for smaller objects such as tumors where the surface-to-volume ratio is higher. Recovery coefficients merely augment the overall scale of the activity in the region without redistributing the activity. Point spread function modeling does a better job, but even a single miss-registered or miss-corrected voxel out of hundreds or thousands of voxels in a tumor will significantly impact the value of the EUD. As such, it is difficult to trust the values of EUD calculated from clinical data.

21.5.2 Normal Organ Nonuniformity: Small-Scale Modeling

In RPT, unlike in EBRT, the absorbed dose to the normal organs is generally quite uniform, at least as measured by clinical imaging devices. The reason for this is that Avogadro's number of molecules are being taken up inside the organ, an uptake that is driven by physiology. Often the cells that drive the uptake are uniformly distributed throughout the organ, such that the activity distribution is uniform at a macroscopic level, and the uniformity of absorbed dose is enhanced by the range of the energy deposition. The EUD concept does not apply well to normal organs in RPT. As mentioned, the uptake is driven by physiology, and nonuniformities are often below the imaging resolution of the clinical scanners. The EUD assumes equivalent radiobiological cells within a given voxel or, alternatively, the same variability in cell type (radiobiology) across all voxels in a given region. Even when nonuniformities can be detected at the microscopic level—as with cortex vs. medullary uptake of peptide receptors—the PVE effect also creates a high level of uncertainty, which discourages quantification with the EUD.

An alternative approach is needed for sub-organ activity and dose distribution, when whole-organ S values or even voxelized dosimetry fail to provide dosimetry that correlates with outcome. This problem appears not infrequently for kidneys and salivary glands, in particular. Smaller-molecular-weight radiopharmaceuticals may bind to the renal tubules. A possible solution to this problem is to simply look for organ-level RPT-specific MTBED (or MTEQDX) knowing that it will likely be different from the expected (uniform) MTBED or MTEQDX. However, this defeats the purpose of radiobiologically standardized dosimetry and reporting if dosimetry has a different scale and significance depending on the radiopharmaceutical and organ. A more elegant solution—and one more consistent with standardized dosimetry practices—is to model the sub-organ structures using the MIRD schema. The scale of the activity nonuniformity distribution and the range of the decaying particles dictate the scale of the model. In MIRD Pamphlet No. 19 (Bouchet et al. 2003), an age-dependent multi-region stylized model of the kidneys was developed to include source and

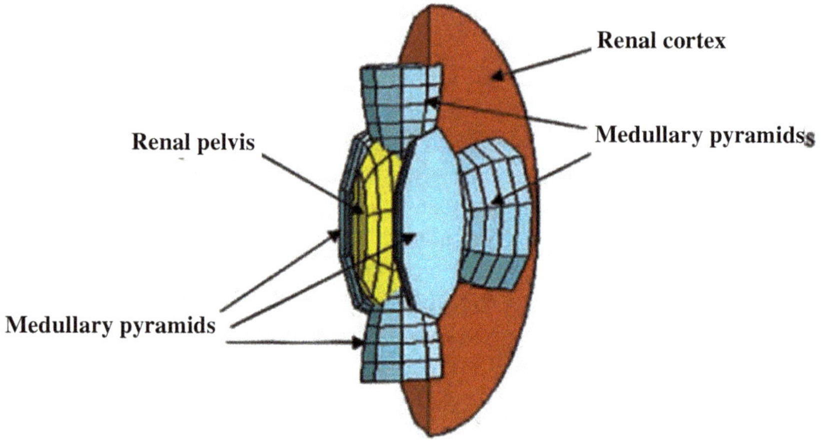

Figure 21.2 Sub-organ model of the kidney developed for peptide receptor RPTs (β particle scale) with cortex, medullary pyramids, and pelvis (Wessels et al. 2008). Reprinted with permission from the *Journal of Nuclear Medicine*.

target regions of the renal cortex, the renal medullary pyramids with papillae (2 vertically and 3 horizontally oriented), and the renal pelvis (see Figure 21.2).

This model was designed to solve the problem of localized uptake of peptide receptor RPTs in the cortex. Photon and electron S values were computed across the model series for 26 radionuclides of diagnostic or therapeutic interest. When in a study by Konijnenberg et al. (2007), a finer scale of localization was observed using ex vivo autoradiography of healthy renal tissue from renal cancer patients, a voxelized model of the MIRD 19 phantom was developed for greater input detail. The scale of these models is still insufficient for alpha-particle therapy. In a model study by Hobbs et al. (2012) a nephron-scale model was developed with stylized glomeruli, proximal and distal tubules with cellular components. Small-scale S values for a range of source and target regions were obtained using Monte Carlo in this model. To apply these microscale S values under the MIRD schema, corresponding values of time-integrated activity at the sub-organ compartment level is needed as input. Since the time-integrated activity can only be measured at the organ level, a component of this methodology, the *macro2micro method*, is to measure both organ-level and small-scale time-integrated activities from *ex vivo* preclinical models in order to establish the apportionment factors (Figure 21.3).

Once these apportionment factors are established, as with the S values, they no longer need to be measured. This user then assigns time-integrated activity to the different compartments based on measured time-integrated activity to the organs and these apportionment factors. The absorbed doses (EQDX) to the different compartments are then obtained by applying S values determined from Monte Carlo simulations run on small-scale anatomical/geometrical models and phantoms.

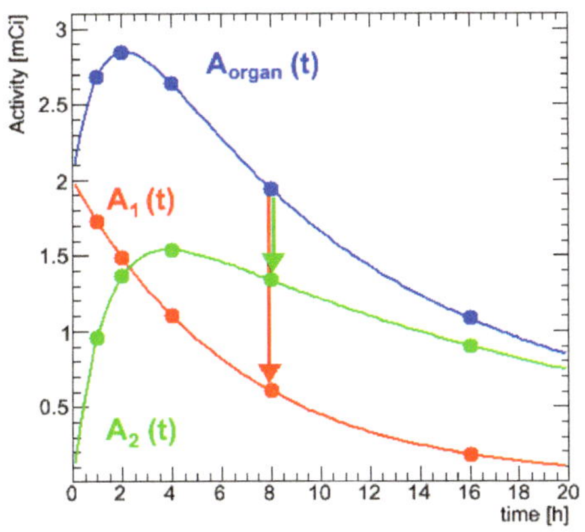

Figure 21.3 Macro2micro apportionment factor determination. The preclinical whole-organ activities are measured at different time points (blue curve). The organs are then imaged at smaller scale using the appropriate modality (auto-radiography, alpha-camera, etc.). Here two sub-compartments are shown in red and green. The apportionment factors assign the amount of activity as a function of time seen in the organ to the compartments.

21.6 Alpha-Particle RPT

Interest in the use of α-particle RPTs is growing at even a more rapid rate than for traditional RPTs. A full treatise or even chapter on αRPT is outside of the scope of this summer school, but we provide a brief overview here to note a few basic concepts and issues relative to the dosimetry.

Alpha particles are helium nuclei and are composed of two protons and two neutrons. As such, they are heavier than β particles by a factor of ~8000. Their energy deposition occurs over a much shorter range, as the greater mass results in α particles plowing through tissue as they deposit energy, rather than scattering through tissue. Alpha particles have energies in the 5 to 10 MeV range, with ranges in tissue on the order of 50 to 100 mm. A number of α-particle emitters are either in use in the clinic, in clinical trials, or being investigated preclinically including, but not limited to, ^{211}At, ^{223}Ra, ^{227}Th, ^{213}Bi, ^{212}Pb, and ^{225}Ac. Dosimetry for these RPTs is generally more complex than for traditional RPTs. There are four major challenges to overcome.

1. ***Small-scale activity and absorbed dose distribution***. As mentioned in the previous section, a small-scale distribution of activity may result in a lack of correlation between organ-level dosimetry and biological outcome. This is particularly true for α particles, as the short range of the energy distribution does little to overcome the nonuniform activity distribution. The macro2micro method has the potential to overcome this issue, but known apportionment factors are still relatively rare.
2. ***The high linear energy transfer (LET) of a particles results in a greater biological damage to tissue per unit of absorbed dose***. This results in a scale

of αRPT absorbed dose that is not consistent with traditional dosimetry. This enhancement relative to traditional (or low-LET) radiation is known as the *relative biological effect* (RBE) and can be quantified, such that multiplying the αRPT absorbed dose by the organ- and RPT-specific RBE provides comparable, standardized dosimetry (see Chapter 22). A number of RBE values have been reported, generally with values in the 3 to 9 range (Konijnenberg et al. 2007); nevertheless, as with the apportionment factors, RBE values are not generally known and are often reported using a non-standard reference.

3. ***Many α-particle emitters have α particles in their decay chain***. After decay of the parent isotope, the daughter particle will generally break the bond with the pharmaceutical and may re-localize independently of the parent RPT. In order to perform complete dosimetry, the localization of the daughters and the resulting absorbed dose contributions from the daughters in each of the different organs must be accounted for; these may be very different for the daughters as compared to the parent isotope.
4. ***Because of the high potency and lack of clear dosimetric understanding, the amount of administered activity in αRPTs is very conservative, with administrations on the order of a few hundred μCi***. As a result, there are relatively few photons being emitted from αRPT, and a low count rate of photons is seen in SPECT/CTs, which makes it challenging to make accurate or precise activity measurements.

Resolutions for all of these issues are currently being worked on, and it is likely that personalized dosimetry-based treatment planning could become an option in the near future.

References

Antipas, V., R. G. Dale, and I. P. Coles (2001). "A theoretical investigation into the role of tumour radiosensitivity, clonogen repopulation, tumour shrinkage and radionuclide RBE in permanent brachytherapy implants of 125I and 103Pd." *Phys. Med. Biol.* 46(10):2557–69.

Baechler, S., et al. (2008). "Extension of the biological effective dose to the MIRD schema and possible implications in radionuclide therapy dosimetry." *Med. Phys.* 35(3):1123–34.

Baechler, S., et al. (2010). "Predicting hematologic toxicity in patients undergoing radioimmunotherapy with 90Y-ibritumomab tiuxetan or 131I-tositumomab." *J. Nucl. Med.* 51(12):1878–84.

Barendsen, G. W. (1982). "Dose fractionation, dose rate and iso-effect relationships for normal tissue responses." *Int. J. Radiat. Oncol. Biol. Phys.* 8(11):1981–97.

Barone, R., et al. (2005). "Patient-specific dosimetry in predicting renal toxicity with (90)Y-DOTATOC: relevance of kidney volume and dose rate in finding a dose-effect relationship." *J. Nucl. Med.* 46 Suppl 1:99S–106.

Bartlett, R. M., et al. *MIRD Primer 2022: A Complete Guide to Radiopharmaceutical Dosimetry.* Reston, VA: Society of Nuclear Medicine and Molecular Imaging, 2022.

Bentzen, S. M., et al. (2010). "Quantitative Analyses of Normal Tissue Effects in the Clinic (QUANTEC): an introduction to the scientific issues." *Int. J. Radiat. Oncol. Biol. Phys.* 76(3 Suppl):S3–9.

Bentzen, S. M., et al. (2012). "Bioeffect modeling and equieffective dose concepts in radiation oncology—terminology, quantities and units." *Radiother. Oncol.* 105(2):266–68.

Besse, I. M., et al. (2009). "Modeling combined radiopharmaceutical therapy: a linear optimization framework." *Technol. Cancer Res. Treat.* 8(1):51–60.

Bodey, R. K., G. D. Flux, and P. M. Evans (2003). "Combining dosimetry for targeted radionuclide and external beam therapies using the biologically effective dose." *Cancer Biother. Radiopharm.* 18(1):89–97.

Bodey, R. K., P. M. Evans, and G. D. Flux (2004). "Application of the linear-quadratic model to combined modality radiotherapy." *Int. J. Radiat. Oncol. Biol. Phys.* 59(1):228–41.

Bouchet, L. G., et al. (2003). "MIRD Pamphlet No 19: absorbed fractions and radionuclide S values for six age-dependent multiregion models of the kidney." *J. Nucl. Med.* 44(7):1113–47.

Brenner, D. J. and E. J. Hall (1991). "Conditions for the equivalence of continuous to pulsed low dose rate brachytherapy." *Int. J. Radiat. Oncol. Biol. Phys.* 20(1):181–90.

Dale, R. G. (1996). "Dose-rate effects in targeted radiotherapy." *Phys. Med. Biol.* 41(10):1871–84.

Fowler, J. F. (1990). "Radiobiological aspects of low dose rates in radioimmunotherapy." *Int. J. Radiat. Oncol. Biol. Phys.* 18(5):1261–69.

Garin, E., et al. (2006). "Effect of a ^{188}Re-SSS lipiodol/^{131}I-lipiodol mixture, ^{188}Re-SSS lipiodol alone or ^{131}I-lipiodol alone on the survival of rats with hepatocellular carcinoma." *Nucl. Med. Commun.* 27(4):363–69.

Goodhead, D. T. (1989). "The initial physical damage produced by ionizing radiations." *Int. J. Radiat. Oncol. Biol. Phys.* 56(5):623–34.

Gustafsson, J., P. Nilsson, and K. S. Gleisner (2013a). "On the biologically effective dose (BED)-using convolution for calculating the effects of repair: I. Analytical considerations." *Phys. Med. Biol.* 58(5):1507–27.

Gustafsson, J., P. Nilsson, and K. S. Gleisner (2013b). "On the biologically effective dose (BED)-using convolution for calculating the effects of repair: II. Numerical considerations." *Phys. Med. Biol.* 58(5):1529–48.

Hobbs, R. and G. Sgouros (2009). "Calculation of the Biological Effective Dose (BED) for Piece-Wise Defined Dose-Rate Fits." *Med. Phys.* 36(3):904–07.

Hobbs, R. F., et al. (2011a). "A model of cellular dosimetry for macroscopic tumors in radiopharmaceutical therapy." *Med. Phys.* 38(6):2892–903.

Hobbs, R. F., et al. (2011b). "A treatment planning method for sequentially combining radiopharmaceutical therapy and external radiation therapy." *Int. J. Radiat. Oncol. Biol. Phys.* 80(4):1256–62.

Hobbs, R. F., et al. (2012). "A nephron-based model of the kidneys for macro-to-micro alpha-particle dosimetry." *Phys Med Biol.* 57(13) 4403–24.

Hobbs, R. F., et al. (2013). "Radiobiologic optimization of combination radiopharmaceutical therapy applied to myeloablative treatment of non-Hodgkin lymphoma." *J. Nucl. Med.* 54(9):1535–42.

Howell, R. W., D. V. Rao, and K. S. Sastry (1989). "Macroscopic dosimetry for radioimmunotherapy: nonuniform activity distributions in solid tumors." *Med. Phys.* 16(1):66–74.

Howell, R. W., S. M. Goddu, and D. V. Rao (1994). "Application of the linear-quadratic model to radioimmunotherapy: further support for the advantage of longer-lived radionuclides." *J. Nucl. Med.* 35(11):1861–69.

Kerr, C. P., et al. (2022). "Developments in Combining Targeted Radionuclide Therapies and Immunotherapies for Cancer Treatment." *Pharmaceutics* 15(1):128.

Konijnenberg, M., et al. (2007). "Radiation dose distribution in human kidneys by octreotides in peptide receptor radionuclide therapy." *J. Nucl. Med.* 48(1):134–42.

Kotzerke, J., D. Bunjes, and D. A. Scheinberg (2005). "Radioimmunoconjugates in acute leukemia treatment: the future is radiant." *Bone Marrow Transplant* 36(12):1021–26.

Kunikowska, J., et al. (2011). "Clinical results of radionuclide therapy of neuroendocrine tumours with ^{90}Y-DOTATATE and tandem ^{90}Y/^{177}Lu-DOTATATE: which is a better therapy option?" *Eur. J. Nucl. Med. Mol. Imaging* 38(10):1788–97.

Leichner, A., et al. (2008). "Targeted radionuclide therapy: theoretical study of the relationship between tumour control probability and tumour radius for a ^{32}P/^{33}P radionuclide cocktail." *Phys. Med. Biol.* 53(7):1961–74.

Madsen, M. T., et al. (2006). "Potential increased tumor-dose delivery with combined 131I-MIBG and 90Y-DOTATOC treatment in neuroendocrine tumors: a theoretic model." *J. Nucl. Med.* 47(4):660–67.

Niemierko, A. (1997). "Reporting and analyzing dose distributions: a concept of equivalent uniform dose." *Med. Phys.* 24(1):103–10.
O'Donoghue, J. A., M. Bardies, and T. E. Wheldon (1995). "Relationships between tumor size and curability for uniformly targeted therapy with beta-emitting radionuclides." *J. Nucl. Med.* 36(10):1902–09.
O'Donoghue, J. A. (1999). "Implications of nonuniform tumor doses for radioimmunotherapy." *J. Nucl. Med.* 40(8):1337–41.
Prideaux, A. R., et al. (2007). "Three-dimensional radiobiologic dosimetry: application of radiobiologic modeling to patient-specific 3-dimensional imaging-based internal dosimetry." *J. Nucl. Med.* 48(6):1008–16.
Sgouros, G., J. C. Roeske, M. R. McDevitt, et al. (2010). "MIRD pamphlet No. 22 (Abridged): Radiobiology and dosimetry of α-particle emitters for targeted radionuclide therapy." *J. Nucl. Med.* 51:311–28.
Sgouros, G., et al. (2021). "ICRU Report 96: Dosimetry-guided radiopharmaceutical therapy." *J. ICRU* 21(1):1–212.
Solanki, J. H., et al. (2017). "Cellular Response to Exponentially Increasing and Decreasing Dose Rates: Implications for Treatment Planning in Targeted Radionuclide Therapy." *Radiat. Res.* 188(2):221–34.
Strigari, L., et al. (2011). "Dosimetry in nuclear medicine therapy: radiobiology application and results." *Q. J. Nucl. Med. Mol. Imaging* 55(2):205–21.
Villard, L., et al. (2012). "Cohort Study of Somatostatin-Based Radiopeptide Therapy With [90Y-DOTA]-TOC Versus [90Y-DOTA]-TOC Plus [177Lu-DOTA]-TOC in Neuroendocrine Cancers." *J. Clin. Oncol.* 30(10):1100–06.
Wessels, B.W., et al. (2008). "MIRD pamphlet No. 20: the effect of model assumptions on kidney dosimetry and response—implications for radionuclide therapy." *J. Nucl. Med.* 49(11):1884–99.

Chapter 22

Standardization and ICRU Report 96[1]

Roger W. Howell, Ph.D., FSNMMI

ICRU Commissioner
Department of Radiology
Rutgers New Jersey Medical School
Newark, NJ

1. To ensure accuracy of information from ICRU Report 96 (Sgouros et al. 2021), excerpts of the report are reproduced herein with minor modifications. Sage permits reuse of the Accepted Manuscript in a book authored or edited by its authors. Students are expected to refer to ICRU Report 96 for explanations and complete details. Access to ICRU Report 96 and other ICRU reports is an AAPM member benefit.

List of Symbols

A	activity
A_0	administered activity
$\tilde{A}$	time-integrated activity
a	fraction of the administered activity in source region
$\tilde{a}$	time-integrated activity coefficient
D	absorbed dose
$\dot{D}$	absorbed dose rate
C	activity concentration
d	absorbed dose coefficient
Δ	mean energy emitted per decay
E_i	initial energy of the ith radiation
fr_S	fraction of administered activity taken up by a single identified source region
G	Lea-Catcheside time factor
L	the unrestricted linear energy transfer
λ	rate constant
λ_e	effective rate constant
λ_{ec}	effective clearance rate constant
λ_{eu}	effective uptake rate constant
λ_p	physical decay constant
M	mass
μ	linear attenuation coefficient; also used for DNA repair rate assuming exponential repair
ϕ	absorbed fraction
Φ	specific absorbed fraction
r_S	source region
r_T	target region
S	mean absorbed dose per decay
T_b	biological half-time
T_{bc}	biological clearance half-time
T_{bu}	biological uptake half-time
T_e	effective half-time
T_{ec}	effective clearance half-time
T_{eu}	effective uptake half-time
T_p	physical half-life (also $T_{1/2}$)
T_μ	repair half-time
τ	dose-integration period
u	atomic mass unit
u	uncertainty
U	uncertainty tolerance limit
Y_i	yield (mean number of ith radiation emitted per nuclear transformation)
Z	charge number of an atomic nucleus

22.1 Introduction

22.1.1 International Standards for Communication and Patient Safety in Radiopharmaceutical Therapy

Radiopharmaceutical therapy is a rapidly growing modality for cancer therapy. There have been several recent U.S. FDA approvals of radiopharmaceuticals for therapy, including Xofigo® ($^{223}RaCl_2$), SIR-Spheres® (^{90}Y resin microspheres), TheraSphere™ (^{90}Y glass microspheres), LUTATHERA® (^{177}Lu DOTATATE), and PLUVICTO™ (^{177}Lu vipivotide tetraxetan). These agents follow their predecessors Zevalin® (^{90}Y ibritumomab tiuxetan), AZEDRA® ($Na^{131}I$), and HICON™ ($Na^{131}I$). These agents, and others under development, comprise different classes of radionuclides and pharmaceuticals.

To ensure that these radiopharmaceuticals are developed and implemented so that tumor response is maximized and normal-tissue toxicity is minimized, ICRU Report 96 recommends that dosimetry-based treatment strategies similar to those used in external beam radiation therapy should be implemented (Sgouros et al. 2021). Report 96 establishes standardized terminology and nomenclature for radiopharmaceutical therapy, an important step to facilitate the common ground that is needed for efficient and accurate communication of treatment plans and maturation of the field. Planning and verifying the absorbed doses and equieffective doses received by both tumor and normal tissues is required to assure optimal treatment for each patient. ICRU Report 96 recommends that a record of this information be maintained for use in planning retreatments. Not all treatment centers are equipped with the same level of technology for radiopharmaceutical therapy, therefore the ICRU has established reporting levels for prescribing, recording, and reporting of treatments with increasing levels of sophistication (Level 1, Level 2, Level 3) (Sgouros et al. 2021). Treatment centers are encouraged to elevate their standards to the highest level possible, and patients can benefit from seeking treatment at the highest level accessible to them.

Due to the unique properties of each radiopharmaceutical, the requirements to achieve a given level are specific to the radiopharmaceutical and will likely evolve as new technology develops. This chapter will review some of the key recommendations in ICRU Report 96. Readers are encouraged to study and implement the report, which is available free for AAPM members at the link below.

22.1.2 Required Course Materials

Summer school attendees are expected to download a copy of ICRU Report 96, *Dosimetry-Guided Radiopharmaceutical Therapy* (free for AAPM members) at https://www.aapm.org/pubs/ICRU/detail.asp?docid=94.

22.1.3 Optional Course Materials

ICRU Report 85, *Fundamental Quantities and Units for Ionizing Radiation* (free for AAPM members): https://www.aapm.org/pubs/ICRU/detail.asp?docid=72.

Other ICRU reports: https://www.aapm.org/pubs/ICRU/?d=d.

22.2 ICRU Quantities, Units, Symbols, and Definitions

The vernacular of radiopharmaceutical therapy is varied and inconsistent, and this often causes confusion among the general population, students, patients, and even professionals in radiology, radiation oncology, and nuclear medicine. As radiopharmaceutical therapy grows into a mainstream therapeutic modality, it important that a common language be adopted. Accordingly, the ICRU recommends that care be taken to ensure the use of internationally accepted quantities, units, and definitions that are specified in ICRU Report 96 (Sgouros et al. 2021). This section begins with Table 22.1 which describes some common misuses of terminology, followed by internationally accepted definitions specified by the ICRU. Table 22.1 also lists the plethora of phrases used for radiopharmaceutical therapy. While some of them seek to specify a type of radiopharmaceutical therapy, others are redundant. ICRU Report 96 recommends that this medical technology be referred to as "radiopharmaceutical therapy" (Sgouros et al. 2021).

22.2.1 Absorbed Dose

The *absorbed dose*, D, is defined by:

$$D = \frac{d\bar{\varepsilon}}{dm}, \tag{22.1}$$

where $d\bar{\varepsilon}$ is the mean energy imparted by ionizing radiation to matter in a mass element dm. The unit of absorbed dose is the joule per kilogram (J kg^{-1}), with the special name gray (Gy).

22.2.2 Absorbed-dose Rate

The *absorbed-dose rate,* $\dot{D}$, is defined as

$$\dot{D} = \frac{dD}{dt}, \tag{22.2}$$

where dD is the increment of absorbed dose in the time interval dt. The unit of absorbed dose is the joule per kilogram per second (J kg^{-1} s^{-1}) or gray per second (Gy s^{-1}).

22.2.3 Mean Absorbed Dose

The mean absorbed dose, $D(r_T)$, in a specific tissue region is defined as (ICRU 2011):

$$D(r_T) = \frac{1}{m_T} \int D \, dm, \tag{22.3}$$

where D is the absorbed dose in the mass element. dm, and m_T is the mass of the tissue region.

Table 22.1 Examples of correct uses, common misuses, and preferred uses

Correct	Incorrect
absorbed dose	dose physical dose
administered activity	dose
biologically effective dose (BED)	dose biological dose biologically equivalent dose
biological half-time	biological half-life
equieffective dose	dose biological dose effective dose biologically equivalent dose
gray	Gray
Gy	Gy (RBE)
physical half-life	half-life
time-integrated activity coefficient	residence time
Preferred	**Not Preferred**
radiopharmaceutical therapy (RPT)	alpha radionuclide therapy Auger therapy Auger-electron therapy beta-plus Auger-electron therapy high-LET radionuclide radiation therapy molecular radiotherapy molecularly targeted radiation therapy molecular targeted radiotherapy peptide receptor radionuclide therapy radioactive iodine therapy radioactive targeted therapy radiochemical therapy radioimmunotherapy radioiodine therapy radioisotope therapy radioligand therapy radionuclide therapy somatostatin radionuclide therapy systemic radiation therapy systemic radioisotope therapy targeted alpha therapy targeted α-particle radiation therapy targeted Auger-electron-emitter therapy targeted beta therapy targeted beta radiotherapy targeted molecular radiotherapy targeted radionuclide therapy

22.2.4 MIRD Schema for Absorbed-Dose Computation

Mean Absorbed-Dose Rate

The absorbed-dose rate, $\dot{D}(r_T,t)$, to the target tissue r_T from radioactivity distributed uniformly within source tissue r_S at time t after administration is given as:

$$\dot{D}(r_T,t) = \Sigma_{r_S} A(r_S,t)\, S(r_T \leftarrow r_S,t), \tag{22.4}$$

where $A(rs,t)$ is the time-dependent activity of the radiopharmaceutical in the source tissue region r_S $S(r_T \leftarrow r_S,t)$. The radionuclide-specific S-coefficient, $S(r_T \leftarrow r_S,t)$, (also called the *S value*), is the mean absorbed dose to target tissue per decay in the source tissue at time t.

Mean Absorbed Dose: Time-Dependent Formulation

The mean absorbed dose $D(r_T,\tau)$ to target tissue r_T over a defined dose-integration period τ after administration of the radioactive material to the subject is given as:

$$D(r_T,\tau) = \int_0^\tau \dot{D}(r_T,t)\,dt = \sum_{r_S} \int_0^\tau A(r_S,t) S(r_T \leftarrow r_S,t)\,dt. \tag{22.5}$$

where the time integration limit is τ.

If $A(r_S,t)$ is normalized to a unit administered activity A_0 and denoted as $a(r_S,t)$, then the absorbed-dose coefficient $d(r_T,\tau)$ in target tissue r_T is given as:

$$d(r_T,\tau) = \Sigma_{r_S} \int_0^\tau a(r_S,t) S(r_T \leftarrow r_S,t)\,dt, \tag{22.6}$$

where $a(r_S,t)$ is the fraction of the administered activity in the source tissue r_S at time t post-administration.

S Coefficient

The S coefficient is given as:

$$\begin{aligned} S(r_T \leftarrow r_S,t) &= \frac{1}{m(r_T,t)} \sum_i E_i\, Y_i\, \phi(r_T \leftarrow r_S, E_i, t) \\ &= \frac{1}{m(r_T,t)} \sum_i \Delta_i\, Y_i\, \phi(r_T \leftarrow r_S, E_i, t), \end{aligned} \tag{22.7}$$

where

E_i	=	the mean (or individual) energy of the ith nuclear transition,
Y_i	=	the number of ith nuclear transitions per nuclear transformation,
Δ_i	=	their product (mean energy of the ith transition per nuclear transformation),
$\Phi(r_T \leftarrow r_S, E_i, t$	=	the absorbed fraction (defined as the fraction of radiation energy E_i emitted within the source tissue r_S at time t that is absorbed in the target tissue r_T), and
$m(r_T,t)$	=	the time-dependent mass of the target tissue r_T in the patient tissue model.

Specific Absorbed Fraction

The specific absorbed fraction $\Phi(r_T \leftarrow r_S, E_i, t)$ is defined as the ratio of the absorbed fraction and the target mass:

$$\Phi(r_T \leftarrow r_S, E_i, t) = \frac{\Phi(r_T \leftarrow r_S, E_i, t)}{m(r_T, t)}, \tag{22.8}$$

such that

$$S(r_T \leftarrow r_S, t) = \sum_i \Delta_i \Phi(r_T \leftarrow r_S, E_i, t). \tag{22.9}$$

Mean Absorbed Dose: Time-Independent Formulation

In many instances, the time dependency of S may be neglected. Under such conditions, the time-independent form is:

$$D(r_T, \tau) = \sum_{r_S} \tilde{A}(r_S, \tau)\, S(r_T \leftarrow r_S), \tag{22.10}$$

where $\tilde{A}(r_S, \tau)$ is the time-integrated activity (or total number of nuclear transformations) in source tissue r_S over the dose-integration period τ such that $\tilde{A}(r_S, \tau) = \int_0^{\tau} A(r_S, t)\, dt$. Furthermore,

$$d(r_T, \tau) = \sum_{r_S} \tilde{a}(r_S, \tau) S(r_T \leftarrow r_S), \tag{22.11}$$

where

$$\tilde{a}(r_S, \tau) = \int_0^{\tau} a(r_S, t)\, dt = \frac{1}{A_0} \int_0^{\tau} A(r_S, t)\, dt, \tag{22.12}$$

and the quantity $\tilde{a}(r_S, \tau)$ is termed the time-integrated activity coefficient. The value of $\tilde{a}(r_S, \tau)$ represents the cumulative number of nuclear transformations (Bq s) occurring in source tissue r_S over a dose-integration period τ per administered activity A_0 (Bq).

22.2.5 Equieffective Dose (EQD*X*)

The ICRU recommends a quantity named *equieffective dose* for bioeffect modeling (Bentzen et al. 2012). The equieffective dose, like the BED, is designed to account for differences in absorbed-dose rate. The equieffective dose (EQD*X*) relates different dose delivery schedules to a reference dose delivery schedule and has the unit Gy. The ICRU recommends this quantity based, in part, on the large body of published clinical response data for EBRT with 2-Gy fractions. To relate this quantity to EBRT wherein a total absorbed dose, *D*, is often delivered over time in 2-Gy fractions, while also retaining the flexibility to accommodate other reference absorbed-dose fraction sizes, represented by *X*, the ICRU recommends that the nomenclature and symbol for equieffective dose include reference to the absorbed-dose fraction size and also to the α/β ratio. With these considerations in mind, the symbol for equieffective dose is

$\text{EQD}X_{\alpha/\beta}$, with the recommended standard of $\text{EQD2}_{\alpha/\beta}$, where "2" refers to 2-Gy daily fractions.

This formalism is applicable to any modality, and its advantage is a common reference frame across all radiation modalities. The BED has traditionally been used in nuclear medicine and is equivalent to the "infinitely low absorbed-dose rate" of near 0-Gy fractions or EQD0 (Bentzen et al. 2012). Consequently, EQD2 and BED are related by a simple multiplicative factor that depends on the α/β ratio but does not depend on absorbed dose and, thus, any cause-effect relationships established using the BED can be converted to an equivalent EDQ2 relationship by a simple factor and vice versa.

The equieffective dose, EQD*X*, is defined in ICRU Report 96 as the total absorbed dose delivered by the reference treatment plan (fraction size *X*) that leads to the same biological effect in a test treatment plan conducted with absorbed dose per fraction, *d*, and total absorbed dose, *D* (Bentzen et al. 2012; Sgouros et al. 2021).

$$\begin{gathered}\text{EQD}X_{\alpha/\beta} = D \times \frac{\alpha/\beta + d}{\alpha/\beta + X}, \text{ where} \\ \text{BED} = \text{EQD0}_{\alpha/\beta} = D \times \frac{\alpha/\beta + d}{\alpha/\beta}.\end{gathered} \tag{22.13}$$

The time-varying absorbed-dose rates in nuclear medicine require a modification with the *G* factor (Hobbs et al. 2014; Hobbs et al. 2011; Konijnenberg 2003):

$$\text{EQD}X_{\alpha/\beta} = D \times \frac{[\alpha/\beta + G(\infty) \times D]}{(\alpha/\beta + X)}, \tag{22.14}$$

where $G(\tau)$ is given by (Lea and Catcheside 1942):

$$G(\tau) = \frac{2}{D^2} \times \int_0^{\tau} \dot{D}(t)dt \int_0^{\tau} \dot{D}(w) \times e^{-\mu(t-w)} dw, \tag{22.15}$$

where μ is the exponential repair rate constant, and the second integral represents the time between breaks; w is the time of the first break, t the time of the second break, and τ the duration of irradiation.

22.2.6 Standardized Relative Biological Effectiveness (sRBE*X*)

ICRU Report 96 defines the *standardized relative biological effectiveness* (sRBE*X*) (Hobbs et al. 2014; Sgouros et al. 2021):

$$\text{sRBE}X_{\alpha/\beta} = \frac{\kappa}{(\alpha + \beta X)}, \tag{22.16}$$

where κ is the coefficient in the log-linear cell survival response of cells to high-LET radiation, ($SF_H = e^{-\kappa D_H}$), and D_{H} is the absorbed dose of high-LET radiation. Therefore, for high-LET radiations, $\text{EDQ}X_{\alpha/\beta} = D_{\text{H}} \times \text{sRBE}X_{\alpha/\beta}$ where:

$$\mathrm{sRBE0}_{\alpha/\beta} = \frac{\kappa}{\alpha}, \qquad (22.17)$$

and

$$\mathrm{sRBE2}_{\alpha/\beta} = \frac{\kappa}{\alpha + 2\beta}. \qquad (22.18)$$

The ICRU recommends that reporting should include both the conventional RBE values calculated over the entire range of absorbed dose, as well as an $\mathrm{sRBE}X_{\alpha/\beta}$ value and the values of the parameters used to derive the $\mathrm{sRBE}X_{\alpha/\beta}$.

22.2.7 Equieffective Dose vs. RBE-Weighted Dose

The absorbed-dose dependence of the conventional RBE, and the often-used RBE-weighted dose, is mitigated substantially by the newly defined $\mathrm{sRBE}X_{\alpha/\beta}$. The product of $\mathrm{sRBE}X_{\alpha/\beta}$ and absorbed dose is simply the equieffective dose:

$$\mathrm{EDQ}X_{\alpha/\beta} = D_{\mathrm{sRBEX}} = \mathrm{sRBE}X_{\alpha/\beta} \times D. \qquad (22.19)$$

This provides strong support for the adoption of the newly defined and the equieffective dose for bioeffect modeling in radiopharmaceutical therapy. Accordingly, the ICRU recommends the use of the quantity equieffective dose and discourages the use of the quantity RBE-weighted dose for radiopharmaceutical therapy. The unit to be used for the equieffective dose is the gray (Gy).

22.2.8 Cellular and Multicellular Dosimetry and Bioeffect Modeling

The biological effects of radiation are ultimately the consequence of effects at the cellular level. In the case of radiation delivered by internally distributed radionuclides—where the distributions of the radionuclides and of the energy emitted and deposited are often highly nonuniform—estimation of the absorbed dose to individual cells and to critical radiobiological targets such as the cell nucleus can be performed with *MIRDcell* (Figure 22.1) a downloadable software application at https://mirdcell.njms.rutgers.edu (Katugampola et al. 2022; Vaziri et al. 2014).

22.3 ICRU Nomenclature for Radiopharmaceutical Therapy

The treatment protocols can differ between different kinds of RPTs. While some are given as a single administration, others involve several serial administrations separated by some preset time interval. When more than one administration is prescribed and planned as part of the therapy protocol, then the series of administrations are regarded to be part of the same treatment (or therapy). The treatment thus extends from the time of the first administration to the (unknown) time point when radiobiological effects from the last administration end. ICRU Report 96 refers to each of these administrations as *treatment fractions*.

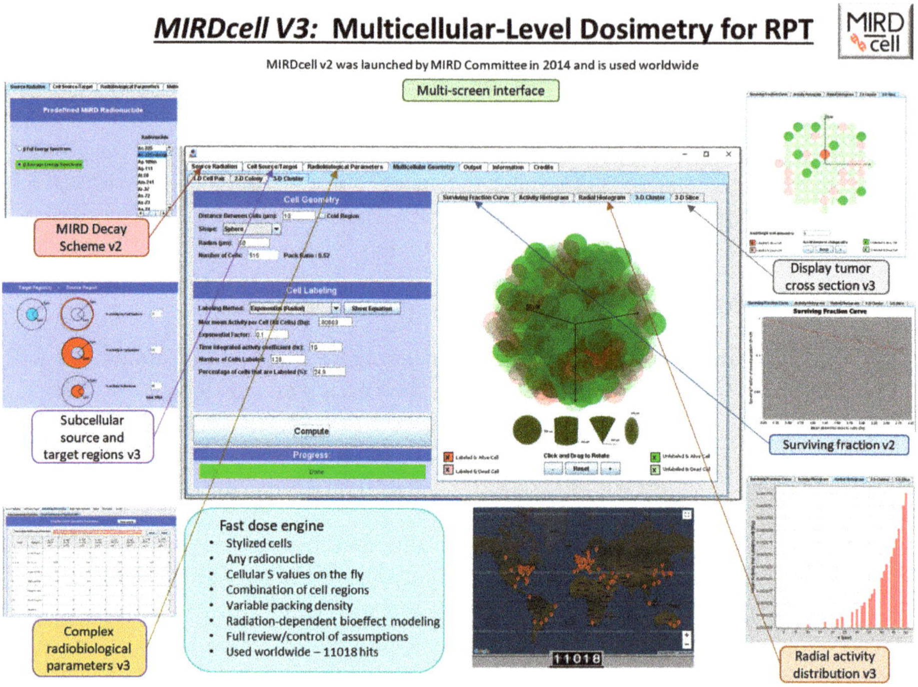

Figure 22.1 Features and graphic user interfaces within MIRDcell V3 (Katugampola et al. 2022; Vaziri et al. 2014).

ICRU Report 96 recommends a new nomenclature system for radiopharmaceutical therapy. Table 22.2 summarizes the suggested nomenclature along with the ICRU nomenclature for external beam radiation therapy (EBRT). Figure 22.2 gives an example of how this suggested nomenclature can be applied

Table 22.2 Summary of the region nomenclature for RPT in comparison with nomenclature from previous external beam ICRU documents

Previous External Beam ICRU Reports	ICRU Report 96
Patient Representation: The patient representation obtained from medical imaging is the volume over which dose from external beams may be computed. Ideally it represents the entire patient anatomy.	**Localization Region:** The organs, tissues, or cells to which the radioactive compound localizes.
Irradiated Volume: Volume over which the primary external beam passes, including the patient, patient support, and patient immobilization systems. All of the irradiated volume of the patient must be in the patient representation with generous superior and inferior borders. The patient representation must contain all of the patient's irradiated volume.	**Source Region (SR):** The organs and tissues for which the activity content as a function of time and the time-integrated activity are estimated. The SR is a subcategory of the Localization Region. The SR is categorized into the Delineable Source Region(s) (SRD) and the Non-Delineable Source Region(s) (SRND), depending on the possibility to perform patient-specific image-based estimation of the time-integrated activity, and the level of uncertainty in the absorbed dose estimate.
Gross Tumor Volume (GTV), Clinical Target Volume (CTV): Region(s) of disease for which the mass and absorbed dose are estimated. The GTV and CTV are subregions of the PTV. The GTV is delineated with clinical and imaging evidence. The CTV is an extension of the GTV (such that the GTV is a subregion of the CTV) delineated to include suspected microscopic extension or may be regions to which disease may have spread away from the GTV, such as draining lymph nodes.	**Clinical Treatment Region (CTR):** Region(s) to treat, including macroscopic disease and microscopic disease (i.e., the total disease burden with margins for delineation uncertainty).
Planning Target Volumes (PTV), Internal Target Volume (ITV): Region(s) to treat, including macroscopic disease and microscopic diseases with margins for delineation, motion, and setup uncertainties.	**Dosimetric Treatment Region (DTR):** Region(s) of disease for which the mass and absorbed dose are estimated. The DTR is a subcategory of the CTR. The DTR is categorized into the Delineable Dosimetric Treatment Region(s) (DTRD) and the Non-Delineable Dosimetric Treatment Region(s) (DTRND), depending on the possibility to perform patient-specific estimation of the tissue mass from images and the level of uncertainty in the absorbed dose estimate. A region of disease included as an SR is generally also included among the DTRs.
Organ at Risk (OAR) and Remaining Volume at Risk (RVR): Critical tissues that if irradiated could suffer significant morbidity or functional loss, and for which the mass and absorbed dose are estimated. The OAR is delineated for organs that may be acutely or chronically damaged by the radiation. The OAR may overlap with CTV or PTV. The RVR is the Patient Representation not included by the CTVs and OARs. The margin between the PTV and CTV would be considered part of the RVR if not otherwise co-occupied by part of an OAR.	**Region at Risk (RAR):** Critical tissues that if irradiated could suffer significant morbidity or functional loss, and for which the mass and absorbed dose are estimated. The RAR is categorized into the Delineable Region(s) at Risk (RARD), the Non-Delineable Region(s) at Risk (RARND), and the Region at Risk for Secondary Effects (RARSE). The categorization into RARD or RARND is determined by the possibility to perform patient-specific estimation of the tissue mass from images and the level of uncertainty in the absorbed dose estimate. The RARSE refers to irradiated body regions for which absorbed-dose estimates may be useful for estimating the long-term risk of stochastic effects. A normal tissue included as an SR is generally included among the RARs.

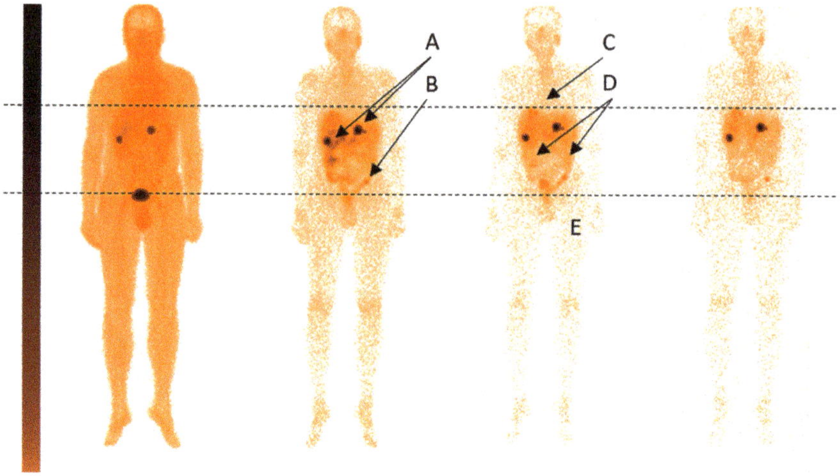

Figure 22.2 Example of the application of the suggested nomenclature for a patient receiving ^{177}Lu-DOTATATE peptide receptor radionuclide therapy. The horizontal dashed lines indicate the limits of the field-of-view of the SPECT/CT scans that were acquired in parallel to these total-body scans, representing the activity distribution at 1 h, 24 h, 96 h, and 168 h after administration. Letters indicate the different types of regions as described below, also including the technique used for quantification of activity and mass.

CTR	Neuroendocrine tumors
Localization Regions:	Body regions that accumulate ^{177}Lu-DOTATATE by the expression of somatostatin receptors or by other mechanisms, including tumors, liver, kidneys, spleen, small intestine, and, in some cases, pituitary gland, salivary gland, and thyroid.
SR_D:	**D:** Left and right kidney (SPECT/CT), **A:** Tumors (SPECT/CT), **E:** Total Body (planar)
SR_{ND}:	**C:** Red bone marrow (whole-blood activity concentration)
DTR_D:	**A:** Tumors (CT or SPECT/CT)
DTR_{ND}:	**B:** Tumors that are either so small so that the relative standard deviation in the absorbed dose estimate is expected to be higher than $U(D)$, or tumors that are located outside the SPECT/CT field-of-view.
RAR_D:	**D:** Left and right kidney (CT)
RAR_{ND}:	**C:** Red bone marrow (mass based on patient weight and literature data)
RAR_{SE}:	**E:** Total Body (patient weight)

22.4 Dosimetry-Based Prescriptions

ICRU Report 96 recommends the following three things with respect to dosimetry-based prescriptions (see the report for full details):

1. ***Dosimetry-based prescriptions should complement, not replace, activity-based prescriptions.***
2. ***Dosimetry-based prescriptions should be specified with the quantity absorbed dose, D, or the equieffective dose, $EQDX_{\alpha/\beta}$, when indicated.*** Equieffective dose prescriptions should always be accompanied by the corre-

sponding absorbed dose and associated parameters. Care should be exercised when the activity distribution is highly nonuniform.

3. ***The prescription absorbed dose (or EQDX) should be constrained firstly by the tolerance of regions at risk (RARs).*** In this case, the maximal activity to be administered is limited by the maximal or mean absorbed dose (or EQD*X*) to RARs. In the absence of constraints from RARs, as is the case for highly efficient uptake of radiopharmaceuticals by the DTR, the administered activity should be high enough to guarantee at least the minimum absorbed dose (or EQD*X*) to the DTR to achieve the stated clinical goals.

22.5 Prescribing, Recording, and Reporting

Robust standards are essential for evaluating clinical dose-response data, for describing the absorbed-dose distribution to DTR and RAR, and for assessing the efficacy of RPT compared to competing therapies. ICRU Report 96 provides a recommended framework to standardize the prescribing, recording and reporting of RPT (Sgouros et al. 2021).

22.5.1 ICRU Reporting Levels

As per previous ICRU reports (Wilke et al. 2019), ICRU Report 96 identifies three reporting levels for RPT. The requirements to achieve a given level depend on the radiopharmaceutical and therapeutic indication. Treating physicians and patients alike should seek the highest level that can be achieved within their constraints.

- ***Level 1 recommendations*** are the minimum standards for prescribing and reporting. Below these standards, any type of radiopharmaceutical therapy should not be performed.
- ***Level 2 recommendations*** apply for the prescribing and reporting of state-of-the-art techniques so that the calculation is patient-specific and that it meets pre-specified uncertainty criteria. A complete *quality assurance* (QA) program is used to ensure that the prescribed treatment is accurately delivered. This translates into collecting the data required to calculate absorbed doses to the DTR and RAR (see Section 22.5 for quantities and nomenclature as they pertain to RPT). Depending upon the agent, clinical circumstances, and ability to accurately collect the needed data, this level should include adjustments to the prescribed activity based on the absorbed dose to the DTR and RAR.
- ***Level 3 recommendations*** are indicated for reporting research and novel developments. They are used for the development of new techniques or approaches for which reporting criteria are not yet standardized by the ICRU (ICRU 2002). The objective here is to develop and implement new techniques that may be implemented and refined in clinical trials to evaluate their applicability for widespread deployment as Level 2 recommendations.

The levels described should be considered the minimum required to achieve that level. The description provided does not preclude prescribing and reporting techniques that are intermediate to the descriptions provided. At all levels, the reports generated should include all input to the calculations used in arriving at the prescribed absorbed dose or administered activity.

22.5.2 Dose Prescription in RPT

Depending on tumor disease and intended treatment, the prescribed absorbed dose is the absorbed dose to the DTR or RAR that will assure the clinical goal. The clinical goal may be overall survival, cancer-specific survival, disease-free survival, or tumor control, depending on the condition of the patient and the treatment plan. Recognizing that in many cases RPT is used to treat disseminated cancer—and that estimation of the absorbed dose to all sites of disease based on activity quantitation with external imaging may not be practically possible—the absorbed dose to DTR is often used as a surrogate of the absorbed dose to the total disease burden (i.e., CTR). In these cases, it is also possible to acquire quantitative information from liquid or solid biopsies that can inform absorbed-dose calculations for *non-detectable* (ND) disease.

When the clinical intent is disease control, the absorbed-dose prescription is based on the probability that there are cancer cells in some anatomically and functionally definable tissues in organs or other areas. These cells, detectable or non-detectable, require an appropriate absorbed dose to reduce the risk of recurrence or later detectable manifestations if not treated adequately in the sub-clinical stage. The estimate of that probability is based on clinical experience from adequately documented treatment. In addition, the intention of treatment can be radical to achieve the tumor control or palliative to relieve patients from discomforting symptoms of cancer (e.g., pain).

The prescribed absorbed dose is a fundamental part of treatment prescription, which includes instructions for treatment delivery to achieve the planned absorbed dose to the DTR or RAR and the technical details to deliver the treatment plan. The responsible physician, in collaboration with the medical physicist, identifies the DTR and RAR as well as their prescribed absorbed doses and gives these aims to the medical physicist as a basis for planning a treatment. The medical decision-making process is based on a detailed evaluation of the patient, baseline diagnostic studies, consultation(s) when indicated, oncological concepts, and delineation of treatment regions. The use of one or more therapeutic modalities is also taken into consideration.

22.5.3 Reporting in RPT

Several variables govern the absorbed-dose distribution. These broadly fall into categories concerning the therapeutic radiopharmaceutical and the method of administration. The following information should be reported for all the levels to assure adequate recordkeeping for the treatment plan and permit retrospective evaluation of dose effect relationships:

- The patient's name and an identification code with birth data, which needs to be verified before actual administration.
- The intended treatment: palliative, curative, locoregional, etc. The treatment indication comes from a medical decision-making process, based on the

detailed evaluation of the patient, baseline diagnostic studies, consultation(s) when indicated, oncological concepts, and delineation of treatment regions.

- The use of one or more therapeutic modalities taken into consideration during the patient evaluation.
- In cases where dosimetry-guided treatment is applied, the intended or prescribed absorbed dose to the DTR and RAR, which includes instructions for treatment delivery to achieve the planned absorbed-dose distribution and the technical details to deliver the treatment plan.
- The instructions for treatment delivery, including the number of treatment fractions, the administered activity, and the time interval between fractions.
- The treatment regions and the RAR. In particular, the following quantities should be reported:
    - radionuclide localization regions
    - treatment regions (CTR, DTR_D, DTR_{ND})
    - regions at risk (RAR_D, RAR_{ND}, RAR_{SE})
    - prescribed $D(DTR_D)$, $D(DTR_{ND})$, $D(RAR_D)$, $D(RAR_{ND})$ as per the treatment goal
- The names of the medical physicist, dosimetrist, and the physician responsible for approval of the treatment plan.
- The planning parameters that govern the absorbed-dose calculations should be reported, including:
    - radiopharmaceutical for prospective dosimetry
    - imaging methods for prospective dosimetry
    - radiopharmaceutical for treatment
    - imaging methods for verification dosimetry
    - administered activities
    - # of fractions
    - time interval between fractions
    - $D(DTR_D)$, $D(DTR_{ND})$, $D(RAR_D)$, $D(RAR_{ND})$, $D(RAR_{SE})$
    - SR_D, SR_{ND}
    - biological and effective half-times in SR_D and SR_{ND}
    - $\tilde{A}(SR_D)$, $\tilde{A}(SR_{ND})$
- The absorbed-dose calculation algorithm and software.
- The prescribed administered activity or calculated administered activity to deliver the prescribed absorbed dose.

Recommendation for Reporting at Level 1

In radiopharmaceutical therapy, Level 1 is typically implemented based on administered activity. The requirement at this level is that the prescribed administered activity be confirmed using a dose calibrator (activity meter) that has been calibrated relative to a national metrology lab. The standard uncertainty in the administered activity must be within 10% to meet this level. However, in certain cases (i.e., certain radiopharmaceuticals and clinical scenarios) absorbed-dose calculations are required even for Level 1.

The mean absorbed dose to the DTR and RAR are reported, as appropriate to the treatment intent. The time activity curve and the fitted function for determining the activity biodistribution (if applicable) are to be indicated in the patient report. When possible, the absorbed dose pertinent to RAR_{SE}. should be reported.

Recommendation for Reporting at Level 2

The methodology for prescribing and reporting at Level 2 should be chosen so that the combined standard uncertainty (one standard deviation) in the absorbed dose for the DTR and RAR are within a specified tolerance level $U(D)$. Both quantities will depend on the agent and clinical circumstances. If the treatment objective is to avoid toxicity, then the absorbed dose to the RAR should be compared with available dose constraints specific to the RAR. If the objective is tumor control, then the absorbed dose to the DTR should be one that is expected to yield a clinical response. Depending on the specific clinical circumstances and RPT agent, this may require patient-specific 3D imaging as input into the DTR absorbed-dose calculation. The RAR and DTR selected for the calculations need to be those that are most likely to predict biological outcome to assess safety and efficacy of the treatment. The time activity curve and the fitted function for determining the activity bio-distribution (if applicable) are to be indicated in the patient report. To account for stochastic effects, the absorbed dose to the RAR_{SE} should be calculated. The key distinctions between Level 1 and Level 2 are (1) the degree to which the calculation is patient-specific and (2) the degree to which uncertainty in the absorbed dose for a given patient is reduced.

Recommendations for Reporting at Level 3

Level 3 is appropriate in a research setting where the objective of the research is to identify dosimetry methodologies that may incorporate bioeffect modeling or biomarkers to account for prior treatment history to better predict response or toxicity. Bioeffect modeling would, for example, include absorbed-dose-rate considerations (e.g., EQD2) and incorporate DVH dependent values (e.g., EUD) and examine their reliability relative to resolution constraints of current imaging instrumentation. In addition, the biological input-parameters and the possibility of measuring these in patients using, for example, PET agents that could report on signaling pathways relevant to DNA damage repair and radiosensitivity at a resolution that is higher than a whole organ or tumor. Since long-term response is often driven by microscopic disease, the possibility of using liquid or solid biopsies to assess the absorbed-dose distribution to relevant tumor cell populations is also a potential level 3 endeavor.

In addition, dosimetric studies based on novel radiopharmaceuticals/radionuclides or the adoption of novel devices are to be considered at the reporting Level 3. These studies may lead to an improved characterization of DTR and RAR features that could produce novel dose constraints or more accurate dose effect relationships.

Although criteria for Level 3 reporting are not standardized by the ICRU, one could envision a reporting scheme that would include images overlaid with isodose contours and associated dose-volume histograms and metrics such as the absorbed dose to a specific point or the absorbed dose to 95% of the DTR or RAR (e.g., D95). Dose-volume constraints on RAR can be visualized by isodose contouring.

Motivation and Rationale for the Different Levels

The Level 1 strategy is a methodology generally not recommended, with the exception of cases in which we have a nonmeasurable treatment region or region at risk, a wide therapeutic window, or for those radiopharmaceuticals for which dosimetry is always required. Adoption of the prescribed absorbed dose can be based on available diagnostic images or patient-specific information such as *body mass index* (BMI) and *body surface area* (BSA) to guarantee at least an optimization to reduce the risk of normal tissue toxicity.

Prescribing and reporting at Level 2 should reflect an effort to incorporate treatment planning, either for an individual or a patient population. The absorbed dose to the DTR and RAR must be reported under Level 2. The extent to which these values impact the administered activity to match desired absorbed doses to these regions will depend upon the agent and the clinical circumstances; for example, administered activity adjustments based on RAR absorbed dose should be considered essential for pediatric patients and when the therapeutic window is narrow. Since absorbed dose to RAR is generally not reported currently for RPT patients, there is concern that patients receiving RPT are undertreated. Clinical scenarios where this is a substantial concern should implement Level 2 reporting so that data are collected to assess this concern and potentially adjust the administered activity.

The main motivation for Level 3 reporting would be to capture the results of advanced methodologies so that the potential clinical impact of these may be assessed to help decide which of these should be adopted as a Level 2 recommendation.

A list of radiopharmaceutical therapeutics, their level as prescribed in 2020, and how to move to a higher-level approach to prescribing each RPT is provided in Tables 22.3 and 22.4.

Table 22.3 .ICRU Report 96: 2021 recommendations for dosimetry assessment for federally approved radiopharmaceutical therapies

Radiopharmaceutical Therapy	Treatment Indication	Posology	Prospective Dosimetry (Pretreatment)	Verification Dosimetry (Pretreatment)	Treatment Driving Quantity (RAR or DTR)
^{131}I NaI	Benign thyroid disease Unifocal disease	*L1*: Administered activity as per package insert *L2*: Dosimetry guided treatment. Plan for 300 to 400 Gy to DTR, following relevant guidelines *L3*: TCP/NTCP- optimization	*L1*: None *L2*: ^{131}I NaI probe* *L3*: ^{131}I NaI quantitative imaging *L3*: ^{131}I NaI collar* *L3*: ^{124}I NaI quantitative PET/CT * Imaging for thyroid mass estimation using ^{99m}Tc pertechnetate or ^{123}I NaI or ultrasound	*L1*: None *L2*: ^{131}I NaI probe* *L3*: ^{131}I NaI quantitative imaging *L3*: ^{131}I NaI collar* *L3*: ^{124}I NaI quantitative PET/CT * Imaging for thyroid mass estimation using ^{99m}Tc pertechnetate / ^{123}I NaI / ultrasound	*L1*: None *L2*: DTR=thyroid *L3*: RAR = salivary glands
^{131}I NaI	Benign thyroid disease Multifocal disease	*L1*: as above. *L2*: Dosimetry guided treatment. Plan for 150 to 200 Gy to DTR, following relevant guidelines *L3*: TCP/NTCP- optimization			
^{131}I NaI	Benign thyroid disease Graves' disease	*L1*: as above. *L2*: Dosimetry guided treatment. Plan for 200 Gy to DTR, following relevant guidelines *L3*: TCP/NTCP- optimization			
^{131}I NaI	Differentiated thyroid cancer ablation	*L1*: Administered activity as per package insert, 1.11-7.4 GBq	*L1*: ^{99m}Tc-pertechnetate or ^{123}I NaI quantitative imaging *L2*: ^{131}I NaI quantitative imaging and quantification of activity concentration in blood *L3*: ^{124}I NaI PET/CT lesions and lung dose	*L1*: Quantitative planar imaging or SPECT/CT *L2*: Multiple time-point WB probe and blood sampling * Imaging for thyroid mass estimation using ^{99m}Tc pertechnetate / ^{123}I NaI / ultrasound	*L1*: None *L2*: DTR=thyroid *L3*: RAR = lung, bone marrow, salivary glands
^{131}I NaI	Differentiated thyroid cancer metastatic recurrent disease	*L1*: 3.7-11.1 GBq *A*(Lung, t=48 h)< 3 GBq *L2*: *D*(blood) < 2 Gy			
^{131}I MIBG	Unresectable, locally advanced or metastatic pheochromocytoma and paraganglioma, neuroendocrine tumors (age ⩾ 12 y)	*L1*: Administered activity as per package insert with thyroid blockade (*e.g.* Lugol's). Adjust administered activity based on absorbed dose to RAR. If *D*(RAR) > D_{limit}(RAR), calculate reduced activity as per package insert. *L3*: TCP/NTCP- optimization	*L1*: ^{131}I MIBG quantitative WB planar imaging at three time points as per package insert. BW>50 kg, 185-222 MBq BW<50 kg 3.7 MBq/kg Total kidney volume (left +right) via CT scan (density = 1.03 g cm^{-3}) *L2*: Use patient specific organ masses estimated from MRI or CT for RAR dosimetry ($^{131/123}$I MIBG) *L3*: Quantitative SPECT/CT or PET/CT (^{124}I MIBG) imaging of RAR		*L1*: RAR= marrow, lungs, kidneys, liver, small intestine *L3*: DTR = tumor
^{131}I MIBG	Neuroblastoma (children and young adults)	*L1*: Administered activity based on BW as per package insert with thyroid blockade (e.g., Lugol's) *L2*: as above *L3*: *D*(lesions)	*L1*: None *L2*: ^{131}I MIBG (5 MBq) or ^{123}I mIBG uptake (Level 1) *L3*: ^{124}I MIBG	*L1*: None *L2*: WB probe for WB/BM dose *L3*: Planar WB or SPECT/CT	*L1*: None *L2*: RAR= marrow, lungs, kidneys, liver, small intestine *L3*: DTR = tumor

(continued)

Table 22.3 continued

Radiopharmaceutical Therapy	Treatment Indication	Posology	Prospective Dosimetry (Pretreatment)	Verification Dosimetry (Pretreatment)	Treatment Driving Quantity (RAR or DTR)
^{177}Lu DOTATATE	Neuroendocrine tumors	*L1*: Administered activity and amino acids as per package insert *L2*: *D*(lesions) and RAR *L3*: TCP/NTCP- optimization	*L1*: None *L2*: ^{177}Lu DOTATATE quantitative imaging for treatment fractions following the first *L2*: ^{68}Ga DOTATATE may be used for mass estimation *L3*: ^{68}Ga DOTATATE used for mass estimation	*L1*: None *L2*: ^{177}Lu DOTATATE quantitative imaging and dosimetry for one, several, or preferably all treatment fractions *L3*: ^{177}Lu DOTATATE quantitative imaging and dosimetry for all treatment fractions	*L1*: None *L2*: RAR_D = kidneys RAR_{ND} = bone marrow *L3*: DTR = tumors
^{90}Y ibritumomab tiuxetan	Non-Hodgkin's lymphoma	*L1*: Administered activity and rituximab as per package insert	*L1*: ^{111}In ibritumomab tiuxetan. Qualitative WB planar imaging at 48-72 h as per package insert *L2*: Quantitative WB planar imaging *L3*: Quantitative SPECT/CT imaging	*L1*: None *L2*: Bremsstrahlung SPECT/CT or PET/CT	*L1*: Intense localization in RAR_D = liver, spleen, or marrow. Increased uptake in normal organs *L2*: RAR_D = bone marrow *L3*: DTR = lymphoma lesions
$^{89}SrCl_2$	bone pain palliation	*L1*: Fixed administered activity as per package insert (148 MBq) *or* Administered activity based on BW as per package insert: 1.5-2.2 MBq/kg	*L1*: Qualitative ^{99m}Tc MDP/HDP WB planar imaging	None	None
^{153}Sm-EDTMP	bone pain palliation	*L1*: Administered activity based on BW as per package insert: 37 MBq/kg	*L1*: Qualitative ^{99m}Tc MDP/HDP WB planar imaging	*L1*: None *L2*: Planar or SPECT/CT	None
^{90}Y microspheres	Primary liver tumors or metastases	*L1*: None *L2*: 80-150 Gy whole liver *L3*: TCP/NTCP- optimization	*L1*: None *L2*: ^{99m}Tc- MAA	*L1*: None *L2*: Bremsstrahlung SPECT/CT PET/CT	*L1*: DTR_D = tumors *L1*: RAR_D = lung *L3*: DTR_D = lesions
^{223}Ra dichloride	bone pain palliation	*L1*: Administered activity based on BW as per package insert: 6´55 kBq/kg at 4-wk intervals	*L1*: Qualitative ^{99m}Tc MDP/HDP WB planar imaging	*L1*: None *L2*: Planar	None

KEY TO ABBREVIATIONS: See the key at the bottom of the table on the next page.

Table 22.4 ICRU Report 96: 2021 recommendations for dosimetry assessment for several RPT therapies not yet approved

Radiopharmaceutical Therapy	Treatment Indication	Posology	Prospective Dosimetry (Pretreatment)	Verification Dosimetry (Pretreatment)	Treatment Driving Quantity (RAR or DTR)
^{90}Y DOTATOC	Neuroendocrine tumors	Not registered *L1*: 1.85-3.7 GBq/m^2 BSA amino acids *L2*:	*L1*: None *L2*: ^{86}Ga DOTATOC ^{111}In DTPA-octreotide	*L1*: None *L2*: Bremsstrahlung SPECT/CT or PET/CT	*L1*: None *L2*: RAR_D: kidneys *L3*: DTR = tumors
^{177}Lu PSMA ligands	Metastatic castration-resistant prostate cancer	Not registered *L1*: 4´6.0–7.4 GBq at 8-w. interval	*L1*: None *L2*: ^{68}Ga –PSMA ^{18}F- PSMA	*L1*: None *L2*: Planar WB or SPECT/CT	*L1*: None *L2*: DTR = lesions
^{225}Ac PSMA ligands	mCRPC	Not registered *L1*: 4´100 kBq/kg at 8-week interval	*L1*: None *L2*: ^{68}Ga –PSMA or ^{18}F- PSMA or ^{177}Lu-PSMA	*L1*: None *L2*: Planar gamma camera	*L1*: None *L2*: DTR = lesions
^{166}Ho microspheres[§]	Primary liver tumors or metastases	*L1*: None *L2*: 60 Gy liver dose *L3*: TCP/NTCP- optimization	*L1*: None *L2*: ^{99m}Tc- MAA ^{166}Ho - microspheres	*L1*: None *L2*: SPECT/CT MRI	*L1*: None *L2*: RAR_D: non-target liver *L3*: DTR = lesions
^{90}Y, and ^{186}Re colloid, ^{169}Er citrate	Radiation synovectomy Rheumatoid arthritis	*L1/L2*: 10-40 MBq ^{169}Er (hand & wrist) 10-190 MBq ^{186}Re (hip) 180-200 MBq ^{90}Y (knee)	*L1*: None *L2*: ^{99m}Tc MDP/HDP and/or ^{99m}Tc-HIG	*L1*: None *L2*: Planar imaging, dicentric chromosomes counts	*L1*: None *L2*: RAR_D: non-target bone tissues

A = activity
CTR = clinical treatment region. Region(s) to treat, the total disease burden.
D = absorbed dose
DTR = dosimetric treatment region. The part of CTR for which the mass and absorbed dose are estimated. DTR_D and DTR_{ND} refers to delineable versus non-delineable.
LR = localization region. The organs, tissues, or cells to which the radioactive compound localizes.
L1 = Level 1
L2 = Level 2
L3 = Level 3
RAR = region at risk. Critical tissues that could suffer significant morbidity or functional loss, and for which the mass and absorbed dose are estimated. RAR_D and RAR_{ND} refers to delineable versus non-delineable. RAR_{SE} refers to RAR at risk for long-term stochastic effects.
SR = source region. The organs and tissues for which the activity content as a function of time and the time-integrated activity are estimated. SR_D and SR_{ND} refers to delineable versus non-delineable.
TDQ = treatment driving quantity
VOI = volume of interest
TB = total body
BW = body weight

22.6 Clinical Example

ICRU Report 96 contains several clinical examples with detailed explanations (Sgouros et al. 2021). The summary of one example is provided below.

<table>
<tr><th colspan="5">Report of treatment of prostate cancer using ^{177}Lu-PSMA-617</th></tr>
<tr><td colspan="5">Date of report: Dec 3, 2020
Patient name: NN (patient 5)
Gender: M
Indication: Low volume prostate cancer, GS 9, PSA 7.8/5 months, metastatic (PET 5)
Treatment prescription: Fixed activity first fraction (3000 MBq) to determine tolerance for second fraction (3000 or 6000 MBq).
Total number of fractions administered (planned): 2 (2)
Reporting level: Level 2</td></tr>
<tr><td>Time of administration</td><td>Aug 12, 2019
14:36</td><td>July 10, 2019
14:40</td><td></td><td></td></tr>
<tr><td>Administered activity (MBq)</td><td>3017</td><td>5977</td><td></td><td></td></tr>
<tr><td>Localization and Source Regions</td><td>LR
SR_D

SR_{ND}</td><td colspan="3">Regions with high PSMA expression density, including tumors, liver, kidneys, salivary glands.
Salivary glands, Left kidney, Right kidney, Tumors (n=5): (SPECT/CT: fraction 1 & 2: d 0,1,2,3,7)
Total Body: (planar scintigraphy: fraction 1: d 0,1,2,3,7)
Red marrow: (blood samples: fraction 1&2, n=9)</td></tr>
<tr><td>Treatment Regions</td><td>CTR
DTR_{D1}</td><td colspan="3">Tumors: (^{68}Ga-PSMA-I&T PET/CT, -15 d)
Tumors (n=1):(SPECT/CT: fraction 1, d 1)</td></tr>
<tr><td>Regions at risk</td><td>RAR_D
RAR_{ND}
RAR_{SE}</td><td colspan="3">Salivary glands and Kidneys: (CT: fraction 1, d 1)
Red marrow: (whole-blood concentration: fraction 1)
Total body absorbed dose, as a surrogate for absorbed dose to tissues at risk for SE</td></tr>
<tr><td colspan="5">Complementary information:
Abbreviations: d = days, m = months, SE = secondary effects</td></tr>
</table>

- Reporting level is the lowest level reported for any category (posology, prospective dosimetry, verification dosimetry).
- Localization Region (**LR**): the organs, tissues, or cells to which the radioactive compound localizes.
- Source Region (**SR**): the organs and tissues for which the activity content as a function of time and the time-integrated activity are estimated. SR_D and SR_{ND} refers to delineable versus non-delineable.
- Clinical Treatment Region (**CTR**): region(s) to treat, the total disease burden.
- Dosimetric Treatment Region (**DTR**): the part of CTR for which the mass and absorbed dose are estimated. DTR_D and DTR_{ND} refers to delineable versus non-delineable.
- Region at Risk (**RAR**): critical tissues that could suffer significant morbidity or functional loss, and for which the mass and absorbed dose are estimated. RAR_D and RAR_{ND} refer to delineable versus non-delineable. RAR_{SE} refers to RAR at risk for long-term stochastic effects.

Summary Dosimetry Report (^{177}Lu-PSMA-617)

Absorbed doses (Gy) to RAR based on within-therapy imaging				
Fraction number	**Kidneys**	**Salivary Glands**	**Red Marrow**	**Total Body**
1	1.7 Gy	2.1 Gy	0.06 Gy	0.14 Gy
2	4.1 Gy	2.9 Gy	0.15 Gy	0.30 Gy
3				
4				
Total	**5.8 Gy**	**5.0 Gy**	**0.21 Gy**	**0.44 Gy**

Absorbed doses (Gy) to DTR based on within-therapy imaging					
Fraction number	**DTR_{D1}**	**DTR_{D2}**	**DTR_{D3}**	**DTR_{D4}**	**DTR_{D5}**
1	33.2 Gy	10.1 Gy	9.8 Gy	4.2 Gy	7.4 Gy
2	40.8 Gy	9.7 Gy	45.5 Gy	19.2 Gy	6.4 Gy
3					
4					
Total	**77.2 Gy**	**19.8 Gy**	**55.3 Gy**	**23.4 Gy**	**13.8 Gy**

Detailed Dosimetry Report per Treatment Fraction (^{177}Lu-PSMA-617)

Fraction number: 1

Source Regions	**Data(time)*** ***A* = activity** **$\dot{D}$ = dose rate** ***C* = activity concentration**	**Assumed biokinetics and integration method**	**Effective half-times and relative amplitude of biokinetic phases (if applicable)**	**Time-integrated activity**
$\mathbf{SR_{D}}$: Kidneys	A(23.6 h) = 76 MBq $\dot{D}$(23.6 h) = 21 mGy h^{-1}	Monoexponential, analytical	30.8 h	6010 MBq h
$\mathbf{SR_{D}}$: Salivary glands	A(24.1 h) = 28.8 MBq $\dot{D}$(23.6 h) = 29 mGy h^{-1}		35.6 h	2105 MBq h
$\mathbf{SR_{D}}$: Total body	A(24 h) = 226 MBq		195 h	44,192 MBq h
$\mathbf{SR_{D1}}$: Tumor 1	A(23.6 h) = 1.22 MBq $\dot{D}$(23.6 h) = 593 mGy h^{-1}	Trapezoidal from 0 to 24 h. Monoexponential from 24 h to ∞, analytical, provided fit $R^2 < 0.7$. Otherwise, trapezoidal between all points and then monoexponential tail.	29.0 h	68.5 MBq h
$\mathbf{SR_{D2}}$: Tumor 2	A(23.6 h) = 0.08 MBq $\dot{D}$(23.6 h) = 63 mGy h^{-1}		31.4 h	13.7 MBq h
$\mathbf{SR_{D3}}$: Tumor 3	A(23.6 h) = 0.52 MBq $\dot{D}$(23.6 h) = 62 mGy h^{-1}		62.3 h	81.9 MBq h
$\mathbf{SR_{D4}}$: Tumor 4	A(23.6 h) = 0.21 MBq $\dot{D}$(23.6 h) = 29 mGy h^{-1}		25.2 h	11.6 MBq h
$\mathbf{SR_{D5}}$: Tumor 5	A(23.6 h) = 0.21 MBq $\dot{D}$(23.6 h) = 264 mGy h^{-1}		13.4 h	6.2 MBq h
$\mathbf{SR_{ND}}$: Red marrow	C(23 h) = 13.8 kBq g^{-1}	Tri-exponential, analytical.	11 min (55%) 2.8 h (31%) 12.6 h (14%)	1.25 MBq h g^{-1}

* Data considered relevant to be able to trace the dosimetry calculation

Fraction number: 2				
Source Regions	**Data(time)*** ***A* = activity** **$\dot{D}$ = dose rate** ***C* = activity concentration**	**Assumed biokinetics and integration method**	**Effective half-times and relative amplitude of biokinetic phases (if applicable)**	**Time-integrated activity**
SR_D: Kidneys	*A*(23.5 h = 131 MBq $\dot{D}$(23.5 h) = 36 mGy h^{-1}	Monoexponential, analytical	44.9 h	14670 MBq h
SR_D: Salivary glands	*A*(23.9 h) = 35.1 MBq $\dot{D}$(23.9 h) = 35 mGy h^{-1}		42.1 h	2911 MBq h
SR_D: Total body	*A*(24 h) = 2020 MBq		51.0 h	148529 MBq h
SR_{D1}: Tumor 1	*A*(23.5 h) = 2.55 MBq $\dot{D}$(23.5 h) = 364 mGy h^{-1}	Trapezoidal from 0 to 24 h. Monoexponential from 24 h to ∞, analytical, provided fit $R^2 < 0.7$. Otherwise, trapezoidal between all points and then monoexponential tail.	41.2 h	252.0 MBq h
SR_{D2}: Tumor 2	*A*(23.5 h) = 0.35 MBq $\dot{D}$(23.5 h) = 57 mGy h^{-1}		86.8 h	48.2 MBq h
SR_{D3}: Tumor 3	*A*(23.5 h) = 3.08 MBq $\dot{D}$(23.5 h) = 50 mGy h^{-1}		72.1 h	406.8 MBq h
SR_{D4}: Tumor 4	*A*(23.5 h) = 1.60 MBq $\dot{D}$(23.5 h) = 260 mGy h^{-1}		55.1 h	166.2 MBq h
SR_{D5}: Tumor 5	*A*(23.5 h) = 0.17 MB $\dot{D}$(23.5 h) = 27 mGy h^{-1}		24.0 h	39.3 MBq h
SR_{ND}: Red marrow	*C*(23 h) = 13.8 kBq g^{-1}	Bi-exponential, analytical.	9.2 min (71%) 8.7 h (29%)	3.20 MBq h g^{-1}
* Data considered relevant to be able to trace the dosimetry calculation				

Detailed Dosimetry Report per Treatment Fraction (^{177}Lu-PSMA-617)

RAR	Absorbed-dose calculation method	Mass	Absorbed dose Fraction 1	Absorbed dose Fraction 2
RAR_D: Kidneys	MIRD Formalism, S coefficients OLINDA adult male phantom, self-dose	0.189 kg (left) 0.182 kg (right)	1.7 Gy	4.1 Gy
RAR_D: Salivary glands		N/A*	2.1 Gy	2.9 Gy
RAR_{ND}: Red marrow	Blood-based dosimetry (BM concentration = blood concentration). MIRD Formalism, S coefficient OLINDA adult male phantom	N/A*	0.061 Gy	0.148 Gy
RAR_{SE}: Total body	MIRD Formalism, S coefficient OLINDA adult male phantom	Patient whole-body weight: 78 kg	0.14 Gy	0.30 Gy
* Absorbed dose calculated from activity concentration.				

DTR	Absorbed-dose calculation method	Mass	Absorbed dose Fraction 1	Absorbed dose Fraction 2
DTR_{D1}: Tumor 1	Voxel-based Monte Carlo calculation (mean dose-rate in VOI)	0.16 g	36.4 Gy	40.8 Gy
DTR_{D2}: Tumor 2		0.11 g	10.1 Gy	10.7 Gy
DTR_{D3}: Tumor 3		0.68 g	9.8 Gy	55.8 Gy
DTR_{D4}: Tumor 4		0.59 g	1.6 Gy	20.8 Gy
DTR_{D5}: Tumor 5		0.06 g	8.0 Gy	6.4 Gy

References

Bentzen, S. M., W. Dorr, R. Gahbauer, et al. (2012). "Bioeffect modeling and equieffective dose concepts in radiation oncology—terminology, quantities and units." *Radiother. Oncol.* 105:266–68. 10.1016/j.radonc.2012.10.006.

Hobbs, R. F., T. McNutt, S. Baechler, et al. (2011). "A treatment planning method for sequentially combining radiopharmaceutical therapy and external radiation therapy." *Int. J. Radiat. Oncol. Biol. Phys.* 80:1256–62.

Hobbs, R. F., R. W. Howell, H. Song, S. Baechler, and G. Sgouros (2014). "Redefining relative biological effectiveness in the context of the EQDX formalism: implications for alpha-particle emitter therapy." *Radiat. Res.* 181:90–98. 10.1667/RR13483.1.

ICRU (2002). "ICRU Report 67: Absorbed-dose specification in nuclear medicine." *J. ICRU* 2:1–110.

ICRU (2011). "ICRU Report No. 85: Fundamental quantities and units for ionizing radiation." *J. ICRU* 11:1–42.

Katugampola, S., J. Wang, A. Rosen, and R. W. Howell (2022). "MIRD Pamphlet No. 27: MIRDcell V3, a Revised Software Tool for Multicellular Dosimetry and Bioeffect Modeling." *J. Nucl. Med.* 63:1441–49. 10.2967/jnumed.121.263253.

Konijnenberg, M. W. (2003). "Is the renal dosimetry for [90Y-DOTA0,Tyr3]octreotide accurate enough to predict thresholds for individual patients?" *Cancer Biother. Radiopharm.* 18:619–25. 10.1089/108497803322287718.

Lea, D. E. and D. G. Catcheside (1942). "The mechanism of the induction by radiation of chromosome aberrations in tradescantia." *J. Genet.* 44:216–45.

Sgouros, G., W. E. Bolch, A. Chiti, et al. (2021). "ICRU REPORT 96, Dosimetry-Guided Radiopharmaceutical Therapy." *J. ICRU* 21:1-212. 10.1177/14736691211060117

Vaziri, B., H. Wu, A. P. Dhawan, P. Du, and R. W. Howell (2014). "MIRD Pamphlet No. 25: MIRDcell V2.0 Software Tool for Dosimetric Analysis of Biologic Response of Multicellular Populations." *J. Nucl. Med.* 55:1557–64. 10.2967/jnumed.113.131037.

Wilke, L., N. Andratschke, O. Blanck, et al. (2019). "ICRU report91 on prescribing, recording, and reporting of stereotactic treatments with small photon beams: Statement from the DEGRO/DGMP working group stereotactic radiotherapy and radiosurgery." *Strahlentherapie Und Onkologie* 195:193–98. 10.1007/s00066-018-1416-x.

Chapter 23

Radiopharmaceutical Therapy Reimbursement: an Evolving Landscape in the United States

Russell Tarver[1], M.S., DABR, FAAPM and Gerald White[2], M.S., FAAPM

[1]Director, Radiation Services
Department of Radiation Oncology
The Center for Cancer and Blood Disorders
Fort Worth, TX
[2]Colorado Associates in Medical Physics
Colorado Springs, CO

23.1 Reimbursement Fundamentals

23.1.1 Healthcare System Models

World-wide, there are four common schemes for reimbursement for the provision of medical care. These models are nicely described by the journalist T. R. Reid (2010), and we repeat his categorization here.

- ***The Bismarck model***. Named for the Prussian Chancellor Otto von Bismarck, this system features private providers of healthcare services, coupled with private payers, typically insurance companies of one type or another. Here we use the "insurance" descriptor with a wide reach to include HMOs and similar endeavors. The cost of the insurance is often split between employers and their employees, but individuals may be able to purchase insurance independently of employment. This results in a model with multiple payers, but in many industrialized countries services and prices are tightly regulated. They may be for-profit entities, but (except for the United States) are typically non-profit.

- ***The Beveridge model***. William Beveridge was a British economist who articulated a vision for much of the social services system currently in place in Great Britain (Beveridge 1942). From his work, the plan for the National Health Service was created where medical procedures were to be provided as a government service, just like defense, fire protection, police, and streets. Patients don't receive bills for medical services, and the providers are government-owned, government-employed, or government contractors.
- ***The National Health Insurance model***. Saskatchewan (under the leadership of Premier Tommy Douglas) set the standard for a model that has similarities with both the Bismarck and Beveridge models. Medical care is provided by private entities (or perhaps government-owned hospitals and clinics), but the payment system is a government-run single payer system. Examples are Canada (provincial payer systems) and Taiwan.
- ***Out-of-pocket personal payment model***. In the absence of one of the above-described models, one is left with the market-based system in which individuals pay for medical care out of pocket. They seek care based on need and ability to pay, typically with a mismatch between these two features; the need frequently exceeds the ability to pay.

In the United States, we find all four of these models in play. A great many Americans have the Bismarck model of employer-based insurance, but without the more common European feature of a regulatory system that requires uniformity of benefits. Veterans, members of the military, and many Native Americans find themselves in a Beveridge model healthcare system, in which the providers are employed by the government and medical services are funded by taxes. For those over 65 or eligible due to disability, the Medicare program provides a National Health Insurance payment model, and Medicaid serves a similar function for low-income Americans. Lastly, approximately 10.2% of the United States population were uninsured in 2021 (Tolbert et al. 2022), finding themselves embedded in the out-of-pocket personal payment model.

We describe these models, at least in general terms, because reimbursement processes for medical procedures in the United States—and for radiopharmaceutical therapy procedures in particular—depend on the reimbursement model in which we find our patient. Too frequently, reimbursement discussions are focused on Medicare patients and Medicare policy. The reader is cautioned on the necessity to fully understand the payment processes of the system particular to each patient. Private payers in the United States may have very different payment policies than Medicare; the provider must be familiar with the relevant particulars.

In the discussion to follow, we'll try to focus on characteristics that are common to the Bismarck and National Health Insurance models, with qualified comments on other situations. In the United States, most reimbursement is based, at least in part, on the description of individual procedures that a patient has received.

23.1.2 Nomenclature and Description of Procedures (AMA)

In the United States, the most common system for a taxonomy of the universe of medical procedures is the Current Procedural Terminology (CPT®) system. It is owned and maintained by the American Medical Association (AMA) and is intended to be "*... a listing of descriptive terms and identifying codes for reporting medical services and procedures performed by physician and other health care professionals.*" (AMA 2023). Not only does the CPT® endeavor include a procedure and associated numerical code, it also includes, via the code application process, a detailed description of each individual procedure. The text of these code applications is not generally available, but the AMA regularly publishes guidance documents with direction and advice as to the proper use of CPT codes for medical procedures. There are currently about 10,000 CPT codes described. Note that the CPT codes are specialty-agnostic. While the codes may be grouped by specialty category in the AMA listing, any code may be reported by the physician who performs the procedure, regardless of specialty. These codes are widely accepted by private payers and are integrated into the Medicare payment system as well (see below).

The AMA CPT system is supplemented in the Medicare payment process with the Healthcare Common Procedure Coding System (HCPCS). Level 1 HCPCS codes are the AMA CPT codes described above. Level II HCPCS codes are generally related to supplies and patient-specific medical equipment, ambulance services, and other costs not classified in the CPT system.

Relevant to our discussion, HCPCS "A" codes are used to reimburse for radiopharmaceuticals and HCPCS "G" codes may be used for various procedures for which Medicare believes CPT codes are not available or do not have adequate descriptors. The "G" codes are intended to be temporary, but they may persist for many years.

23.1.3 Physician Work and Practice Expense

Both the private payer system and the Medicare system are structured around concepts promulgated in the United States during the formative years of the health insurance enterprise. Hospitals began the process by offering insurance or prepayment plans, some of which became consolidated under the Blue Cross brand in the 1930s. Physicians were initially opposed to the concept, preferring to have a direct financial relationship with their patients. After viewing the success of the Blue Cross concept, physician opposition weakened and insurance schemas to cover physician services were begun, finding utility in an umbrella organization, Blue Shield (more on this division in Section 23.1.5).

This consolidation of insurance entities encouraged efforts toward uniformity in payment levels, eventually resulting (in the 1980s) in the currently near-universal Resource Based Relative Value System for determining payment levels (Hsiao 1979). Just as the AMA maintains control over procedure descriptions in the CPT system, they exercise considerable influence on payment levels through the AMA Relative Value Update Committee (RUC). This committee determines the relative value of physician work for each of the 10,000 CPT codes, as well as the practice expense (equipment, staff, supplies, etc.) associated with the physician work (Laugesen 2016).

23.1.4 Payer Mix and Influence of CMS

Individual medical practices considering implementing or expanding a radiopharmaceutical therapy program will certainly evaluate the reimbursement aspects. When doing so, it is essential to understand the payment policies of the various payers for the patient population the practice will serve. The four major categories of payment systems described above are only the beginning of the granularity levels that must be considered.

It is noteworthy that, in the United States the Center for Medicare and Medicaid Services (CMS) develops payment policies that may guide or be adopted by private payers in the employer-based or privately purchased insurance systems. The private payers may further qualify payments based on the particular insurance company, and further by the particular type of insurance plan within the spectrum of offerings for an individual insurer. Further, insurance companies may engage benefit management companies; these entities provide an additional level of review for the eligibility of various medical procedures or drugs, making recommendations to the insurer for a decision to allow or deny payment. These "pre-authorization" barriers may be particularly intrusive for high-cost care processes such as may be seen in radiopharmaceutical therapy. Medicare Advantage insurers have been cited by the Office of Inspector General for restrictions to medically necessary care (Grimm 2022).

While CMS provides fairly transparent policies regarding payment eligibility and rates, they contract with a handful of intermediaries (Medicare Administrative Contractors) for the actual payment review and processing and, thus, there may be regional differences in some aspects of coverage. Also, many Medicare-eligible patients have chosen private insurance coverage under the "Medicare Advantage" branding. These are commercial insurance policies, and although they nominally provide coverage equivalent to Medicare, their actual policies and procedures may be much more restrictive.

An individual practice or hospital may have a hundred or more payment schemas to deal with. A detailed knowledge of the payer mix, the patient mix, and the payers' policies will be necessary when considering the financial viability of a radiopharmaceutical therapy program.

23.1.5 RUC and CMS Valuations—HOPPS and MPFS

As mentioned above, the AMA RUC exercises substantial influence on actual reimbursement levels in the United States. Their role is to assign a relative value to physician work and determine practice expense costs for each of the 10,000 CPT codes. As part of this process, specialty societies present to the RUC their opinion as to the time and intensity of physician work for a procedure. Included in this submission, the specialty societies typically conduct a survey of physicians in the relevant specialty to determine best estimates of physician work and practice expense. In a rapidly evolving care process such as radiopharmaceutical therapy, providers may well receive a survey associated with either a new CPT code or the revaluation of an existing code. It is very important to respond to such a survey carefully and completely, as the aggregate results will determine the reimbursement level for years to come. Following a survey, a society will offer a "recommendation" to CMS for consideration; in practice, the vast majority of recommendations to the RUC are accepted as proposed or with minor revisions and adopted into the Medicare Physician Fee Schedule (MPFS).

For hospital outpatient reimbursement (the technical component only; physician reimbursement is described in the MPFS), the AMA does not play a direct role in valuation. Rather, CMS examines actual reported costs for procedures in the Hospital Outpatient Prospective Payment System (HOPPS). This is done via a convoluted process; CMS examines the charges submitted for a particular CPT code or group of CPT codes from a providing facility. Recognizing that the charges (from the hospital chargemaster) are typically wildly inflated (Bai and Anderson 2016), they have the facility report costs on a less granular level, say departmental, and then deflate the charges by a charge to cost ratio.

Aggregating costs determined in this way across the nation, CMS then determines a payment level to apply to various procedures in the HOPPS system. In an effort at simplification, payment bins designated as Ambulatory Payment Classifications (APCs) are established. CPT codes with similar cost valuations are placed into an APC, and all procedures in a specific APC are reimbursed at the same rate. This process is important to understand when establishing a radiopharmaceutical therapy program in a hospital outpatient department paid under the HOPPS. If chargemaster values for various procedures are inappropriately created (recall the charge to cost ratio described above), then contributing to a national aggregate, the eventual CPT and APC values will be skewed. For example, if a department believes a particular procedure costs $200 to deliver and the procedure has a chargemaster value of $200, but the department's charge-to-cost ratio is 5, CMS will believe the actual cost to deliver the procedure to a patient is $40. If repeated nationally, eventually the CMS reimbursement would become $40.

Private payers and providers frequently use CMS values as a baseline to anchor prices negotiated in various health insurance plans. It is often useful when estimating potential revenues for a procedure to examine CMS valuations. But care must be taken to understand payment levels of major third-party payer entities with whom the practice or facility has contracts. Providers with a substantial Medicaid patient population will need to carefully consider payment rates and allowed services, as they vary from state to state.

If patients in the out-of-pocket personal payment model (described above) are considered in the mix, care must be taken in applying the chargemaster values as a "rack rate" or "list price." Recent legislative and judicial actions may limit the use of chargemaster prices for actual patient billing.

CMS valuations can be used as a baseline for reimbursement discussions, but it is important to recognize that actual payments can vary from payer to payer, and even from plan to plan, in a particular payer's portfolio (Kliff and Katz 2021).

23.1.6 Drugmaker Price Policies

Most of this chapter is devoted to descriptions of physician work and practice expense. For many radiopharmaceutical treatment courses, these costs are a minor portion of the total expense and will be dwarfed by the drug costs. Drug costs vary by orders of magnitude. A single treatment for hyperthyroidism using ^{131}I sodium iodide might incur a drug cost of about \$600, while the drug cost for a course of therapy using ^{177}Lu DOTATATE could approach \$300,000. There are a variety of mechanisms for coping with such large costs, and each practice/institution will need to review the array of payer and drugmaker policies applicable to their individual situation.

For patients insured through non-government entities, there are programs that can assist with out-of-pocket patient costs associated with deductibles or co-pays. This assistance can be substantial, and providers of radiopharmaceutical therapy should be familiar with options in their locality. Providers offering radiopharmaceutical therapy in certain settings may be eligible for reduced prices for radiopharmaceutical drugs through a program sponsored by the United States' Health Resources and Service Administration. The program is known colloquially as the "340b Drug Pricing Program." The eligibility requirements and program details are beyond the scope of this chapter. This program deals with a wide variety of drugs, so if your facility is eligible, it is highly likely that your revenue team or pharmacy team has details of the program and can assist you with implementation in the radiopharmaceutical realm.

Lastly, purchasing agreements between providers and vendors may involve rebate programs that reward providers with "cash back" rebates based on purchasing volume. These rebates, while generally not large in the context of the drug cost, can add up and represent a significant dollar amount returned to the provider or practice. One could argue that these programs give providers an incentive to deliver more care, so negotiating equivalent net rates that do not rely on rebates is a preferred drug cost model.

23.2 Process of Care and Procedure Coding

In determining the appropriate procedure codes to report in a process of care, one must keep in mind a few basic principles.

- ***The provider should only report codes for procedures that were actually performed***. While this seems to be too obvious to restate, there is sometimes an overreliance on "coding guidance" that lists a series of CPT codes that might be reported for a particular process of care. For example, one might see a list of codes associated with a course of IMRT for prostate cancer, superficial radiation therapy for skin cancer, or, relevant to this discussion, administration of radiopharmaceutical therapy. In any of these scenarios, the universe of potentially applicable procedure codes may well be larger than the list of procedures actually performed on an individual patient.
- ***The provider is obligated to report the CPT code that most closely describes the actual procedure provided to the patient***. This match may often be clear,

but not always. Further, the provider must avoid substituting one code for another to avoid payer problems related to procedure-to-procedure edits, date-of-service edits, or payer policy restrictions.

- ***Individual payers may require a specific code reporting process***. For example, for a particular procedure, some payers might require a CPT code, while others might require an HCPCS reporting process.
- ***CMS reimburses for drugs delivered to the patient, and in some cases, for waste associated with unit dosage packaging***. The nature of radiopharmaceuticals prohibits keeping stock on hand, so dosages are generally ordered for a specific patient and time. If a patient cancels after a dosage has been delivered or the procedure is aborted, the cost of the radiopharmaceutical is born by the provider and is not reimbursable. Private payers have similar policies. Many vendors have policies that allow for these situations, and they either waive the pharmaceutical cost or charge a reduced amount. The wise administrator is encouraged to fully understand this aspect of their purchasing contract before embarking on a new program. Better yet, the administrator should consider this aspect during negotiations of purchasing contracts.

23.2.1 Evaluation and Consultation

A *radiation medicine physician* (RMP) may receive a referral from another provider to consider a patient for radiopharmaceutical therapy. Alternately, the RMP may want to consider radiopharmaceutical therapy for a patient already under their care. In either case, the initial step may be an office or other outpatient visit for the evaluation of the patient. The CPT formalism lists a series of *evaluation and management* (E/M) procedures that may be reported for this care (CPT 99029–99215). The procedures are stratified by complexity intensity, time spent with the patient, status as a new or established patient, and inpatient or outpatient status. Providers are advised to carefully review the definitions for each in this set of codes to select the one that most closely describes the E/M service provided. The CPT Professional 2023 Codebook (American Medical Association 2022) has nine pages of guidelines to assist in selecting the appropriate E/M code.

Following the E/M process, the RMP may provide a Clinical Treatment Planning service. This work will encompass review and interpretation of tests, prescribing additional tests or imaging procedures, and determining the applicability and appropriateness of available treatment options, some of which may include radiopharmaceutical therapy. There are three levels of complexity reportable for Therapeutic Radiology Treatment Planning, and the procedure is described using the CPT codes 77261 (simple), 77262 (intermediate), or 77263 (complex).

23.2.2 Preparation for Treatment

Once the decision has been made to proceed with radiopharmaceutical therapy, the RMP may identify the need for imaging studies to assist with planning for the activity of radioactive material to be administered. This process may involve quantitative

Table 23.1 CPT codes and descriptions of common RPT studies during the early stages of the treatment planning process

CPT® Code	Descriptor
78580	Pulmonary perfusion imaging (e.g., particulate)
78800	Radiopharmaceutical localization of tumor or distribution of radiopharmaceutical agent(s); limited area
78801	Radiopharmaceutical localization of tumor or distribution of radiopharmaceutical agent(s); multiple areas
78802	Radiopharmaceutical localization of tumor or distribution of radiopharmaceutical agent(s); whole body, single day imaging
78803	Radiopharmaceutical localization of tumor or distribution of radiopharmaceutical agent(s); tomographic (SPECT)
78804	Radiopharmaceutical localization of tumor or distribution of radiopharmaceutical agent(s); whole body, requiring 2 or more days imaging
78812	PET imaging, skull base to mid-thigh
78814	Positron emission tomography (PET) with concurrently acquired computed tomography (CT) for attenuation correction and anatomical localization imaging; limited area (e.g., chest, head/neck)
78830	Radiopharmaceutical localization of tumor, inflammatory process or distribution of radiopharmaceutical agent(s) (includes vascular flow and blood pool imaging, when performed); tomographic (SPECT) with concurrently acquired computed tomography (CT) transmission scan for anatomical review, localization and determination/detection of pathology, single area (e.g., head, neck, chest, pelvis), single day imaging
78831	Radiopharmaceutical localization of tumor, inflammatory process or distribution of radiopharmaceutical agent(s) (includes vascular flow and blood pool imaging, when performed); tomographic (SPECT), minimum 2 areas (e.g., pelvis and knees, abdomen and pelvis), single day imaging, or single area imaging over 2 or more days
78832	Radiopharmaceutical localization of tumor, inflammatory process or distribution of radiopharmaceutical agent(s) (includes vascular flow and blood pool imaging, when performed); tomographic (SPECT) with concurrently acquired computed tomography (CT) transmission scan for anatomical review, localization and determination/detection of pathology, minimum 2 areas (e.g., pelvis and knees, abdomen and pelvis), single day imaging, or single area imaging over 2 or more days
78835	Radiopharmaceutical quantification measurement(s) single area (List separately in addition to code for primary procedure)

imaging studies and subsequent dosimetric analysis predictions for the therapy. Table 23.1 describes procedures that might be used in this process. Additional procedures that might be performed during preparation for treatment are listed in Table 23.2.

23.2.3 Treatment Administration

Administration of the radiopharmaceutical may occur in a variety of settings: hospital outpatient, a free-standing practice, or within a hospital department variously named nuclear medicine, radiation oncology, endocrinology, or radiology. Whatever the setting, a variety of CPT codes are available for reporting treatment procedures (Table 23.3). As mentioned above, the codes are specialty-agnostic, and the appropri-

Table 23.2 CPT codes and descriptions of additional RPT studies in preparation for patient treatment

CPT® Code	Descriptor
77300	Basic radiation dosimetry calculation, calculation of absorbed dose to a point or region based on planned activity to be administered or actual deposited activity
77280 77285 77290	Simulation (simple, intermediate, or complex), possibly imaging procedures associated with microsphere deposition, or positioning and immobilization processes in connection with the need for reproducible patient positioning for a series of pre-administration and post-administration imaging procedures
77295	Should a dosimetric analysis prior to administration rise to the level of a three-dimensional isodose plan, this code might be the closest to describing the procedure (at the time of this writing)
77370	An RMP may request a special medical physical consult for unusual circumstances. This code would not be routinely used
77470	A special treatment procedure might be appropriate for situations involving unusual complexity. The code would not be routinely used

Table 23.3 CPT codes and descriptions of various RPT treatment procedures

CPT® Code	Descriptor	Comments
77750	Infusion or instillation of radioelement solution	Includes 3-month follow-up care
79005	Radiopharmaceutical therapy, by oral administration	
79445	Radiopharmaceutical therapy, by intra-arterial particulate administration	
79403	Radiopharmaceutical therapy, radiolabeled monoclonal antibody by intravenous infusion	
79101	Radiopharmaceutical therapy, by intravenous administration	
79200	Radiopharmaceutical therapy, by intracavitary administration	
79300	Radiopharmaceutical therapy, by interstitial radioactive colloid administration	
79440	Radiopharmaceutical therapy, by intra-articular administration	
79999	Radiopharmaceutical therapy, unlisted procedure	
S2095	Transcatheter occlusion or embolization for tumor destruction, using Y-90 microspheres	May be required by certain payers

ate code is determined by the procedure that is performed and not the specialty of the RMP administering the procedure.

Two of these codes have potential overlap of definitions:

- CPT 77750, Infusion or instillation of radioelement solution and
- CPT 79101, Radiopharmaceutical therapy, by intravenous administration.

Note that 77750 has a 90-day global period, meaning all subsequent related care is included in this procedure for 90 days. This may cause difficulties for radiopharmaceutical therapy that is fractionated over a number of administrations.

23.2.4 Follow-up

The post-administration period may involve additional patient procedures. There is an increasing awareness of the clinical significance of dosimetric analysis for radiopharmaceutical therapy. As mentioned in Section 23.2.2 above, dose calculations may be performed and described by CPT codes 77300 or 77295. Additional quantitative imaging procedures will be common for use in patient-specific dosimetric analysis. They are also listed in Section 23.2.2.

Follow-up visits, management of side effects, surveillance for disease progression, etc. can be coded using the E/M code set described earlier (with the exception of the 90-day post-treatment period if CPT 77750 was the administration procedure).

23.3 Future Developments in Procedure Definitions and Coding

One hesitates to put into print predictions about future developments. However, it would not seem overly speculative to take note that many of the existing codes were developed, described, and valued for processes that differ from our current spectrum of radiopharmaceutical therapy activities. In particular, dosimetric calculation and isodose planning codes developed for external beam or sealed-source brachytherapy may not be good fits for the dosimetric analysis necessary for radiopharmaceutical therapy. Administration codes perhaps do not have appropriate descriptions to adequately differentiate the levels of physician work and practice expenses associated with the current (and on the horizon) array of radiopharmaceutical agents. We should look forward to the professional societies addressing these needs within the CPT/RUC process in the near future, with the hope that such changes would migrate into the private payer and HOPPS world shortly afterwards.

References

American Medical Association. "CPT® Professional 2023 Codebook." AMA, 2022.

Bai, G. and G. F. Anderson (2016). "US Hospitals Are Still Using Chargemaster Markups To Maximize Revenues." *Health Affairs* 35(9):1658–64. https://doi.org/10.1377/hlthaff.2016.0093.

Beverage, W. (1942). "Social Insurance and Allied Services." Reprinted in *J. Bull. World Health Org*. 78(6):847–55. Available at https://www.ncbi.nlm.nih.gov/pmc/articles/PMC2560775/pdf/10916922.pdf.

Grimm, C. A. (2022). "Some Medicare Advantage Organization Denials of Prior Authorization Requests Raise Concerns About Beneficiary Access to Medically Necessary Care." U. S. Department of Health and Human Services Office of Inspector General, OEI-09-18-00260: April 2022.

Hsiao, W. C. and W. B. Stason (1979). "Toward Developing a Relative Value Scale for Medical and Surgical Services." *Health Care Financing Rev.* 1(2):23–38.

Kliff, S. and J. Katz (2021). The Upshot: "Hospitals and Insurers Didn't Want You to See These Prices. Here's Why." *New York Times*, 22 Aug. 2021.

Laugesen, M. *Fixing Medical Prices: How Physicians Are Paid.* Cambridge, MA: Harvard University Press, 2016.

Reid, T. R., ed. *The Healing of America: A Global Quest for Better, Cheaper, and Fairer Health Care.* New York: Penguin Books, 2010.

Tolbert, J., J. Drake, and A. Damico (2022). "Key Facts About the Uninsured Population." Kaiser Family Foundation website. Accessed 17 Feb. 2023. https://www.kff.org/uninsured/issue-brief/key-facts-about-the-uninsured-population/.